APPLICATIONS OF MATHEMATICAL PROGRAMMING TECHNIQUES

APPLICATIONS OF MATHEMATICAL PROGRAMMING TECHNIQUES

A conference held at CAMBRIDGE, U.K. in June 1968
under the aegis of the N.A.T.O. Scientific Affairs Committee

Edited by E. M. L. BEALE
Technical Director, Scientific Control Systems Ltd., and
Visiting Professor, Imperial College of Science and Technology

THE ENGLISH UNIVERSITIES PRESS LTD
St Paul's House, Warwick Lane, London E.C.4

First printed 1970

ISBN 0 340 05395 X

Printed and bound in Great Britain for
The English Universities Press Limited

PRINTED BY Unwin Brothers Limited
THE GRESHAM PRESS OLD WOKING SURREY ENGLAND

Produced by 'Uneoprint'

A member of the Staples Printing Group **(UCO4847)**

FOREWORD

A conference on applications of mathematical programming techniques was held at the University of Cambridge from 24th to 28th June, 1968. This conference was sponsored by the NATO Science Committee, and was attended by about 140 people from 14 countries. These included Professor George Dantzig, the founder of mathematical programming, and other eminent mathematicians, as well as many people with substantial experience of mathematical programming applications.

This book will not replace any of the many textbooks on mathematical programming now available. But it can supplement any of them, since it contains many descriptions of applications—often in novel fields—of linear, nonlinear, integer and dynamic programming written by people who have been personally involved in them. Aspects covered range from mathematical methods and computer problems to model building and discussions on the acceptability of the work to management.

The contents are organized in the same way as the conference. Section 1 consists of a survey by Professor Dantzig of the historical development of mathematical programming. Section 2 contains papers on two developing fields of application: engineering and transport scheduling. Section 3 covers applications to national economic problems. Section 4 covers a vital aspect of the task of taking mathematical programming applications out of the field of research into routine operation: matrix generators and output analysers, i.e. methods of using computers to assemble and process both the input data and the results of a mathematical programming application, as well as performing the optimization.

The next part of the book extends beyond purely linear programming. In Section 5 Professor Balas discusses project scheduling with resource constraints, a field that has so far been tackled in practice almost entirely by heuristic methods, but where genuine optimization by mathematical programming is just beginning to be practically feasible. Section 6, devoted to methods for non-convex problems, is introduced by a general survey from Professor Zoutendijk. Section 7 is devoted to applications of integer programming, while Section 8 consists of an introductory paper on geometric programming, a new approach to some nonlinear programming problems.

Section 9 is devoted to detailed accounts of work, both in the United States and the United Kingdom, on applications of mathematical programming to the military problems of strategic deployment. Although the context of this work is a military one, the methodology will be of much wider interest.

Finally, Section 10 discusses two applications of nonlinear programming.

This book follows the conference in emphasizing relatively new and growing fields of application of mathematical programming, and particularly of nonlinear and integer programming. The quantitatively most significant applications, which are still of purely linear programming to industrial production problems, are nevertheless represented by the papers of A. E. Clark and L. Pessina in Section 4.

Those who took part in the discussions at the conference were invited to produce a written version of their remarks for publication, and the speakers to produce written replies. These are reproduced in the text. Other discussion has not been preserved, but one incident seems worth recalling: a purely terminological discussion about the apparent contrast between maximizing a concave function and solving a convex problem was forestalled by Professor Vajda, the chairman at the time, with the remark 'I am not surprised that something that looks concave from one side of the Atlantic should look convex from the other.'

E. M. L. BEALE

Conference Director

CONTENTS

SECTION 1 Survey Paper

G. B. Dantzig

Linear Programming and its Progeny *

GEORGE B. DANTZIG
Stanford University, U.S.A.

1. ORIGINS AND INFLUENCES

In the years following its conception in 1947 in connection with planning activities of the military, linear programming has come into wide use. In academic circles, mathematicians and economists have written books on the subject. Our purpose is to give a brief account of its origins and to point out those influences which brought about its development. Interestingly enough, in spite of its now known wide applicability to everyday problems, linear programming was unknown before 1947. Fourier may have been aware of its potential in 1823, and it is true that in the U.S.S.R., in 1939, Kantorovitch made proposals that were neglected there during a period that witnessed its discovery and rapid development elsewhere.

2. INFLUENCE OF MILITARY PLANNING

The following statement of M. K. Wood and M. A. Geisler is pertinent:

> 'It was once possible for a Supreme Commander to plan operations personally. As the planning problem expanded in space, time, and general complexity, however, the inherent limitations in the capacity of any one man were encountered. Military histories are filled with instances of commanders who failed because they bogged down in details, not because they could not eventually have mastered the details, but because they could not master all the relevant details in the time available for decision. Gradually, as planning problems became more complex, the Supreme Commander came to be surrounded with a General Staff of specialists which supplemented the Chief in making decisions. The existence of a General Staff permitted the subdivision of the planning process and the assignment of experts to handle each part. The function of the Chief then became one of selecting objectives, co-ordinating, planning, and resolving conflicts between staff sections.'

World War II witnessed the development of staff planning on a gigantic scale in all parts of the U.S. military establishment and in such civilian counterparts as the War Production Board. A nation's military establishment, in wartime or in peace, is a complex of economic and military activities requiring almost unbelievably careful coordination in the implementation of plans produced in its many departments. If one such plan calls for equipment to be designed and produced, then the rate of ordering equipment has to be coordinated with the capabilities of the economy to relinquish men, material, and productive capacity from the civilian to the military sector. These development and support activities should dovetail into the military program itself. To be coordinated between the major activities are hundreds of personnel subtypes and thousands of supply subtypes.

During the war, the planning process itself became so intricate, lengthy, and multipurposed that a 'snapshot' of the military staff at any one time showed it to be working on many different programs—some in early phases of development and based on earlier ground rules and facts. To cut the time of the planning process, a patchwork of several of these programs was often thrown together based on necessarily inconsis-

---

* Reprinted from Naval Research Reviews, June 1966

tent facts and rules. To coordinate this work better, the Air Staff, for example, around 1943, created the program-monitoring function under Professor E. P. Learned of Harvard. The entire program was started off with a war plan in which were contained the wartime objectives. From this plan, by successive stages, the wartime program specifying unit deployment to combat theaters, training requirements of combat personnel and technical personnel, supply and maintenance, etc., was computed. To obtain consistent programming, the ordering of the steps in the schedule was so arranged that the flow of information from echelon to echelon was only in one direction, and the timing of information availability was such that the portion of the program prepared at each step did not depend on any following step. Even with the most careful scheduling, it took about seven months to complete the process.

After the war, it became clear that efficient coordination of the energies of whole nations in the event of a total war would require scientific programming techniques. Undoubtedly, this need had occurred many times in the past, but this time there were two concurrent developments that had a profound influence: (a) the development of large-scale electronic computers and (b) the development of the inter-industry model proposed by Wassily Leontief.

The potential attraction of the inter-industry model was its simple linear structure. In some ways it was too simple. It was not dynamic, and it assumed that each industry had a unique technology which produced only one product. Another limitation of the model was that it was not possible to have alternative feasible programs.

It was necessary, therefore, to generalize the inter-industry approach. The result was the development of the linear-programming model. Intensive work began in June 1947 in an Air Force group under the Comptroller General Ed Rawlings that later was given the official title of Project SCOOP (Scientific Computation of Optimum Programs). Principals in the group included Marshall Wood, Murray Geisler, John Norton, and the author.

The simplex computational method for choosing the optimal feasible program was developed by the end of the summer of 1947. Interest in linear programming began to spread quite rapidly. During this period, the military sponsored work at the Bureau of Standards on electronic computers and on mathematical techniques for solving such models.

Early contacts with Tjalling Koopmans, of the Cowles Commission, who then was at the University of Chicago and who now is at Yale; and Robert Dorfman, then of the Air Force, now at Harvard; and the interest of such economists as Paul Samuelson, of the Massachusetts Institute of Technology, initiated an era of intense reexamination of classical economic theory based on the results and ideas of linear programming.

Early contact with John von Neumann at the Institute for Advanced Study gave fundamental insight into the mathematical theory and sparked the interest of A. W. Tucker of Princeton University and two of his former students, David Gale and Harold Kuhn. With Office of Naval Research support, they attacked problems in linear inequality theory and game theory. Princeton became an academic focal point in these related fields.

It was the size of the military planning problem which made it evident in the immediate post-war period that even the best of future computing facilities would not be powerful enough to find an optimal solution to a general detailed military planning model. Accordingly, Project SCOOP modified its approach and, in the spring of 1948, proposed that there be developed special linear programming models, called 'triangular models', whose structure and computational solution would closely parallel traditional stepwise staff procedure to provide a feasible, but not necessarily optimal, solution.

Since 1948, the military has been making more and more active use of mechanically computed programs. The triangular models are in constant use for the computation of detailed programs, while the general linear-programming models have been applied to smaller systems, such as (a) contract bidding; (b) balanced aircraft, crew training, and wing deployment schedules; (c) scheduling of maintenance overhaul cycles; (d) personnel assignment; and (e) airlift routing problems.

During the period from 1948 on, granting agencies—particularly the Office of Naval Research—began to support research seeking to develop efficient methods for finding optimal solutions to larger and larger planning systems.

3. THE INFLUENCE OF ECONOMIC MODELS

The current introduction of linear programming into economics appears to be an anachronism; it would seem more likely to have begun when economists first begun to describe economic systems in mathematical terms. Indeed, a crude example of a linear-programming model can be found in the Tableau Economique of Quesnay around 1758; he attempted to interrelate the roles of the landlord, the peasant, and the artisan. We find that L. Walras proposed, in 1874, a very sophisticated mathematical model which had as part of its structure fixed technological coefficients such as are assumed in linear programming. Oddly enough, until the 1930's there was little in the way of exploitation of the linear-type model.

Of particular note is the effort during the 1930's of a group of Austrian and German economists who worked on generalizations of the linear technology of Walras. This work raised some questions that may have stimulated the mathematician Von Neuman, in his paper 'A Model of General Economic Equilibrium', to formulate a dynamic linear-programming model in which he introduced alternative methods of producing given commodities singly or jointly. Von Neumann assumed (a) a constant rate of expansion of the economy and (b) a completely self-supporting economy. While the model did not contain any explicit objective, Von Neumann showed that market forces would maximize the expansion rate and proved a theorem that the maximum rate of expansion was equal to the interest rate on capital invested in production.

The inspiration of the general linear-programming model arose out of the practical planning needs of the military and the possibility of generalizing to this end the simple structure of the Leontief Model. From a purely formal standpoint, one could consider the Leontief Model as a simplification of the Walrasian Model. Actually, theoretical economic models were a kind of ivory tower. According to Leontief:

> 'One hundred and fifty years ago when Quesnay first published his famous schema, his contemporaries and disciples acclaimed it as the greatest discovery since Newton's laws. The idea of general interdependence among the various parts of the economic system has become by now the very foundation of economic analysis. Yet, when it comes to the practical application of this theoretical tool, modern economists must rely exactly as Quesnay did upon fictitious numerical examples.'

Leontief's great contribution (in the opinion of the author) was his construction of a quantitative model of the American economy for the purpose of tracing the impact of government policy and consumer trends upon a large number of industries which were imbedded in a highly complex series of interlocking relationships. To appreciate the difference between a purely formal mathematical model and an empirical model, it is well to remember that the acquisition of data for a real model requires an organization working many months, sometimes years. After the model has been put together, a second obstacle looms—the solution of a very large system of simultaneous linear equations. In the period 1936-1940 there were no electronic computers; the best that one could hope for in general would be to solve 20 equations in 20 unknowns. After this, there is still a third obstacle, the difficulty of 'marketing' the results of such studies. From the onset, the undertaking initiated by Leontief represented a triple gamble.

As a result of the Great Depression and the advent of the New Deal, a serious attempt was made by the government to determine and then support certain activities which it was hoped would speed recovery. This brought about more intensive collections of statistics on costs of living, wages, national resources, productivity, etc. There was a need to organize and interpret this data in order to construct a mathematical model to describe the economy in quantitative terms.

From 1936 on, the scope, accuracy, and area of application of Leontief-type models were greatly extended by the Bureau of Labor Statistics. It was the work there by Duane Evans, Jerome Cornfield and Marvin Hoffenberg that stimulated efforts toward seeking a mathematical generalization suitable for dynamic military applications.

A few words about the Leontief model itself are in order. The focal point of input-output analysis is an array of coefficients variously called the 'input-output' matrix or 'tableau economique'. A column of this matrix represents the input requirements of various commodities for the production of one dollar's worth of particular commodity. There is exactly one column for each commodity produced in the economy. Thus the production of a commodity corresponds to the concept of an activity in the linear-programming model. If the input factors appearing in a row of the matrix are multiplied by the corresponding buying industry's total output, the totals represent the distribution of the dollar value of purchases among the selling industries. Thus, the model makes it possible not only to determine each industry's rate of output to meet specified direct demand by civilians and the military, but also to trace the indirect effect on each industry of government expenditures in, say, military programs.

In 1947, T. C. Koopmans took the lead in bringing to the attention of economists the potentialities of the linear-programming models. His rapid development of the economic theory of such models was due to the insight he gained during the war with a special class of linear-programming models, called 'transportation models', which he applied to Allied shipping problems. In 1949, he organized the historic Cowles Commission conference on 'linear programming', which was attended by such well-known economists as K. Arrow, R. Dorfman, N. Georgescu-Roegen, L. Hurwicz, A. Lerner, J. Marschak. O. Morgenstern, S. Reiter, P. Samuelson, and H. Simon; such mathematicians as G. W. Brown, M. M. Flood, D. Gale, H. W. Kuhn, C. B. Tompkins, and A. W. Tucker; and by government statisticians, including W. D. Evans, M. A. Geisler, M. Hoffenberg, and M. K. Wood. The papers presented there were later collected into the book Activity Analysis of Production and Allocation.

The following is an interesting quotation from that book's introduction, written by Koopmans, the editor, which serves to characterize the linear-programming model:

> The adjective in 'linear model' relates only to (a) assumption of proportionality of inputs and outputs in each elementary productive activity and (b) the assumption that the result of simultaneously carrying out two or more activities is the sum of the results of the separate activities. In terms more familiar to the economist, these assumptions imply constant returns to scale in all parts of the technology. They do not imply linearity of the production function .... Curvilinear production functions ... can be obtained from the models ... by admitting an infinite set of elementary activities ....
>
> Neither should the assumption of constant returns to scale ... be regarded as essential to the method ... although new mathematical problems would have to be faced in the attempt to go beyond this assumption. More essential to the present approach is the introduction of ... the elementary activity, the conceptual atom of technology into the basic postulates of the analysis. The problem of efficient production then becomes one of finding the proper rules for combining these building blocks. The term 'activity analysis' ... is designed to express this approach.

The number of practical economic applications is continually growing. Linear programming is being used by economists to study in detail the economics of specific industries, such as metal working, petroleum refining, and iron and steel, and to yield long-range plans for electricity generation in an entire economy.

For a better appreciation of the economic implications, the reader is referred to Linear Programming and Economic Analysis, by Dorfman, Samuelson, and Solow, and Economic Theory and Operations Analysis, by W. J. Baumol.

4. MATHEMATICAL HISTORY

The linear-programming model, when translated into purely mathematical terms, requires a method for finding a solution to a system of simultaneous linear equations and linear inequalities which minimizes a linear form. This central mathematical problem was not known to be important until the advent of linear programming in 1947.

We are all familiar with methods for solving linear equation systems, which start with our first course in algebra. The literature of mathematics contains thousands of papers concerned with techniques for solving linear equation systems with the theory of matrix algebra (an allied topic), with linear approximation methods, etc. On the other hand, the study of linear inequality systems excited virtually no interest until the advent of game theory in 1944 and linear programming in 1947. For example, T. Motzkin, in 1936, in his doctoral thesis on linear inequalitites, was able to cite after diligent search only some 30 references for the period 1900-1936 and about 42 in all. In the 1930's, four papers dealt with the building of a comprehensive theory of linear inequalities and with an appraisal of earlier works. These were by R. W. Stokes, Dines McCoy, H. Weyl, and T. Motzkin. As evidence that mathematicians were unaware of the importance of the problem of seeking a solution to an inequality system that also minimized a linear form, we may note that none of these papers made any mention of such a problem, although there had been earlier instances in the literature.

The famous mathematician Fourier, while not going into the subject deeply, appears to have been the first to study linear inequalities systematically and to point out their importance to mechanics and probability theory. He was interested in finding the least maximum deviation fit to a system of linear equations, which he reduced to the problem of finding the lowest point of a polyhedral set; he suggested a solution by a vertex-to-vertex descent to a minimum, which is the principle behind the simplex method used today. This is probably the earliest known instance of a linear-programming problem. Later, another famous mathematician, de la Vallee Poussin, considered the same problem and proposed a similar solution.

4.1 The Work of Kantorovitch

The Russian mathematician L. V. Kantorovitch has for a number of years been interested in the application of mathematics to programming problems. He published an extensive monograph in 1939 entitled 'Mathematical Methods in the Organization and Planning of Production'.

In his introduction, Kantorovitch states:

> There are two ways of increasing efficiency of the work of a shop, an enterprise or a whole branch of industry. One way is by various improvements in technology, that is, new attachments for individual machines, changes in technological processes, and the discovery of new, better kinds of raw materials. The other way, thus far much less used, is by improvement in the organization of planning and production. Here are included such questions as the distribution of work among individual machines of the enterprise or among mechanisms, orders among enterprises, and the correct distribution of different kinds of raw materials, fuels and other factors.

Kantorovitch should be credited with being the first to recognize that certain important broad classes of production problems had well defined mathematical structures which, he believed, were amenable to practical numerical evaluation and could be numerically solved. If Kantorovitch's earlier efforts had been appreciated at the time they were first presented, it is possible that linear programming would be more advanced today. However, his early work in this field remained unknown both in the Soviet Union and elsewhere for nearly two decades while linear programming became a highly developed art. According to The New York Times,

> The scholar, Professor L. V. Kantorovitch, said in a debate in 1959 that Soviet economists had been inspired by a fear of mathematics that left the Soviet Union far behind the United States in applications of mathematics to economic problems. It could have been a decade ahead.

During the summer of 1947 Leonid Hurwicz, well-known econometrician associated with the Cowles Commission, worked with the author on techniques for solving linear-programming problems. This effort and some suggestions of T. C. Koopmans resulted in the 'simplex method'. The obvious idea of moving along edges from one vertex of a convex polyhedron to the next (which underlies the simplex method) was rejected earlier on intuitive grounds as inefficient. In a different geometry it seemed efficient, and so, fortunately, it was tested and is now accepted as the standard solution procedure.

4.2 The Work of Von Neumann

Credit for laying the mathematical foundations of this field goes to John von Neumann more than to any other man. During his lifetime, he was generally regarded as the world's foremost mathematician. He played a leading role in many fields. Perhaps in the long run his stimulation of electronic-computer development during World War II will prove to be his most significant contribution. In 1944, John von Neumann and Oskar Morgenstern published their monumental work on the theory of games, a branch of mathematics that aims to analyze problems of conflict by use of models termed 'games'. A theory of games was first broached in 1921 by Emile Borel and was first established in 1928 by von Neumann with his famous 'Minimax Theorem'. The significance of this effort for us is that game theory, like linear programming, has its mathematical foundation in linear inequality theory.

Von Neumann, at the first meeting with the author in October 1947, was able immediately to translate basic theorems in game theory into their equivalent statements for systems of linear inequalities. He introduced and stressed the fundamental importance of duality and conjectured the equivalence of games and linear-programming problems. Later, he made several proposals for the numerical solution of linear-programming and game problems.

New computational techniques and variations of older techniques are continuously being developed in the United States and abroad. The well-known econometrician Ragnar Frisch at the University of Oslo has done extensive research work on his 'multiplex method'. Investigations in Great Britain have been spearheaded by S. Vajda and M. Beale. A number of important variants of the simplex method have been proposed by C. Lemke, W. Orchard-Hays, E. M. L. Beale, P. Wolfe and others.

5. ELECTRONIC COMPUTER CODES

A special variant of the simplex method that was developed for the transportation problems was first coded in 1950 for the National Bureau of Standards SEAC computer. The general simplex method was coded in 1951 under the general direction of A. Orden of the Air Force and A. J. Hoffman of the Bureau of Standards (later, Hoffman was with ONR and IBM). In 1952, W. Orchard-Hays of the RAND Corporation worked out a simplex code for the IBM—C.P.C. and later for the IBM 701, 704, etc.

His code turned out to be practical for commercial applications. As a result, the use of electronic computers by business and industry grew by leaps and bounds. Many of the digital computers which are commercially available provide, as part of their software, codes of the simplex technique. In fact, computer companies are now spending close to a half million dollars on software development of a complete linear-programming system which is free to customers who buy or rent their equipment.

6. EXTENSIONS OF LINEAR PROGRAMMING

If we distinguish, as indeed we must, between those types of generalizations in mathematics that have led to existence proofs and those that have led to constructive solutions of practical problems, then current developments mark the beginning of several important constructive generalizations of linear-programming concepts to allied fields.

6.1 Network Theory

A remarkable property of one very special class of linear programs, namely, the transportation or the equivalent network flow problem, is that their solutions are always in integers. This has been a key fact linking certain combinatorial problems found in mathematical topology with certain continuous problems of network theory. The field has many contributors. Of special mention is the work of Kuhn (for finding a permutation of ones in a matrix composed of zeros and ones) and the related work of Ford and Fulkerson of RAND (for network flows).

6.2 Nonlinear Programming

A natural extension of linear-programming occurs when the linear part of the inequality constraints and the objective are replaced by convex functions. Early work by Barankin and Dorfman centered about a quadratic objective and culminated in an elegant procedure developed independently by Beale, Houthakker, and Wolfe. Wolfe showed how a minor variant of the simplex procedure could be used to solve such problems. Duality concepts first proposed by von Neumann have been successfully extended to certain classes of nonlinear programs. As a result of these investigations, a new uniform procedure has been developed which solves linear programs, quadratic programs, and general matrix games. The research of C. Lemke and Howson, H. Scarf, and the joint work of R. Cottle and the author should be mentioned. Large-scale systems with mixed linear and nonlinear parts are of great practical importance in industry today, and interest in this field is growing rapidly.

6.3 Integer Programming

Important classes of nonlinear, nonconvex, discrete, combinatorial problems can be shown to be formally reducible to a linear-programming problem with the additional restriction that some or all of the variables must be integer valued. The linear-programming approach was used in 1954 by Fulkerson and Johnson and the author to construct an optimal tour for a salesman visiting Washington, D.C., and 48 State capitals of the United States. Our theory was incomplete, however. The foundations for a rigorous theory were first developed by Ralph Gomory in 1958 under an Office of Naval Research contract with Princeton. Many important planning problems are integer programming problems. The best location of warehouses, the optimal routing of a fleet of supply ships, optimal provisioning under space and weight limitations, the optimal sequencing of jobs on machines, the assignment of crews to meet an airline routing schedule, and optimal ways to cut out patterns from stock materials are some examples.

6.4 Programming under Uncertainty

It has been pointed out that programming under uncertainty cannot be usefully stated as a single problem. One important class is a multistage one in which the technological matrix of input-output coefficients is assumed known and the values of the constant terms uncertain, but the joint probability distribution of their possible values is assumed to be known. Research in this field is still in its infancy, and practical planners will continue to resort for some time to heuristic schemes to cover stochastic events.

7. LARGE-SCALE SYSTEM DEVELOPMENT

Of all the progeny of linear programming, perhaps the most fruitful development at the present time is that of techniques for solving linear programs with special structures. Many groups have been developing linear-programming models of their firms for over a decade. As the years go by, the trend is toward larger and more comprehensive corporate models. These are multistaged and dynamic and exhibit a hierarchal structure. Although many proposals have been made, little in the way of practical codes was developed to handle such problems until recently.

In 1959, Philip Wolfe and the author proposed the decomposition principle, an approach which decomposes a model into smaller parts which can be independently optimized. The solution of each of the parts are treated as proposals which are modified to be consistent with total system resources and demands. Several companies, such as C—I—E—R, Mathematica, and Bonner-and-Moore, which specialize in developing computer programs, have written decomposition codes. In one application, that of the National Biscuit Company, a system of nearly a million variables and thirty-two thousand equations has been successfully solved.

8. INDUSTRIAL APPLICATIONS OF LINEAR PROGRAMMING

The history of the first years of linear programming would be incomplete without a brief survey of its use in business and industry. These applications began in 1951 but have had such a remarkable growth that the commercial offspring has overtaken its military parent.

Linear programming has been serving industrial users in several ways. First, it has provided a novel view of operations; second, it induced research in the mathematical analysis of the structure of industrial systems; and third, it has become an important tool for business and industrial management for improving the efficiency of their operations. Thus, the application of linear programming to a business or industrial problem has required the mathematical formulation of the problem and an explicit statement of the desired objectives. In many instances, such rigorous thinking about business problems has clarified aspects of management decision-making which previously had remained hidden in a haze of verbal arguments. As a partial consequence, some industrial firms have started educational programs for their managerial personnel in which the importance of the definition of objectives and constraints on business policies is being emphasized. Moreover, scheduling industrial production traditionally has been based on intuition and experience, a few rules, and the use of visual aids, as has scheduling in the military. Linear programming has induced extensive research in developing quantitative models of industrial systems for the purpose of scheduling production. Of course, many complicated systems have not as yet been quantified, but sketches of conceptual models have stimulated widespread interest.

The first and most fruitful industrial applications of linear programming have been to the scheduling of petroleum refineries. As noted earlier, Charnes, Cooper, and Mellon started their pioneering work in this field in 1951. Two books have been written on the subject, one by Gifford Symonds and another by Alan Manne. So intense has been the development that a survey by Garvin, Crandall, John, and Spellman in

1957 showed that there are applications by the oil industry in every phase of its activities from exploration, production, and refining to final distribution and sales.

The food-processing industry is perhaps the second most active user of linear programming. In 1953, a major producer first used it to determine shipping schedules for catsup from six plants to 70 warehouses.

In the iron and steel industry, linear programming has been used for the evaluation of various iron ores and of the pelletization of low-grade ores and used to determine the shape of the hole to be dug in open-pit mining. Additions to coke ovens and shop loading of rolling mills have provided additional applications. A linear-programming model of an integrated steel mill has been developed. The British steel industry has used linear programming to decide what products their rolling mills should make in order to maximize profit.

Metalworking industries use linear programming for shop loading and for deciding whether to make a part in a shop or to buy it outside. Paper mills use it to decrease trim losses and to decide which of their several mills should be assigned to respond to an order for paper.

The optimal routing of messages in a communication network, contract-award determinations, and the routing of aircraft and ships are problems that were first considered for application of linear-programming methods by the military but that are now applied in industry.

One overall measure of the use of linear-programming and its extensions is the dollars spent on computer time in the United States. This is known to run in the millions.

9. A FEW WORDS ABOUT THE FUTURE

One of the most startling developments in recent years is the penetration of the electronic computer and mathematics into almost every phase of human activity.

If there is a library, then someone is at work representing (inside the memory of a computer) a book number, its title, its shelf location, who has it on loan, the date due, the author, its call number, its cross references, its frequency of use, and so on. A library is like a population that does not bury its dead. Out of this rather straightforward effort of getting some of the present information about a library into a more manipulatable form will emerge the 'information storage and retrieval system' of tomorrow, in which the old physical book and printed paper page may well become a relic like an ancient scroll.

Wherever one finds a system for processing insurance premiums, for keeping track of bank deposits and withdrawals, for recording airline reservations, or for any other type of inventory control, someone is at work simulating such a system in an electronic computer and forging the links whereby the real world supplies information to the computers and the orders of the computer are translated into real actions.

It is correct to regard much of what has been done so far as a vast tooling up, a preparation for new ways to do old tasks. It is the exponential improvement in electronic hardware, the availability of new machine languages and special machine programs that now permit the practical implementation of these ideas. We are witnessing an accelerated trend toward automation of simple human control tasks.

Operations research refers to the science of decision and its application. In its broad sense, the word 'cybernetics', the science of control, may be used in its place. This science is directed toward those tasks that humans have not yet delegated to machines. Tasks involving human energy and (as we have seen) those involving simple human control already have been conceded to machines even though they have not been taken over fully by them. It is the automation of higher order human decision processes that is the last citadel.

At the lowest level of these higher order tasks is the human ability to recognize patterns in sight, sound, touch, smell, and taste. Although these tasks may elicit simple responses (such as turn the wheel to the right or left), human presence is needed because a complex mental recognition process is involved. It would be easy to get a machine to mechanically separate returned Coke and Pepsi bottles if it were smart enough to recognize which is which.

At the next level of complexity is the human ability to observe and to adapt to physical movement; for example, to observe a dial or a car's angle to the road direction and to manipulate certain controls to change the physical movement in some preferred way. Here, again, it is easy to get a machine to make the physical movement of the controls if the machine is smart enough to adapt to trends in the observed movements as changes are made in the controls.

Although pattern recognition is by no means a solved problem, banks do have machines that recognize account numbers on checks, and machines do exist that give change for a dollar bill but not for a blank piece of paper. Automatic feed-back controls in simple situations have been known for a long time. The governor invented by Watt to control the speed of a steam engine is such a device. Closed-loop controls that rely on computers to analyze input data are now a reality in certain large-scale operations, such as oil refineries, chemical plants, and power-distribution systems.

At still a higher level of complexity are those decision processes that involve many alternative courses of action. An industrial complex may have at its disposal many types of equipment and a variety of raw materials and personnel skills. The complex could manufacture a vast variety of products by means of many alternative process sequences. If the wrong decisions are made in the scheduling of the various processes, labor and machines are idle, through-put is reduced, and in-process inventories are increased. If the wrong decisions are made in raw-material selection, the procedure for manufacture, or the choice of final product, labor and machines are over-worked, expensive materials are purchased when cheap ones will do, and unwanted products are dumped on the market.

In the last two decades, great strides have been made to effectively use electronic computers as part of the planning process. As we have noted already, a pioneering effort of this kind was first initiated by the military around 1947 in Project SCOOP. Part of the ramifications of that project included a 400-sector inter-industry model of the national economy. Except for the preparation of input data, the calculation of various planning programs was completely mechanized. The size of systems handled was truly enormous. A program typically stated by month (for 36 months) the level of each type of activity for thousands of activity types. The balanced flows of some tens of thousands of input and output items (necessary to support these activities) were also stated as a function of time. As ground rules, appropriations, or international conditions changed, these programs were recalculated rapidly again and again.

This early pioneering effort at mechanization of the planning process showed that it was possible to describe in mathematical terms the interdependence of various activities, such as training, the work of a combat unit, an engine overhaul, steps in an industrial process, and the shipment of goods from various places of origin to numerous destinations. The approach, as we saw earlier, is to make each activity elementary enough so that its inputs and outputs are proportional to the level of the activity. The resulting mathematical system is one of linear inequalities called a linear program. By incorporating this mathematical approach, planning staffs have been relieved of much drudgery and have been able to concentrate more and more on overall objectives.

True optimization is the revolutionary contribution of modern research to decision processes. In the entire history of mankind, a great gulf has always existed between man's aspirations and his actions. He may have wished to state his wants in terms of objectives, but there were so many possible different ways to go about it, each with its own good and bad, that it was impossible to compare them and to say which was the best. Man invariably turned to a leader, a manager, a governor, or a commanding officer, whose 'experience' and 'mature judgement' would point the way. Inevitably, 'the way' became the new objective. This substitution of the means for the objective is the history of mankind. The slogan 'the end justifies the means' perhaps could be better stated as 'the end might conceivably justify the means if one could ever remember what the original objective was'.

Because man did not have the ability to select the best among infinite alternatives, his planning was characterized by many ground rules and policies dictated by men with 'mature judgement'. It seemed impossible, therefore, that planning could ever be done by computer without constantly stopping the machine to await decisions from the experts. The habits of centuries are not easily overcome, but planning staffs freed from the drudgery of computing one or possibly two alternatives now are beginning to express themselves in terms of overall objectives and to ask the computers to find them the 'best'.

As the power of computing machinery has increased and the power of methods proposed by mathematicians has grown, the achievement of optimal solutions to large-scale complex planning problems has become a reality.

The mathematical systems developed for planning in industry and the military are truly among the largest in the world. Typical problems run from 300 to 800 equations. Some codes are designed to solve practical problems involving as many as 4000 equations. In all cases, the number of possible activities (variables) run in the thousands. We have already mentioned a method, called the decomposition principle, for handling extremely large systems.

Let us now turn to cybernetics developments in the U.S.S.R. In 1939, Leonid Kantorovitch proposed that mathematical methods be applied to planning problems. As we pointed out earlier, for one and a half decades his work was unknown to the Western world as well as in Russia (Communist policy had shelved it as dangerous). In 1959, the Supreme Soviet decreed cybernetics top priority. Scientific circles in Russia, as a result, have begun a vast tooling up. Computers are not as plentiful there as they are in the United States, but this does not prevent education on their design nor work on the development of machine languages nor work on the design of complex control systems. It is stated by one authority that there are ten times as many engineers and mathematicians working on control theory in the U.S.S.R. as in the United States.

It is reported that Russians feel that through their efforts in cybernetics they will outdistance the productivity of the Western world. According to economists who have studied developments there, the Soviets feel that the weak point of their planning system lies at the middle management level, the factory managers. This group apparently is unable to exercise initiative and modify its given production plans. On the other hand, their top management (at industrial and governmental levels) appears to be the equal of ours. The same is true at their lower management levels. Their hope is to replace middle-management planning with a highly flexible scheme which they believe is inherently superior to ours.

We should not dismiss too quickly their idea of overtaking us. It should be noted that, historically, it has been difficult to change the direction of large-scale enterprises once they have gathered momentum. These momentums have been the underlying cause of our business cycles and depressions. It would thus appear that timely, balanced, optimal programs encompassing a broad spectrum of enterprises could have enormous payoffs. They believe that excellent detailed planning and rapid flow of information could avoid these problems completely. Soviet Academician Vasili Nemchinov sums it up this way:

> The Communist system alone gives sufficient room to apply the combination of

> mathematics and cybernetics to the national economy. Only under the system of public ownership is it possible to introduce into the economy a single automated electronic system of planning. In a private enterprise system the use of cybernetics is restricted by the framework of companies, corporations, and syndicates.

Whether computers and programming can accomplish this miracle of putting the Soviet Union ahead of everyone else is, therefore, a challenging question. We have seen that every step necessary to fulfill this aim has already been service-tested on a large scale in this country, particularly in military planning and in industry. We have seen the rapid improvements in computer technology and how computers are penetrating into record keeping and process control so that input data can be made available instantly to the planning system and the results of the computation can be prepared in a form for necessary feed-back action.

Can computers be programmed to solve the truly immense systems characteristic of a national economy, particularly dynamic systems involving optimization? Here again we note that by use of the decomposition principle of linear programs, systems of the order of 3×10^4 equations and 10^6 variables have already been solved. Ponder the statement made by the Russian, Malkov: 'Half of the mathematicians in the Moscow Computing Center are working on decomposition problems.' Even if total system optimization is presently impossible—and it is—there are all kinds of schemes involving partial aggegation that permit near-optimal solutions.

We are witnessing a computer revolution in which nearly all tasks of man—be they manual labor or simple control, pattern recognition or complex higher order decision making—all are being reduced to mathematical terms and their solution delegated to computers. It is in the latter development that linear programming (and its extensions) is playing a key role.

REFERENCES

DANTZIG, GEORGE B., and PHILIP WOLFE, Decomposition principle for linear programs, Oper. Research, 8, (January-February 1960), 101-111; also Econometrica, 29, (October 1961), 767-778. These are different papers on the same subject. Also in Rand Symposium on Mathematical Programming (P. Wolfe, Ed.) (March 1959), RAND R-351, p. 5.

DANTZIG, GEORGE B., Large-scale linear programming, Technical Report 67-8 (November 1967), Department of Operations Research, Stanford University; also in Proceedings of Sixth International Symposium on Mathematical Programming, Princeton University, (August 1967) published by Princeton University Press, also in Proceedings of American Mathematical Society Summer Seminar on Mathematics of the Decision Sciences, published by American Mathematical Society (1968).

DANTZIG, GEORGE B., Linear Programming and Extensions, Princeton University Press, Princeton, New Jersey, (August 1963), 621 pages, revised edition Fall, 1966; Fourth Printing, 1968.

DANTZIG, GEORGE B., Linear programming under uncertainty, Management Science, 1, (1955), 197-206; also in Mathematical Studies in Management Science (A. F. Veinott, Jr., ed.) The MacMillian Co., New York (1965), pp. 330.

DANTZIG, GEORGE B., Operations research in the world of today and tomorrow, Operations Research Center, University of California, Berkeley, 65-7 (January 1965); also submitted to Technion Yearbook; also in commemorative volume for Professor Kreller, Institut fur Okonometric; presidential address TIMS 1966 entitled Management Science in the World of Today and Tomorrow in Management Science, Vol. 13, No. 6, February 1967.

FORD, LESTER R., Jr. and DELBERT R. FULKERSON, 1954-1. 'Maximal Flow through a network', The RAND Corporation, Research Memorandum RM-1400, November 19, 1954; also The RAND Corporation, Paper P-605, November 19, 1954. Published in Canad. J. Math., Vol. 8, No. 3, 1956, pp. 399-404.

GOMORY, R. E., 1965-1. 'On the Relation Between Integer and Noninteger Solutions to Linear Programs,' Proc. Natl. Acad. Sci., U.S.A., Vol. 53, No. 2, pp. 260-265.

HOFFMANN, A. J., 1960-1, 'Some Recent Applications of the Theory of Linear Inequalities to Extremal Combinatorial Analysis', in R. Bellman and Marshall Hall, Jr. (eds.), Proceedings of Symposia in Applied Mathematics, Vol. X, Combinatorial Analysis, American Mathematical Society, Providence, Rhode Island, 1960, pp. 113-127.

KANTOROVICH, L. V., 1939-1. 'Mathematical Methods in the Organization and Planning of Production', Publication House of the Leningrad State University, 1939, 68 pp. Translated in Management Sci., Vol. 6, 1960, pp. 366-422.

KOOPMANS T. C., (ed.) 1951-1. Activity Analysis of Production and Allocation, John Wiley and Sons, Inc., New York, 1951, 404 pp.

LEONTIEF, WASSILY 1951-1. The Structure of American Economy, 1919-1931, Oxford University Press, New York, 1951.

von NEUMANN, JOHN, 1928-1, 'Zur Theorie de Gesellschaftsspiele', Math. Ann., Vol. 100, 1928, pp. 295-320. Translated by Sonya Bargmann in A. W. Tucker and R. D. Luce (eds.) Contributions to the Theory of Games, Vol. IV, Annals of Mathematics Study No. 40, Princeton University Press, Princeton, New Jersey, 1959, pp. 13-42.

von NEUMANN, JOHN, 1937-1. 'Uber ein okonomisches Gleichungssystem und ein Verallgemeinerung des Brouwerschen Fixpunktsatzes', Ergebnisse eines mathematischen Kolloquiums, No. 8, 1937. Translated in Rev. Econ. Studies, Vol. 13, No. 1, 1945-46, pp. 1-9.

von NEUMANN, JOHN, and OSKAR MORGENSTERN, 1944-1. Theory of Games and Economic Behavior, Princeton University Press, Princeton, New Jersey, 1944; 2nd ed. 1947; 3rd ed. 1953.

WILLIAMS, N., Linear and Nonlinear Programming in Industry, Pitman Press, 1967.

WOLFE, P., 'Methods of Nonlinear Programming', in Nonlinear Programming, J. Abadie, (ed.) pp. 99-134, North-Holland Press, 1967.

WOOD, Marshall K., and M. A. GEISLER, 1951-1, 'Development of Dynamic Models for Program Planning', in T. C. Koopmans (ed.), Activity Analysis of Production and Allocation, John Wiley & Sons, Inc., New York, 1951, pp. 189-192.

DISCUSSION

M. Arbabi

When should one use decomposition codes to solve large linear programs, assuming that the present LP code on third generation hardware can solve an LP having up to 8000 rows (e.g. MPS 360 on IBM 65 with I core can solve an 8192 row-LP)?

Is decomposition code faster even on small problems?

G. B. Dantzig

I recommend that a special large-scale technique be used to solve linear programs of size greater than one thousand equations. It is my opinion that if this is done the computation time will be a fraction of that needed using a general simplex approach (particularly for problems of the size of 8000 equations). A great deal more experimentation is needed, however, to determine which of the various proposals for solving large-scale systems (partitioning, decomposition principles, etc.) would probably be best for a particular application.

A. M. Dunsmuir

I should like to report very encouraging results with a 4000 row problem containing some 2000 common rows using the MPS/360 agendum BASIC, rather than true decomposition. Running on a 360-65/50 ASP system we have found near-optimal solutions in some $4\frac{1}{2}$ hours, which includes 10 minutes preliminary optimisation for each of six subproblems.

For formulations such as this, which contain too many common rows for true decomposition to provide a feasible approach but for which intelligent guesses can be made as to the 'boundary values' of the subproblems, BASIC does seem to offer a promising line of attack.

G. B. Dantzig

As I understand it, your method allocates the RHS of the common rows to the various subs in order to get a good starting solution. In general it appears to be a good idea to begin with a good guess. I believe you would get markedly better results with BASIC (with very little extra effort) if you next change the amount of allocation of each row when the prices on the row as obtained from various subs differ.

SECTION 2 Linear Programming

P. Demonsablon
R. Razani
N. Williams

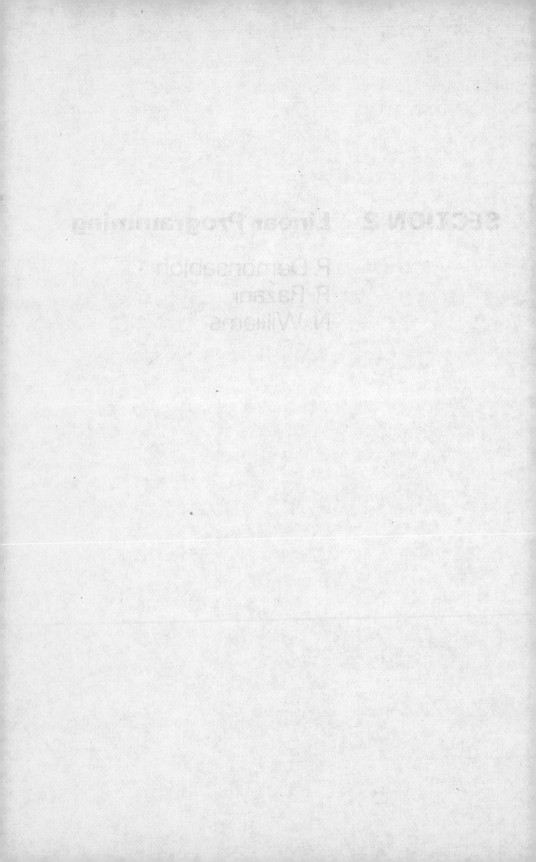

Application de la Programmation Linéaire au Calcul à la Rupure des Fondations sur Pieux

P. DEMONSABLON
Serequip, France

SUMMARY

This paper is based on Nökkentved's model for the determination of the reactions in a group of piles. The model enables one to define a coefficient of safety, either by an elastic expression for which one ascertains the limitations, or, alternatively, by a calculation of the load at which failure occurs.

The plastic failure of an isolated pile can take two forms: through disruption of the equilibrium between the pile and the ground (first type), or fracture of the constituent material of the pile (second type).

The theorem of the uniqueness of the state of fracture (which is applicable to failures of the first type when the resultant of the exterior forces is known) is reviewed, and this property is extended to more general types of forces. The determination of the state of fracture from these properties corresponds to the optimisation of a linear program of which the details relating to the problem are brought out. The problem is discussed with the help of a computer program; some examples of applications are given.

The dual form of the problem results from the theorem of virtual work.

The method is only applied to failures of the first type, but does not bear any restriction as far as the external forces and their transmission to the foundations are concerned.

1. SCHEMAS STATIQUES DE DETERMINATION DES REACTIONS DANS UN GROUPE DE PIEUX

Les efforts reportés sur les pieux d'un groupe liés à une superstructure se répartissent entre ceux-ci en fonction

- de la déformabilité de la superstructure, des pieux et du terrain
- des liaisons des pieux avec la superstructure et le terrain

Différentes hypothèses élémentaires peuvent ainsi être envisagées, comme indiqué dans le tableau ci-dessous

	Hypothèse concernant	1	2	3
a	La superstructure	Indéformable	Déformable	
b	La liaison pieu—superstructure	Articulation	Encastrement	Encastrement élastique
c	La liaison pieu—terrain, à l'égard des déplacements latéraux	pas de liaison	Réaction proportionnelle au déplacement	Réaction tenant compte de la résistance du sol au cisaillement
d	La liaison pieu—terrain, à l'égard des rotations axiales	pas de liaison	Réaction proportionnelle au déplacement	Réaction tenant compte de la résistance du sol au cisaillement

Les lois régissant la répartition des réactions sur un groupe de pieux, sous l'effet d'un effort extérieur donné, ont été formulées pour un certain nombre de modèles élastiques qu'il est commode de représenter par une suite ordonnée de quatre chiffres, correspondant aux quatre hypothèses élémentaires explicitées ci-dessus. Par exemple:

Modèle de Nökkentved:	1. 1. 1. 1.	[Nökkentved, 1924]
Modèle d'Hetenyi:	1. 1. 2. 1.	
Modèle d'Asplund:	1. 2. 2. 1.	[Asplund, 1956]
Modèle de Paduart:	2. 3. 1. 1.	[Paduart, 1949]

Ces différentes hypothèses appellent les remarques suivantes:

<u>Hypothèses a et b</u>

La prise en compte de l'encastrement des pieux dans la superstructure ne s'impose généralement pas. On le constate a posteriori en calculant les moments en têtes de pieux résultant de l'application à ceux-ci des déplacements estimés selon le modèle de Nökkentved; ces moments sont faibles dans la plupart des cas, et l'on en conclut que l'hypothèse de l'articulation en tête ne conduit pas à surestimer sensiblement la déformabilité de la fondation.

On ne pourra toutefois plus négliger l'encastrement:

— lorsque les pieux auront des diamètres importants (de l'ordre du mètre)

— lorsque la superstructure sera déformable: dans le cas des appontements ou des semelles souples sur pieux, il conviendra d'adopter le modèle de Paduart.

<u>Hypothèses c et d</u>

A notre connaissance, la réaction du terrain à la rotation du pieu autour de son axe n'a jamais été prise en compte. Elle devrait, dans un modèle cohérent, faire l'objet des mêmes hypothèses que la réaction au déplacement latéral.

Il est prudent de négliger la réaction latérale du terrain:

— lorsque les pieux sont encaissés dans un sol susceptible de disparaître ou d'être mis temporairement en suspension (lits affouillables)

— lorsque les réactions latérales risquent de provoquer dans le terrain des déformations permanentes (efforts de longue durée appliqués à des fondations en terrain cohérent).

Les réactions latérales des terrains, lorsqu'elles sont prises en compte, sont généralement assimilées à l'action de ressorts de raideur constante (terrains cohérents) ou proportionnelle à la profondeur (terrain pulvérulents). Ce modèle néglige par conséquent:

— l'intéraction des diverses couches horizontales rencontrées le long du fût d'un même pieu, c'est-à-dire la résistance du terrain au cisaillement dans le sens horizontal

— l'interaction des différents pieux sur les réactions latérales qu'ils peuvent exercer sur le terrain, problème particulièrement complexe, même en terrain homogène, par suite de la variation des distances entre pieux selon la profondeur.

En définitive, nous estimons que, dans le cas des superstructures indéformables, on pourra généralement s'en tenir aux hypothèses de l'articulation pieu—superstructure et de l'absence de réaction pieu—terrain

— soit pour des raisons de sécurité

— soit parce que les modèles plus raffinés, qui sont d'un maniement plus lourd, ne peuvent toutefois représenter de façon plus satisfaisante les phénomènes physiques ni, par conséquent, fournir des résultats plus sûrs.

Nous nous référons donc au modèle de Nökkentved, caractérisé comme suite (Fig. 1):

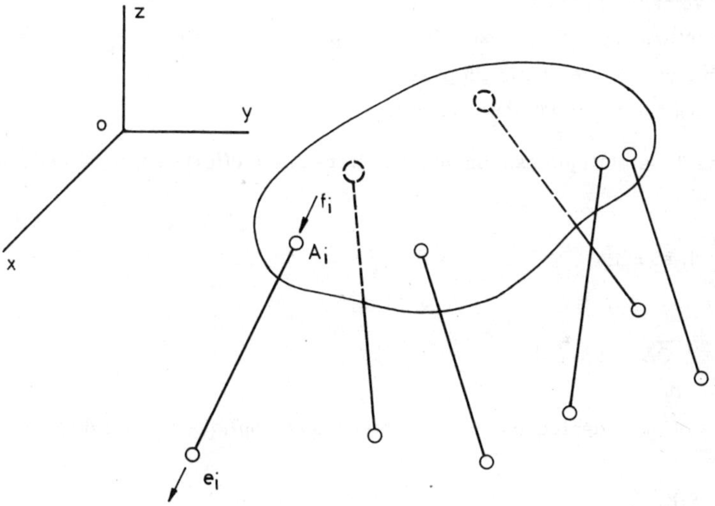

Fig. 1

— les pieux sont des barres élastiques subissant des variations de longueur proportionnelles aux efforts appliqués suivant leur axe

— ces barres sont articulées en tête à un solide indéformable, le massif, qui peut subir des déplacements compatibles avec ses liaisons, et en pied à un solide fixe, indéformable, le sol.

Ce dernier point de la définition semble exclure à la fois les pieux flottants et les pieux portant en pointe sur un sol non rocheux. En fait, ces deux cas rentrent dans le domaine de validité du modèle, à condition que l'on remplace le module d'élasticité du matériau constitutif du pieu par un module d'élasticité apparent.

Soient, pour le pieu i d'un groupe de n pieux repérés dans un système orthonormé oxyz:

$\vec{e_i}$ le vecteur unitaire porté par l'axe du pieu

$\vec{u_i}$ le moment en o de $\vec{e_i}$

A_i la position de la tête du pieu

l_i la longueur du pieu

S_i la section du pieu

E_i le module d'élasticité apparent du pieu

soit d'autre part un système d'efforts appliqués à la superstructure et défini par ses éléments de réduction en o:

$\vec{R}$ résultante

$\vec{M}$ moment

Ce système provoque

— relativement à la superstructure, un déplacement dont les éléments de réduction en o sont

$\vec{t}$ translation

$\vec{\omega}$ rotation

— relativement au pieu i:

f_i réaction (algébrique) exercée sur le pieu

$\overrightarrow{\delta A_i}$ déplacement de la tête du pieu

δl_i allongement (algébrique) du pieu

Les équations de la statique exprimant l'équilibre entre efforts appliqués et réactions sont

$$(1)\quad \begin{cases} \sum_{i=1}^{n} f_i\, \vec{e_i} = \vec{R} \\ \\ \sum_{i=1}^{n} f_i\, \overrightarrow{OA_i} \wedge \vec{e_i} = \vec{M} \end{cases}$$

Les relations efforts—déplacements sont obtenues en appliquant la loi de Hooke à chacun des pieux:

$$(2)\quad f_i = \frac{S_i E_i}{l_i}\,(\vec{t}\,.\,\vec{e_i} + \vec{\omega}\,.\,\vec{m_i})$$

soit, en posant

$$\vec{\mathcal{E}} = \begin{array}{r} 1 \\ \\ 6 \end{array} \left| \begin{array}{c} \vec{R} \\ \\ \vec{M} \end{array} \right| , \qquad \vec{u_i} = \begin{array}{r} 1 \\ \\ 6 \end{array} \left| \begin{array}{c} \vec{e_i} \\ \\ \vec{m_i} \end{array} \right| , \qquad \vec{d} = \begin{array}{r} 1 \\ \\ 6 \end{array} \left| \begin{array}{c} \vec{t} \\ \\ \vec{\omega} \end{array} \right|$$

$$P = \begin{array}{r} \\ 1 \\ i \\ \\ n \end{array} \left| \begin{array}{c} 1 \qquad 6 \\ \cdots\cdots \\ \hline \vec{u_i} \\ \hline \cdots\cdots \\ \cdots\cdots \end{array} \right| \; . \; \vec{f} = \begin{array}{r} 1 \\ i \\ n \end{array} \left| \begin{array}{c} \vdots \\ f_i \\ \vdots \end{array} \right| , \qquad K = \begin{array}{r} \\ 1 \\ i \\ n \end{array} \left| \begin{array}{ccc} 1 & i & n \\ & o & \\ & o & \\ & oo S_i E_i oo & \\ & o & \\ & o & \end{array} \right|$$

(1) : $\boxed{P^t . \vec{f} = \vec{\mathcal{E}}}$

(2) : $\boxed{\vec{f} = K . P . \vec{d}}$

d'où l'on déduit le déplacement $\vec{d}$ résultant du système d'efforts $\vec{\mathcal{E}}$:

(3) $\boxed{\vec{d} = (P^t . U . P)^{-1} . \vec{\mathcal{E}}}$

2. INSUFFISANCE DU CALCUL ELASTIQUE—UTILITE D'UN CALCUL A LA RUPTURE

Les formules (3) et (2) permettent de calculer les réactions exercées par les pieux sur le terrain sous l'effet d'un système donné d'efforts extérieurs $\vec{\mathcal{E}}$, pourvu que les réactions ainsi définies soient intérieures à un domaine élastique, constitué de l'intersection des domaines élastiques suivants:

- domaine élastique de la superstructure
- domaine élastique du pieu (matériau constitutif du pieu)
- domaine élastique du terrain (liaison pieu—terrain)

Dans le modèle de Nökkentved, ces deux domaines ne sont considérés que sous efforts longitudinaux aux pieux. Le lecteur adaptera de lui-même aux autres modèles les considérations développées dans ce chapitre.

Dans un calcul élastique classique, le coefficient de sécurité est défini à partir de la comparaison entre les contraintes réelles et les contraintes-limites en toute section de l'ouvrage étudié. Dans le problème qui nous intéresse, si l'on admet que la superstructure est dimensionnnée de telle sorte qu'elle ne conditionne pas la sécurité de l'ensemble, le coefficient de sécurité sera défini par

$$\sigma = \inf_i \left[\sup_j (|f_{ij}/F_i|) \right]$$

où i identifie les différents pieux

j identifie les différents cas de charge

f_{ij} est l'effort appliqué au pieu i sous le cas de charge j

F_i est l'effort limite applicable au pieu i, eu égard aux domaines élastiques du pieu et du terrain

f_{ij} et F_i sont considérés pour les deux sens d'application des efforts (traction et compression)

Or, la sortie du domaine élastique précité, pour l'une ou plusieurs des causes de plastification qui ont été énumérées, n'entraîne la ruine de l'ouvrage que dans la mesure où les éléments non plastifiés de la fondation ne suffisent plus à s'opposer aux déplacements d'ensemble de la superstructure.

Tel n'est généralement pas le cas lorsqu'apparaissent les premiers éléments plastifiés, c'est-à-dire lorsqu'est atteinte la limite du domaine élastique. Par exemple on peut montrer que, dans le modèle de Nökkentved, le nombre de pieux suffisant pour empêcher un système de pieux de se transformer en mécanisme articulé à au moins un degré de liberté est en général de

3 pour les systèmes plans

6 pour les systèmes spatiaux

Les fondations d'ouvrages importants (piles, culées) comportent couramment plusieurs dizaines de pieux: on voit ainsi que la définition du coefficient de sécurité par la méthode élastique ne fournit qu'une appréciation insuffisante de la capacité de résistance offerte par un groupe de pieux à l'égard d'un mode de rupture donné.

De plus, il est facile de concevoir des configurations de pieux et des systèmes d'efforts extérieurs tels que la suppression d'un ou plusieurs pieux fasse disparaître certains états de plastification, alors que la suppression de ces liaisons ne peut que diminuer la résistance du groupe à la rupture: dans ce cas, le modèle élastique fournit une appréciation erronée de la résistance à la rupture.

C'est pourquoi certains auteurs [D. Vandepitte, 1953, Ch. Massonnet et H. Maus, 1962] ont proposé de substituer le calcul à la rupture au calcul élastique.

Les méthodes exposées ci-après s'appuient, en les prolongeant, sur les travaux de ces auteurs.

3. BASES DU CALCUL A LA RUPTURE

3.1 Données relatives à la rupture d'une fondation sur pieux

Deux types de rupture des pieux peuvent être envisagés:

(a) les ruptures d'équilibre entre le pieu et le terrain encaissant

Les diagrammes effort-déformation revêtent alors l'un ou l'autre aspect indiqué sur la Fig. 2:

— à l'arrachement (traction), phase élastique et linéaire suivie d'un palier de plasticité correspondant à un écoulement indéfini sous charge constante

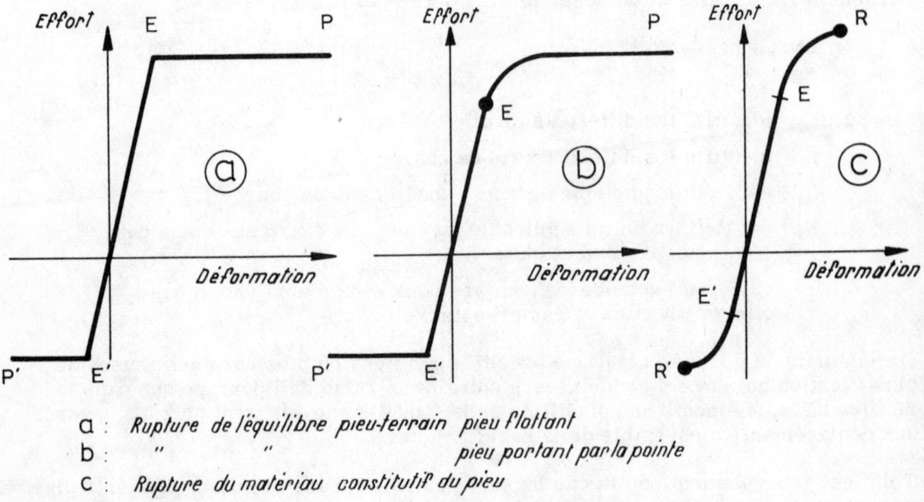

Fig. 2

— au poinçonnement (compression), diagramme analogue pour les pieux flottants, dont la force portante est assurée seulement par frottement latéral; si la force portante du pieu est assurée au moins en partie par la pointe, le diagramme effort-déformation comporte une zone de transition, qui correspond à la plastification progressive du terrain sous le pointe.

(b) les ruptures d'équilibre du matériau constitutif du pieu.

Ces ruptures sont plus rares mais peuvent se produire:

— à la compression, pour les pieux reposant sur le rocher: il y a alors rupture du béton

— à la traction, pour les pieux (généralement précontraints) ancrés dans le rocher. Le processus de rupture comprend alors les phases de décompression, fissuration du béton, rupture des aciers de précontrainte.

Dans les deux cas, le diagramme effort-déformation présente une phase élastique et linéaire, une phase de transition et un arrêt correspondant à la rupture du corps du pieu.

Le processus de rupture d'un groupe de pieux solidarisés par une superstructure rigide est le suivant.

Lorsque les efforts extérieurs appliqués et le déplacement de la superstructure sont tels que la réaction exercée atteigne la limite de proportionnalité (point E ou E'des diagrammes effort-déformation), le pieu continuera à subir les déplacements qui lui sont imposés par la superstructure, et cet accroissement de déplacement s'effectuera

— sous effort constant si E(ou E') est un point de passage direct de la phase de proportionnalité au palier de plasticité P (ou P')

— sous effort pouvant croître encore si E (ou E') correspond à la fin de la phase de proportionnalité, sans passage direct à un palier de plasticité éventuel.

Dans ce cas, nous remarquerons que la présence, dans la fondation, d'autres pieux n'ayant pas atteint l'état de sollicitation E (ou E') limitera très sensiblement l'accroissement des efforts exercés sur les pieux ayant atteint l'état de sollicitation en question. C'est seulement lorsque tous les pieux auront atteint ou légèrement dépassé cet état de sollicitation que la déformabilité du système permettra de développer dans les pieux des efforts correspondant au palier de plasticité P (ou P') ou au point de rupture R (ou R'). De tels états ne sont atteints qu'au prix de grands déplacements, qui sont généralement incompatibles avec la conservation des ouvrages portés, surtout si ceux-ci sont hyperstatiques.

Nous retiendrons donc, dans tous les cas, le critère de rupture suivant: un groupe de pieux entre dans une phase de rupture lorsque les efforts extérieurs appliqués sont tels que le nombre de pieux n'ayant pas atteint l'état de sollicitation E (ou E') ne suffise plus à empêcher le système de se transformer en mécanisme.

C'est un critère de début de rupture. Le système de pieux pourrait supporter des efforts supérieurs au seuil ainsi défini, mais tout effort supérieur à ce seuil entraînerait des désordres importants dans l'ouvrage, parcequ'il laisserait subsister des déformations après suppression de l'effort.

Pour l'étude de l'état de rupture, nous remplacerons donc le diagramme efforts-déformations de la Fig. 2 par ceux de la Fig. 3. En définitive, l'effort f_i appliqué au pieu i doit satisfaire aux conditions:

(4) $T_i \leqslant f_i \leqslant C_i$

La Fig. 3 fait apparaître la différence des deux types de plastification.

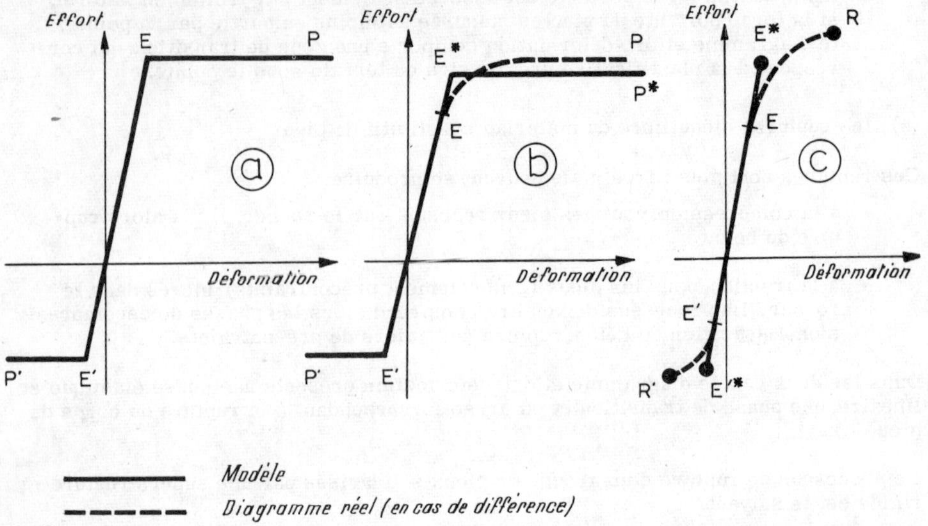

Fig. 3

Dans la rupture de l'équilibre pieu-terrain, l'existence d'un palier de plasticité entraîne la réversibilité de l'état de rupture. Un pieu plastifié pourra rentrer dans le domaine élastique à toute étape ultérieure du chargement si le déplacement de la superstructure tend à diminuer l'intensité de l'effort appliqué alors au pieu considéré; un exemple banal de ce phénomène de plastification suivie d'une rentrée dans le domaine élastique est le battage des pieux.

En revanche, dans la rupture du matériau constitutif du pieu, la plastification est irréversible et un pieu plastifié le reste définitivement, quels que soient les déplacements ultérieurs de la superstructure.

Nous verrons ultérieurement comment il convient de tenir compte de cette différence dans l'application de la méthode de calcul proposée.

3.2 Propriétés de l'état de rupture

Nous considérerons tout d'abord un état de rupture dans lequel le support de l'effort-limite applicable au groupe de pieux est donné. Soient

$\vec{\epsilon}$ le vecteur unitaire associé à ce support

$\lambda\vec{\epsilon}$ l'effort de rupture correspondant.

Les principes du calcul à la rupture, appliqués au problème qui nous intéresse, s'expriment comme suit [Ch. Massonnet et M. Save 1961]

— aucun effort extérieur $\check{\lambda}\vec{\epsilon}$ équivalent à une distribution de réactions statiquement admissible (c'est-à-dire satisfaisant toutes aux conditions (4)) n'a une intensité supérieure à l'intensité de l'effort de rupture:

$$\check{\lambda} \leqslant \lambda$$

— aucun effort extérieur $\hat{\lambda}\vec{\epsilon}$ correspondant à une distribution de réactions cinématiquement admissible (c'est-à-dire entraînant la transformation du système en mécanisme articulé) n'a une intensité inférieure à l'intensité de l'effort de rupture:

$$\hat{\lambda} \geqslant \lambda$$

D. Vandepitte [1953] en a déduit un théorème d'unicité que, en égard aux restrictions que nous expliciterons ultérieurement, nous exprimerons comme suit:

Théorème d'unicité

Lorsque la plastification des pieux se produit par rupture de leur équilibre dans le terrain, la force portante-limite d'un groupe de pieux dépend uniquement du support de la résultante et non du mode de chargement.

En d'autres termes, pour des plastifications du premier genre et pour un support donné S de l'effort de rupture, l'intensité R de cet effort est indépendante de la loi de chargement f (D) associant à une famille de supports D tendant vers S des intensités de l'effort extérieur f tendant vers R, sous réserve évidemment qu'aucune intensité f_i ne soit supérieure à l'effort de rupture R_i associé au support D_i (Fig. 4).

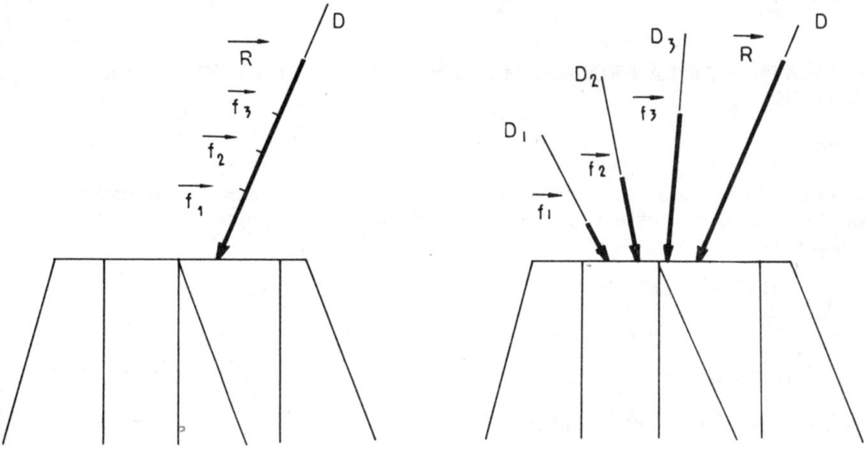

Fig. 4

En outre, la distribution des réactions sur les pieux dans l'état de rupture est généralement unique. Toutefois, dans certains cas de dégénérescence (dans lesquels la configuration du groupe de pieux est telle que trois d'entre eux, s'il s'agit d'un système plan, ou six d'entre eux, s'il s'agit d'un système spatial, ne suffisent pas à empêcher le système de se transformer en mécanisme), plusieurs distributions de réactions sont statiquement admissibles dans l'état de rupture: mais ces différentes distributions ont alors la même résultante et le même moment résultant.

Dans les états de rupture envisagés par Vandepitte, le support de la résultante est donné a priori. Il peut être intéressant, pour les applications pratiques, de considérer d'autres états de rupture correspondant aux éventualités suivantes:

— accroissement différencié des efforts extérieurs en service, en fonction de leur nature (poids mort, surcharges, poussée des terres, effet du vent)

— introduction d'efforts exceptionnels, de support donné, et se superposant aux efforts en service (choc d'un corps flottant sur une pile, effort d'accostage, séisme)

On définira, en conséquence:

— un état de chargement en service ($\mathcal{E}$) caractérisé par les six composantes des efforts extérieurs en service

— une base d'accroissement des efforts (ϵ) correspondant au mode de rupture envisagé et caractérisée, à un facteur constant près, par les six composantes des efforts extérieurs supplémentaires provoquant la rupture.

L'état de rupture associé à $(\mathcal{E}, \epsilon)$ correspondra à un système d'efforts extérieurs $\vec{\mathcal{E}} + \lambda\vec{\epsilon}$, le paramètre λ caractérisant la résistance du groupe de pieux à l'égard du mode de rupture envisagé.

Le théorème de Vandepitte concerne les modes de chargement dans lesquels $\mathcal{E} = o$. Il peut être étendu aux modes de chargement qui viennent d'être décrits, grâce à la propriété suivante [P. Demonsablon, 1965].

Théorème des états correspondants

Si T_i et C_i sont les seuils de plastification (du premier genre) des pieux constituant une fondation, et f_i les efforts qui leur sont appliqués dans l'état de chargement $(\mathcal{E})$, l'état de rupture associé à $(\mathcal{E}, \epsilon)$ coïncide avec l'état de rupture associé à (o, ϵ) pour un groupe de pieux de même configuration, et dont les seuils de plastification seraient $T_i - f_i$, et $C_i - f_i$.

4. APPLICATION DE LA PROGRAMMATION LINEAIRE DE CALCUL A LA RUPTURE

4.1 Principe

La détermination de l'état de rupture associé, pour un groupe donné de n pieux, au système d'efforts $(\mathcal{E}, \epsilon)$ revient donc à définir un ensemble de n réactions f_i satisfaisant aux conditions suivantes:

$$(5)\quad \begin{cases} \text{(a)}\ [P^t] \cdot \vec{f} = \vec{\mathcal{E}} + \lambda\vec{\epsilon} & \text{(6 équations)} \\ \text{(b)}\ T_i \leqslant f_i \leqslant C_i & \text{(2n inéquations)} \\ \text{(c)}\ \lambda \text{ maximum} & \end{cases}$$

Les six équations (5a) se développent en

$$(6)\quad \sum_{j=1}^{n} P_{ji} f_j = \mathcal{E}_i + \lambda\epsilon_i \qquad \forall i \in [1, 6]$$

L'une au moins des composantes ϵ_i est différente de zéro. Soit k l'indice correspondant; en éliminant ϵ_k entre cette équation et les cinq autres, on obtient:

$$(7)\quad \begin{cases} \text{(a)}\ \sum_{j=1}^{n} \left(P_{ji} - \frac{\epsilon_i}{\epsilon_k} P_{jk}\right) f_j = \mathcal{E}_i - \frac{\epsilon_i}{\epsilon_k} \mathcal{E}_k & \text{(5 équations)} \quad \forall i \neq k \\ \text{(b)}\ T_i \leqslant f_i \leqslant C_i & \text{(2n inéquations)} \\ \text{(c)}\ \lambda \text{ maximum} & \end{cases}$$

L'ensemble des conditions (7) définit la distribution des réactions dans l'état de rupture comme la solution optimale d'un programme linéaire.

Pour mettre celui-ci sous forme standard, nous substituerons aux variables f_i les variables non négatives:

$$f'_i = f_i - T_i$$

Nous introduirons en outre les variables d'écart f''_i permettant de transformer en équations les inéquations (7b):

$$f'_i + f''_i = C_i - T_i$$

$$f'_i \geqslant o, f''_i \geqslant o$$

On obtient ainsi le tableau des coefficients et des seconds membres:

Tableau I

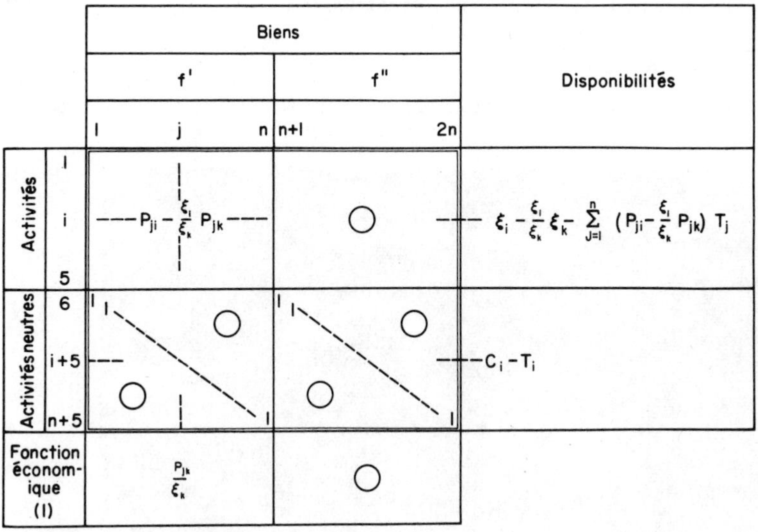

(1) A la forme linéaire en f′ définissant la valeur optimale de la fonction économique, il conviendra naturellement d'ajouter le terme constant $1/\epsilon_k\ [\Sigma P_{jk}T_j - \mathcal{E}_k]$ pour obtenir la valeur du paramètre caractérisant l'effort limite.

Le programme comporte ainsi 5 activités non neutres. D'après une propriété connue [R. Dorfman, 1951], le programme optimum comportera seulement 5 activités à un niveau non nul.

En d'autres termes, 5 seulement dans 2n variables (f′, f″) seront strictement positives dans l'état de rupture

Or, $(f'_i = o) \Rightarrow (f_i = T_i)$ = le pieu correspondant est plastifié en traction

$(f''_i = o) \Rightarrow (f'_i = C_i - T_i) \Rightarrow (f_i = C_i)$ = le pieu correspondant est plastifié en compression

On retrouve ainsi une propriété de l'état de rupture déjà énoncée au § 2.

La méthode exposée fait l'objet d'un programme de calcul automatique, le programme 'OBERON', qui existe en deux versions utilisant l'une et l'autre l'algorithme du Simplexe dans sa forme révisée:

— une version en Fortran IV pour calculatrice I.B.M 7094/II

— une version en Fortran II pour calculatrice I.B.M 360/75

Le programme OBERON fournit les résultats suivants:

— dans tous les cas, le paramètre multiplicateur λ caractéristique de l'état de rupture

— sauf cas d'indétermination (cas détectés par le programme), la distribution des réactions sur les pieux dans l'état de rupture, avec indication des pieux plastifiés; vérification est faite, en outre, de l'équivalence entre le système des réactions et le système des efforts extérieurs.

Les performances du programme sont très intéressantes et rendent possible l'étude approfondie du comportement d'un groupe de pieux sous divers cas de charge, ou la comparaison, pour un ensemble donné de cas de charge, de plusieurs configurations de pieux. Nous en donnons ci-après quelques exemples.

4.2 Exemples d'application

Le premier exemple concerne un poste d'amarrage pour navires petroliers (ce sont les organes rectangulaires apparaissant sur la Fig. 5)

Fig. 5

Ce poste est fondé sur des pieux métalliques tubulaires, présentant

— à l'arrachement une résistance de 120 t

— au poinçonnement une résistance de 580 t

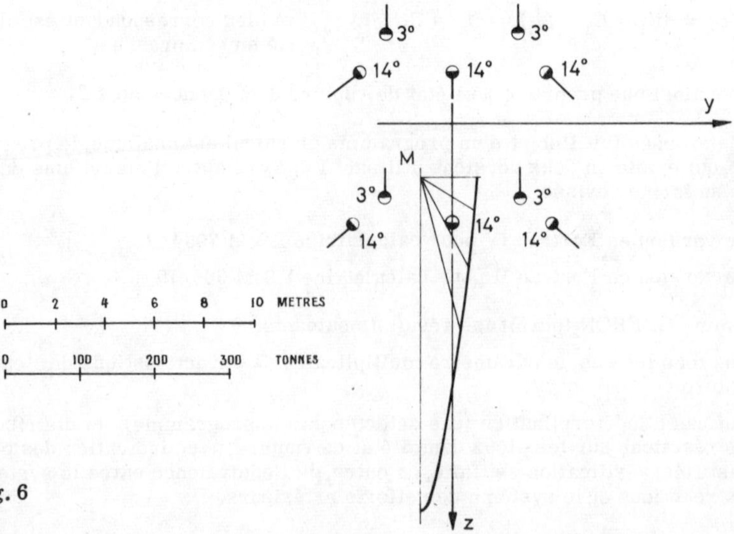

Fig. 6

La configuration des pieux est donnée sur la Fig. 6 qui indique la vue en plan du massif et l'inclinaison des pieux sur la verticale.

L'état de charge en service correspond au poids de massif: 1415 t.

L'état de rupture a été étudié sous l'effet d'un effort de traction du navire, effort appliqué au point M (—5, 30; —1, 22; 2, 13 m.) sous divers azimuts.

Les valeurs de l'effort de rupture, indiquées sur le tableau II ci-après, sont reportées sur la Fig. 6

Tableau II

Azimut sur Oz, degrés	Effort de rupture tonnes
0	441, 3
15	194, 0
30	122, 8
60	80, 5
90	74, 9

Les calculs correspondants ont demandé 6/1000 h sur I.B.M 7094/II dont 2/1000 h d'exécution, pour 5 cas de charge.

Dans le second exemple on a cherché à mettre en évidence l'influence de la position du support de la résultante sur la résistance limite d'un groupe de pieux, dont la configuration est donnée sur la Fig. 7. Ces pieux, dont l'inclinaison sur la verticale est indiquée, présentent:

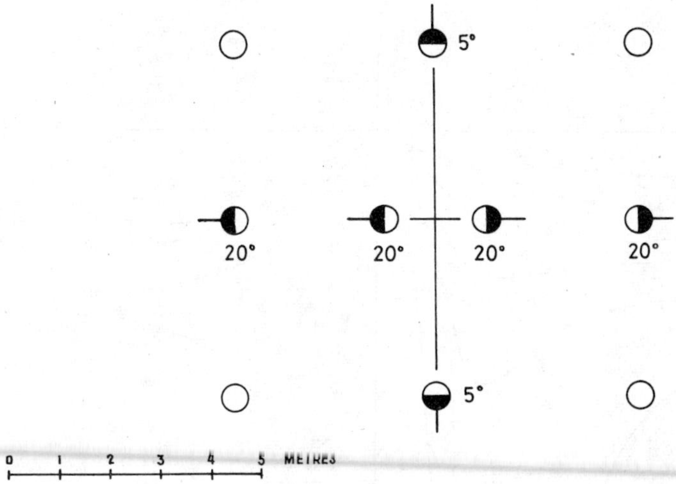

Fig. 7

— à l'arrachement une résistance de 70 t

— au poinçonnement une résistance de 420 t

Les états de chargement en service correspondent à $\mathcal{E} = o$.

L'état de rupture a été étudié pour différents supports de la résultante:

- supports passant par 0 sous divers azimuts: 13 cas étudiés
- supports verticaux: 16 cas étudiés
- supports horizontaux: 6 cas étudiés
- autres supports: 50 cas étudiés

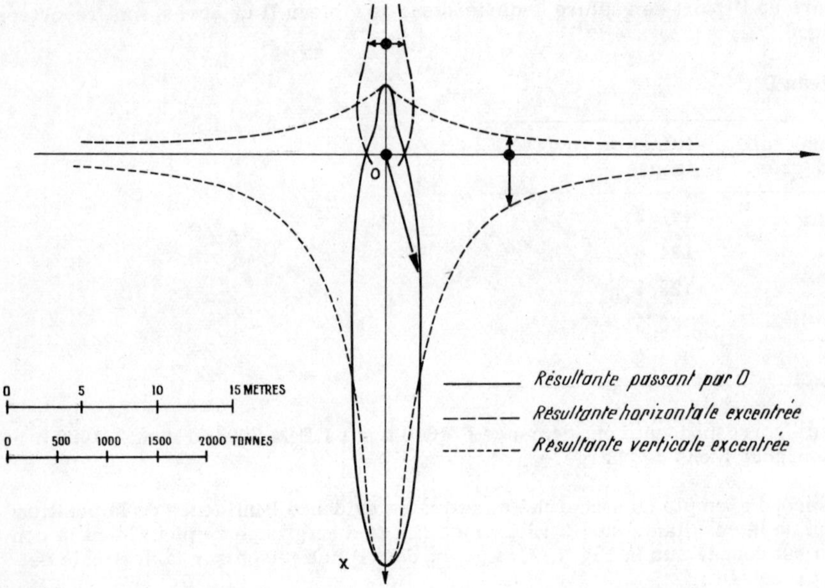

Fig. 8

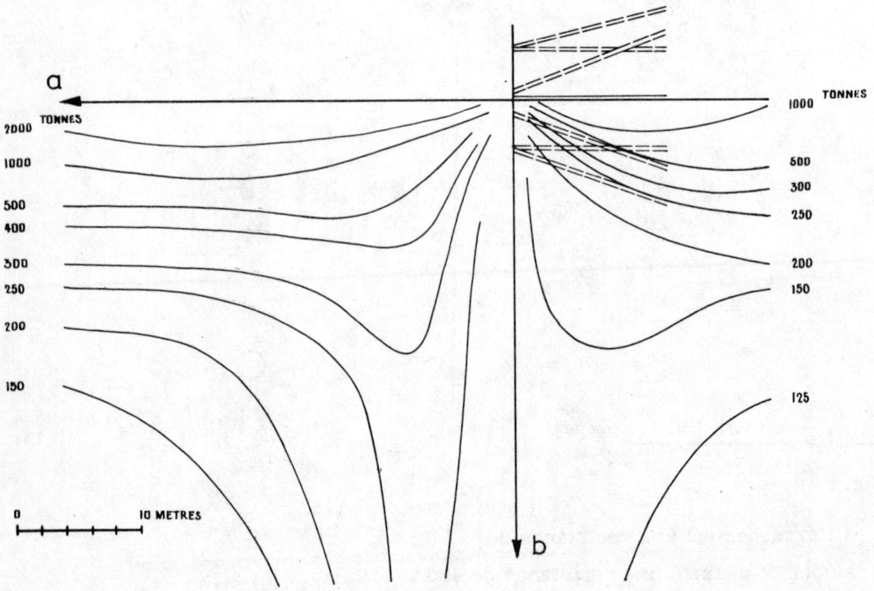

Fig. 9

Les valeurs de l'effort à la rupture sont données dans les tableaux III à VI ci-après, et sur Fig. 8 et 9, qui représentent respectivement

- Fig. 8: valeurs de l'effort de rupture en fonction de divers paramètres (résultats des tableaux III, IV et V)
- Fig. 9: courbes d'égales valeurs de l'effort-limite $\lambda\|\vec{\epsilon}\|$ en fonction de a et b, pour des résultantes dont le support a pour equation

$$\frac{x}{a} + \frac{y}{b} = 1 \text{ (tableau VI).}$$

Dans le tableau VI, qui correspond à des efforts dirigés vers le bas, les efforts sont exprimés en tonnes et les paramètres a et b en mètres.

L'ensemble des cas de charge étudiés dans cet exemple a été traité 21/1000 h. sur I.B.M. 7094, dont 13/1000 h. pour l'exécution proprement dite.

Tableau III

Supports passant par O

Azimut sur la verticale, degrés	Effort de rupture Tonnes	
	Vers le bas	Vers le haut
0	4095	683
15	1299	368
30	546	264
45	323	218
60	241	195
75	203	189
90	187	

Tableau IV

Supports verticaux

Excentrement mètres	Effort de rupture Tonnes	
	Vers le bas	Vers le haut
0	4095	683
1	3276	546
2	2455	455
4	1325	341
8	534	228
12	331	171
16	228	137
20	171	114

Tableau V

Supports horizontaux

Distance au plan des têtes de pieux, mètres	Effort de rupture Tonnes
0	187
2	229
4	248
6	203
8	170
10	137

4.3 Forme duale du problème

Nous avons formulé le problème en utilisant le principe statique du calcul à la rupture, et en recherchant l'état extrême (maximal) des états statiquement possibles.

Nous pouvons utiliser le principe cinématique et rechercher l'état extrême (minimal) des états cinématiquement possibles.

Nous serons ainsi amenés à rechercher:

— 2n déplacements virtuels des têtes de pieux selon leur axe

δ en compression, δ' en traction,

avec (δ_i ou δ_i') = o $\forall i$

— 6 déplacements virtuels de la superstructure,

Les pieux étant soumis aux efforts C_i (si $\delta_i' = o$)

ou T_i (si $\delta_i = o$) et la superstructure aux efforts $\vec{\mathcal{E}} + \lambda\vec{\epsilon}$

Nous exprimons:

— n conditions de compatibilité des déplacements virtuels de la superstructure au droit des têtes de pieux

— la condition de minimum de la somme des travaux virtuels des réactions despieux (principe des travaux virtuels)

Cette formulation conduira à la forme duale du programme linéaire précédemment posé, et définira le paramètre λ caractéristique de l'état de rupture comme un minimum.

5. DISCUSSION DE LA VALIDITE DE LA METHODE

Nous avons été amenés, en énonçant le théorème d'unicité qui fonde la méthode précédemment exposée, à frapper celui-ci de certaines restrictions quant à son domaine d'application.

Nous avons en effet limité cette application aux plastifications dites 'du premier genre', c'est à dire correspondant à une rupture de l'équilibre pieu-terrain. Il nous reste à justifier cette restriction.

Tableau VI

Efforts de rupture associés à des supports d'équation $\frac{x}{a} + \frac{y}{b} = 1$

a \ b	1	2	3	4	5	6	8	10	12	16	20	24	32	36	40
–10					632			236			130				96
–8				768			280			150			110		
–6			960			343			178			128		118	
–4		1149		417			218			154		141	136		
–2		509		265			191		175	168					
2	1379	746		416			287		260	250					
4		1379		746			474			321		290	278		
6			1379			746			358			251		229	
8				1379			564			287			206		
10					1264			452			240				169

La méthode de programmation utilisée consiste à examiner différentes 'bases' possibles (combinaisons de 2n valeurs de f_i' et f_i'' satisfaisant aux équations de la statique, et comprenant sauf cas de dégénérescence (n-5) valeurs nulles) et, par un processus d'itération, à définir la base correspondant à la valeur maximale du paramètre λ.

En d'autres termes, on examine successivement divers états de rupture statiquement possibles, dont un seul, qui est aussi possible cinématiquement, réunit les conditions de réalité: aucun autre état, parmi ceux que le procédé d'itération a permis de construire, ne pourrait être atteint par un processus physique de chargement.

Or, le processus physique d'épuisement progressif de la force portante du groupe, tel que nous l'avons décrit en 3. 1; modifie la déformabilité de la fondation au fur et à mesure qu'apparaissent de nouvelles plastifications de pieux et les déplacements de la superstructure peuvent très bien évoluer dans un sens qui soulage certains pieux déjà plastifiés. Cette possibilité est illustrée par la Fig. 10, qui montre quelques types d'évolution de la réaction sur les pieux au cours de processus physique de chargement, étudié comme indiqué ci-après.

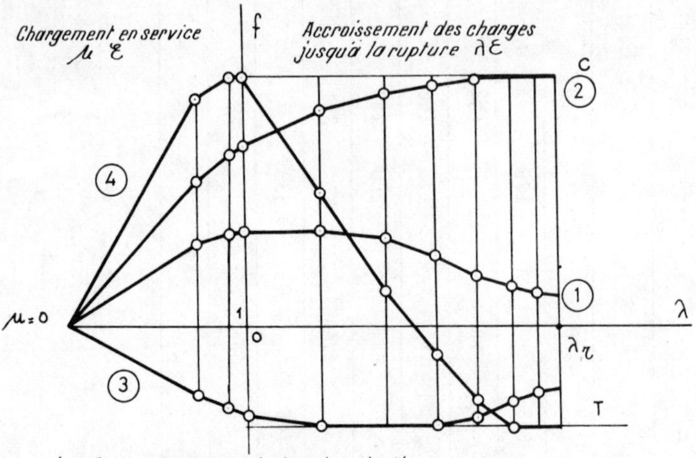

1 : *Pieu restant dans le domaine elastique*
2 : *Pieu plastifié restant plastifié*
3 : *Pieu plastifié rentrant ultérieurement dans le domaine elastique*
4 : *Pieu plastifié, deplastifié puis plastifié à nouveau*

Fig. 10

Encore faut-il que cette plastification ne soit pas définitive, c'est à dire corresponde à une rupture d'équilibre pieu-terrain, laquelle est toujours réversible.

Ainsi, pour que l'état de rupture ne dépende que du support de la résultante et non du mode de chargement (nous pouvons en effet, par le théorème des états correspondants, nous ramener au type de chargement envisagé par Vandepitte), il faut que l'application réelle des chargements destinés à provoquer la rupture n'entraîne pas, avant épuisement de la fondation, d'états plastiques irreversibles: il suffit, pour cela, que les pieux ne soient pas susceptibles de plastification du second genre.

Cette distinction est illustrée sur la Fig. 11. Celle-ci symbolise, dans l'espace $\mathbb{R}^n$ des réactions sur les pieux:

— les conditions de la statique (segments de type 1)

— les conditions de borne des variables (segments de type 2)

Tous les points strictement intérieurs au domaine *D* délimité par ces conditions correspondent à des distributions de réactions admissibles sans qu'il y ait ruine de la fondation. L'état de rupture correspond à un point R du contour.

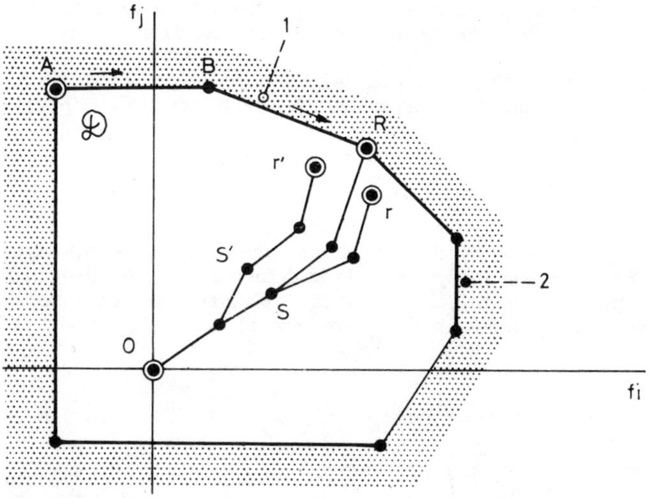

Fig. 11

Dans la méthode de programmation linéaire, on chemine (de manière discontinue) sur le contour de $\mathcal{D}$ à partir d'une base quelconque: par exemple, en A, B, R.

Dans le processus réel de chargement, on chemine au contraire (de manière continue) à l'intérieur de $\mathcal{D}$: par exemple, selon OSR, où O est l'état non chargé et S l'état de chargement en service.

Si les pieux sont susceptibles d'une plastification du second genre, le processus réel de chargement pourra conduire à des états de rupture intérieurs au domaine $\mathcal{D}$: par exemple OSr ou OS 'r' (si l'état de chargement en service comporte des pieux plastifiés). Eventuellement le même état R sera atteint; mais en général, pour les plastifications du second genre, l'unicité de l'état de rupture associé à $(\mathcal{E}, \epsilon)$ ne peut être assurée. Il convient en outre de spécifier le mode de chargement pour définir l'état de rupture et celui-ci ne peut être atteint par la méthode de programmation linéaire.

Pour l'étude des fondations comportant des pieux susceptibles de plastification du second genre, il convient donc d'adopter un algorithme permettant de suivre le processus physique d'épuisement de la fondation, tel que:

- dèfinition de la matrice de déformabilité du système initial
- calcul de la charge la plus faible provoquant une plastification
- identification du (ou des) pieu(x) plastifié(s) et suppression de ce(s) pieu(x) dans le système
- définition de la matrice de déformabilité du système réduit
- calcul de l'accroissement de charge le plus faible provoquant une plastification

et ainsi de suite jusqu'à l'état de rupture.

Cet algorithme peut également s'appliquer aux pieux susceptibles de plastification du premier genre et c'est pourquoi nous le mentionnons ici car il a permis de vérifier les résultats fournis par la méthode de programmation linéaire; le programme OBERON comporte en effet, en option, cette méthode 'pas à pas' dont les résultats concordent (à 10^{-8} près) avec ceux de la méthode de programmation linéaire.

Il faut prendre garde, toutefois, que tout pieu plastifié peut alors rentrer dans le domaine élastique; l'algorithme décrit ci-dessus est donc complété, dans le cas de plastification du premier genre, par un 'test de tendance' qui permet de déterminer, dans l'ensemble des pieux plastifiés, ceux qu'il convient de réintroduire dans le système avant d'en définir l'étape suivante de plastification.

C'est par cette méthode qu'ont été établies les courbes de la Fig. 10, et reconnue l'existence du phénomème de 'déplastification', existence établie théoriquement d'autre part.

6. EXTENSION DE LA METHODE

Nous avons jusqu'à présent traité le problème dans lequel le système d'efforts $(\mathcal{E}, \epsilon)$ appliqué à la fondation est indépendant de la déformabilité de celle-ci, c'est à dire de son état de plastification. Tel ne sera généralement pas le cas des ouvrages hyperstatiques (portiques encastrés élastiquement, arcs, etc...) dont la Fig. 12 donne un exemple.

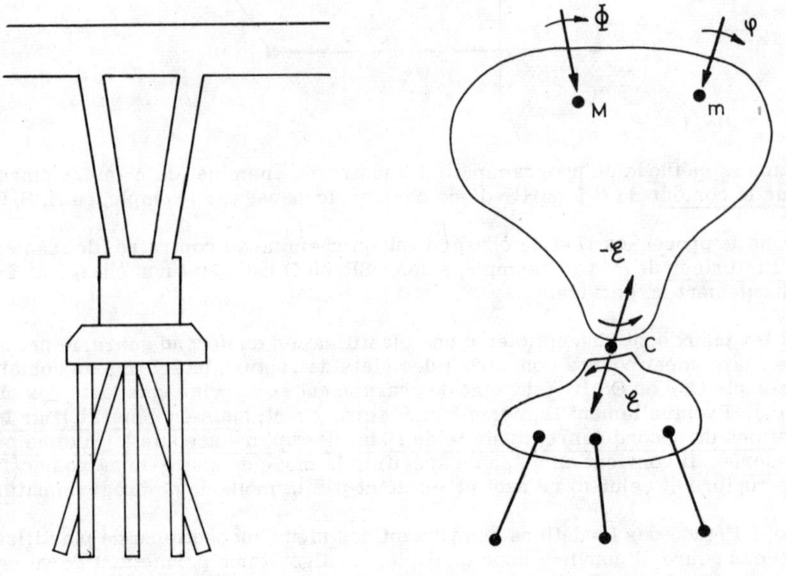

Fig. 12

Dans ces ouvrages, on donnera non le système d'efforts $(\mathcal{E}, \epsilon)$ appliqué à la fondation, mais un système (Φ, φ) appliqué à l'ouvrage lui-même. Pour déterminer les efforts $(\mathcal{E}, \epsilon)$ à prendre en compte, nous pouvons utiliser une <u>méthode de coupure</u>. Créons en C une coupure rendant indépendants les déplacements de la superstructure et de la fondation. Pour rétablir la liaison, il suffira d'introduire en C les systèmes d'efforts $(\mathcal{E}, \epsilon)$ rendant compatibles les déplacements de part et d'autre de la coupure.

Soit $D_C(M)$ la matrice permettant d'exprimer le déplacement $d_C(M)$ tout point M de la superstructure sous l'effet d'un système d'efforts appliqué en C. D'après le théorème de Maxwell:

$$D_M(C) = D_C(M)$$

Soit M le point d'application de $\vec{\Phi}, \vec{\mathcal{E}}$ l'effort correspondant reporté <u>sur</u> la fondation. Le déplacement en C est

$$\vec{d} = D_C(M)\,\vec{\Phi} - D_C(C)\,\vec{\mathcal{E}}$$

$$= D_M(C)\,\vec{\Phi} - D_C(C)\,\vec{\mathcal{E}}$$

Le déplacement de la fondation est d'autre part

$$\vec{d} = (P^t . K . P)^{-1} \vec{\mathcal{E}}$$

$$\text{d'où } \vec{\mathcal{E}} = [(P^t . K . P)^{-1} + D_C(C)]^{-1} . D_M(C) \vec{\Phi}$$

$$= \mathcal{T}_M . \vec{\Phi}$$

$\mathcal{T}_M$ est une matrice de transfert exprimant les efforts transmis à la fondation lorsque des efforts sont appliqués au point M de l'ouvrage porté.

Bien entendu, l'ouvrage porté n'est généralement pas un solide indéformable. On ne peut donc pas remplacer le système des efforts appliqués par leur résultante et leur moment résultant, et l'on écrira:

$$\vec{\mathcal{E}} = \sum_M \mathcal{T}_M \vec{\Phi}_M$$

M désignant les points d'application des différents efforts élémentaires $\vec{\Phi}_M$

De même, si $\vec{\varphi}$ est la base d'accroissement des efforts, appliqués au point m de l'ouvrage, on voit que les conditions (6) deviendront:

$$\begin{cases} P^t . \vec{f} = \sum_M \mathcal{T}_M \Phi_M + \lambda \sum_m \mathcal{T}_m \vec{\varphi}_m \\ T_i \leqslant f_i \leqslant C_i \\ \lambda \text{ max.} \end{cases}$$

La formulation du problème est donc identique à la formulation déjà étudiée. On en déduit:

(a) que l'état de rupture de la fondation associé au système d'efforts (Φ_M, φ_m) appliqués à l'ouvrage porté est unique (pour les plastifications du premier genre)

(b) que la méthode de programmation linéaire précédemment décrite s'applique dans les mêmes conditions.

BIBLIOGRAPHIE

ASPLUND, S. O. (1956), 'Generalized Theory for pile-groups'. Mémoires de l'A.I.P.C., vol. 16, p. 1/23

DEMONSABLON, P. (1965), 'Le Calcul à la rupture des fondations sur groupes de pieux—Principe et traitement sur ordinateur'. Construction, T. XX, No. 7. 8.

DORFMAN, R. (1951), Application of linear programming to the theory of the firm, University of California Press, Berkeley.

MASSONNET, Ch., MAUS, M. (1962), 'Force portante plastique des systèmes de pieux'. Comptes rendus du Symposium sur l'utilisation des calculatrices dans le Génie Civil, Lisbonne—Vol. 1, p. 12. 1/12. 15.

MASSONNET, Ch., SAVE, M. (1961), Jl. Calcul plastique des Constructions, C.B.L.I.A., Bruxelles.

NÖKKENTVED, Ch. (1928), Berechnung von Pfahlrosten, Ernst, Berlin.

PADUART, A. (1949), 'Calcul des semelles de fondation de raideur finie reposant sur pieux verticaux et inclinés'. Annales des Travaux Publics de Belgique, No. 4, p. 407.

VANDEPITTE, D. (1953), 'Charge admissible des fondations sur pieux' Annales des Travaux Publics de Belgique, Nos. 4, p. 567/627; 5, p. 755/93; 6, p. 901/53.

VANDEPITTE, D. (1957), 'Charge admissible des fondations sur pieux' Annales des Travaux Publics de Belgique, No. 2, p. 5/46.

DISCUSSION

J. Abadie

Les liaisons entre la programmation linéaire et la plasticité ont derrière elles une longue histoire. Les applications de la théorie de la dualité à la plasticité ont, en particulier, fait le sujet d'un certain nombre de publications (voir plus loin), la plus ancienne datant, à ma connaissance, de 1951 (Charnes et Greenberg). Ces méthodes ont même été récemment étendues par la considération de la programmation non linéaire (Gavarini, Sacchi et Buzzi Ferraris). Vos recherches ont-elles quelques points communs avec les précédentes?

CHARNES, GREENBERG, Plastic collapse and linear programming. Preliminary Report, Bull. Amer. Math. Soc. Nov. 1951.

DORN, GREENBERG, Linear programming and plastic limit analysis of structures An. Appl. Math., Vol. 15, 1957.

CHARNES, LEMKE, ZIENKIEWICZ, Virtual work, linear programming and plastic limit analysis, Proc. Roy. Soc., Series A, Math. and Phys. Sciences, May 1959.

PRAGER, Lineare Ungleichungen in der Baustatik, Schweizerische Bauzeitung, 10 Mai 1962.

WOLFENSBERGER, THURLIMANN, Untere Grenzwerte des Traglast von Platten. Bulletin C.E.B. N. 43-Sept. 1964.

CERADINI, GAVARINI, Calcolo a rottura e programmazione lineare, Giornale del Genio Civile, gennaio, febbraio 1965.

KOOPMAN, LANCE, On linear programming and plastic limit analysis, J. Mech. Phys. Solids, March 1965.

GAVARINI, Fundamental plastic analysis theorems and duality in linear programming, Ingegneria Civile, N. 18, 1966.

GAVARINI, Plastic analysis of structures and duality in linear programming, Meccanica, N. 3/4, 1966

P. G. HODGE, Yield-point load determination by nonlinear programming. Proceedings of the 11th Int. Congress of Ap. Mech. Munich 1964 (printed 1966).

SACCHI, Contribution à l'analyse limite des plaques minces en béton armé à l'aide des solutions statiquement admisibles. Istituto Lombardo 1966.

R. H. LANCE, D. C. A. KOOPMAN, Limit analysis of shells of revolution by linear programming, Cornell University Report, NSF GK 687/1, Jan. 1967.

R. H. LANCE, Duality in the finite difference method of plastic limit analysis, Cornell University Report, NSF GK 687/2, April 1967.

P. G. HODGE, Elastic-plastic torsion as a problem in nonlinear programming. Int. J. Solid Structures. 1967.

G. SACCHI, G. BUZZI FERRARIS, Sul criterio cinematico di calcolo a rottura di piastre inflesse mediante programmazione non lineare, Istituto Lombardo, 1967.

C. GAVARINI, Calcolo a rottura e programmazione non lineare, Istituto Lombardo, 1967.

G. CERADINI, C. GAVARINI, Calcolo a rottura e programmazione lineare Continui bi-e tridimensionali. Nota I: Fondamenti teorici, Giornale del Genio Civile, 1968.

G. CERADINI, C. GAVARINI, Calcolo a rottura e programmazione lineare. Continui bi-e tridimensionali. Nota II: Applicazioni a piastre e volte di rivoluzione, Giornale del Genio Civile, 1968.

G. MAIER, Programmazione quadratica e teoria delle strutture elasto-plastiche, V Convegno Nazionale sulla Plasticità-Stresa, 1968.

W. PRAGER, Mathematical Programming and Theory of Structure, J. Soc. Indust. Appl. Math., Vol. 13, No. 1, March 1965.

P. Demonsablon

Les travaux de Charnes et de Prager concernent, je crois, des systèmes continus, alors que les travaux de Massonnet, aux quels les miens se réfèrent, concernent des systèmes discrétisés.

Le phénomène de rupture comporte un aspect commun aux deux types de systèmes, et un aspect spécifique à chacun d'entre eux.

Aspect commun d'abord. Du point de vue physique, la rupture d'un système à liaisons surabondantes, qu'il soit ou non continue, est progressive: c'est ce que l'on appelle généralement l'adaptation des structures. Du point de vue mathématique, l'état de rupture est associé à l'extremum d'une fonction des variables d'état de la structure (état qu'on peut définir soit par les contraintes, soit par les déformations), les variables étant d'autre part liées par des conditions qui définissent un domaine convexe.

Aspect spécifique maintenant. Dans le système discrétisé, les éléments susceptibles de plastification (les pieux dans la communication que j'ai présentée) sont en nombre fini et identifiables. Dans le système continue, la plastification peut atteindre tout point ou toute section de la structure, et c'est seulement par approximation qu'on peut exprimer en nombre fini les conditions de l'équilibre-limite.

D'autre part, la plastification des structures se produit par épuisement du matériau constitutif. Elle est donc irréversible et, comme je l'ai signalé, l'emploi d'une méthode de programmation linéaire peut, dans ce cas, conduire à définir comme état de rupture un état non susceptible d'être atteint par le processus physique de chargement, donc à surestimer la résistance à la rupture du système.

A Decentralized Programming Method Using Iterative Smoothing

REZA RAZANI
Pahlavi University, Iran

1. SUMMARY

The cost optimization of girders and some problems dealing with decentralization of management activities can be formulated as mathematical programming problems where the objective function represents the costs relating to central and regional activities, where regional costs consist of operating and fixed costs for each region. The unknowns are the global variables representing central activities, the number and the size of the decentralized regions, and the level of regional activities necessary to fulfil regional demands. Also, various other central and regional constraints must be met.

For determinate systems where regional demands are constant within each region, an iterative solution method is presented which decomposes the above optimization problem into two suboptimization problems. The first one determines the number and the size of the regions for given global variables using a smoothing method. The second one determines the global and regional variables for regions obtained in the first problem. The newly obtained global variables are resupplied to the first problem and the process is repeated until a converged solution is obtained. Proof of convergence is given and the optimality condition is discussed. Due to non-convexity of the problem the converged solution is not necessarily optimum.

For indeterminate systems where the demand within each region depends upon the level of activities of the system, the iterative method is modified by analysing and adjusting the demand in each cycle of iteration. For indeterminate systems with small sensitivity the convergence and optimality is discussed. Various applications of the method to minimum-cost design of girders is presented.

2. GENERAL STATEMENT OF THE PROBLEM

Determine variables $W = X, K, Y, Z$ where $X = (x_1, x_2, \ldots x_n) \geqslant 0$, $K \geqslant 1$ is an integer, $Y = (y_1, y_2, \ldots y_K) \geqslant 0$ and $Z = (z_1, z_2, \ldots z_K) \geqslant 0$ so as to minimize the function

$$f(W) = f_1(X) + \sum_{i=1}^{i=K} [\Phi_i(X, y_i, z_i) + \Psi_i(X, y_i, y_{i-1})] \tag{2.1}$$

where y_0 is known, subject to

$$g_i(X) \geqslant 0 \qquad i = 1, 2, \ldots m \tag{2.2}$$

$$\sum_{i=1}^{i=K} z_i = L \tag{2.3}$$

$$H_i(y_i) \geqslant 0 \qquad i = 1, 2, \ldots K \tag{2.4}$$

$$S(X, y_i) \geqslant 0 \qquad i = 1, 2, \ldots K \tag{2.5}$$

$$R[X, y_i, U_i(W)] \geqslant 0 \qquad i = 1, 2, \ldots K \tag{2.6}$$

where variables X are called global variables, variables Y are called regional variables, and variables Z are region-size variables. K represents the total number of different regions.

Function $f_1(X)$ represents the costs relating to the level of activity of global variables, functions $\Phi_i(X, y_i, z_i)$ represent the costs relating to regional activities, and function $\Psi_i(X, y_i, y_{i-1})$ generally represents a fixed charge relating to the existence of each region, and in some cases may vanish if the level of activity of two adjacent regions are identical and at the same time it may also be a function of global variables.

The constraint set (2. 2) represents limitations on global variables. The equality constraint set (2. 3) expresses that the sum of the size of all the regions is equivalent to the size of the entire region L. Inequality constraint sets (2. 4) and (2. 5) represent relationships between regional activities and regional and global activities within each region respectively.

Inequality constraint sets (2. 6) represent relationships between the level of activities within each region and the so-called demand $U_i(W)$ within that region, where the level of demand within each region may depend upon the behavior of the entire system. These types of constraints are called behavior constraints. The demand U(W) within each region may have three forms.

(a) A determinate system where the demand within each region is a constant value for that region and is not a function of the level of activities of the system.

$$U_i(W) = c_i \tag{2. 7}$$

where c_i is constant for the ith region.

(b) A normal indeterminate system where the demand within each region is little affected by the level of activities of the system.

(c) A hybrid indeterminate system where the demand within each region is highly sensitive to the level of activities of the system.

In the subsequent sections, first as an example, the problem of minimum-cost design of plate girders is formulated; then the method of solution is developed which converges to a solution of the above problem for the case of determinate demand. The case of indeterminate demand is also discussed briefly.

2. 1 Formulation of the minimum-cost design of plate girders

As an example a simple version of this problem is presented here. More comprehensive examples are given by Razani (1966).

Consider the plate girder shown in Figure (1). The design parameters, a, which are the known data of the problem and are given apriori are the length of the spans, the

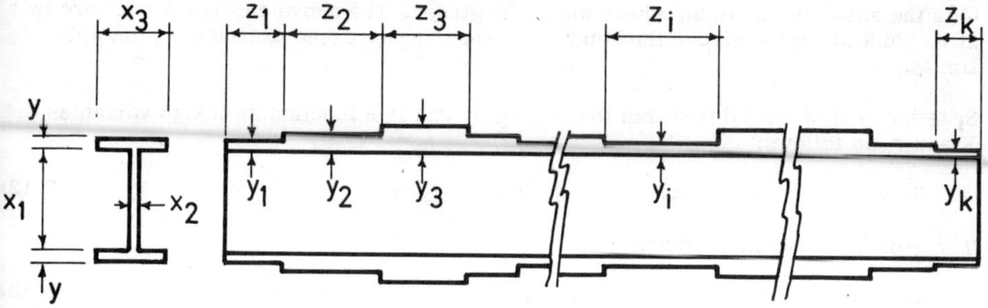

Fig. 1 Plate Girder Arrangement and Design Variables

types of steel used, the allowable stresses, and the limiting value for the dimension of the component plates and their width-thickness ratios.

The global design variables are web height x_1, web thickness x_2, and flange width x_3. The value of these variables are assumed to be constant along the entire length of the girder. The flange thickness varies along the girder and is uniform within various regions of the girder as shown in Figure (1).

It is assumed that there are K of such regions where K is a variable itself. The length of these regions are $(z_1, z_2, \ldots z_K)$, respectively, which are called region-size variables. The corresponding flange thickness within these regions are $(y_1, y_2, \ldots y_K)$, respectively, which are called regional variables. Thus, the total number of design variables N is

$$N = 3 + 2K \tag{2.8}$$

where K is a design variable itself.

The objective function is

$$f(W) = x_1 x_2 L c_1 + \sum_{i=1}^{i=K} (2 x_3 z_i y_i c_1) + \sum_{i=1}^{i=K-1} c_2(1 - \delta_{y_i, y_{i+1}}) \tag{2.9}$$

where c_1 represents the cost of the unit volume of a steel plate. For simplicity here, it is assumed that c_1 is constant and c_2 represents the cost of a flange splice and generally consists of two terms. The first term is a fixed charge representing the cost of preparation, set-up, and x-rays of the welded area of each splice. The second term may be a function of volume of welding material used. In this example for the sake of simplicity, the second term is neglected. Thus, there is only a fixed charge c_2 for each splice. The splices exist only if the flange thickness of the two adjacent regions are different, otherwise, no flange splice is used. This is shown by the function $(1 - \delta_{y_i, y_{i+1}})$ where

$$\delta_{y_i, y_{i+1}} = 0 \qquad \text{if } y_i \neq y_{i+1}$$

$$\delta_{y_i, y_{i+1}} = 1 \qquad \text{if } y_i = y_{i+1}$$

The behavior constraints are as follows:

$$S_i \sigma_b - M_i \geq 0 \qquad i = 1, 2, \ldots K \tag{2.10}$$

$$x_1 x_2 \sigma_s - Q \geq 0 \tag{2.11}$$

where M_i and S_i are the absolute maximum design moment and section modulus within region i of the girder. σ_b and σ_s are the allowable bending and shearing stresses respectively. Here for simplicity, it is assumed the value of these stresses are constant. Q is the absolute maximum shear along the girder. The above constraints ensure that at no point along the girder the bending or shearing stresses exceed the allowable limits.

S_i is the section modulus within the ith region and is a function of design variables X and y_i as follows:

$$S_i = [x_3(x_1 + 2y_i)^3 - x_1^3 (x_3 - x_2)]/[6(x_1 + 2y_i)] \tag{2.12}$$

The geometric constraints are:

$$x_j - x_{j\,min} \geq 0 \qquad j = 1, 2, 3 \tag{2.13}$$

$$y_i - y_{min} \geq 0 \qquad i = 1, 2, \ldots K \tag{2.14}$$

$$a_1 x_2 - x_1 \geq 0$$

$$x_2 - x_3 + 2a_2 y_i \geq 0 \qquad i = 1, 2, \ldots K \qquad (2.15)$$

where $x_{j\,min} > 0$ and $y_{min} > 0$ are the lower bounds on the size of variables x_j and y_i respectively. a_1 and a_2 represent the assumed maximum acceptable ratios of width thickness for web plates and outstanding portions of the flange plates respectively to prevent local buckling of plate elements.

3. METHOD OF SOLUTION

In order to prepare the foundation for the solution method, the definition and solution of the following two problems is necessary:

3.1 Problem 1

Given variables $X = \overline{X} \geq 0$ such that,

$$g_i(\overline{X}) \geq 0 \qquad i = 1, 2, \ldots m \qquad (3.1)$$

find variables K, $Z = (z_1, z_2, \ldots z_K) \geq 0$ and $Y = (y_1, y_2, \ldots y_K) \geq 0$ so as to minimize

$$f(W) = f_i(\overline{X}) + \sum_{i=1}^{i=K} [\Phi_i(\overline{X}, z_i, y_i) + \Psi_i(\overline{X}, y_i, y_{i-1})] \qquad (3.2)$$

where $y_0 = \overline{y}_0$ and subject to the following constraints:

$$K = \text{an integer} \qquad (3.3)$$

$$\sum_{i=1}^{i=K} z_i = L \qquad (3.4)$$

$$H_i(y_i) \geq 0 \qquad i = 1, 2, \ldots K \qquad (3.5)$$

$$S(\overline{X}, y_i) \geq 0 \qquad i = 1, 2, \ldots K \qquad (3.6)$$

$$R(\overline{X}, y_i, c_i) \geq 0 \qquad i = 1, 2, \ldots K \qquad (3.7)$$

If the nonvariable terms are eliminated from the above formulation, the problem becomes:

$$\text{Min } f'(K, Z, Y) = \sum_{i=1}^{i=K} [\Phi_i(z_i, y_i) + \Psi_i(y_i, y_{i-1})] \qquad (3.8)$$

subject to constraints (3.3) and (3.4), and

$$T_i(y_i) \geq 0 \qquad i = 1, 2, \ldots K \qquad (3.9)$$

where constraint sets (3.9) represent constraint sets (3.5), (3.6), and (3.7).

The type of problem presented above is a complex mathematical problem where the number of variables is unknown. Usually, this problem can be cast in the form of smoothing problems in one or more dimensions, where, in general, two types of conflicting costs exist.

The first type of costs are represented by:

$$c_1 = \sum_{i=1}^{i=K} \Phi_i(z_i, y_i) \tag{3.10}$$

where Φ_i, in general, represents the costs dealing with activities within each region. Usually, these costs reduce if the size of the region becomes small and the number of regions increases. In other words, these costs reduce if the level of activities, y_i, within each region i is chosen so that the demand level along the entire region is met with as small an amount of waste as possible. If one wants to fulfill this condition, he has to subdivide the entire region into as many subregions as possible and within each subregion adjust the level of his activities to the level of demands as close as possible; thus, the value of K must increase.

The second type of costs is represented by function

$$c_2 = \sum_{i=1}^{i=K} \Psi_i(y_i, y_{i-1}) \qquad \text{given } y_o = \bar{y}_o \tag{3.11}$$

where Ψ_i, in general, represents either the fixed cost dealing with the existence of each subregion or the change-over costs due to the difference between levels of activities of two adjacent subregions. Usually, these costs can be minimized by reducing either the number of regions or the variation between level of activities of regions which is equivalent to choosing K as small as possible.

In order to obtain a solution for the problems of the above nature where there are two types of conflicting costs, it is expedient to pursue a middle path, balancing one type of cost against another so as to minimize the over all costs. These types of problems are usually called smoothing problems.

A solution of the smoothing problems has been under study by many investigators. Bellman (1962) gives an efficient method for the solution of various types of one dimensional smoothing problems using dynamic programming. Other methods also have been used. When the dynamic programming method is used for the solution of one dimensional smoothing problems, the entire region L will be subdivided into many, say N, small subregions. The size of each subregion is given by ΔL_i, such that,

$$\sum_{i=1}^{i=N} \Delta L_i = L \tag{3.12}$$

The regional variables Y are selected among discrete values which represent closely the entire range of variation of these types of variables. The number and the size of these subregions and the number of the discrete values representing regional variables influence the length of computation, the speed of convergence of the iterative solution presented in section (3.3), and the accuracy of the results.

For the formulation and solution of the above smoothing problem, it is necessary to solve the following sub-problem:

3.1.1 Sub-Problem 1

Given variables $X = \bar{X} \geqslant 0$, $K = N$, $Z = (\Delta L_1, \Delta L_2, \ldots \Delta L_N) \geqslant 0$ satisfying constraints (3.1) and (3.12), find variables $Y = (y_1, y_2, \ldots y_N) \geqslant 0$ so as to minimize

$$f'(Y) = \sum_{i=1}^{i=N} \Phi_i(\Delta L_i, y_i) \tag{3.13}$$

subject to constraint

$$T_i(y_i) \geqslant 0 \qquad i = 1, 2, \ldots N \tag{3.14}$$

This problem is completely decentralized and can be reduced to the following problem:

$$\underset{y_i}{\text{Min}}\{\Phi_i(y_i) \mid T_i(y_i) \geqslant 0\} \qquad i = 1, 2, \ldots N \tag{3.15}$$

It is assumed that the above problem is feasible and a method of solution exists which can obtain the global optimum y_i' within each subregion. Also, it is assumed that the nature of the above problem is such, that for each subregion, $y_i \geqslant y_i'$ is feasible and $y_i < y_i'$ is infeasible.

3.1.2 Smoothing Problem

Let y_i' be the minimum required level of activity within the ith element, and y_i be the actual level of activity within that element, such that,

$$y_i \geqslant y_i' \qquad i = 1, 2, \ldots N \tag{3.16}$$

There is a cost relating to the deviation of the actual level of activity from the minimum required level of activity. This function is defined as:

$$\eta_i(y_i, y_i') = \Phi_i(y_i) - \Phi_i(y_i') \tag{3.17}$$

The smoothing problem becomes:

Choose variables $Y = (y_1, y_2, \ldots y_N) \geqslant 0$, such that,

$$\text{Min } C(y_1, y_2, \ldots y_N) = \sum_{i=1}^{i=N} [\eta_i(y_i, y_i') + \Psi_i(y_i, y_{i-1})] \tag{3.18}$$

where $y_0 = y_0'$ is given, subject to constraint (3.16).

Using the dynamic programming method, Bellman (1962) shows that the above problem can be embedded within the family of problems requiring minimization of the function

$$C_k = \sum_{i=k}^{i=N} [\eta_i(y_i, y_i') + \Psi_i(y_i, y_{i-1})] \tag{3.19}$$

over the region defined by (3.16) for $i = k, k + 1, \ldots N$ and $k = 1, 2, \ldots N$.

Assume that the proposed level of activity at the $(k-1)^{th}$ subregion is ξ in which

$$\xi \geqslant y_{k+1} \tag{3.20}$$

The minimum of $C(y_1, y_{i+1}, \ldots y_N)$ over the region depends upon the starting subregion k and the value of the proposed level of activity ξ at the $(k-1)^{th}$ subregion. This dependence can be shown specifically by introducing

$$F_k(\xi) = \text{Min}\,(C_k) \qquad i = k, k + 1, \ldots N \quad k = 1, 2, \ldots N \tag{3.21}$$

subject to constraint (3.16) in which $F_k(\xi)$ is the minimum wasted cost relating to the furnished level of activities, y_i, between subregions k and N, if at subregion k-1 the proposed level of activity is ξ.

In the theory of dynamic programming these types of problems are solved by determining a recurrence relationship connecting $F_k(\xi)$ and $F_{k+1}(\xi)$ for arbitrary feasible values of k and ξ. This recurrence relationship is

$$F_k(\xi) = \text{Min}\{\eta_k(y_k, y_k') + \Psi_k(y_k, \xi) + F_{k+1}(y_k)\} \tag{3.22}$$

$$y_k \geqslant y_k', \qquad \xi \geqslant y_{k-1}', \qquad k = 1, 2, \ldots N$$

where, as initial condition, it is assumed y_0' is given and $F_{N+1}(\xi) = 0$ for all values of ξ.

The efficiency of computation increases if the values of y_k, y_k' and ξ are chosen among a discrete set and if the specific nature of each individual problem is utilized.

As a result of solving the above problem, the optimum values of $y_1, y_2, \ldots y_N$ within the subregions 1, 2, ... N are obtained. If the corresponding sizes of the subregions are $\Delta L_1, \Delta L_2, \ldots \Delta L_N$, then the sizes of the adjacent subregions having an equal level of activity can be combined. In this manner, two matrices are obtained; Matrix Z = $z_1, z_2, \ldots z_K$ in which each of its elements represents the sum of the sizes of the adjacent subregions having uniform levels of activity (thus, it can be administered as a unit), and Matrix Y = $\bar{y}_1, \bar{y}_2, \ldots \bar{y}_K$ which represents the uniform level of activities corresponding to the above Z regions. K represents the total number of various regions obtained above.

The nature of the solution using dynamic programming is such, that in most cases it obtains a unique global optimum value of variables Z and Y among the assumed discrete values.

3.2 Problem 2

Given variables $K = \bar{K}$ and $Z = (\bar{z}_1, \bar{z}_2, \ldots \bar{z}_k)$ where K is an integer and $\sum_{i=1}^{K} \bar{z}_i = L$, find variables X and Y, such that

$$f(W) = f_1(X) + \sum_{i=1}^{\bar{K}} [\Phi_i(X, \bar{z}_i, y_i) + \Psi_i(X, y_i, y_{i-1})] \tag{3.23}$$

subject to

$$g_j(X) \geqslant 0 \qquad j = 1, 2, \ldots m \tag{3.24}$$

$$H_i(y_i) \geqslant 0 \qquad i = 1, 2, \ldots \bar{K} \tag{3.25}$$

$$S(X, y_i) \geqslant 0 \qquad i = 1, 2, \ldots \bar{K} \tag{3.26}$$

$$R(X, y_i, c_i) \geqslant 0 \qquad i = 1, 2, \ldots \bar{K} \tag{3.27}$$

This is a usual programming problem with a known number of variables. The structure of the problem and the existence of regional activities presents the problem as a multi-stage system where a number of smaller problems are tied together by some coupling variables, constraints, and an objective function. As an example, if for a special case the objective function and constraints become linear, the problem can be cast into the form of the following ordinary linear programming problem:

$$\begin{aligned}
&c_1y_1 + c_2y_2 + \ldots + c_Ky_K + cX = Z(\text{Min}) \\
&\bar{A}_1y_1 + \bar{A}_2y_2 + \ldots + \bar{A}_Ky_K + \bar{A}X \geqslant \bar{b} \\
&A_1y_1 \qquad\qquad\qquad\qquad \geqslant b_1 \\
&\qquad A_2y_2 \qquad\qquad\qquad \geqslant b_2 \\
&\qquad\qquad\qquad A_Ky_K \qquad \geqslant b_K
\end{aligned} \tag{3.28}$$

where A_i and $\overline{A}_i$ are matrices, and y_i, c_i, X, and b_i are vectors. The problems of the above type can be solved either by usual linear programming methods or it may be more efficient to solve them by decomposition methods, such as the one given by Dantzig (1963). In such a case there exists sufficient proof that the solution converges to a global optimum in a finite number of steps.

In some special cases it may also be possible to cast problem (3.23-3.27) into the form of a convex partition programming problem similar to the one given by Rosen (1963) having the following form:

$$\min_{X,y_i} \left\{ CX + \sum_{i=1}^{i=K} C_i^T y_i \mid A_i^T y_i \geqslant b_i(X) \quad i = 1, 2, \ldots K \right\} \tag{3.29}$$

where, if we can assume that $b_i(X)$ are convex functions, then there is sufficient proof that the solution converges to a global optimum in a finite number of steps.

It may happen that in some cases, problem (3.23-3.27) has no solution or the solution may not be a global optimum. We assume that we are dealing with those cases where problem (3.23-3.27) has a global optimum solution. Thus, the result of applying problem II is to find the global optimum values of variables X and Y for given values of $\overline{K}$ and region-size variables Z.

3.3 Iterative method of solution

A solution to the problem (2.1-2.6) can be found using the following steps iteratively until convergence.

Step I Initialization: Select a set of global variables $X = X_0 \geqslant 0$ as an initial design such that:

$$g_j(X_0) \geqslant 0 \quad j = 1, 2, \ldots m$$

Step II Application of Problem I: For the given value of global variable $X = X_0$ obtained from the previous step, problem I as stated in section (3.1) is solved and the optimum values of the number of regions $\overline{K}$, region-size variables $\overline{Z}$, and regional activities $\overline{Y}'$ are obtained. The solution to problem (2.1-2.6) is feasible during and at the end of this step.

Step III Application of Problem II: For the given value of the number of regions $K = \overline{K}$, and region-size variables $Z = \overline{Z}$ as obtained from Step 2, Problem II as stated in Section (3.2) is solved; and the optimum values of global variables $\overline{X}$, and regional variables $\overline{Y}$ are obtained. The solution to problem (2.1-2.6) is feasible during and at the end of this step.

Step IV Test for Convergence: If the convergence condition is satisfied, then, the final solution is $X^* = \overline{X}$, $Y^* = \overline{Y}$, $Z^* = \overline{Z}$, $K^* = \overline{K}$. Otherwise, a transfer is made to Step II assuming $X = \overline{X}$ as a new initial value of global variables. The procedure is repeated until the convergence condition is satisfied.

A schematic flow chart for the above iterative process is shown in Figure (2).

3.4 Definition of convergence

The iterative process of Section (3.3) is defined as converged if the value of design variables W = X, Y, Z and the objective function f(W) remains unchanged after the application of one complete cycle. At the jth cycle the optimum value of variables Z_j obtained from the solution of problem I depends only on the value of global variables X_{j-1} of the previous cycle. The optimum value of variables X_j, Y_j and objective func-

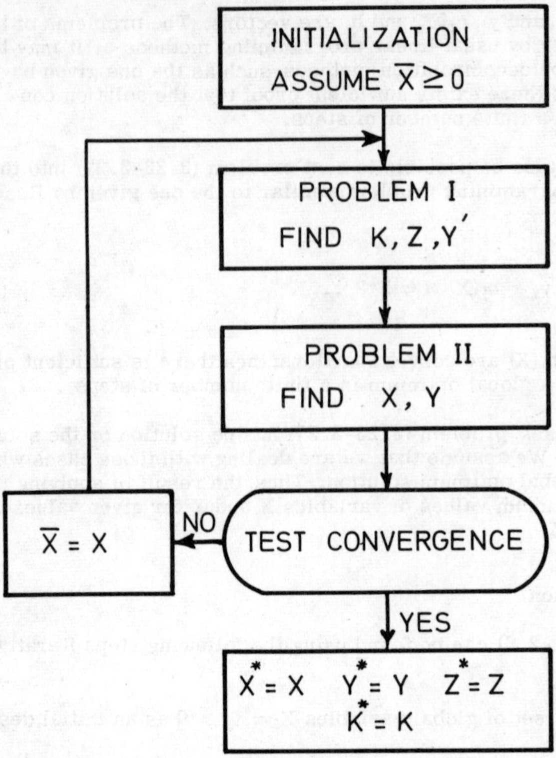

Fig. 2 Schematic Flow Chart for Iterative Smoothing Method for Systems with Determinate Behavior

tion $f(W)_j$, obtained from the solution of problem II depends only on the value of variables Z_j. Therefore, it is evident that if $X_j = X_{j-1}$, then, at the j + 1 cycle $Z_{j+1} = Z_j, X_{j+1} = X_j, Y_{j+1} = Y_j$, and $f(W)_{j+1} = f(W)_j$. Thus, a suitable test for convergence is to stop the iteration when

$$X_j = X_{j-1} \tag{3.30}$$

With the same reasoning, it follows, that if $Z_j = Z_{j-1}$, then $X_j = X_{j-1}$, $Y_j = Y_{j-1}$, and $f(W)_j = f(W)_{j-1}$. Thus, another suitable convergence test is to stop the iteration when

$$Z_j = Z_{j-1} \tag{3.31}$$

3.5 Proof of convergence

Theorem I:

If the solution method of each of the problems I and II obtains the global optimum solution for each of the problems, then, the iterative algorithm of Section 3.3 will converge to a feasible solution of problem (2.1-2.6) in a finite number of cycles. The converged solution has the least value of the objective function among all other solutions obtained during the iterative cycles.

Proof:

Consider the jth cycle of iteration: At the start of the jth cycle, there is a feasible vector X_{j-1}, Y_{j-1}, Z_{j-1} and a corresponding value of $f(W)_{j-1}$. As a result of applying problem I at the jth cycle, a new feasible design is obtained giving a vector Z_j and Y'_j which constitutes a global optimum for problem I on the basis of the given value of global variable X_{j-1}. Let it be assumed that the value of the objective function at the end of the application of problem I is $f(W')_j$. Because the solution $W_{j-1} = X_{j-1}, Y_{j-1}, Z_{j-1}$ is a feasible solution for problem I at the jth cycle, then, the application of problem I leads to a new feasible design $W'_j = X_{j-1}, Y'_j, Z_j$ where

$$f(W')_j \leqslant f(W)_{j-1} \tag{3.32}$$

Applying problem II of the iterative scheme gives a new feasible solution, $W_j = X_j$, Y_j, Z_j which constitutes a global optimum for problem II for the given value of region-size variable Z_j. If the value of the objective function at the end of problem II is $f(W)_j$, then, because the solution W'_j is a feasible solution for problem II, the application of this problem leads to the feasible solution W_j where

$$f(W)_j \leqslant f(W')_j \tag{3.33}$$

From inequality (3. 32) it can be concluded that

$$f(W)_j \leqslant f(W)_{j-1} \tag{3.34}$$

Thus, during each cycle the solution remains feasible and the objective function in general improves.

To show that the solution converges, consider two cases:

Case 1: Problems I and II have only unique optimal bases. At the start of the jth cycle, the feasible vector $W_{j-1} = X_{j-1}, Z_{j-1}, Y_{j-1}$ and the corresponding value of the objective function $f(W)_{j-1}$ is available. From the solution of problem I, during the jth cycle the value of variables Z_j and Y'_j, and the objective function $f(W')_j$ is obtained. These values depend upon the value of X_{j-1}.

From the solution of problem II, during the jth cycle the value of variables X_j, Y_j, and the value of the objective function $f(W)_j$ is obtained. These values depend only upon the value of Z_j.

Case A: If $X_j = X_{j-1}$, then the iteration will converge at the next cycle as defined and discussed in Section 3. 4.

Case B: If $X_j \neq X_{j-1}$, then

$$f(W')_{j+1} \leqslant f(W)_j < f(W')_j \leqslant f(W)_{j-1} \tag{3.35}$$

Two cases may happen as follows:

Case B1: When $Z_{j+1} = Z_j$ at the j + 1 cycle.

This case is possible because the value of Z variables are selected among discrete sets and it is always possible that a small change in variables X during the jth cycle will not be able to force the Z variables to change from the former basis to another new basis. In such a case the design will converge at the j + 1 cycle as defined in Section 3. 4 and

$$f(W)_{j+1} = f(W)_j \tag{3.36}$$

But $f(W)_{j+1} \leqslant f(W')_{j+1}$, therefore, from (3.35) and (3.36) it follows that:

$$f(W)_{j+1} = f(W')_{j+1} = f(W)_j$$

Case B2: When $Z_{j+1} \neq Z_j$ at the j + 1 cycle.
In this case:

$$f(W)_{j+1} < f(W)_j \tag{3.37}$$

Again two possibilities exist; either $X_{j+1} = X_j$ which leads to Case A and the solution converges at the j + 2 cycle, or $X_{j+1} \neq X_j$ which leads to Cases B1 or B2. At Case B1 the solution converges at the j + 2 cycle. At Case B2 we have $Z_{j+2} \neq Z_{j+1}$ and the process will continue again. As shown by inequality (3.37) during each continuation of the process, the value of the objective function decreases. Because the region-size variables Z can only be selected among a finite number of discrete values forming a limited number of bases for problem I, then, the number of cycles must be finite.

Case II: Problems I and II may have alternate optimal bases.

It is assumed that the solution methods of problem I and II will furnish all the existing alternate optimal bases. If at the jth cycle for problem I a few alternate optimal bases were found, then each of these bases will be tried in turn. If a basis, Z_j, is found for which the complete problem converges, i.e., $Z_{j+1} = Z_j$, then, we move to the next basis, Z_j'. Trying this basis, it is found that either the complete problem converges, i.e., $Z_{j+1}' = Z_j'$, or $Z_{j+1}' \neq Z_j'$ where, in the latter case $f(W)_{j+1} < f(W)_j$ and the problem moves to a new Z basis which is different from the Z basis of the previous cycle. Since there are only a finite number of possible problem I bases, then, the number of cycles is finite.

3.6 Discussion of optimality

The iterative process presented in Section 3.3 converges to a feasible solution of problem (2.1-2.6). This solution has the least value of the objective function obtained during the iterative cycles. The solution obtained at the point of convergence satisfies the optimality conditions of problems I and II simultaneously.

So far the optimality conditions of the converged solution for the entire problem has not been found. This has been mainly due to the difficult situation governing this problem, the fact being, that the number of variables is also a variable and the problem is a non-convex problem. For the iterative problems having a fixed number of variables and when certain convexity properties exist, it may be possible to obtain certain optimality conditions in a similar manner as was done in the case of convex partition programming by Rosen (1963). As Beale shows (Abadie (1967)), for non-convex partition programming there is no guaranteed optimality.

In the proof of the condition of optimality of problem II, when applicable, the Kuhn-Tucker optimality condition or those of linear programming may be used. In the proof of the optimality of problem I when using dynamic programming, the principle of optimality as discussed by Bellman (1963) is used. In problem I the number of variables $Z = z_1, z_2, \ldots z_K$ and their value is unknown. Therefore, no condition of optimality in a form compatable to that of problem II is obtainable. This makes it theoretically very difficult to obtain a general condition for optimality of the entire problem.

In his review of this paper, Beale presented the following simple example to show that the iterative method of solution of section 3.3 is not always guaranteed to converge to a global optimum solution.

Example of non-optimality

Consider a problem with N = 3, i.e. with 3 elements in the smoothing problem, with one global variable x_1 restricted to lie between 1 and 2, and 'elementary' local variables y_1, y_2, and y_3 subject to the constraints

$$3x_1 + y_1 \geqslant 7$$
$$2x_1 + y_2 \geqslant 5$$
$$x_1 + y_3 \geqslant 4.$$

Let the objective function be

$$23x_1 + 4(y_1 + y_2 + y_3) + (1 - \delta_{y_1y_2}) + (1 - \delta_{y_2y_2}).$$

Then the suggested iterative procedure could be applied as follows:

Try $x_1 = 1$.

Then the minimum feasible values of y_1, y_2, and y_3 are 4, 3, and 3 respectively.

Conditional on x_1 being equal to 1, we find that these are the best values to the smoothing problem. (The total cost increases by 7 if we put $y_1 = y_2 = y_3$.)

So we put K = 2, $z_1 = 1$, $z_2 = 2$, and Problem 2 becomes:

Minimize

$$23x_1 + 4y_1 + 8y_2 + (1 - \delta_{y_1y_2})$$

subject to the constraints:

$$1 \leqslant x_1 \leqslant 2$$

$$3x_1 + y_1 \geqslant 7$$

$$x_1 + y_2 \geqslant 4.$$

If we now keep y_1 and y_2 distinct, then we obviously put

$$y_1 = 7 - 3x_1$$
$$y_2 = 4 - x_1$$

and the total cost is $60 + 3x_1 + 1$, which takes its minimum value of 64 when $x_1 = 1$.

If on the contrary we make y_1 and y_2 equal, then both must equal the larger of $7 - 3x_1$ and $4 - x_1$, so the total cost is the larger of $84 - 13x_1$ and $48 + 11x_1$. These quantities both equal $64\frac{1}{2}$ if $x_1 = 1\frac{1}{2}$, and the larger of the two is larger for any other value of x_1.

So the procedure terminates with this initial solution with $x_1 = 1$ and a cost of 64.

On the other hand we can easily verify that if $x_1 = 2$, $z_1 = 2$, $z_2 = 1$, $y_1 = 1$, $y_2 = 2$, then the total cost becomes:

$$46 + 16 + 1 = 63.$$

So the procedure does not converge to the global solution of the problem.

Further theoretical investigations are needed to verify in a mathematically sound form the conditions of the optimality of the solution obtained from the convergence of the iterative process of Section 3.3.

In order to increase the confidence in the optimality of the solution obtained from the convergence of the iterative process, some numerical experiments have been carried out. For example, the problem of the minimum-cost design of plate girders shown in Figure (3), and as formulated in Section 2.1 has been studied. The numerical solution of this problem is discussed in Section 5. For verification of optimality, two methods are used; in the first method at the neighborhood of the point of convergence of iterative process, the global variables X are varied in a prescribed discrete manner. Then, each choice of variable X used as a starting point, and the iterative method of Section 3.3, is applied and the final converged solution is obtained. It was found that for all these alternate starting points, the final solution converges to the same original solution.

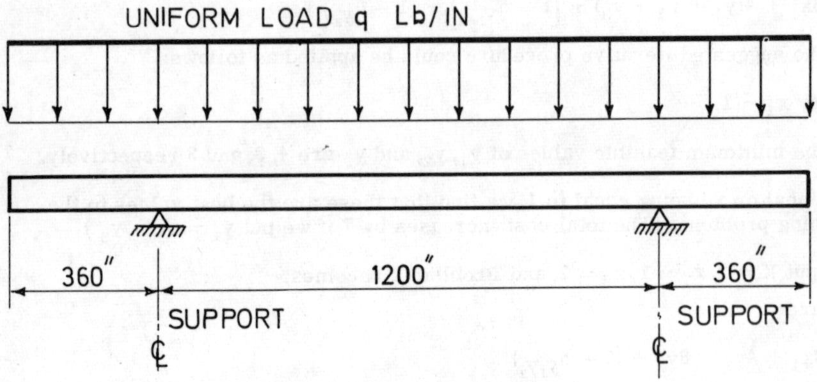

Fig. 3 Determinate Girder with Overhangs

In the second method the iterative process of Section 3.3 is started using various starting values of global variables X which were very different from each other. Again for almost all cases the solutions converged to the same original point; however, the number of cycles differed for various starting points.

The above limited experiments indicates that for the above problem the solution at least converges to a local minimum point. More numerical experiments are necessary to give a stronger confidence. In general, the optimality is not guaranteed.

4. EXTENSION OF THE SOLUTION METHOD TO INDETERMINATE SYSTEMS

In the case of determinate systems the behavior (or demand) variables $U_i(W)$, which are present in constraint set (2.6) and correspond to each region i, are constant for that region: that is,

$$U_i(W) = c_i \tag{4.1}$$

For indeterminate cases these variables are explicit functions of variables W = X, Y, Z of the entire system. For an engineering structure the behavior variables may be the forces (or the stresses) in various members, displacements, accelerations, or temperature magnitude at different points of the structure under the action of applied static, dynamic, or thermal loading. These behavior variables are functions of the geometry of the structure, the value of design variables W, and the applied loading condition.

For a company having interdependent managerial or economical activities within a country, the behavior variables within each region of that country may be the demand for the company's products, the prices of natural resources, or the supply and cost of labor within each region. These behavior variables may be explicitly dependent upon the level of various activities of that company within the entire country.

For the determination of behavior variables U_i within each region i, the following problem is solved:

4.1 Analysis problem

For engineering structures the determination of the behavior of structure U for a given set of variables W under an applied loading condition is called analysis of the structural system and is obtained from the solution of the proper analysis equations represented by

$$\phi(W, U) = S(P) \tag{4.2}$$

where S(P) is a function of the applied loads.

In most methods of formulating analysis problems, the set of equations (4.2) can be subdivided into two sets of equations, namely, the equilibrium equations representing the equilibrium between internal and external forces which are represented by

$$\phi'(W, U) = S'(P) \tag{4.3}$$

and the compatability equations representing the compatibility of the behavior of the structural system under the applied loads which are represented by

$$\phi''(W, U) = S''(P) \tag{4.4}$$

Equilibrium and compatibility equations similar to equations (4.3) and (4.4) can be written for various other physical or managerial systems.

Here, it is assumed that relevant technologies for the solution of an analysis problem represented by equations (4.2) exist. On this basis, for a given set of variables W, the value of behavior variables within each subregion can be obtained from the solution of equation (4.2) in the following form for the ith region

$$U_i = U_i(W) \tag{4.5}$$

4.2 Iterative method of solution

The iterative process of Section 3.3 is extended to the case of indeterminate systems in the following manner:

Step I Initialization: Select a set of initial variables $W_0 = X_0, Y_0, Z_0 \geqslant 0$, such that, constraints (2.2), (2.3), (2.4), and (2.5) are satisfied.

Step II Applying an Analysis Problem: For the given value of variables $W = W_0$ obtained from the previous step, solve analysis equations (4.2). Compute $U = U(W)$ for the entire system.

Step III Apply Problem I: This step is the same as Step II, Section 3.3 and computes $\overline{K}$ and $\overline{Z}$.

Step IV Apply Problem II: This step is the same as Step III, Section 3.3 and computes $\overline{X}$ and $\overline{Y}$.

Step V Test for Convergence: The convergence condition is defined in Section 4.3. If the condition of convergence is satisfied, then the final solution is: $W = \overline{W} = \overline{X}, \overline{Y}, \overline{Z}, \overline{K}$. Otherwise, a transfer is made to Step II assuming $W_0 = \overline{W}$ as a new set of initial variables. The cycles are repeated until the convergence condition is satisfied.

A schematic flow chart for the above iterative process is shown in Figure (4).

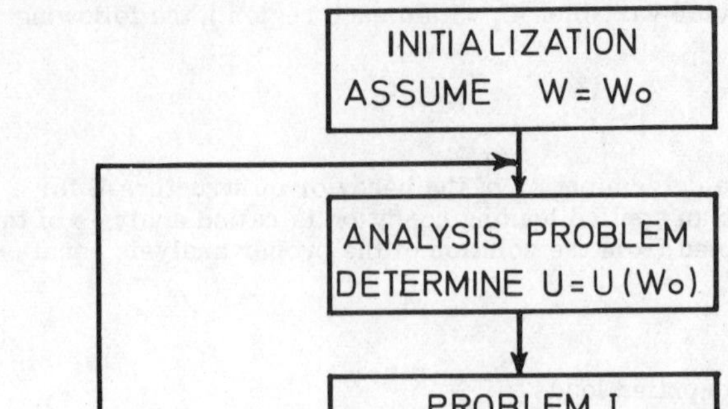

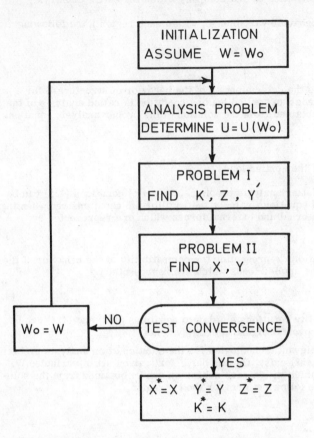

Fig. 4 Schematic Flow Chart for Iterative Smoothing Method for Systems with Indeterminate Behavior

4. 3 Definition of convergence

The iterative process in Section 4. 2 is defined as converged if the value of variables W = X, Y, Z and the objective function f(W) remains unchanged after the application of one complete cycle.

4. 4 Discussion of convergence and optimality of the iteration

Consider the jth cycle of iteration: At the start of this cycle, there is a vector $W_{j-1} = X_{j-1}, Y_{j-1}$, and Z_{j-1} satisfying all the constraints but not necessarily the constraint set (2. 6). Vector W_{j-1} satisfies the constraint set (2. 6) with the value of behavior (demand) variables obtained on the basis of W_{j-2}; this means

$$R[X_{j-1}, Y_{j-1}, U_i(W_{j-2})] \geqslant 0 \quad i = 1, 2, \ldots K \tag{4. 6}$$

Applying the analysis problem, the value of behavior variables $U_i(W_{j-1})$ is obtained which may be different from $U_i(W_{j-2})$. Therefore, the constraint sets (2. 6) are not necessarily satisfied, and thus, during the iteration the solution may not remain fea-

sible. If the iterative process converges at the nth cycle as defined in Section 4.3, it is evident that constraint set (2.6) is satisfied because $W_{n+1} = W_n$ and

$$\bar{U}_i(W_{n+1}) = U_i(W_n) \quad i = 1, 2, \ldots K \tag{4.7}$$

Thus, the final converged solution is feasible.

The rate of convergence and the relative optimality of the solution of the iteration depends upon the characteristics of the analysis equations and behavior variables. If these variables are very sensitive to the variation of variables W and the magnitude of their variation due to a change in variable W is large, then there is no way to ensure convergence and relative optimality.

The problem of convergence and optimality can be better understood by discussing an example from methods of structural design, namely, the fully-stressed design method. This method operates in the following manner:

If analysis shows that a certain member is overstressed (demand exceeds supply), the fully-stressed design method increases the area (level of activity) of that member to such an extent as to remove the over stress. It does the opposite if the member is understressed (supply is more than demand).

Statically indeterminate structures have a tendency to attract more force to any member whose area or stiffness has increased. Therefore, a few cycles of iterative analysis and fully-stressed design are necessary to obtain a feasible design.

For the structures designated by Cross (1936) as having normal action, the iterative cycles of analysis and fully-stressed design converges rapidly, and generally, the magnitude of forces in members is little affected by relative variations in the design of other members. For structures with the so-called hybrid action, variation in the size of a member changes the magnitude of forces in other members; thus, each member must be designed with due consideration to its effect on other members. For this type of structure the convergence is generally slow and the repetition of analysis and fully-stressed design may tend to eliminate some members of the structure and may result in a simpler structure.

The relationship between an iterative fully-stressed design and a design with minimum design-moment-volume has been discussed by Razani (1965a + b), where minimum design-moment-volume is defined as

$$\text{Min}\left\{\int_0^L |M|\,d\xi \cong \sum_{i=1}^{i=N} |M_i|\,\Delta L_i\right\} \tag{4.8a}$$

For the general problem this is equivalent to minimum demand volume

$$\text{Min}\left\{\int_0^L U d\xi \cong \sum_{i=1}^{i=N} U_i \Delta L_i\right\} \tag{4.8b}$$

He has shown that for constant depth girders an iterative fully-stressed design converges if the spectral norm of Matrix B is less than unity, i.e.,

$$\|B\| < 1 \tag{4.9}$$

B is an N × N influence matrix whose elements b_{ij} reprcsent the rate of change in the design moment (or demand) in the ith region due to a unit increase in the design moment (or demand) in the jth region, i.e.,

$$B = \partial\,|M_i|\,/\partial\,|M_j| \tag{4.10a}$$

or in general

$$B = \partial U_i / \partial U_j \tag{4.10b}$$

The speed of convergence depends upon the smallness of the spectral radius $\rho(B)$ of Matrix B.

Razani (1965b) also has used the Kuhn-tucker (1951) optimality condition for nonlinear programming to show that for a girder the condition of optimality of fully-stressed design is satisfied if

$$\lambda = (I - B^T)^{-1} \Delta L > 0 \tag{4.11}$$

where I is an $N \times N$ unit Matrix, B^T is the transpose of B, and $\Delta L > 0$ is a vector representing the length of the members or subregions. For a converging system where (4.9) holds, the above optimality condition can be expanded in the form of the following converging series:

$$\lambda = (I + B^T + B^{T^2} + \ldots) \Delta L > 0 \tag{4.12}$$

For indeterminate structures Matrix B has generally the following properties:

(1) The diagonal elements of B are always positive. This is due to the fact that for this type of structures more forces (demand) are induced in any member whose area or stiffness (or level of activity) has increased. This can be verified from close inspection of analysis equations for any indeterminate structures.

(2) The absolute value of the off diagonal elements of Matrix B become smaller as their relative location from diagonal elements increases.

Certain relationships exist between the spectral radius of Matrix B and the size of its element. Varga (1962) shows that

$$\rho(B) \leqslant \mathrm{Min}(V, V') \tag{4.13}$$

where

$$V = \mathrm{Max} \sum_{j=1}^{j=N} |b_{ij}| \qquad 1 \leqslant i \leqslant N \tag{4.14}$$

$$V' = \mathrm{Max} \sum_{i=1}^{i=N} |b_{ij}| \qquad 1 \leqslant j \leqslant N \tag{4.15}$$

From the above discussion it is seen that for Matrix B, whose elements have small absolute values, the spectral radius is small; thus, the rate of convergence is large. Also, the fast converging iteration indicates a small spectral radius which indicates a Matrix B having many elements with a small absolute value.

Considering other properties of Matrix B, it can be argued that for a rapidly converging fully-stressed design, such as in the case of structures with normal action, for vector $\Delta L > 0$, the inequality (4.12) is satisfied and the iterative cycles of analysis and redesign lead to a design with minimum design-moment-volume (or minimum demand volume) as defined by Equations (4.8a) and (4.8b). This situation is evident in Figure (5) where a three span girder is designed. The initial design is shown by broken lines and the value of global variables are shown. The regional variables are assumed to be uniform along the girder. The design moment (demand) is shown by a broken line for initial design. After three cycles of iteration the design at the end of the third cycle is shown by solid lines. The design moment (demand) is also shown by a solid line. It is seen that this structure has a normal action and, even though the design variables W have changed sharply, the change in design moment is very small. Also, it is observed that the area under the design moment shown by the solid line representing the latest demand is less than the area under the design moment shown by a broken line which represents the initial demand.

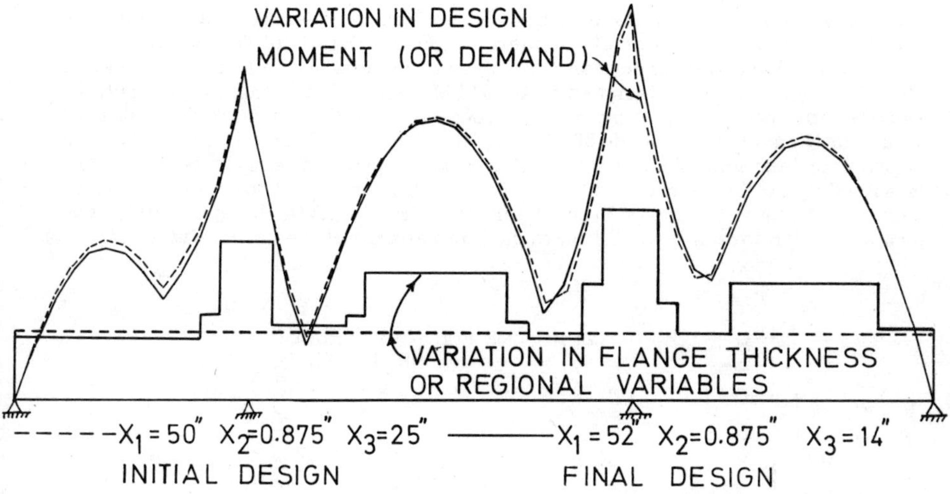

Fig. 5 Variation of Behavior (Demand) Variables Due to Variation in Design Variables

For many problems there is a direct relationship between the level of demand, level of activities, and costs, such that, if the volume of demand increases, the level of activities must increase to meet these demands and correspondingly the costs increase also. For this type of problem a solution having a minimum volume of demand provides a basis for a minimum-cost solution, where determination of this type of solution is possible by an iterative process of analysis and optimum design.

Finally, it can be concluded that if the indeterminate system has normal action, and if the problem has properties similar to that discussed in the previous paragraph, then successive cycles of the iteration process of Section 4.2 reduces the demand volume and indirectly reduces the objective function (costs), i.e.,

$$\int_0^L U(W_j)d\xi \leq \int_0^L U(W_{j-1})d\xi$$

and $\qquad j = 0, 1, \ldots n$

$$f(W_j) \leq f(W_{j-1})$$

The solution may become infeasible during the iteration; however, at the point of convergence the solution is feasible and the objective function has the least value along the path followed to the convergence point.

The effect of problems I and II on convergence and optimality was discussed in Sections 3.1 and 3.2. For indeterminate systems with normal behavior using the same line of reasoning as in Section 3.5 it is possible to prove that the iteration converges in a finite number of cycles. The optimality of the procedure is not guaranteed, however, the converged solution has the least value of the objective function among all other feasible solutions obtained during the iterative cycle, and thus, is an improvement over the starting solution.

5. NUMERICAL EXAMPLES

Minimum-cost design of plate girders was formulated in Section 2.1 and was solved by Razani (1965a) and (1966) using the so-called Iterative Flange Smoothing Method (or IFSM). This method uses an early version of the general algorithm of the decentralized programming method using iterative smoothing.

The global variables x_1, x_2, and x_3 are chosen from the discrete values which represent the available dimension of commercial plates. Details of data and results is given in the above references for many case examples. As an example, the girder shown in Figure (3) was designed by the IFSM method. The length of computation required for the solution of problem I, problem II, and the entire IFSM problem for values of N equal to 15, 30, 45, 60, 75, and 90 is shown in Figure (6). For the solution of problem I the smoothing method using dynamic programming is used. Problem II is solved by direct search on discrete values of x_1 and x_3. The value of x_2 and the values of regional variables Y are determined from a particular property of the problem which forces the solution to certain vertices of the constraint set.

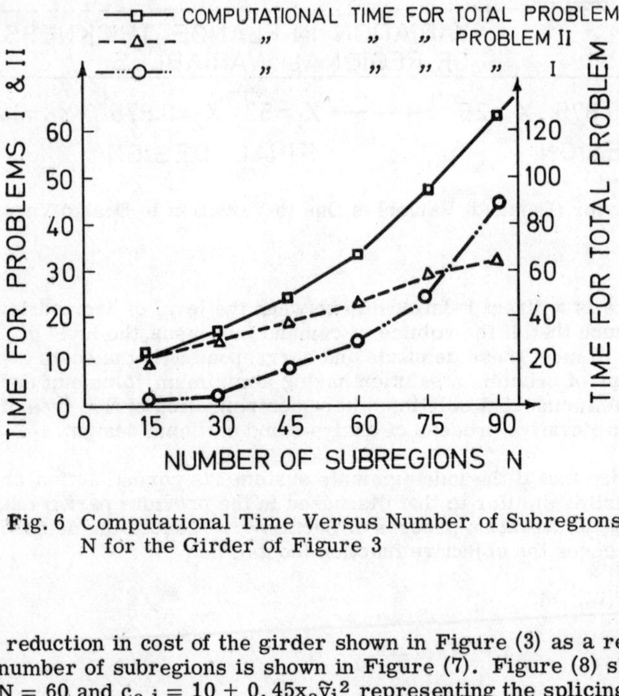

Fig. 6 Computational Time Versus Number of Subregions N for the Girder of Figure 3

The reduction in cost of the girder shown in Figure (3) as a result of the increase in the number of subregions is shown in Figure (7). Figure (8) shows the final design for $N = 60$ and $c_{2,i} = 10 + 0.45x_3\tilde{y}_i^2$ representing the splicing costs between the ith and $i + 1$ regions where $\tilde{y}_i = \text{Min}(y_i, y_{i+1})$. Figure (9) shows the final design for $N = 60$ and $c_{2,i} = 10 + 0.15x_3\tilde{y}_i^2$. From a comparison of Figures (8) and (9) it is seen that the less expensive splicing costs cause more variation in flange thickness, and increases the number of regions ($K = 5$ for Figure (8) and $K = 9$ for Figure (9)). It is seen that the value of global variables in the two above cases are the same; this is probably due to the discreteness of the value of the global variables.

Figure (10) shows the final design for $N = 30$ and $c_{2,i} = 10 + 0.15x_3\tilde{y}_i^2$. From a comparison of Figures (9) and (10), it is seen that a reduction in the number of subregions increases the costs of the girder by forcing the formation of splices at less suitable points. It is seen that the value of global variables are different for the two above cases. For the above problems the iteration converged after three cycles.

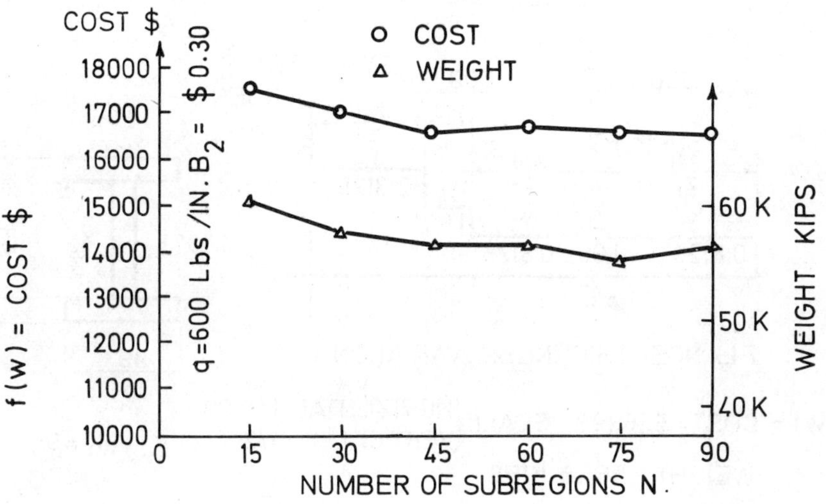

Fig. 7 Variation of Optimum Cost and Weight Versus Number of Subregions N

$f(w) = COST$ $^\${}17235$ SCALE $\begin{cases} HORIZONTAL & 1/200 \\ VERTICAL & 1/2 \end{cases}$ $K = 5$

WEIGHT 58.5 KIPS

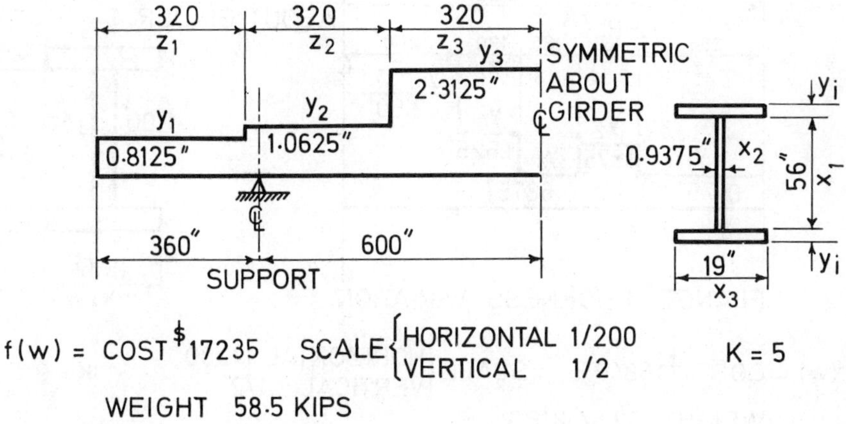

Fig. 8 Optimum Design for N = 60, q = 600 Lb/in, and $B_2 = \$0.45$

FLANGE THICKNESS VARIATION

$f(w) = COST = \$16398$ SCALE $\begin{cases} \text{HORIZONTAL } 1/1200 \\ \text{VERTICAL} \quad\quad 1/2 \end{cases}$ K = 9

WEIGHT 56.48 KIPS

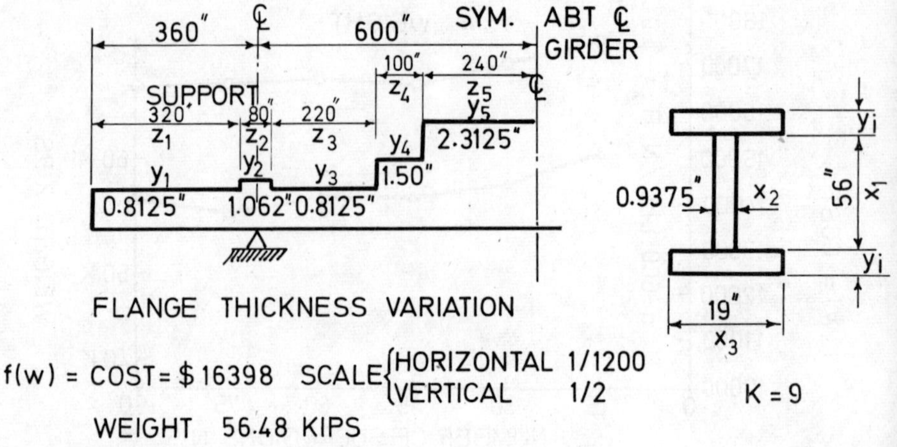

Fig. 9 Optimum Design for $N = 60$, $q = 600$ Lb/in, and $B_2 = \$0.15$

FLANGE THICKNESS VARIATION

$f(w) = COST = \$16840$ SCALE $\begin{cases} \text{HORIZONTAL } 1/200 \\ \text{VERTICAL} \quad\quad 1/2 \end{cases}$ K = 9

WEIGHT 57.54 KIPS

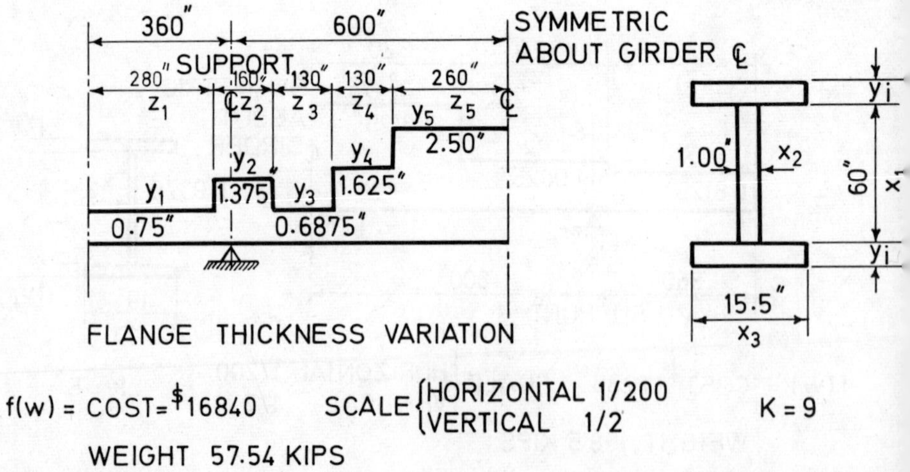

Fig. 10 Optimum Design for $N = 30$, $q = 600$ Lb/in, and $B_2 = \$0.15$

Figures (11) and (12) compare two designs obtained for a three span continuous (indeterminate) girder under a moving load. The number of subregions for both cases is $N = 40$. The splicing cost for the design shown in Figure (11) is $c_{2,i} = 10.0 + 0.10x_3\tilde{y}_i^2$ and the number of final regions is $K = 14$. For the design shown in Figure (12): $c_{2,i} = 14.0 + 0.25x_3\tilde{y}_i^2$. The number of final regions for this case is $K = 10$. The value of global variables is the same for both cases except for x_3 which is different.

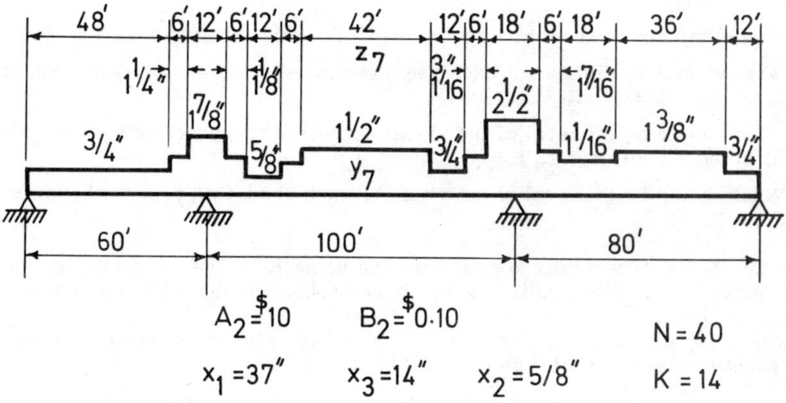

Fig. 11 Three-Span Girder Design, Variation of Flange Thickness

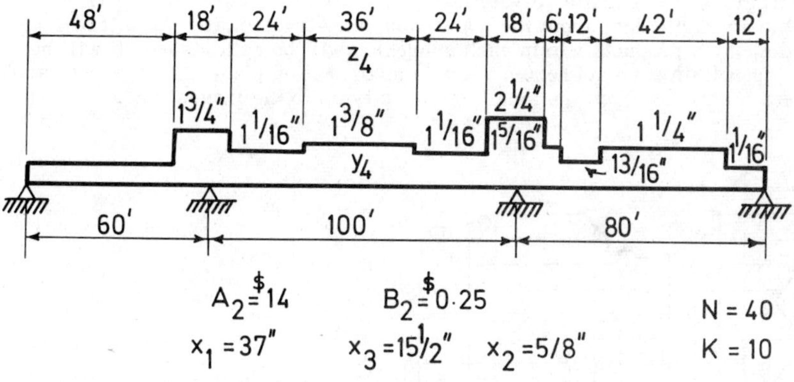

Fig. 12 Three-Span Girder Design, Variation of Flange Thickness

For the two above indeterminate cases, the iteration converged in four cycles which took approximately three minutes of computer time using the 1107 Univac computer. More details are given in the above mentioned references.

6. POTENTIAL APPLICATIONS TO MANAGEMENT PROBLEMS

Consider a company which is planning to locate manufacturing plants or sales offices for producing or selling some products at different locations within a country. Within each plant there are two types of activities and constraints. The first types are directed from the central headquarter and is common for all the plants. The level of these activities are shown by global variables X. As an example, these variables

may represent the prices of raw materials, or finished products furnished by the central headquarter, advertisement policies, general specifications for products, technology, and management of all the plants. The second type is local in nature and includes the size of each plant, the level of its local activities, such as its local purchasing, manufacturing, and sales policies. These activities are shown by regional variables y_i for the ith plant.

The problems, facing the central management who want to optimize the main objective of the company, are:

(1) How many plants should be established within the country, i.e., $K = ?$

(2) What should be the size of each region which should be dependent upon each plant, i.e., $Z = z_1, z_2, \ldots z_K = ?$

(3) What should be the level of the global activities representing the centrally directed activities, i.e., $X = ?$

(4) What should be the level of activities of each plant, i.e., $y_i = ?$ where $i = 1, 2, \ldots K$.

The algorithm of the decentralized programming using iterative smoothing as presented in Section 3. 3 can be applied to the above problem in the following manner:

(1) The central management assumes an initial set of policies, procedures, prices, etc., representing a set of initial global variables $X_0 \geqslant 0$.

(2) For the given set of global variables the problem I is solved in the following manner:

The entire range of the country under consideration is assumed to be subdivided into small subregions (say counties or districts) as shown in Figure (13) for Iran. On the basis of the available data about the population, industry, etc., the magnitude of demand for the company's products within each subregion will be established. It will be assumed that within each subregion there is an operating plant. These plants are of various sizes starting from the smallest capacity up to the largest one.

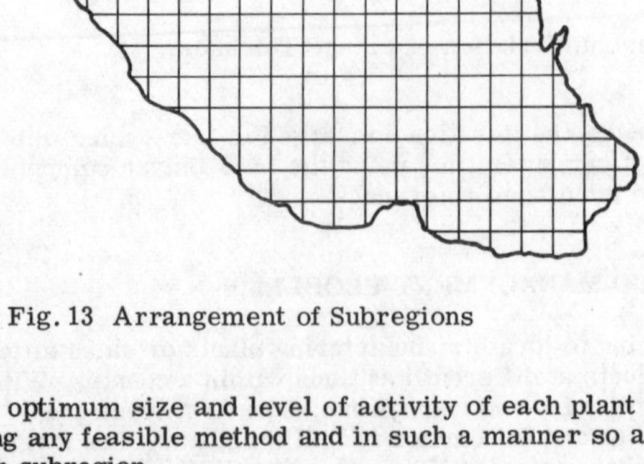

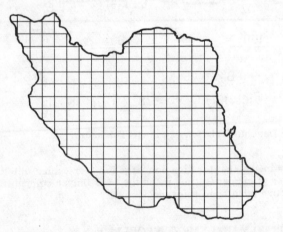

Fig. 13 Arrangement of Subregions

The optimum size and level of activity of each plant within each subregion is obtained using any feasible method and in such a manner so as to satisfy the demand within each subregion.

Generally, there is a fixed cost relating to the establishment and the size of a plant, and an operating cost corresponding to the size of the plant, the size of the region served by the plant, and its level of activity.

The operating cost generally increases when the size of the region served by the plant increases. These costs are small for plants operating in small size regions. However, it is not generally economical for the company to locate small plants in each community due to the many fixed costs relating to the establishment, maintenance, and management of each plant. If the company tries to minimize the operating costs by putting efficient small sized plants in small subregions he increases his fixed costs. If he tries to minimize his fixed costs by establishing giant plants within large provinces he may increase his operating costs. A feasible method of smoothing can be used to balance these two conflicting costs and minimize the total costs.

Using a smoothing technique the central management will determine the optimum number, size, and arrangements of the regions which will minimize the total costs. Figure (14) represents a possible map of regions obtained at this stage.

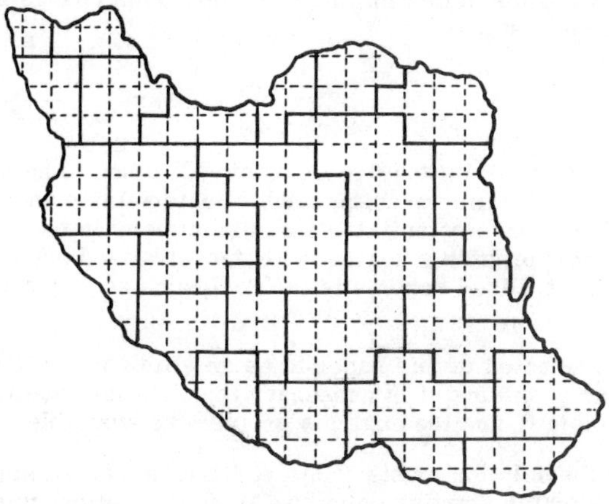

Fig. 14 Arrangement of Smoothed Subregions

(3) Now, assuming that the regions obtained in stage 2 are temporarily fixed, problem II must be solved. In this case the central management tries to find the optimum value of the global variables, and the corresponding optimum level of the regional activities corresponding to each plant within each region. This can be accomplished by various programming methods, such as the decomposition method of Dantzig (1963), the convex partition programming of Rosen (1963), or other applicable methods.

As a result of the above stage a new set of global variables are found which constitute an optimum solution on the basis of the assumed regionalization. However, the regionalization obtained in stage 2 on the basis of the global variable of the previous cycle may not be an optimum arrangement for the newly obtained set of global variables if the new set is different from the set of the previous cycle. In such a case the new set of global variables will be substituted in stage 2 as initial global variables and the cycles are repeated until convergence.

By analogy to the structural design problem, three cases can happen.

(1) The determinate behavior where the activities of the company will not at all change the pattern of the demand in any part of the country.

This is the case with small industries such as post offices, chain restaurants, etc.

(2) The case of normal indeterminate behavior where the allocation of plants and their activities within various regions of the country will change the pattern of demand within the country only by a small magnitude.

This is the case with small industries such as chain stores, light manufacturing, banks, etc., where the volume of the industry under study is small compared with the total volume of other industries.

(3) The hybrid indeterminate behavior, where the establishment of plants in various zones of the country will appreciably change the demand pattern in all the regions of the country.

This happens when a major manufacturing or administrative industry allocates its plants within some regions of the country. As an example, the establishment of an automobile manufacturing plant within a region so as to satisfy the prescribed need of that region will bring a large number of people and other industries from other regions of the country toward that region. This will increase the value of the demand for automobiles in that region much more than the initially assumed demand. However, it may decrease the demand for automobiles in other regions. This change in demand may take a long time to become stable.

7. CONCLUSION

The decentralized programming method using iterative smoothing is found to be very efficient when applied to the solution of minimum-cost design of determinate and indeterminate girders. This method may have potential useful applications in the solution of administrative and management problems dealing with decentralized activities, where determination of the best decentralized scheme is the goal, such as in locating plants, departments, etc.

In order for the iterative scheme presented in this paper to be an efficient and feasible tool for the solution of various problems, it is necessary that efficient methods for the solutions of problem I, problem II, and the analysis problem be available.

No guarantee for the optimality of the solution exists. However, the converged solution is generally an improvement over the starting solution. More theoretical work is needed to verify the optimality of the converged solution of the iterative method, and in general, to obtain the conditions of optimality for cases of non-convex programming where the number of variables are unknown. Also, more theoretical work in the area of convergence and optimality of systems with indeterminate behavior is needed.

ACKNOWLEDGEMENTS

The author wishes to thank his wife Sally for typing and proof reading the manuscript and Mr. Parviz Emami for preparing the illustrations.

The author also wishes to thank Mr. E. M. L. Beale for critically reviewing this paper, his valuable suggestions, and presenting a simple numerical example which shows that the converged solution is not necessarily the global optimum.

REFERENCES

BELLMAN, R. E. and S. E. DREYFUS (1962) Applied Dynamic Programming, Princeton University Press, Princeton, N. J.

CROSS, H. (1936) 'The Relation of Analysis to Structural Design' Trans. Amer. Soc. of Civil Engr., Volume 62, No. 8, Part 2, pp. 1363-1408.

DANTZIG, GEORGE B. (1963) Linear Programming and Extensions, Princeton University Press, Princeton, N.J., pp. 448-470.

KUHN, H. W. and A. W. TUCKER (1951) 'Non-linear programming' Proceedings of the 2nd Berkeley Symposium on Maths, Statistics, and Probability, Univ. of California Press, pp. 481-492.

RAZANI, R. (1965) The Iterative Smoothing Method and Its Application to Minimum Cost Design of Highway Bridge Girders, Ph.D. Thesis, Case Institute of Technology, Cleveland, Ohio.

RAZANI, R. (1965) 'Behavior of Fully-Stressed Design of Structures and Its Relationship to Minimum-Weight Design' Journal, Amer. Inst. of Aeronautics and Astronautics, New York, N.Y. Volume 3, No. 12.

RAZANI, R. and G. G. GOBLE (1966) 'Optimum Design of Constant Depth Plate Girders', Journal of the Structural Division, ASCE, Volume 92, No. ST2, Proc. paper 4787, April 1966, pp. 253-281.

ROSEN, J. B. (1963) Recent Advances in Mathematical Programming, McGraw-Hill Book Co., pp. 159-176.

VARGA, RICHARD S. (1962) Matrix Iterative Analysis, Prentice-Hall, Inc., Englewood Cliffs, N.J.

ABADIE, J. (1967) Nonlinear Programming, North-Holland Publishing Company, Amsterdam, pp. 200-202.

DISCUSSION

E. M. L. Beale

The approach to optimization problems in which different sets of variables are held constant in turn, while optimal values are found for the other variables, can be very powerful, as Dr. Razani's paper illustrates. One might think that if the process terminates it must produce at least a locally optimal solution. But, as Dr. Razani was careful to point out, this is not necessarily so when the objective function is not a differentiable function of the independent variables; and inequality constraints usually make the objective function non-differentiable in this sense.

R. Razani

As I have discussed in the paper the condition of optimality is not theoretically obvious. However, I have shown that the iterative process converges to a feasible solution in a finite number of steps. Also I have shown that the cost selected to the converged solution is the least cost relating to the path of solution starting from an initial situation so if the initial situation is an existing design or an existing organization so the converged solution will be either a better solution or at least equal to the starting solution.

The reason for considering a set of variables constant and varying the rest of the variable and then keeping the variables constant and varying the previously held constant variables is to avoid the dimensionality of the problem. As discussed in the paper the number of variables is not known therefore this decentralization follows a natural path of first defining a set of variables, then obtaining their optimum value and repeating this process iteratively until convergence.

P. Demonsablon

Dr. Razani pointed out the question of the process convergence. In my opinion it cannot be answered from a mathematical point of view (at least regarding the statistically indeterminate structures) but from a technological one. The reason is that the convergence depends on the algorithms used for the shape alteration in the optimization process.

In the continuous girder case, certain constraints were impressed (such as minimum wall thickness, minimum flange width), which eliminated the risk of a divergence to a cantilever-girder.

In the case of an arch, where support conditions have a larger influence on the stress distribution, the shape alteration algorithm might finally result into a fixed arch or a hinged arch as well, depending only on the initial data (i.e. the approximate shape).

To sum up, I consider that a good shape alteration algorithm is essential in solving the structure optimization problem.

R. Razani

The problem of convergence of the iterative cycle is proved in a rigorous mathematical form in the paper. It is shown that the cycles of iteration converge in a finite number of cycles.

In many design situations there are certain types of variables which are held constant and some other ones may vary for optimization. Engineering designs have various levels. The first level is generation of concepts. This level is generally a non-quantifiable process and depends upon the so-called engineering judgement. As computational methods become more effective and comprehensive more problems of the non-quantifiable type can become quantifiable. In the case of problems discussed in the paper certain assumptions are made a priori such as the girders have constant meh height and meho thickness, or constant flange width. The rest of the variables are obtained for optimization. The problem is not to obtain the absolute minimum cost girder (because this girder may have a variable height). The objective is to obtain the least cost girder of the type assumed.

Liner Scheduling

N. WILLIAMS
Unilever Ltd., U.K.

1. DESCRIPTION OF THE PROBLEM

This paper describes an application of L.P. to the problem of scheduling cargo ships so as to provide a liner service between nominated ports. Estimates are made of the cargo that will be available for shipment from the Northern ports to the Southern ports, and from the Southern ports to the Northern ports. The shipping company issues a time-table which shows when vessels will be sailing and where they will be going. Problems of cargo stowage restrict voyage possibilities because, as a general rule, cargo for discharge at the first port-of-call must not be underneath cargo for the second port-of-call, and the vessel must have good trim both when full, and when partly loaded or unloaded.

For even a small number of ports the problem soon becomes large, so this application is described by means of a very small model of the real situation—one interesting feature of the problem is that we must not lose a ship, i.e. it must not be allowed to sail into one port and emerge from another.

The map is shown below

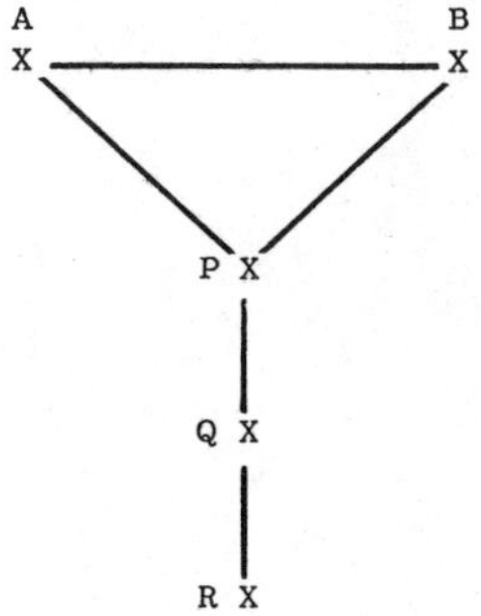

A southbound voyage starts (with the ship empty) at A or B, where cargo may be loaded. The ship then sails to P and/or Q and/or R where cargo may be unloaded. A southbound voyage ends when a ship is in P, Q or R with empty holds, but the last of the cargo may have been taken out at the previous port of call. A northbound voyage starts (with the ship empty) at P, Q or R where cargo may be loaded. The ship then sails for A or B, but may call at intermediate ports (but not more southerly ports) to pick up additional cargo. A northbound voyage ends when a ship is in A or B with an empty hold. Ships may sail from A to B, or B to A, in ballast.

2. POSSIBLE VOYAGES

The following voyages are possible:-

Table I

Voyage No	Direction	Start at	Call at	Finish at
(i)		(j)	(k)	(k)
1	S	A	—	P
2	S	A	—	Q
3	S	A	—	R
4	S	A	P	Q
5	S	A	P	R
6	S	A	Q	R
7	S	A	P, Q	R
8	S	B	—	P
9	S	B	—	Q
10	S	B	—	R
11	S	B	P	Q
12	S	B	P	R
13	S	B	Q	R
14	S	B	P, Q	R

(for southbound voyages the starting port is a 'loading' port, and the calling and finishing ports are 'unloading' ports).

(i)		(j)	(j)	(k)
15	N	P	—	A
16	N	P	—	B
17	N	Q	—	A
18	N	Q	—	B
19	N	Q	P	A
20	N	Q	P	B
21	N	R	—	A
22	N	R	—	B
23	N	R	Q	A
24	N	R	Q	B
25	N	R	P	A
26	N	R	P	B
27	N	R	Q, P	A
28	N	R	Q, P	B

(for northbound voyages the starting and calling ports are 'loading' ports, and the finishing port is an 'unloading' port).

29	E	A	—	B
30	W	B	—	A

(The two voyages 29 and 30 can only be made in ballast—i.e. no cargo.)

3. VARIABLES

At this stage it is convenient to define the variables which will be used in this problem:-

Variable Ref.	Definition
v	number of vessels used
i	number of times vessels sail on voyage i
ijk	tons of cargo loaded at j for deliver to k by vessels sailing on voyage i

4. TIMES AND COSTS

With every voyage is associated a fixed and a variable time, and a fixed and a variable cost. The fixed time (d_i) is the total time (in days) spent at sea steaming from the starting port to the intermediate and finishing ports, plus the time to enter the intermediate and finishing ports. The variable time (d_{jk}) is the time spent in the various ports loading or unloading cargo.

d_{jk} consists of two parts:-

(a) the number of days required to load 1 ton of cargo at port j, <u>plus</u>

(b) the number of days required to unload 1 ton of cargo at port k.

The fixed cost (c_i) is the cost of time spent at sea plus the port charges, and the variable cost (c_{jk}) is the cost of working the ship while the cargo is being loaded or unloaded.

c_{jk} consists of two parts:-

(a) the cost of working the ship while 1 ton of cargo is loaded at port j, <u>plus</u>

(b) the cost of working the ship while 1 ton of cargo is unloaded at port k.

On the credit side, of course, are the freights earned (f_{jk}).

f_{jk} is the freight earned for carrying 1 ton of cargo from port j to port k.

We wish to maximize

$$Z = \sum_i [g_{jk}x_{ijk} - c_i x_i]$$

where $g_{jk} = f_{jk} - c_{jk}$.

If we have to hire the ships, and/or have the possibility of hiring out unused ships we may wish to include an additional term

$$Z = \sum_i [g_{jk}x_{ijk} - c_i x_i] - vx_v$$

where v is the cost of hiring a ship for a year.

The time available is a function of the number of ships. x_v denotes the number of vessels employed; then in one year we shall have available $365x_v$ days, subject to

$$x_v \leqslant V \tag{1}$$

where V is the maximum number of vessels available. We thus get the constraint

$$\sum_i [d_i x_i + d_{jk} x_{ijk}] \leqslant 365x_v \tag{2}$$

5. CARGO CAPACITY

If we assume that all the available ships are the same size, and can each carry T tons of cargo, then every time a ship sails on a specified voyage it makes available T tons of capacity. x_i denotes the number of times a ship sails on voyage i, so the total capacity will be Tx_i. The cargo on board at any one time must, of course, be less than or equal to T; however, as we insist on (a) a southbound ship loading only at A or B and (b) a northbound ship loading only at R, Q and/or P, and unloading only at A or B, then this restriction becomes:-

total cargo loaded on voyage i $\leqslant Tx_i$

x_{ijk} denotes the amount of cargo loaded at port j for discharge at port k when we are using voyage i. Then for any one voyage

$$\sum x_{ijk} \leqslant Tx_i \tag{3}$$

where this summation takes place over all unloading ports (k) for a southbound voyage, or all loading ports (j) for a northbound voyage.

6. CARGO AVAILABLE

The total quantity of cargo to be shipped from every port to every other port in the year is given. Let the quantity to be shipped from j to k be denoted by M_{jk}

Then $$\sum_i x_{ijk} \leqslant M_{jk} \tag{4}$$

(or this may be an equality).

7. FREQUENCY OF CALL

If the amount of cargo to be shipped from any one port to any other is small then there would be, as the model has been developed so far, a temptation to pick it all up at one time. This would be unrealistic, and it is assumed that cargo arises evenly over the year; if ships call too often they will tend to sail only partly loaded, but if they only call infrequently then some competitor is likely to lift this cargo. It is, therefore, desirable to build into the model some conditions that ships are to call with at least some minimum frequency. There appear to be two ways in which this could be expressed:-

(a) we could say that only some fraction of the year's cargo could be lifted at any one call—say $\frac{1}{n}$ th. This gives $x_{ijk} \leqslant \frac{1}{n} M_{jk} \cdot x_i$ (5a)

(b) we could insist that certain voyages or combinations of voyages should be made some minimum number of times. This gives $x_i \geqslant n$ (5b)

where the summation is over voyages which include specified pairs of ports.

For example, if we specify that we must call at P to collect cargo for A at least n times we get:-

$$x_{iPA} \leq \frac{1}{n} \cdot M_{PA} \cdot x_i \quad \text{for } i = 15, 19, 25 \text{ and } 27 \tag{5a}$$

or $$x_{15} + x_{19} + x_{25} + x_{27} \geq n \tag{5b}$$

(5a) is the more flexible, as if we are not insisting on picking up all the cargo we can get a solution in which uneconomic ports are not called at. On the other hand, if we insist on picking up all the cargo, (5b) is a simpler constraint, especially if we wish to explore the effect of changes in the tonnage availabilities (M_{jk}'s).

8. TURNROUND PORTS

A final condition which must be imposed is that every ship starts a voyage from the same port as the one where the previous voyage finished.

Table II (6)

finishing/ starting port	constraint
A	$x_1 + x_2 + x_3 + x_4 + x_5 + x_6 + x_7 + x_{29} - x_{15} - x_{17} - x_{19} - x_{21} - x_{23} - x_{25} - x_{27} - x_{30} = 0$
B	$x_8 + x_9 + x_{10} + x_{11} + x_{12} + x_{13} + x_{14} + x_{30} - x_{16} - x_{18} - x_{20} - x_{22} - x_{24} - x_{26} - x_{28} - x_{29} = 0$
P	$x_1 + x_8 - x_{15} - x_{16} = 0$
Q	$x_2 + x_4 + x_9 + x_{11} - x_{17} - x_{18} - x_{19} - x_{20} = 0$
R	$x_3 + x_5 + x_6 + x_7 + x_{10} + x_{12} + x_{13} + x_{14} - x_{21} - x_{22} - x_{23} - x_{24} - x_{25} - x_{26} - x_{27} - x_{28} = 0$

and it will be seen that (A) + (B) = (P) + (Q) + (R).

9. THE OPENING MATRIX

A section of the opening matrix is shown in Appendix I. Constraints (5a) or (5b) would be in the matrix, but not both.

This is an application where the matrix can soon become very large. If the matrix is subject to modification before each run there would be a case for generating it by a matrix generator program as the rules for constructing the matrix are fairly straightforward.

10. INTEGER VALUES OF VARIABLES

It is desirable that the variable x_v (number of ships employed) should be an integer in the solution. This can be achieved quite simply by a second calculation with x_v fixed at an integer near to the value found in the first solution. It is desirable that the x_i should also be integers, and perferably equal to kn where k = 0, 1, 2... A straightforward linear programming calculation is unlikely to give this desirable result (although the constraints (5a) and (5b) do help to achieve it). The nature of the data and the assumptions in the model may be such that an ordinary L.P. solution is a sufficient guide as to the best way of deploying the ships.

REFERENCE

RAO, M. R. and ZIONTS, S., 'Allocation of Transportation Units to alternative trips—a column generation scheme with out-of-kilter sub-problems', Operations Research, Vol. 1, Jan.-Feb. 1968.

DISCUSSION

E. Koenigsberg

Mr. Williams's model is related to the paper by Sullivan and Koenigsberg and to a paper by Conley, Farnsworth et alia of Matson Research Corp. which will shortly be published in Transportation Science. We consider the model as one for Liner Allocation rather than Liner Scheduling. In the paper to appear in Transportation Science we have used a scheme similar to that of Williams to designate routes. In our case we had a fleet of about 60 vessels of 6 types moving between five parts on one coast and 12 particular ports on the opposite coast. To reduce the number of route combinations we introduced a mythical port (Atlantis ?) or funnel through which all vessels pass without interchange of cargoes. Routes are round trip voyages north of the funnel or south of the funnel; the northern and southern routes are linked by vessel continuity and cargo continuity equations. Both of our papers differ from Williams in that we deal with a mixed fleet and with mixed cargoes. Further we include the cost of vessel layup in our objective function. Our own computational experience has been good (rather less good in the case of the mixed integer problem) and are discussed in the papers.

E. M. L. Beale

It is perhaps worth noting that problems can arise with constraints of the form (5b) if one wants to determine the composition of the fleet as well as its allocation to routes. This is because a 'solution' in which all the cargo for the year is carried in one big ship, while rowing boats visit the ports from time to time to make up the required number of calls, could be unacceptable.

N. Williams

With a mixed fleet the problem immediately becomes more complicated. I agree that constraints of type 5(b) would then need to be reconsidered. I would prefer not to comment on the proposal to add a mythical port as I have not yet seen the paper indicated.

APPENDIX 1: (Section of) Opening Matrix:

Constraint Type	V	1	1AP	7	7AP	7AQ	7AR	26	26RB	26PB	30		
	V	c_1	g_{AP}	c_7	g_{AP}	g_{AQ}	g_{AR}	c_{26}	g_{RB}	g_{PB}	c_{30}	$-z$	
(1)	1											$\leq V$	
(2)	−365	d_1	d_{AP}	d_7	d_{AP}	d_{AQ}	d_{AR}	d_{26}	d_{RB}	d_{PB}	d_{30}	≤ 0	
(4)			1		1							$\leq M_{AP}$	
(4)						1						$\leq M_{AQ}$	
(4)							1					$\leq M_{AR}$	
(4)									1			$\leq M_{RB}$	
(4)										1		$\leq M_{PB}$	
(3)		$-T$	1									≤ 0	
(5a)		$-\frac{M_{AP}}{n}$	1									≤ 0	
(3)				$-T$	1	1	1					≤ 0	
(5a)				$-\frac{M_{AP}}{n}$	1							≤ 0	
(5a)				$-\frac{M_{AQ}}{n}$		1						≤ 0	
(5a)				$-\frac{M_{AR}}{n}$			1					≤ 0	
(3)								$-T$	1	1		≤ 0	
(5a)								$-\frac{M_{RB}}{n}$	1			≤ 0	
(5a)								$-\frac{M_{PB}}{n}$		1		≤ 0	
(5b)		1		1								$\geq n$	(AP)
(5b)						1						$\geq n$	(AQ)
(5b)							1					$\geq n$	(AR)
(5b)									1			$\geq n$	(RB)
(5b)										1		$\geq n$	(PB)
(5b)		1		1								-0	(A)
(6)								−1				-0	(B)
(6)		1										-0	(P)

SECTION 3 Economic Applications

J. Y. Jaffray
D. J. Clough, M. B. Bayer
Frances E. Hobson

Programmation Linéaire d'un Modèle de Léontief Dynamique de l'Economie Française

JEAN-YVES JAFFRAY
C.I.R.O., France

SUMMARY

In the study of the medium-term evolution of an economy by means of a dynamic Leontief model, one is faced with the problem of the choice of an objective function such that the optimal solution obtained by the linear program is realistic and, in particular, at the end of the period leaves the economy in a state which will allow subsequent harmonious development.

It is here demonstrated that, in such a model, there exists a succession of production vectors (the 'path of balanced growth') which is optimal in the long-term; and it is proposed to impose, as a goal for the economy, the rejoining of this optimum path in the medium term at the highest possible level.

The study of an application serves as a means of assessing this formula.

1. INTRODUCTION

Devant étudier les possibilités de développement à moyen terme d'une économie dans des hypothèses précises, nous avons été amenés à la représenter par un modèle de Leontief dynamique; des premiers essais ont fait apparaitre que le choix pour fonction économique de la valeur du capital final (ou de toute fonction analogue) conduisait à une simulation irréaliste, le capital obtenu n'ayant pas une structure harmonieuse et ne permettant pas un développement ultérieur satisfaisant. Nous avons alors cherché si l'on ne pouvait pas justifier certaines contraintes supplémentaires et une fonction économique particulière.

Or, on connait pour certains modèles de Leontief l'existence de chemins de croissance équilibrée dans le voisinage desquels passe nécessairement, à long terme, tout chemin efficace [Mc KENZIE (1963), TSUKUI (1966)]; nous établissons pour un modèle de Léontief avec consommation finale, mais une hypothèse restrictive sur l'évolution de celle-ci, l'existence d'un chemin de croissance équilibrée généralisée, dont tout chemin efficace doit être initialement proche si les matrices d'input-output et d'investissement vérifient certaines conditions.

Il est alors naturel d'imposer à la solution du modèle à moyen terme d'amener l'économie sur le chemin optimal, et au niveau le plus élevé possible; la recherche du niveau maximum accessible se fera par tâtonnement en résolvant successivement un certain nombre de programmes linéaires.

Dans une première partie, nous ferons l'étude théorique des propriétés d'un modèle de Léontief dynamique; nous montrerons ensuite, dans une deuxième partie, que ce modèle s'applique bien à notre problème et traiterons un exemple.

2. ETUDE THEORIQUE

Nous commencerons par décrire un modèle du type 'Leontief dynamique', c'est-à-dire un modèle économique global, à périodes, linéaire, plurisectoriel, avec capital; nous montrerons que, sur une longue période, moyennant certaines restrictions sur l'évolution possible de la consommation finale, la suite des productions donnant une croissance optimale sera nécessairement, aux premières étapes, voisine d'une suite particulière.

2.1 Description du modèle

Le modèle comprend 16 secteurs; le temps est divisé en périodes indicées par t ($t = 0, 1, \ldots, T-1, T$).

2.1.1 Notations des données et variables

Les données sont:

$A \geqslant 0$: matrice 16 × 16 des coefficients d'input-output;
$B \geqslant 0$: matrice 16 × 16 des coefficients de capacité de production;
$D \geqslant 0$: matrice 16 × 16, diagonale, à éléments diagonaux strictement positifs, des coefficients d'amortissement du capital;
$C_o > 0$: vecteur 1 × 16 des capacités de production initiales;
$Y_t \geqslant 0$: vecteur 1 × 16 des consommations finales l'année t.

Les variables sont:

X_t : vecteur 1 × 16 des productions (outputs) l'année t;
Z_t : vecteur 1 × 16 des investissements l'année t;
C_t : vecteur 1 × 16 des capacités de production l'année t;
δC_t : vecteur 1 × 16 des accroissements bruts des capacités de production l'année t.

2.1.2 Relations

Le modèle est décrit par les relations suivantes:

A chaque période, la production se répartit entre input, investissement et consommation finale,

$$X_t \geqq AX_t + Z_t + Y_t; \tag{1}$$

l'investissement est non-négatif et fournit l'accroissement brut de capacité de production,

$$Z_t = B \quad \delta C_t \geqq 0, \delta C_t \geqq 0 \tag{2}$$

l'accroissement net de capacité de production s'obtient en retranchant de l'accroissement brut l'amortissement de la capacité de la période précédente.

$$C_{t+1} - C_t = \delta C_t - DC_t; \tag{3}$$

la production, non-négative, est limitée par la capacité de production,

$$0 \leqq X_t \leqq C_t. \tag{4}$$

(Nous discuterons dans la troisième partie de l'adéquation de ce modèle au traitement de notre problème).

2.1.3 Evolution de la consommation finale Y_t

La suite $\{Y_t\}$ des consommations finales est une donnée exogène du modèle; pour un même vecteur des capacités initiales C_o, le système des relations du type (1), (2), (3), (4) aura une solution réalisable pour certaines suites $\{Y_t\}$, n'en aura pas pour d'autres.

S'il existait une relation d'ordre sur ces suites, on pourrait chercher la suite maximale $\{Y_t\}$ admissible; comme il n'en existe pas, nous allons nous restreindre à un sous-ensemble de l'ensemble des suites:

— soit celui des suites de la forme

$$Y_t = \lambda^t Y_o$$

où λ est un scalaire.

— soit celui des suites de la forme

$$Y_t = \widetilde{Y} + \lambda^t V$$

où λ est un scalaire, $\widetilde{Y}$ et V des vecteurs 16×1.

Dans l'un et l'autre cas, une suite de consommations finales Y_t sera préférée à toutes celles correspondant à des valeurs inférieures du paramètre λ.

Nous pouvons alors étudier, dans l'une ou l'autre hypothèse, à quelle condition une suite de production X_t permet, sur une longue période, de réaliser une suite de consommations Y_t maximale. Nous nous placerons d'abord dans la première hypothèse; nous montrerons ensuite que les résultats obtenus se transposent aisément dans la deuxième hypothèse.

Nous montrerons dans la troisième partie de cette note que cette deuxième hypothèse peut nous aider dans notre problème.

2.2 Existence et propriétés optimales de la croissance équilibrée, lorsque $Y_t = \lambda^t Y_o$

2.2.1 Définition

Nous dirons qu'il y a croissance équilibrée de l'économie s'il existe $\lambda > 1$ et des suites

$$Y_t = \lambda\, Y_{t-1};\ \overline{X}_t = \lambda\, \overline{X}_{t-1}$$

telles que dans les relations (1), (2), (3) et (4), les inégalités autres que celles de signe des variables, soient vérifiées en tant qu'égalités.

Autrement dit, il y a croissance proportionnelle, et au même taux, de la production et de la consommation finale; toute la capacité de production est utilisée à plein; aucune production ne reste inemployée.

2.2.2 Existence d'une croissance équilibrée

Les relations (1), (2), (3) et (4) montrent alors qu'il y aura croissance équilibrée s'il existe $\lambda > 1$ et $X_t \geqq 0$ tels que:

$$\{E - [A + B((\lambda - 1)E + D)]\}\, X_t = Y_t$$

pour un certain t; la solution pour chaque t s'en déduit aussitôt.

Faisons sur les matrices A, B et D les hypothèses suivantes:

— (A + BD) a son rayon spectral, maximum du module de ses valeurs propres μ_i, inférieur à l'unité:

$$\rho = \operatorname*{Max}_i |\mu_i| < 1$$

Cette hypothèse est naturelle, car l'on montre [VARGA (1962)] que sinon, il n'existerait aucun $X \geq 0$ tel que

$$(A + BD)X \leqq X$$

c'est-à-dire que le modèle ne permettrait aucune croissance.

— (A + B) est indécomposable, c'est-à-dire qu'il n'existe pas de matrice de permutation P telle que

$$P(A + B)P' = \left|\begin{array}{c|c} M_{11} & M_{12} \\ \hline 0 & M_{22} \end{array}\right| \qquad M_{11} \text{ et } M_{22} \text{ carrées.}$$

L'interprétation de cette condition est qu'il n'existe pas de sous-ensemble de secteurs n'utilisant, comme inputs ou investissements que des produits du même sous-ensemble.

Ces hypothèses faites, par continuité, il existera un intervalle $[1, \lambda^*[$ tel que $[A + B((\lambda - 1)E + D)]$ ait également son rayon spectral inférieur à 1(λ^* correspondant au rayon 1);il en résulte [VARGA (1962)] que

$$\{E - [A + B((\lambda - 1)E + D)]\}^{-1} \geq 0$$

et même >0, car $[A + B((\lambda - 1)E + D)]$ est indécomposable quand (A + B) l'est.

Donc, étant donné Y_0, à tout $\lambda \epsilon [1, \lambda^*]$ correspond une croissance équilibrée possible:

$$\overline{X}_0 = \{E - [A + B((\lambda - 1)E + D)]\}^{-1} Y_0;$$

$$\overline{X}_t = \lambda^t \overline{X}_0; Y_t = \lambda^t Y_0; \qquad (5)$$

plus le taux de croissance λ est élevé et plus $\overline{X}_0$ est grand par rapport à Y_0, c'est-à-dire que l'investissement augmente au détriment de la consommation finale.

2. 2. 3 Capacité initiale nécessaire à la réalisation de la croissance $Y_t = \lambda^t Y_0$

La croissance équilibrée exige d'après (4):

$$\overline{C}_0 = \overline{X}_0$$

Nous allons chercher une condition nécessaire pour que la capacité initiale C_0 permette de satisfaire une consommation $Y_t = \lambda^t Y_0$ pendant un nombre indéfini de périodes, c'est-à-dire pour T aussi grand que l'on veut.

Nous verrons apparaitre une pondération de C_0, permettant de définir sa 'valeur'; la condition cherchée sera que la valeur de C_0 doit être supérieure ou égale à celle de $\overline{C}_0$.

Nous montrerons ensuite, sous une hypothèse supplémentaire, que si la 'valeur' initiale de C_0 est égale à celle de $\overline{C}_0$, la suite de consommations $Y_t = \lambda^t Y_0$ ne pourra être satisfaite sur un grand nombre T de périodes, si la production initiale X_0 (et donc C_0) n'est pas suffisamment voisine de $\overline{X}_0 = \overline{C}_0$.

Nous procéderons de la façon suivante:

Nous écrirons d'abord en détail le système (I), ensemble des relations reliant les variables du modèle sur T périodes, à l'exception des conditions de signes; si ce système n'a pas de solution, a fortiori, le système complet n'en a pas.

Or, d'après le lemme de Farkas, pour qu'un système d'inéquations n'ait pas de solution, il faut et il suffit qu'un certain système 'dual' ait une solution*. Nous formerons donc ce système II et montrerons que pour certaines valeurs de C_0 il y a une solution dès que T est assez grand.

(a) Calcul du système II

Système I: Multiplicateurs associés aux contraintes

$$\begin{cases} -(E - A)X_0 + Z_0 \leqq -Y_0 & R_0 \geqq 0 \\ BC_1 - Z_0 \leqq B(E - D)C_0 & S_0 \geqq 0 \\ X_0 \leqq C_0 & U_0 \geqq 0 \end{cases}$$

- - - - - - - - - - - - - - -

$$t = 1, 2, \ldots, (T - 1) \quad \begin{cases} -(E - A)X_t + Z_t \leqq -Y_t & R_t \geqq 0 \\ B(C_{t+1} - C_t + DC_t) - Z_t \leqq 0 & S_t \geqq 0 \\ X_t - C_t \leqq 0 & U_t \geqq 0 \end{cases}$$

- - - - - - - - - - - - - - -

$$\begin{cases} -(E - A)X_T \leqq -Y_T & R_T \geqq 0 \\ X_T - C_T \leqq 0 & U_T \geqq 0 \end{cases}$$

Système II: il se compose des contraintes de signe des multiplicateurs et de:

$$\begin{cases} -R_0(E - A) + U_0 = 0 \\ R_0 - S_0 = 0 \end{cases}$$

- - - - - - - - - - - - - - -

$$t = 1, 2, \ldots, (T = 1) \quad \begin{cases} -R_t(E - A) + U_t = 0 \\ R_t - S_t = 0 \\ S_{t-1}B - S_tB(E - D) - U_t = 0 \end{cases}$$

- - - - - - - - - - - - - - -

$$\begin{cases} -R_T(E - A) + U_T = 0 \\ S_{T-1}B - U_T = 0 \end{cases}$$

$$\sum_0^T R_t Y_t > U_0C_0 + S_0B(E - D)C_0$$

---

* X, Y vecteurs $n \times 1$, A matrice $m \times n$, R vecteur $1 \times m$: $\{X: AX \leqq Y\}$ vide $\Leftrightarrow \{R \geqq 0: RA = 0, RY < 0\}$ non vide

En éliminant les variables S_t et U_t, le système II se met sous la forme équivalente suivante

<u>Système II (forme simplifiée)</u>

$$\begin{cases} R_0 \geqq 0 \\ R_0(E - A) \geqq 0 \end{cases}$$

- - - - - - - - - - - - - - - - - - -

$$t = 1, 2, \ldots, T-1 \quad \begin{cases} R_t \geqq 0 \\ R_t(E - A) \geqq 0 \\ R_t(E - A) = R_{t-1}B - R_tB(E - D) \end{cases}$$

- - - - - - - - - - - - - - - - - - -

$$\begin{cases} R_T \geqq 0 \\ R_T(E - A) = R_{T-1}B \end{cases}$$

$$\sum_0^T R_t\, Y_t > R_0(E - A)C_0 + R_0B(E - D)C_0 \tag{6}$$

(b) <u>Recherche de la capacité C_0 rendant le système II réalisable</u>

Par définition de λ^*, $[A + B((\lambda^* - 1)E + D)]$ a son rayon spectral égal à 1; elle est indécomposable, sa transposée l'est donc également; en appliquant à celle-ci le théorème de Perron-Frobénius [VARGA (1962)], on en déduit qu'il existe $\overline{R} > 0$ tel que:

$$\overline{R}\{E - [A + B((\lambda^* - 1)E + D)]\} = 0 \tag{7}$$

Cherchons pour le système II une solution notée $\{\overline{R}_t\}$ de la forme

$$\begin{cases} \overline{R}_0 = \overline{R} \\ - - - - - - - \\ \overline{R}_t = \lambda^* \overline{R}_{t+1} \\ - - - - - - - \\ \overline{R}_T = \overline{R}_{T-1}B(E - A)^{-1} \end{cases}$$

On vérifie facilement que cette suite $\{\overline{R}_t\}$ vérifie les égalités du système II, ainsi que les inégalités larges puisque:

$$\overline{R}(E - A) = \overline{R}\, B[(\lambda^* - 1)E + D] > 0 \Rightarrow R_t(E - A) > 0$$

Reste l'inégalité stricte (6); or

$$\overline{R}_tY_t = \left(\frac{\lambda}{\lambda^*}\right)^t \overline{R}_0Y_0 = \left(\frac{\lambda}{\lambda^*}\right)^t \overline{R}\, Y_0 \qquad (t = 1, 2, \ldots, T-1)$$

$$\sum_0^T R_tY_t = \sum_0^{T-1}\left(\frac{\lambda}{\lambda^*}\right)^t \overline{R}\, Y_0 + \left(\frac{\lambda}{\lambda^*}\right)^{T-1} \overline{R}B(E - A)^{-1}\, Y_0;$$

le 1er terme du second membre tend lorsque T tend vers l'infini, vers:

$$\frac{\lambda^*}{\lambda^* - \lambda}\overline{R}\, Y_0$$

et le 2ème terme vers zéro; pour que (6) soit vérifié pour T grand, il suffit donc que C_0 soit tel que:

$$\frac{\lambda^*}{\lambda^* - \lambda} \overline{R}\, Y_0 > \overline{R}[E - A + B(E - D)]\, C_0$$

or (5) et (7) impliquent:

$$\overline{R}\, Y_0 = \overline{R}\{E - A - B((\lambda - 1)E + D)\}\overline{X}_0 = (\lambda^* - \lambda)\overline{R}B\, \overline{X}_0$$

et $\overline{R}\{E - A + B(E - D)\} = \lambda^*\, \overline{R}\, B\, C_0$

l'inégalité (6) s'écrit donc finalement:

$$\overline{R}\, B\, \overline{X}_0 > \overline{R}\, B\, C_0. \tag{8}$$

Pour tout C_0 vérifiant (8), le système II aura une solution et le système I sera impossible dès que T sera assez grand. En d'autres termes, pour une telle capacité initiale, aucune évolution ne permet de satisfaire indéfiniment à la suite de consommations $\{Y_t\}$.

2.2.4 Attraction par la croissance équilibrée

Etant donné une capacité initiale C_0, appelons dorénavant $\overline{X}_0$ le vecteur croissance équilibrée dont l'intensité est telle que:

$$\overline{R}\, B\, \overline{X}_0 = \overline{R}\, B\, C_0 \tag{9}$$

$\overline{X}_0$ permet la réalisation de consommations finales $Y_t = \lambda^t\, Y_0$.

Nous allons montrer qu'une suite de productions $\{X_t\}$ assurant les mêmes consommations $\{Y_t\}$ est nécessairement voisine au départ de la direction $\overline{X}_0$, d'autant plus voisine que T sera plus grand. Comme

$$X_0 \leqq C_0 \text{ et } \overline{R}\, B\, \overline{X}_0 = \overline{R}\, B\, C_0$$

il en résultera que la capacité initiale C_0 devra elle-même être voisine de $\overline{X}_0$.

Considérons, en effet, la quantité

$$Q_t = \left\{ \sum_{\theta=t}^{\infty} \overline{R}_\theta\, Y_\theta - \overline{R}_t[E - A + B(E - D)]C_t \right\}.$$

Q_t est un indice d'impossibilité du système I, en ce sens que si Q_t est positif, le système I est impossible. (9) entraîne $Q_0 = 0$; montrons que Q_t ne peut que croître:

$$Q_t - Q_{t-1} = \overline{R}_{t-1}[E - A + B(E - D)]C_{t-1} - \overline{R}_t[E - A + B(E - D)]C_t - \overline{R}_{t-1}Y_{t-1}$$

De (7) et de $\overline{R}_t = \left(\frac{1}{\lambda^*}\right)^t \overline{R}$, on déduit

$$(\lambda^*)^{t-1}\, (Q_t - Q_{t-1}) = \overline{R}[E - A + B(E - D)]C_{t-1} - \overline{R}\, B\, C_t - \overline{R}\, Y_{t-1}$$

$$= \overline{R}\{(E - A)X_{t-1} - B(C_t - C_{t-1}) - BDC_{t-1} - Y_{t-1}\} + \overline{R}(E - A)\, \{C_{t-1} - X_{t-1}\}$$

Les quantités entre parenthèses sont nécessairement positives ou nulles et $\overline{R}$ et $\overline{R}(E - A)$ sont strictement positives; il en résulte donc que

$$Q_t \geqq Q_{t-1}$$

Autrement dit, puisque $Q_0 = 0$, seule une évolution où:

$$Q_t = Q_{t-1} = 0$$

peut éviter que le système I soit impossible; nécessairement donc:

$$\begin{cases} C_{t-1} = X_{t-1} \\ (E - A)X_{t-1} = B(C_t - C_{t-1}) + BDC_{t-1} + Y_{t-1} \end{cases}$$

ce qui donne entre X_t et X_{t-1}, la relation:

$$Y_{t-1} = (E - A)X_{t-1} - B(X_t - X_{t-1}) - BDX_{t-1}.$$

La même relation vaut entre $\overline{X}_t$ et $\overline{X}_{t-1}$; en les soustrayant membre à membre, on obtient:

$$(E - A)(X_{t-1} - \overline{X}_{t-1}) = B[(X_t - \overline{X}_t) - (X_{t-1} - \overline{X}_{t-1})] + BD(X_{t-1} - \overline{X}_{t-1}).$$

D'où, en posant

$$X_t - \overline{X}_t = (\lambda)^t V_t,$$

la relation:

$$[E - A + B(E - D)]V_{t-1} = \lambda B V_t$$

Nous allons supposer que $[E - A + B(E - D)]$ est inversible, d'où

$$V_{t-1} = \lambda[E - A + B(E - D)]^{-1} B V_t$$

et que le rayon spectral de $\lambda[E - A + B(E - D)]^{-1} B$ est inférieur à un, propriété qui n'est pas toujours vérifiée par A, B et D; cependant, si

$$[E - A + B(E - D)]^{-1} B > 0^{(1)} \text{ (ou } \geqslant 0 \text{ et indécomposable)}$$

on peut le démontrer comme suit: Par définition de λ^*, il existe $X > 0$ tel que:

$$\{E - [A + B(D + (\lambda^* - 1)E)]\} \quad X = 0$$

(1) C'est le cas, par exemple, pour l'économie française sur laquelle ont porté nos essais.

d'où l'on déduit que:

$$[E - A + B(E - D)]^{-1} BX = \frac{1}{\lambda^*} X$$

Une matrice à éléments positifs (ou indécomposable à éléments non-négatifs) ayant un seul vecteur propre positif, qui est associé à sa valeur propre positive maximum, la matrice ci-dessus a pour rayon spectral $\frac{1}{\lambda^*}$ et:

$$\lambda[E - A + B(E - D)]^{-1} B$$

a pour rayon spectral $\frac{\lambda}{\lambda^*}$, inférieur à un.

Or, lorsque $\lambda[E - A + B(E - D)]^{-1} B$ a un rayon spectral inférieur à un, $\{\lambda[E - A + B(E - D)]^{-1} B\}^h$ tend en norme vers zéro lorsque h tend vers l'infini [VARGA (1962)]. Or (10) entraîne que:

$$\| V_0 \| \leqq \| \{\lambda[E - A + B(E - D)]^{-1} B\}^h \| \; \| V_t \|$$

et $\| V_T \| = \left\| \frac{X_T}{(\lambda)^T} - \overline{X}_0 \right\|$ est borné puisque

$$Q_T = 0 \Longleftrightarrow \overline{R}BX_T = \overline{R}B\overline{X}_T \Longleftrightarrow \frac{\overline{R}BX_T}{(\lambda)^T} = \overline{R}\ B\ \overline{X}_0$$

et que $X_T \geqq 0$; donc $\| V_0 \|$ sera aussi petit que l'on veut pourvu que T soit assez grand.

En résumé, nous venons de montrer que seuleune suite de productions $\{X_t\}$ de directions initiales voisines de $\overline{X}_0$ peut permettre de satisfaire pendant une période T aux mêmes consommations $\{Y_t\}$ que la croissance équilibrée vérifiant initialement

$$\overline{R}\ B\ \overline{X}_0 = \overline{R}\ B\ C_0$$

Comme $X_0 \leqq C_0$, il en résulte que C_0 lui-même doit être voisin de $\overline{X}_0$.

2. 3 Généralisation à une consommation finale $Y_t = Y + \lambda^t V$

Si l'on cherche maintenant à satisfaire une consommation finale de la forme

$$Y_t = \widetilde{Y} + \lambda^t V$$

un changement de variable va nous permettre d'utiliser les résultats du paragraphe précédent. Soit, en effet, $\widetilde{X}$ le vecteur positif tel que:

$$[E - A - BD]\ \widetilde{X} = \widetilde{Y}$$

et posons

$$\widetilde{C} = \widetilde{X};\ \delta\widetilde{C} = D\widetilde{C};\ \widetilde{Z} = B\ \delta\widetilde{C}$$

$$X'_t = X_t - \widetilde{X};\ C'_t = C_t - \widetilde{C};\ \delta C'_t = \delta C_t - \delta\widetilde{C};\ Z'_t = Z_t - \widetilde{Z}$$

Les nouvelles variables X'_t, C'_t, $\delta C'_t$ et Z'_t vérifieront les relations:

$$X'_t \geqq AX'_t + Z'_t + V_t$$

$$Z'_t = B\ \delta C'_t$$

$$C'_{t+1} - C'_t = \delta C'_t - DC'_t$$

$$X'_t \leqq C'_t$$

Pour ce système, d'après les résultats précédents, il existe pour tout λ tel que $\lambda \epsilon]1, \lambda^*[$ un vecteur $\overline{X}'_0$ tel que

$$\{E - A - B((\lambda - 1)E + D)\}\overline{X}'_0 = V$$

et la suite de productions $X'_t = \lambda^t\ \overline{X}'_0$ permet d'assurer sans excédents la consommation $V_t = \lambda^t\ V$, et les propriétés d'optimalité sont encore valables.

Si l'on revient au système des X_t, on peut donc affirmer qu'il existe une suite de productions $\overline{X}_t = \widetilde{X} + \lambda^t\ \overline{X}'_0$ permettant de satisfaire sans excédents la consommation finale $Y_t = \widetilde{Y} + \lambda^t\ V$; aucune capacité initiale C_0 telle que

$$\overline{R}BC_0 < \overline{R}B\overline{X}_0\ (\Longleftrightarrow \overline{R}BC'_0 < \overline{R}B\overline{X}'_0 \text{ car } \overline{R}B\widetilde{X} = \overline{R}B\widetilde{C})$$

ne permet cette même consommation; de plus, dans le cas de l'égalité

$$\overline{R}BC_0 = \overline{R}B\overline{X}_0$$

seule une production initiale X_0 voisine de $\overline{X}_0 = \widetilde{X} + \overline{X}'_0$ (et donc une capacité C_0 voisine de $\overline{X}_0$) peut autoriser cette même consommation sur un grand nombre T de périodes.

Nous appellerons la suite $\overline{X}_t$ suite de 'croissance équilibrée généralisée'.

3. APPLICATIONS

Nous allons maintenant exposer notre problème, montrer comment les résultats de la partie théorique peuvent s'y appliquer et enfin donner les résultats d'un essai numérique servant de test.

3.1 Le problème

Il peut se poser de la façon suivante: quelles sont les possibilités de développement à moyen-terme d'une économie, à capacité de production initiale des différents secteurs donnée, lorsque tout est mis en oeuvre (capital efficacement utilisé, main d'oeuvre éventuellement suremployée, consommation finale rationnée) pour assurer une croissance maximale. Dans un tel problème, la donnée essentielle est le capital initial de chacun des secteurs; aussi le choix d'un modèle de Leontief dynamique s'impose-t-il. La nécessité d'effectuer de nombreux essais où l'on fait varier les données initiales ou les consommations nous a conduits à limiter le nombre de secteurs aux seize figurant dans la comptabilité nationale française;

Mais quel objectif fixer? Il doit être tel que l'économie soit conduite, en cinq années, à un état qui offre des conditions favorables à un développement ultérieur normal, c'est-à-dire qu'on veut qu'à la phase I, économie de restriction, succède une phase II, économie de consommation, où l'on cherche à accroître la consommation finale.

C'est pourquoi, il n'est pas possible de retenir comme fonction économique la valeur globale du capital terminal; la structure de ce capital doit entrer en ligne de compte; pour cela on maximisera une variable λ telle que

$$C(t_1 + 1) \geqq \lambda C^*$$

où t_1 est l'année terminale* et C^* la structure souhaitée (qu'il faudra définir). On reste ainsi dans le domaine de la programmation linéaire.

3.2 Adéquation du modèle au problème posé

3.2.1 Les hypothèses du modèle de Leontief

Le modèle est décrit par les relations (1), (2), (3) et (4). Nous laissons de côté la discussion des hypothèses inhérentes à tout modèle de Leontief (linéarité, fixité des coefficients) dont les inconvénients, inévitables, sont bien connus.

Nous avons écrit les relations (1) sous forme d'inégalités; l'interprétation des variables d'écart de (1) comme stocks ne sera pas valable ici car nous avons fait entrer les stocks, en tant que fonds de roulement, dans le capital; (3) montre bien, en revanche, que le capital inutilisé ne disparait pas pour autant.

Nous n'avons pas introduit de contrainte de main d'oeuvre; cette hypothèse se justifie dans la phase I, période d'efforts sévères (par exemple, période de guerre, ou de reconstruction d'après guerre), où la main d'oeuvre est extensible et où l'on doit même plutôt craindre le chomage.

---

* En effet, si t_1 est l'année terminale, l'investissement Zt_1 est connu, et donc la capacité de production $C_{(t_1+1)}$.

Dans cette phase I, rien n'empêche de faire varier les matrices A et B d'une année sur l'autre si l'on dispose de données suffisantes.

En revanche, dans la phase II, il y a normalement plein emploi, et la croissance n'est possible que par l'effet simultané de l'accroissement démographique et de la variation des coefficients de main d'oeuvre et des matrices A et B. Or, l'existence et les propriétés optimales de la croissance équilibrée, généralisée ou non, supposent la fixité de A et B sur une longue période; A et B cependant ne varient que lentement, et les résultats du paragraphe II pourront être utilisés en première approximation.

Les hypothèses: '$(A + BD)$ de rayon spectral inférieur à un', '$(A + B)$ indécomposable' sont vraies dans toute économie réelle; '$[E - A + B(E - D)]^{-1} B$ de rayon spectral inférieur à $\frac{1}{\lambda^*}$' qui n'est pas évident, sera vérifiée parce que $[E - A + B(E - D)]^{-1} B$ se révèle être positive.

3.2.2 L'hypothèse d'évolution de la consommation finale

Lorsque l'intensité du vecteur consommation finale $Y(t)$ varie, sa direction varie également et à $\| Y \|$ donné on peut supposer qu'il correspond un seul Y; l'ensemble des extrémités des vecteurs Y possibles dépend donc d'un paramètre et décrit une courbe dans R^n. Soit Y_0 un point de cette courbe et V la direction de la tangente en ce point (nous admettons qu'elle existe); la courbe des Y peut être approchée par la droite

$$Y = Y_0 + uV$$

On utilise cette approximation lorsque l'on cherche une suite de consommations finales

$$Y_t = \tilde{Y} + \lambda^t V$$

Mais peut-on trouver $\tilde{Y}$ et V? Posons

$$\lambda = 1 + \epsilon$$

et calculons

$$Y_0 = \tilde{Y} + V; Y_1 - Y_0 = \epsilon V; Y_{t+1} - Y_t \sim \epsilon V$$

cette dernière approximation étant valable tant que t n'est pas trop grand.

On voit que si l'on possède les données des consommations finales d'une économie sur un certain nombre de périodes et se fixe $\lambda = 1 + \epsilon$, on peut facilement estimer V et en déduire $\tilde{Y}$. On obtiendra ainsi une formule permettant d'extrapoler les consommations futures.

En résumé, l'hypothèse

$$Y_t = \tilde{Y} + \lambda^t V$$

est assez réaliste tant que l'on ne considère pas un nombre trop élevé de périodes et se prête à des applications pratiques.

3.2.3 Conclusion de la discussion: choix du vecteur C*

Il résulte de ce qui précède que le modèle étudié s'applique de façon satisfaisante à l'étude de l'évolution de l'économie en phase I.

En phase II, en revanche, ce n'est que sur un petit nombre de périodes que l'on peut admettre les hypothèses du modèle (fixité de A et B; consommation de la forme $Y_t = \tilde{Y} + \lambda^t V$). Les propriétés optimales de la croissance équilibrée mises en évidence dans la deuxième partie, étant des propriétés à long terme, ne sont donc pas vraies en toute rigueur. Si cependant, la modification des coefficients de A et B est assez faible et l'approximation linéaire de Y_t assez bonne, la convergence de la production X_t vers la production $\overline{X}_0$ de croissance équilibrée généralisée (calculée à l'instant initial de la phase II) aura encore lieu dans tout développement optimal.

Ceci nous amène au raisonnement suivant: puisque tout développement optimal en phase II part d'une structure de la capacité de production voisine de la structure de croissance équilibrée il est raisonnable de choisir cette même structure comme structure finale C* de la phase I. D'où:

3.2.4 Procédé itératif de détermination de l'évolution optimale pendant la phase I

— Soit (t_0, t_1) l'intervalle de durée de la phase I; on estime les matrices A_1 et B_1 l'année $(t_1 + 1)$.

— On choisit un point Y_1 sur la droite, (estimée à partir de données antérieures) $Y_0 + uV$.

Connaissant Y_1, V, on prend $\lambda(=1 + \epsilon)$ tel que $\overline{X}_1$, vecteur de croissance équilibrée, que l'on sait alors calculer, et Y_1 donnent au taux

$\frac{\Sigma Y_1^i}{\Sigma \overline{X}_1^i}$ la valeur désirée.

— On résoud alors le programme linéaire correspondant à l'évolution de l'économie pendant la période (t_0, t_1) lorsque l'on a les contraintes

$$C(t_1 + 1) \geqq \mu \overline{X}_1$$

— Si la valeur maximum $\hat{\mu}$ de μ est voisine de un, on a trouvé l'évolution optimale en phase I; sinon, suivant que $\hat{\mu}$ est inférieure ou supérieure à un, on recommence les calculs avec un vecteur Y_1 de norme plus faible ou plus forte.

3.3 Essais: Résultats et Commentaires

Nous avons procédé à des essais du programme linéaire défini précédemment afin de nous assurer:

(a) qu'en cinq années, le vecteur capacités C_t a effectivement le temps de se rapprocher de la direction C* imposée,

(b) qu'en sens inverse, on ne passe pas par des capacités intermédiaires C_t très éloignées de C*, c'est-à-dire qu'il y a une certaine stabilité de l'évolution.

Le plus simple était de choisir pour période (t_0, t_1) une période passée de l'économie française où les valeurs effectivement prises par les variables étaient connues, de prendre pour C* la capacité de production effectivement atteinte l'année $(t_1 + 1)$ et de comparer évolution simulée et évolution réelle.

3.3.1 Estimation des données du modèle

Nous avons choisi la période 1959-63, un tableau d'échanges interindustriels ayant été publié pour l'année 1959.

L'unité de valeur est le franc 1959; toutes les données recueillies ont été converties pour être exprimées en cette unité.

L'existence d'une ligne 'commerce' dans le tableau d'échanges interindustriels (T.E.I.) nous a conduits à introduire un secteur 'commerce', secteur 16 (nous avons, en revanche, regroupé dans le secteur 15 les secteurs 'service du logement' et 'autres services' du T.E.I.); cela nous a obligés à prendre pour prix les prix du marché:

prix du marché = prix à la production + marges commerciales + droits et taxes sur les importations.

(a) estimation de A: soit A_{ij} les élements de A; alors

$$A_{ij} = \frac{X_{ij}}{X_j}$$

— où X_{ij} est la consommation du produit i par la branche j donnée par le T.E.I. 59 (Etudes et Conjoncture—Mars 1966),

— où X_j est la production disponible, somme de la production des Entreprises Non Financières et des Ménages.

(b) estimation de B: B se calcule par la formule

$$B = \beta \,.\, K^{-1}$$

où β est la matrice de ventilation des investissements par branches entre les investissements par produits et K la matrice diagonale des coefficients de capital.

— estimation de β: pour chaque branche, les investissements utilisent deux produits, ceux des secteurs

09 : 'Industrie méchanique et électricité'
13 : 'Bâtiment et Génie Civil'.

Nous avons calculé pour chaque année, de 1959 et 1963, dans chaque secteur, le pourcentage des investissements en 'Bâtiment et Génie Civil', et pris la moyenne des résultats pour ligne 13 de β; la ligne 09 est constituée des compléments à l'unité de ces chiffres.

— estimation de K: nous avons pris pour K_i, $i^{\text{ème}}$ composante de K le rapport

$$\frac{(\Delta\, X_i)_{59}^{63}}{(\Sigma\, Z_i')_{59}^{62}}$$

où . $(\Delta X_i)_{59}^{63}$ est la variation de la production du secteur i entre 1959 et 1963

. $(\Sigma\, Z_i')_{59}^{62}$ est la somme des investissements nets de 1959 à 1962, ce que revient à supposer un décalage d'un an entre investissement et production.

(c) estimation de D: de diverses estimations, après essais, nous avons retenu l'inverse de la durée de vie du capital, n, où n est calculée comme la moyenne harmonique des durées de vie des 'Bâtiments et Travaux Publics' et de 'Matériels', les coefficients de pondération étant les pourcentages moyens des investissements du secteur en B.T.P. et Matériels.

(d) estimation de C_0: on a supposé que l'année 1959, il y avait en plein emploi des capacités d'où

$C_0 = X_{1959}$, production en 1959

(e) estimation des consommations finales Y_t: ces essais ayant pour but de simuler l'évolution réelle de l'économie française entre 1959 et 1963, nous avons pris pour consommations finales les estimations des consommations réelles.

Les importations et exportations ne figurent pas explicitement dans le modèle; elles sont implicitement incorporées dans la consommation finale* (dont les composants peuvent donc être négatives; c'est le cas pour le secteur 08 'Minerais et Métaux non ferreux').

---

* C'est pourquoi, dans les possibilités de développement en phase I, on peut être amené à étudier de nombreux cas, pour différents Y_t correspondant à des échanges extérieurs plus ou moins importants.

La consommation finale est donc la somme des consommations des Ménages, Administrations et Institutions Financières, de leur Formation brute de capital fixe, des Exportations et des Variations de Stocks, diminuée des Importations et des Ventes.

Les données ont été tirées:

— pour 1959 et 1960 'd'Etudes et Conjonctures' (Mars 1966 et Août 1963),

— pour 1960 des 'Comptes de la Nation' (1965).

(f) Choix de C*

Nous avons vu que dans notre cas, il fallait prendre pour C* la valeur de la production réelle l'année 1964; nous ne disposons malheureusement de données que jusqu'à 1963. Nous avons donc pris C* égal à X_{1963}; la direction du vecteur X_{1964} doit être assez proche, son intensité étant supérieure; il faut donc s'attendre à trouver une valeur maximale de λ supérieure à un.

On trouvera en annexe la valeur de A, B, D, Y_t, C_{1959} et C*.

3.3.2 Les essais

Le modèle a été programmé sur UNIVAC 1108.

Les premiers passages effectués avec les seconds membres Y_{1959} à Y_{1963} estimés plus haut ne nous ont pas donné de solution réalisable. Nous avons fait de nouveaux passages en multipliant tous les seconds membres par un scalaire ρ inférieur à un, que nous avons fait croître progressivement; la valeur maximum pour laquelle le programme est réalisable est $\rho = 0,980$ ce qui est satisfaisant étant donné la précision des estimations. Les résultats que nous donnons ci-dessous sont ceux qui correspondent à ce second membre.

3.3.3 Tableau des résultats et commentaires

Le tableau des résultats se trouve aux pages 23 et 24.

La valeur de λ à l'optimum est $\hat{\lambda} = 1,183$, ce qui est un peu élevé, l'économie réelle ne permettant d'atteindre que la capacité de production X_{1964} que l'on peut grossièrement évaluer à 1,05 C*. Le modèle, bien qu'étant à la limite de l'impossibilité pour $\rho = 0,98$, permet un taux de croissance plus élevé que le taux réel; l'explication tient sans doute à ce que l'économie doit produire un gros effort initial, les estimations de C_{1959} et Y_{1959}, faites indépendamment l'une de l'autre ne concordant pas exactement.

Si l'on regarde les excédents des capacités sur les productions on voit, qu'à part la troisième et la quatrième année dans le secteur 15, tous les excédents ont lieu la première année; il se produit donc la première année un réajustement des capacités de manière à satisfaire au mieux au but imposé; de trop grands excédents indiqueraient un désaccord entre le modèle et la réalité, par exemple, des estimations irréalistes de A et B. En fait, ils ne sont importants que dans les secteurs 09 et 13, qui sont les secteurs d'investissement; cela confirme l'explication du paragraphe précédent: la première année, l'économie a du mal à satisfaire la consommation finale et s'y consacre au détriment de l'investissement.

On constate, d'autre part, que la capacité C_{1964} obtenue n'est pas exactement égale à 1,183 C*; il y a, en effet, des excédents dans les secteurs 08 et 13; ce sont des secteurs importants en ce qui concerne les investissements (13 directement, 08 indirectement); qu'ils soient proportionnellement moins importants dans C* que dans les vecteurs capacités intermédiaires calculés est en rapport avec le fait que la solution donnée par le modèle accorde une plus large part à l'investissement et une moindre à la consommation que la réalité.

La croissance des productions et des capacités de production pendant les cinq années n'est pas régulière; on ne pouvait guère l'espérer avec un programme linéaire; et cependant, si l'on considère les accroissements bruts de capacité δC_t qui étant des

Tableaux Des Resultats

Année	1959				1960				1961			
Variable	X	C	C − X	δ C	X	C	C − X	δ C	X	C	C − X	δ C
Secteurs	42895	43932	1037	4204	46160	46160			44083	44083		4408
	52425	53612	1187	5244	56766	56766		2272	56824	56824		4161
	7231	7231			6992	6992		234	6996	6996		1388
	6042	6042			5848	5848		850	6512	6512		1108
	10393	10393		73	10123	10123		2013	11802	11802		1206
	5337	6650	1312		6370	6370		1490	7593	7593		1097
	10482	10482			10010	10010		2420	11980	11980		1799
	1616	1908	291		1831	1831			1758	1758		812
	37296	53755	16459	1944	53442	53442		10428	61626	61626		11841
	15217	16308	1091	2326	17868	17868		1811	18840	18840		2511
	25300	25300		3533	27669	27669		1987	28383	28383		3072
	19680	21107	1427	2048	22185	22185		1964	23129	23129		2694
	24296	35882	11586	1454	35615	35615		8528	42434	42434		6764
	19705	19815	110		18983	18983		3309	21496	21496		2585
	45651	47638	1987	4996	51444	51444			44402	50158	5756	
	40302	43271	2969	4163	45401	45401		3312	46579	46579		5582

Tableaux Des Resultats — continué

Année	1962				1963				1964		
Variable	X	C	C − X	δ C	X	C	C − X	δ C	X	C	$C - \hat{\lambda}C$
Secteurs	46507	46507		3300	47715	47715		14454		60022	
	58770	58770		4062	60540	60540		15011		73100	
	8153	8153		1133	9018	9018		587		9308	
	7412	7412		1216	8391	8391		2147		10269	
	12620	12620		2185	14388	14388		5194		19108	
	8372	8372		2549	10569	10569		481		10607	
	13240	13240		2729	15374	15374		107		14789	
	2500	2500		690	3090	3090				2967	242
	70879	70879		17649	85551	85551		6114		88072	
	20465	20465		3290	22794	22794		6092		27816	
	30151	30151		4523	33287	33287		10305		42061	
	24760	24760		3389	27011	27011		7577		33346	
	47161	47161		18555	63453	63453				60407	4455
	23179	23179		2970	25176	25176		5114		29233	
	47128	48904	1776	3607	51289	51289		21669		71676	
	49973	49973		6801	54425	54425		8651		60519	

Maximum de λ: $\hat{\lambda} = 1,1831$

variables marginales ont les évolutions les plus irrégulières, on constate que la deuxième, troisième et quatrième année il y a investissement dans tous les secteurs (sauf le secteur 01, la deuxième année), c'est-à-dire qu'une certaine régularité de développement s'instaure pendant les années intermédiaires.

3.3.4 Conclusion des essais

On pouvait craindre que la simulation de l'évolution réelle d'une économie à l'aide d'un programme linéaire, ou bien soit impossible, le modèle n'étant pas assez souple, ou bien soit possible, mais conduise à des résultats aberrants (développement trop irrégulier).

Nous avons constaté qu'au prix d'une modification minime de la valeur de second membre, il existait des solutions réalisables, et que la solution optimale correspondait à une évolution relativement régulière de l'économie. Cette régularité est l'effet des contraintes

$$C(t_1 + 1) \geqq \lambda C^*$$

qui imposant la structure des capacités de production terminales influencent encore, dans une certaine mesure, les capacités intermédiaires.

BIBLIOGRAPHIE

VARGA, RICHARD S., (1962), Matrix Iterative Analysis, Prentice-Hall, Englewood Cliffs, N.J.

MC KENZIE, LIONEL W., (1963), 'Turnpike theorem for a generalized Leontief Model', Econometrica Vol. 31, no 1-2, pp 165-180

TSUKUI, JINKICHI, (1966), 'Turnpike theorem in a Generalized Dynamic Input-Output System', Econometrica Vol. 34, no 2, pp 396-407

ANNEXE: DONNEES DU MODELE DE L'ECONOMIE FRANCAISE UTILISE POUR LES ESSAIS

1. Liste des Secteurs

Agriculture, forêts	0 1
Industries agricoles et alimentaires	0 2
Combustibles minéraux solides et gaz	0 3
Electricité	0 4
Pétrole, gaz naturel et carburants	0 5
Matériaux de construction et verre	0 6
Mines de fer et sidérurgie	0 7
Minerais et métaux non ferreux	0 8
Industries mécaniques et électricité	0 9
Chimie	1 0
Textile, habillement, cuir	1 1
Bois, papier et industries diverses	1 2
Bâtiment et génie civil	1 3
Transports et télécommunications	1 4
Services	1 5
Commerce	1 6

2. Matrice A

	01	02	03	04	05	06	07	08	09	10	11	12	13	14	15	16
01		0, 393								0, 005	0, 005	0, 042		0, 001	0, 041	
02	0, 037									0, 023	0, 020	0, 003		0, 002	0, 099	0, 001
03		0, 003		0, 141		0, 041	0, 170	0, 026	0, 004	0, 032	0, 005	0, 005	0, 001	0, 016	0, 002	0, 004
04	0, 005	0, 004	0, 032		0, 005	0, 022	0, 029	0, 069	0, 013	0, 026	0, 014	0, 015	0, 001	0, 011	0, 007	0, 005
05	0, 016	0, 007	0, 039	0, 037		0, 044	0, 010	0, 029	0, 010	0, 030	0, 009	0, 017	0, 024	0, 096	0, 018	0, 026
06	0, 001	0, 003	0, 001	0, 001			0, 025		0, 005	0, 009		0, 007	0, 124		0, 001	0, 003
07			0, 062	0, 002		0, 006		0, 010	0, 127	0, 008			0, 036	0, 005		
08						0, 002	0, 059		0, 036	0, 013		0, 005				
09	0, 009	0, 013	0, 025	0, 030	0, 018	0, 010	0, 038	0, 015		0, 026	0, 006	0, 013	0, 125	0, 039	0, 047	0, 002
10	0, 048	0, 004	0, 032	0, 004	0, 017	0, 036	0, 012	0, 020	0, 045		0, 044	0, 058	0, 023	0, 019	0, 015	0, 005
11	0, 004	0, 001		0, 001					0, 007	0, 020		0, 017	0, 005	0, 004	0, 004	0, 004
12	0, 001	0, 012	0, 014	0, 004	0, 005	0, 027	0, 004	0, 011	0, 018	0, 034	0, 024		0, 046	0, 009	0, 029	0, 023
13	0, 007	0, 002	0, 016	0, 070	0, 004	0, 009	0, 012	0, 003	0, 008	0, 007	0, 006	0, 006		0, 043	0, 004	0, 005
14	0, 002	0, 012	0, 020	0, 010	0, 065	0, 019	0, 063	0, 018	0, 023	0, 036	0, 019	0, 029	0, 023		0, 016	0, 099
15	0, 045	0, 015	0, 003	0, 015	0, 010	0, 016	0, 009	0, 004	0, 019	0, 032	0, 025	0, 041	0, 048	0, 031		0, 037
16	0, 251	0, 194	0, 148			0, 146	0, 036		0, 111	0, 169	0, 269	0, 184				

3. Matrice B

09	0, 635	0, 123	1, 580	5, 707	0, 223	0, 332	1, 480	0, 845	0, 211	0, 312	0, 061	0, 117	0, 147	1, 429	0, 061	0, 054
13	0, 128	0, 067	1, 753	9, 676	0, 265	0, 116	0, 360	0, 524	0, 071	0, 111	0, 013	0, 021	0, 013	0, 494	0, 227	0, 036

4. Matrice D D est la matrice diagonale ayant pour éléments diagonaux:

0, 045	0, 039	0, 033	0, 032	0, 033	0, 042	0, 045	0, 040	0, 042	0, 047	0, 046	0, 046	0, 048	0, 042	0, 025	0, 047

5. Vecteur C_0 (ou C_{1959})

1	2	3	4	5	6	7	8	9	10	11	12	13	14	15	16
43933	53613	7231	6042	10393	6650	10482	1908.	53756	16309	25300	21108	35883	19816	47639	43271

6. Vecteur C* (ou X_{1963} réel)

1	2	3	4	5	6	7	8	9	10	11	12	13	14	15	16
50730	61860	7867	8680	16150	8965	12500	2303	74438	23510	35550	28184	47290	24708	60580	51160

7. Vecteurs $Y_{1959}, Y_{1960}, \ldots, Y_{1963}$

Y_t Secteur	Y_{59}	Y_{60}	Y_{61}	Y_{62}	Y_{63}
1	19767	20983	18158	21457	20796
2	46249	49848	50651	52183	53389
3	2381	2376	2155	2396	2415
4	2093	2233	2437	2807	3174
5	2100	2262	2815	2836	3321
6	945	1007	1000	1067	1055
7	2381	2774	2101	1700	1272
8	−1026	−1115	−1189	−1264	−1376
9	23166	23832	25997	28574	31426
10	6028	7122	7530	8108	8717
11	24108	26196	26817	28424	31299
12	13557	14771	15319	16230	16964
13	18907	19478	21025	22565	24347
14	8500	8766	9408	10042	10263
15	37377	41776	33913	35742	38248
16					

DISCUSSION

D. G. Clough

Two questions arise concerning the 'balanced growth' aspects of the model and the fact that the objectives of economic growth and economic stability seem to be incompatible.

First, the model incorporates constraints on capacity which appear to be constraints on physical facilities only. Perhaps in further extensions of the model it would be worthwhile to incorporate such other constraints as limitations on the availability of manpower inputs and limitations on complimentary imports resulting from constraints on foreign exchange reserves. These short-run constraints on manpower resources and foreign exchange reserves seem to knock the indicative planning of Leontieff models into a cocked hat. In Britain, for example, long-run economic growth has been cut short by the exigencies of short-run stabilization policies and full employment policies.

Second, the model does not account for time lags between inputs and outputs or for time lags between changes in capacity constraints (investments) and outputs. If one considers the real storage effect of production processes—that is, the work-in-process effect—it seems clear that time lags between inputs and outputs may range from weeks or months (style clothing) to years (aircraft). The choice of an accounting interval of, say, one year, is tantament to assuming that time lags are discrete and exactly the same for all commodities or industries. If one considers an elementary linear model with different exponential time lags for different commodities, it seems clear that equilibrium growth generally cannot exist. Examination of the eigenvalues for such models indicates that outputs oscillate with different periods and amplitudes for different commodities. One way out of the dilemma of accounting for time lags might be to subdivide the classification of commodities and industries into elementary classes for which all time lags are relatively short and nearly the same.

J. Y. Jaffray

Le modèle exposé n'est pas destiné à mettre en évidence des variations conjoncturelles mais à dégager une tendance de l'évolution à long-terme; l'introduction de considérations de politique monétaire ou d'un délai entre inputs et outputs différencié par secteur aurait pour résultat de créer des oscillations autour de la croisance équilibrée que apparaitrait alors comme une direction movenne du vecteur production.

Le problème de la main-d'oeuvre a été abordé au paragraphe 3. 2. 1; ajoutons que c'est la question de l'introduction du progrès technique qui est soulevée ici; il faudrait sans doute introduire une contrainte main d'oeuvre, dont les coefficients décroitraient avec le temps tandis que les coefficients de capacité de production augmenteraient; mais l'estimation de ces coefficients est si difficile, par manque de données, qu'il est pratiquement impossible de trouver la loi de leur variation, et donc de dire si la propriété établie d'existence d'une croissance équilibrée se généralise à un modèle tenant compte du progrès technique.

Optimal Waste Treatment and Pollution Abatement Benefits on a Closed River System

DONALD J. CLOUGH
University of Toronto and
Systems Engineering Associates Ltd., Canada
M. B. BAYER
University of Toronto, Canada

SUMMARY

This paper describes a non-linear programming model to be used for the selection of an optimal set of waste treatment plant efficiencies on a closed river basin system. Both short-run and long-run objective functions are specified. The short-run objective is to minimize the relevant incremental costs of changing the plants from an existing set of efficiencies to some new set of efficiencies which must satisfy stream water quality constraints at various points in the system. The long-run objective is to minimize the relevant incremental costs of a sequence of irreversible discrete changes in efficiencies over a finite planning horizon. Efficiency is measured in terms of the percentage removal of B.O.D. (biological oxygen demand). The constraints are specified as upper limits on allowable B.O.D. concentrations and lower limits on allowable D.O. (dissolved oxygen) concentrations. The stream dilution effects of multi-purpose storage reservoirs are also described, and a new basis laid for the measurement of the economic pollution abatement benefits of multi-purpose reservoir systems.

1. POLITICO-ECONOMIC CONTEXT

The model described in this paper is being developed for application to river systems in the Province of Ontario, Canada. The model only makes sense if we postulate the existence of an agency which has the legal authority to manage water resources (e.g., storages and stream flows) on an entire river system, and if we postulate the existence of a regulatory agency which has the legal authority to specify and codify stream water quality standards.

In Ontario there exists a unique and highly effective organization of regional Conservation Authorities which have the authority under statutes of Ontario to construct works such as stream channel improvements and storage dams for such multiple purposes as flood control, water supply, pollution abatement, recreation, erosion control, fish and wild-life conservation, and other purposes. (Hydro-electric power developments fall within the jurisdiction of the Ontario Hydro Electric Power Commission, and navigation aspects of streams fall within the jurisdiction of the federal government.) Each Conservation Authority has jurisdiction over one or more complete watershed areas or complete river systems and is organized as a quasi-political body with members appointed by all of the municipal governments within that Authority's area of jurisdiction. Financial support for water resource management programs is obtained from federal, provincial and municipal levels of government under a number of different federal, provincial and municipal statutes.

The Ontario Water Resources Commission holds the legal authority to specify stream water quality standards and standards of waste treatment and water treatment. The O.W.R.C. also has the power to carry out pollution studies and to design and recommend waste treatment works.

The initiative for multi-purpose water conservation works generally comes from a local Conservation Authority. If an Authority wishes to seek financial assistance from the senior levels of government it will generally engage consultant engineers to prepare preliminary engineering designs and benefit-cost analyses, and then submit its proposals and requests in the form of a legal brief.

One of the problems facing consultants at the present time is the lack of a hard-and-fast set of rules or guide-lines for the analysis of economic benefits. Existing guides such as the 'Green Book' (Federal U.S.Govt., 1950) used widely in the United States are not explicit in the specification of underlying economic theories and models on which methods of measurement must be based. It is almost universally assumed that benefits are determined by equilibrium conditions of supply and demand in an unconstrained free market system, but explicit models of such a mechanism are not specified. Existing constraints on supply and demand variables—imposed by legal statutes, physical laws of nature, and engineering design criteria, for example—are almost universally ignored or are never specified in terms of underlying theories and models.

We contend that economic benefits are at least partially determined by various constraints imposed by customs, by prior decisions, by the specification of engineering design criteria, by the specification of legal water quality standards, and not least by the physical laws of nature. Any realistic theory or model should incorporate such constraints explicity, and the measures of economic benefits should be based on an explicit mathematical model.

This paper deals with a special model which has limited usefulness for the measurement of some pollution abatement benefits. To simplify the presentation, we assume first that the river system is a closed economic system so that we do not have to worry about the external diseconomies of internal pollution abatement decisions. We also assume that explicit legal water quality standards are specified and that these standards must be met one way or another. For example, the constraints on stream water quality may be met by improving the efficiencies of some of the waste treatment plants which discharge into the streams. Alternatively, the constraints may be met by building dams and storage reservoirs to augment stream flows and to dilute waste effluents during the low flow periods. In practice there may be an infinity of feasible alternative combinations of stream flow augmentation and waste treatment schemes that will satisfy the stream water quality constraints. The economic pollution abatement benefit of a proposed set of dams (with fixed designs, fixed construction schedules and fixed operating rules) is measured as the difference between the relevant costs of optimal* waste treatment with and without the existence of the dams. (See Dean, 1952, and Clough, 1963, for discussions of relevant costs in managerial economics.) The pollution abatement benefits of a number of discrete alternative sets of proposed dams should be measured and compared in the same way.

It is important to note that <u>economic pollution abatement benefits cannot be properly defined or measured unless stream water quality constraints are specified explicitly. If no water quality standards exist, waste treatment and stream dilution are not required and economic measures of pollution abatement benefits cannot be defined.</u>

(It should be noted that in Canada water quality standards are constitutionally the responsibility of the provincial governments. Water quality standards which are either stated explicitly or imputed by the decisions of a provincial agency such as the Ontario Water Resources Commission are indirect reflections of demands of the provincial body politic and objectives of the provincial government. In this case provincial rather than national demands and objectives dictate the constraints which force the economic variables. This raises some perplexing questions concerning the validity of any of the standard economic theories currently used in the measurement of 'national' benefits of multi-purpose water conservation projects.)

---

* by 'optimal' we mean 'minimum cost'.

2. MATHEMATICAL PROGRAMMING MODELS

The model described in the following sections follows the development of linear programming models by Deininger (1965), and Loucks, Revelle and Lynn (1967, 1968). The model described is different from its predecessors in its objective function, in its water quality constraints, in the matrix representation of multi-branch systems, and in the interpretation of relevant economic costs. It is extended to deal with systems expanding irreversibly over some specified time horizon. It is also extended as a basis for the measurement of the economic benefits of flow augmentation by storage reservoirs. In particular, the authors cited above employ a linear objective function which is restricted to plants with B.O.D. removal efficiencies less than 90 percent, while in this paper we employ a logarithmic objective function which applies up to efficiencies in the neighbourhood of 100 percent. The other authors also employ D.O. as the sole measure of water quality and specify a set of constraints on D.O. concentration. Considering what is relevant in Ontario, in this paper we specify both B.O.D. and D.O. as measures of water quality and define a set of simultaneous constraints on both measures.

3. WATER QUALITY MEASURES AND CONSTRAINTS

In this paper we shall consider only bio-degradable wastes, which make up the largest proportion of municipal and industrial waste effluents and are the most important from a public health point of view. Non-degradable chemical wastes are not considered, but they can be incorporated readily into the model if their effects are independent and approximately additive. Generally speaking, streams have the power to cleanse themselves of bio-degradable waste materials, provided that the waste concentrations are not too high and appropriate conditions of flow, temperature, mechanical agitation, and biochemistry exist.

Two inter-related measures of stream pollution are commonly used. One is the biological oxygen demand (B.O.D.), which measures the stream's concentration of waste material in terms of the quantity of oxygen demanded by the biochemical processes of waste degradation under specified laboratory test conditions. (See Frankel, 1966, for a comprehensive bibliography.) The other is the dissolved oxygen concentration (D.O.), which measures the stream's concentration of dissolved oxygen, and hence its ability to support wildlife, plant life and biochemical processes. The unit of measurement for both B.O.D. and D.O. concentrations is milligrams per litre (mpl).

For the purposes of this paper, we shall use the Streeter-Phelps differential equations which describe the B.O.D. and D.O. concentrations as functions of time (Streeter and Phelps, 1925). Let us denote the B.O.D. concentration by $x(t)$, the D.O. concentration by $y(t)$, and the D.O. saturation concentration by S. The difference $S - y(t)$ is called the oxygen deficit. The Streeter-Phelps equations can then be written in the following form:-

$$d\,x(t)/dt = -K\,x(t) \qquad 0 < x(t) \tag{1}$$

$$d\,[S - y(t)]/dt = K\,x(t) - L\,[S - y(t)],\ 0 < y(t) < S \tag{2}$$

where K is a rate constant for deoxygenation and L is a rate constant for reaeration. The constant K depends on temperature and other factors, and L depends on temperature, mechanical agitation conditions and other factors. The Streeter-Phelps equations do not constitute a comprehensive model of stream pollution and purification processes but only a relatively crude approximation. They have been extended by Phelps (1944), Camp (1963), Dobbins (1964), Thomas (1948), Li (1962), O'Connor (1960) and others to incorporate rate constants for sedimentation, B.O.D. addition caused by runoff and scour, oxygen production or depletion caused by photosynthesis and respiration, and other factors. However these improved equations have forms similar to the Streeter-Phelps equations and can be used in a similar fashion in the following model.

The solution of the pair of equations (1) and (2) is given as follows:-

$$x(t) = x(0) \,.\, \exp(-Kt) \tag{3}$$

$$y(t) = y(0) \,.\, \exp(-Lt) - x(0)[K/(L-K)][\exp(-Kt) - \exp(-Lt)] + S[1-\exp(-Lt)] \tag{4}$$

where x(0) and y(0) are initial values at time t = 0. Considering a unit volume of water travelling downstream, it is clear that the process time t is directly related to the distance travelled (for a fixed set of flow conditions). The concentration of dissolved oxygen may decrease as the unit volume proceeds downstream, and then may increase. The plot of y(t) against t is called the oxygen sag curve.

The Streeter-Phelps equations apply only if the D.O. concentration does not fall to zero. If the D.O. concentration does reach zero, the biochemical process changes from an aerobic reaction to an anaerobic reaction. One of the objectives of pollution control is to prevent oxygen depletion which leads to anaerobic reactions and the production of foul-smelling gases.

Let us denote the point of minimum D.O. concentration by t^{**}, and an intermediate point by t^*, where $0 \leqslant t^* \leqslant t^{**}$. We note that $y(t^*) \geqslant y(t^{**})$. Setting $\dot{y}(t) = 0$, we obtain the minimum point t^{**} as follows:

$$t = \frac{1}{L-K} \ln\left[\frac{L}{K} - \frac{S-y(0)}{x(0)} \cdot \frac{L}{K}\left(\frac{L}{K} - 1\right)\right], \text{if } \frac{S-y(0)}{x(0)} < \frac{K}{L} \tag{5}$$

$$\text{and } t^{**} = 0 \text{ if } [S - Y(0)]/x(0) \geqslant K/L. \tag{6}$$

We obtain $y(t^{**})$ by substituting t^{**} from (5) into (4).

Now suppose that we want to specify a maximum constraint on the B.O.D. concentration at the worst point:

$$x(0) \leqslant x^0_{max} \tag{7}$$

Suppose also that we want to specify a minimum constraint on the D.O. concentration at the worst point:

$$y(t^{**}) \geqslant y^{**}_{min}, \text{ when } x(0) = x^0_{max} \tag{8}$$

From equation (4) we can write

$$y^0_{min} = S - [K/(L-K)][x^0_{max}][1-\exp((L-K)t^{**})]$$
$$- (S - y^{**}_{min}) \exp(Lt^{**}) \tag{9}$$

where

$$t^{**} = \frac{1}{L-K} \ln\left[\frac{L}{K} - \frac{S-y^0_{min}}{x^0_{max}} \cdot \frac{L}{K}\left(\frac{L}{K} - 1\right)\right], \text{if } \frac{S-y^0_{min}}{x^0_{max}} < \frac{K}{L} \tag{10}$$

and

$$t^{**} = 0, \text{if } [S - y^0_{min}]/x^0_{max} \geqslant K/L.$$

Substituting t^{**} from (10) into (9) and rearranging, we obtain

$$y^0_{min} = S - [K/(L-K)\{x^0_{max} - [(K/L)x^0_{max}]^{L/K}[S - y^{**}_{min}]^{(K-L)/K}\} \tag{11}$$

if the right hand side of (11) is greater than $S - (K/L)x^0_{max}$, and $y^0_{min} = y^{**}_{min}$ if the right hand side is equal to or less than $S - (K/L)x^0_{max}$. Then we can rewrite(8) in the equivalent form

$$y(0) \geqslant y^0_{min}, \text{ when } x(0) = x^0_{max}. \tag{12}$$

If we wish to specify a constraint $y(t^*) \geqslant y^*_{min}$ at an intermediate point, $0 < t^* < t^{**}$, we can make a similar transformation to obtain an equivalent constraint such as (12).

The constraints apply under some specified set of critical low-flow conditions. For example, we may use the low seven-day average flow having an estimated mean return period of ten years; we may assume steady-state flow conditions; and we may specify B.O.D. and D.O. measurements applying at some specified point in the diurnal cycle (e.g. the lowest point of the diurnal D.O. cycle, which generally occurs in the middle of the night).

4. SHORT RUN SINGLE BRANCH OPTIMIZATION

Let us consider a single branch of a river system and specify a sequence of points $i = 0, 1, \ldots, n$ running downstream from an origin point 0 to an end point n. We identify a point wherever we want to specify a local water quality constraint and wherever one of the following conditions exists: (1) a flow enters the stream from a tributary on which there are no waste treatment plants, (2) a flow is released into the stream from a storage reservoir, (3) a flow is released from a waste treatment plant on the stream, (4) a flow is withdrawn for consumption or irrigation, or (5) a flow is lost from the stream through evaporation or percolation. (In this discrete-point model we can approximate runoff flows along a stream by discrete tributary flows at points, and we can approximate continuous losses by losses at discrete points.) We call the interval between two points i and i + 1 the reach i.

Let us assume that we are dealing with specified critical low-flow conditions (e.g., steady-state conditions using the 7-day average low flow having an estimated mean return period of 10 years; dissolved oxygen measured at the lowest point in the diurnal cycle). Let us assume that a unit volume of water starting at point 0 at time t_0 passes points $1, 2, \ldots, n$ at times $t_1, t_2, \ldots, t_n$ respectively, where $t_0 < t_1 < \ldots < t_n$. The residence times for the unit volume in reaches $0, 1, \ldots, n\text{-}1$ are $(t_1 - t_0), (t_2 - t_1), \ldots, (t_n - t_{n-1})$ respectively. We define the following fixed parameters and variables.

4.1 Fixed parameters

Q_{Ui} = Flow of water upstream of point i (litres/sec).

Q_{Di} = Flow of water downstream of point i (litres/sec).

Q_i = Inflow at point i from waste treatment plant, tributary, or additional reservoir release (litres/sec).

Q_{Ci} = Outflow at point i to consumption or loss (litres/sec).

S_{Ui} = Saturation concentration of dissolved oxygen upstream of point i (milligrams/litre).

S_{Di} = Saturation concentration of dissolved oxygen downstream of point i (milligrams/litre).

S_i = Saturation concentration of dissolved oxygen in the inflow to point i (milligrams/litre).

K_i = Deoxygenation rate constant for reach i.

L_i = Reaeration rate constant for reach i.

4.2 Variables

x_{Ui} = Concentration of B.O.D. upstream of point i (milligrams/litre).

x_{Di} = Concentration of B.O.D. downstream from point i (milligrams/litre).

x_i = Concentration of B.O.D. in the inflow to point i (milligrams/litre).

y_{Ui} = Concentration of D.O. upstream of point i (milligrams/litre).

y_{Di} = Concentration of D.O. downstream from point i (milligrams/litre).

y_i = Concentration of D.O. in the inflow to point i (milligrams/litre).

4.3 Continuity equations

The continuity equations for flow (litres per sec.), dissolved oxygen (milligrams per sec.) and biological oxygen demand (milligrams per sec.) are based on the assumption that quantities are additive:

$$Q_{Di} = Q_{Ui} + Q_i - Q_{Ci} \qquad , i = 0, \ldots, n \tag{13}$$

$$Q_{Ui} = Q_{D(i-1)} \qquad , i = 1, \ldots, n \tag{14}$$

$$x_{Di}Q_{Di} = x_{Ui}(Q_{Ui} - Q_{Ci}) + x_iQ_i, i = 0, \ldots, n \tag{15}$$

$$y_{Di}Q_{Di} = y_{Ui}(Q_{Ui} - Q_{Ci}) + y_iQ_i, i = 0, \ldots, n \tag{16}$$

It should be noted that at any particular point i one or more of the flows might be zero. For example, if there were no waste effluent, tributary flow or added storage reservoir flow at a point i, we would set $Q_i = 0$ at that point; if there were no withdrawal or loss we would set $Q_{Ci} = 0$.

4.4 Oxygen sag equations

From the basic Streeter-Phelps equations discussed earlier (equations (3) and (4)) we derive the following sets of equations:

$$x_{Ui} = \alpha_{i-1}x_{D(i-1)} \qquad , i = 1, \ldots, n \tag{17}$$

$$y_{Ui} = \gamma_{i-1}y_{D(i-1)} - \beta_{i-1}x_{D(i-1)} + \theta_{i-1}, i = 1, \ldots, n \tag{18}$$

where

$$\alpha_i = \exp[-K_i(t_{i+1} - t_i)] \qquad > 0 \tag{19}$$

$$\gamma_i = \exp[-L_i(t_{i+1} - t_i)] \qquad > 0 \tag{20}$$

$$\beta_i = [K_i/(L_i - K_i)][\alpha_i - \gamma_i] \tag{21}$$

$$\theta_i = S_{U(i+1)} - \gamma_i S_{Di} \tag{22}$$

4.5 Recursion equations in matrix form

Combining equations (15) and (17) we obtain

$$x_{Di} = a_{i-1}x_{D(i-1)} + b_ix_i \quad , i = 1, \ldots, n \tag{23}$$

where

$$a_{i-1} = \alpha_{i-1}(Q_{Ui} - Q_{Ci})/Q_{Di} \qquad > 0 \tag{24}$$

$$b_i = Q_i/Q_{Di} \qquad \geqslant 0 \tag{25}$$

(Note that if $Q_i = 0$ and $Q_{Ci} = 0$ we obtain $a_{i-1} = \alpha_{i-1}$ and $b_i = 0$.) We can rewrite (23) in matrix form as follows:

$$\mathbf{Ax_D} = \mathbf{Bx} + \mathbf{g}$$

$$\mathbf{x_D} = \mathbf{A^{-1}Bx} + \mathbf{A^{-1}g} \tag{26}$$

where

$$\mathbf{A} = \begin{pmatrix} 1 & 0 & \cdots & 0 & 0 \\ -a_1 & 1 & \cdots & 0 & 0 \\ 0 & -a_2 & \cdots & 0 & 0 \\ \cdots & \cdots & \cdots & \cdots & \cdots \\ 0 & 0 & \cdots & -a_{n-1} & 1 \end{pmatrix} \tag{27}$$

$$\mathbf{B} = \begin{pmatrix} b_1 & 0 & \cdots & 0 & 0 \\ 0 & b_2 & \cdots & 0 & 0 \\ 0 & 0 & \cdots & 0 & 0 \\ \cdots & \cdots & \cdots & \cdots & \cdots \\ 0 & 0 & \cdots & 0 & b_n \end{pmatrix} \tag{28}$$

$$\mathbf{x'_D} = (x_{D1} x_{D2} \cdots x_{DN}) \tag{29}$$

$$\mathbf{x'} = (x_1 x_2 \cdots x_n) \tag{30}$$

and where the n-dimensional vector **g** gives initial values for point $i = 0$.

$$\mathbf{g'} = (g_1\ 0 \ldots\ 0) \tag{31}$$

where

$$g_1 = a_0 x_{D0} \tag{32}$$

In a similar fashion we combine equations (16) and (18) to obtain

$$y_{Di} = c_{i-1} y_{D(i-1)} - d_{i-1} x_{D(i-1)} + b_i y_i + f_{i-1},$$

$$i = 1, \ldots, n \tag{33}$$

Here

$$c_{i-1} = \gamma_{i-1}(Q_{Ui} - Q_{Ci})/Q_{Di} \quad > 0 \tag{34}$$

$$d_{i-1} = \beta_{i-1}(Q_{Ui} - Q_{Ci})/Q_{Di} \tag{35}$$

$$f_{i-1} = \theta_{i-1}(Q_{Ui} - Q_{Ci})/Q_{Di} \tag{36}$$

Then we can write (33) in the following matrix form:

$$\mathbf{Cy_D} = -\mathbf{Dx_D} + \mathbf{By} + \mathbf{f} + \mathbf{h}$$

$$\mathbf{y_D} = -\mathbf{C^{-1}Dx_D} + \mathbf{C^{-1}By} + \mathbf{C^{-1}f} + \mathbf{C^{-1}h} \tag{37}$$

Here

$$\mathbf{y'_D} = (y_{D1}\ y_{D2} \cdots y_{Dn}) \tag{38}$$

$$\mathbf{y}' = (y_1\ y_2\ \dots\ y_n) \tag{39}$$

$$\mathbf{C} = \begin{pmatrix} 1 & 0 & \dots & 0 & 0 \\ -c_1 & 1 & \dots & 0 & 0 \\ 0 & -c_2 & \dots & 0 & 0 \\ \dots & \dots & \dots & \dots & \dots \\ 0 & 0 & \dots & -c_{n-1} & 1 \end{pmatrix} \tag{40}$$

$$\mathbf{D} = \begin{pmatrix} 0 & 0 & \dots & 0 & 0 \\ 0 & d_1 & \dots & 0 & 0 \\ 0 & 0 & \dots & 0 & 0 \\ \dots & \dots & \dots & \dots & \dots \\ 0 & 0 & \dots & 0 & d_{n-1} \end{pmatrix} \tag{41}$$

$$\mathbf{f}' = (f_0\ f_1\ \dots\ f_{n-1}) \tag{42}$$

and the n-dimensional vector **h** gives the initial values for point $i = 0$,

$$\mathbf{h}' = (h_1\ 0\ \dots\ 0) \tag{43}$$

where

$$h_1 = c_0 y_{D0} - d_0 x_{D0} \tag{44}$$

Substituting $\mathbf{x}_D$ from (26) into (37) we obtain

$$\mathbf{y}_D = \mathbf{C}^{-1}\mathbf{B}\mathbf{y} - \mathbf{C}^{-1}\mathbf{D}\mathbf{A}^{-1}\mathbf{B}\mathbf{x} + \mathbf{C}^{-1}(\mathbf{f} + \mathbf{h}) - \mathbf{C}^{-1}\mathbf{D}\mathbf{A}^{-1}\mathbf{g} \tag{45}$$

The matrix equations (26) and (45) give the B.O.D. and D.O. concentrations immediately downstream of points i ($i = 1, 2, \dots, n$) as linear combinations of the B.O.D. and D.O. concentrations at point $i = 0$ and the B.O.D. and D.O. concentrations of the entering waste effluents, reservoir releases or tributary flows at points upstream of point i.

4.6 Objective Function

Let us consider an existing waste treatment plant of fixed capacity (design flow). The average annual costs of the plant—including amortized capital costs and operation and maintenance costs—depend on the designed efficiency of the plant, as measured by the percentage of B.O.D. removed from the treated waste flow. The plant efficiency depends on some sequence of treatment processes such as chlorination of raw sewage, primary settling, chemical flocculation, trickling filters, stabilization ponds, oxidation ponds, activated sludge treatment, and tertiary sand filtration and chemical treatment. Most plants today are designed for future changes and are built on sufficiently large areas of land to permit expansion. The efficiency of such a plant can generally be increased in two main ways: (1) by twinning facilities such as settling basins, trickling filters and oxidation ponds to slow the rate of flow and to increase the residence times of wastes in the treatment processes, and (2) by adding new treatment processes in sequence with existing ones. Because of this flexibility it is possible to deal with the incremental costs of changing the efficiency of a plant of fixed capacity.

Let us first consider the average annual value of total treatment costs for a plant of fixed capacity. We denote the B.O.D. concentration in a treatment plant flow Q_i before treatment by z_i and after treatment by $\mathbf{x}_i$. Then we write the proportion of B.O.D. removed (percent efficiency/100) as follows:

$$p_i = 1 - x_i/z_i \tag{46}$$

An examination of extensive statistical data compiled by Frankel (1966) indicates that the average annual value of total costs, m_i, is a smooth monotonically increasing function of p_i in the region $0.30 \leq p_i \leq 1$, yielding the following equation:

$$m_i = K_i - k_i \ln x_i, \; 0 \leq x_i \leq 0.70 z_i \tag{47}$$

where $k_i > 0$ and $K_i > k_i \ln (0{\cdot}70 z_i)$. Equation (47) implies that the cost m_i approaches infinity as the effluent B.O.D. concentration x_i approaches zero (or as efficiency approaches 100 percent). Of course the approximation does not hold in the neighbourhood of $p_i = 1$ since there exists an upper limit of cost corresponding to an extreme treatment such as complete distillation of wastes. (see the numerical example at the end).

Now let us consider the incremental cost of changing a plant of fixed size from some existing efficiency $0.30 \leq p_i^* \leq 1.00$ to some new efficiency $p_i^* \leq p_i \leq 1.00$. The output B.O.D. will change form x_i^* to x_i and total cost will change from m_i^* to m_i. The incremental cost will be the difference

$$m_i - m_i^* = k_i \ln x_i^* - k_i \ln x_i, \; 0 \leq x_i \leq x_i^* \tag{48}$$

A similar incremental cost may be incurred at any point on the river i where a waste treatment plant discharges its effluent. At those points i where waste treatment plants do not exist we set $m_i = 0$. Suppose the objective is to minimize the sum of these incremental costs. Then we can write the objective in terms of the relevant parts of the $m_i - m_i^*$, as follows:

$$\text{Minimize} - \sum_i k_i \ln x_i \tag{49}$$

subject to the constraints $0 \leq x_i \leq x_i^*$ at those points i where waste treatment plants exist. (The minimization is also subject to the water quality constraints and other constraints specified below.)

We assume for this simplified version of the model that once specified, any further reductions in B.O.D. below the constraints will have negligible effects on the treatment costs of water withdrawn for consumption.

Therefore we have not included incremental costs of water treatment in the objective function.

4.7 Constraints

Now suppose that an authorized agency specifies maximum constraints on B.O.D. concentrations and minimum constraints on D.O. concentrations. From (7) and (12) we can write constraints immediately downstream of points $i = 1, 2, \ldots, n$, as follows:

$$\mathbf{x}_D \leq \mathbf{x}_{D,\max} \tag{50}$$

$$\mathbf{y}_D \geq \mathbf{y}_{D,\min} \tag{51}$$

Substituting $\mathbf{x}_D$ and $\mathbf{y}_D$ in equations (26) and (45), we obtain the constraints in terms of $\mathbf{x}$, as follows:

$$\mathbf{A}^{-1}\mathbf{B}\mathbf{x} \leq \mathbf{x}_{D,\max} - \mathbf{A}^{-1}\mathbf{g} \tag{52}$$

$$\mathbf{C}^{-1}\mathbf{D}\mathbf{A}^{-1}\mathbf{B}\mathbf{x} \leq \mathbf{C}^{-1}(\mathbf{f} + \mathbf{h}) - \mathbf{C}^{-1}\mathbf{D}\mathbf{A}^{-1}\mathbf{g} + \mathbf{C}^{-1}\mathbf{B}\mathbf{y} - \mathbf{y}_{D,\min} \tag{53}$$

We note that $\mathbf{A}^{-1}$ and $\mathbf{C}^{-1}$ are both lower diagonal matrices of positive elements. We assume that the elements on the right hand side in (52) and (53) are such that a feasible solution exists (e.g., the value g_1 in $\mathbf{g}$ is small enough that elements on the right side of (52) do not become negative).

For a single-branch system with no waste treatment plants on its tributaries, we assume that **y** is fixed and known—that is, all the D.O. concentrations in the waste plant effluents are fixed independently of the B.O.D. concentrations, and all the D.O. concentrations in tributary inflows are fixed. We also assume that the B.O.D. concentrations in tributary inflows are fixed and known, so that

$$x_i = \mu_i > 0, \text{for some i.} \tag{54}$$

We also introduce some constraints to limit the plant effluent B.O.D. concentrations to the levels of existing plants (assuming first that none of them are operating at efficiencies less than 30%), as follows:

$$x_i \leqslant x_i^*, \text{for some specified i.} \tag{55}$$

Finally, we write the non-negativity conditions

$$x_i \geqslant 0, \text{for all i.} \tag{56}$$

A feasible set of B.O.D. concentrations must satisfy the constraints (52), (53), (54), (55) and (56). An optimal set will also satisfy the cost minimization objective (49). It is important to note that tests for redundancy should be applied and redundant constraints removed. It is also important to note that equality constraints like (54) can be removed by substituting the constants $x_i = \mu_i$ into (52) and (53) and carrying them over to the right hand side. After reductions and eliminations, we can divide through by the elements on the right side to obtain a standard formulation

$$\text{maximize} \sum_{s=1}^{S} k_s \ln x_s \tag{57}$$

subject to the constraints

$$\sum_{s=1}^{S} \phi_{rs} x_s \leqslant 1, r = 1, 2, \ldots, R \tag{58}$$

$$x_s \geqslant 0, s = 1, 2, \ldots, S$$

where $c_s > 0$, and the index i has been relabelled as $s = 1, 2, \ldots, S$ to correspond to those points at which waste treatment plants exist.

5. SHORT RUN MULTI-BRANCH OPTIMIZATION

Let us denote the points on the main branch of a river system by $i = 0, 1, \ldots, n$. Wherever a tributary enters the main branch at a point i we shall denote the points on the tributary by $j(i) = 0, 1, \ldots, n(i)$. The flow, B.O.D. and D.O. concentrations entering the main branch at point i are equal to the same quantities on the tributary at the end point n(i). Thus we can write the continuity equations

$$Q_i = Q_{Dn(i)} \tag{59}$$

$$x_i = x_{Dn(i)} \tag{60}$$

$$y_i = y_{Dn(i)} \tag{61}$$

From tributary equations like (26) and (45) we can write as follows (using a subscript i to denote vectors and matrices related to a tributary i):

$$x_{Dn(i)} = \rho_i'[\mathbf{A}_i^{-1}\mathbf{B}_i\mathbf{x}_i + \mathbf{A}_i^{-1}\mathbf{g}_i] \tag{62}$$

$$y_{Dn(i)} = \rho_i'[\mathbf{C}_i^{-1}\mathbf{B}_i\mathbf{y}_i - \mathbf{C}_i^{-1}\mathbf{D}_i\mathbf{A}_i^{-1}\mathbf{B}_i\mathbf{x}_i + \mathbf{C}_i^{-1}(\mathbf{f}_i + \mathbf{h}_i) - \mathbf{C}_i^{-1}\mathbf{D}_i\mathbf{A}_i^{-1}\mathbf{g}_i] \tag{63}$$

where

$\rho_i' = (0 \ 0 \ldots 0 \ 1)$ is an n(j) — dimensional row vector which is used as an operator to select the bottom rows corresponding to $x_{Dn(i)}$ and $y_{Dn(i)}$. From equations (60), (61), (62) and (63) we obtain x_i and y_i (entering the main branch) as linear combinations of the constants $y_{j(i)}$ and variables $x_{j(i)}$ on tributary i. The expressions for x_i and y_i can be then substituted directly into main branch equations (26) and (45).

5.1 Mathematical programming formulation

The relevant incremental costs of waste treatment plants on tributaries and on the main branch are additive, so that we can formulate the objective function as follows:

$$\text{minimize} - \sum_i k_i \ln x_i - \sum_i \sum_j k_{j(i)} \ln x_{j(i)} \tag{64}$$

Here the summations contain terms with $k_i > 0$ only at those points i on the main branch where a waste treatment plant exists, and terms with $k_{j(i)} > 0$ only at those points on a tributary i where a waste treatment plant exists.

We must write a set of constraints for each tributary i as follows:

$$\begin{aligned} &\mathbf{A}_i^{-1}\mathbf{B}_i\mathbf{x}_i \leq \mathbf{x}_{D,\,max,\,i} - \mathbf{A}_i^{-1}\mathbf{g}_i \\ &\mathbf{C}_i^{-1}\mathbf{D}_i\mathbf{A}_i^{-1}\mathbf{B}_i\mathbf{x}_i \leq \mathbf{C}_i^{-1}(\mathbf{f}_i + \mathbf{h}_i) - \mathbf{C}_i^{-1}\mathbf{D}_i\mathbf{A}_i^{-1}\mathbf{g}_i + \mathbf{C}_i^{-1}\mathbf{B}_i\mathbf{y}_i - \mathbf{y}_{D,\,min,\,i} \\ &x_{j(i)} = \mu_{j(i)}, \text{for some specified i} \\ &x_{j(i)} \leq x^*_{j(i)}, \text{for some specified i.} \end{aligned} \tag{65}$$

There would be a similar set of constraints for each existing tributary i. We must also write a set of constraints for the main branch, as follows:

$$\begin{aligned} &\mathbf{A}^{-1}\mathbf{B}\mathbf{x} \leq \mathbf{x}_{D,\,max} - \mathbf{A}^{-1}\mathbf{g} \\ &-\mathbf{C}^{-1}\mathbf{B}\mathbf{y} + \mathbf{C}^{-1}\mathbf{D}\mathbf{A}^{-1}\mathbf{B}\mathbf{x} \leq \mathbf{C}^{-1}(\mathbf{f} + \mathbf{h}) - \mathbf{C}^{-1}\mathbf{D}\mathbf{A}^{-1}\mathbf{g} - \mathbf{y}_{D,\,min} \\ &x_i = \mu_i, \text{for some specified i} \\ &x_i \leq x_i^*, \text{for some specified i} \end{aligned} \tag{66}$$

where elements x_i and y_i corresponding to tributary inflows are obtained from the continuity equations (60) and (61) through (62) and (63).

The tributary constraints (65) and main branch constraints (66), together with the non-negativity conditions, are given in terms of the B.O.D. concentrations at waste treatment plants on the tributaries, $x_{j(i)}$, and on the main branch, x_i. Wherever waste treatment plants exist we can relabel the points i and j(i) as s = 1, 2, . . . , S. Then we can recast the problem into the standard format given by (57) and (58).

6. LONG RUN MULTI-BRANCH OPTIMIZATION

A waste treatment plant has a specified design capacity and a corresponding design efficiency when operating at that capacity. Suppose that when a plant is built it is designed with enough excess capacity to handle increases in flow corresponding to increases in population up to, say, ten years in the future. Under such conditions a plant can usually be operated with higher efficiency at first because the waste flows can be retained longer in the treatment processes. As the flow increases the efficiency will generally decrease. The critical design efficiency is reached at the end of the period when the plant utilizes its full design capacity.

Suppose that it is necessary to expand the capacity of the plant at the end of the ten-year interval, say to handle forecast increases in flow for another ten-year interval. Suppose also that water quality constraints must not be relaxed, and that there is a policy that the design efficiency of the expanded plant must be at least as high after expansion as it is before expansion. In this case the cost associated with a capacity expansion at the same efficiency is irrelevant since it must be incurred in any event. But the cost of an increase in efficiency at the new capacity is relevant for a decision on the selection of optimal efficiency.

Consider discrete points in time $t = 0, 1, 2, \ldots, T$, not necessarily equally spaced. A plant design capacity increases from time to time. Its design capacity at time t (when it is expanded) is determined by $Q_{i,\,t+1}$ if it is at point i on the main branch or by $Q_{j(i),\,t+1}$ if it is at point j(i) on tributary i. The total cost of a plant at point i over the interval between $t-1$ and t is

$$m_{it} = K_{it} - k_{it} \ln x_{it} \tag{67}$$

The total cost of the plant at the same efficiency (but expanded capacity) over the interval between t and $t+1$ is

$$m^*_{i,\,t+1} = K_{i,\,t+1} + k_{i,\,t+1} \ln x^*_{i,\,t+1} \tag{68}$$

where $x^*_{i,\,t+1} = x_{it}$. The total cost of the plant at a higher efficiency ($x_{i,\,t+1} < x^*_{i,\,t+1}$) over the interval between t and $t+1$ is

$$m_{i,\,t+1} = K_{i,\,t+1} + k_{i,\,t+1} \ln x_{i,\,t+1} \tag{69}$$

The incremental cost of the increase in efficiency is

$$-\,k_{i,\,t+1}[\ln x_{i,\,t+1} - \ln x_{it}] \tag{70}$$

Introducing discount factors δ_t for periods $t = 1, \ldots, T$, the present value P_i of the incremental costs of a series of increases in efficiency is

$$\begin{aligned} P_i &= -\,\delta_1 k_{i1} \ln x_{i1} - \sum_{t=2}^{T} \delta_t k_{it}[\ln x_{it} - \ln x_{i,\,t-1}] \\ &= -\sum_{t=1}^{T-1} [\delta_t k_{it} - \delta_{t+1} k_{i,\,t+1}] \ln x_{it} - \delta_T k_{iT} \ln x_{iT} \end{aligned} \tag{71}$$

Where the following constraints apply

$$x_{it} \leq x_{i,\,t-1},\ t = 2, 3, \ldots, T \tag{72}$$

We can define a similar value P_i for every point i on the main branch where a waste treatment plant exists, and a similar value $P_{j(i)}$ for every point j(i) on a tributary i where a plant exists. As before, we can relabel such points i and j(i) and rename the coefficients in the objective function so that we can cast the problem into a standard format:

maximize

$$\sum_{t=1}^{T} \sum_{s=1}^{S} K_{st} \ln x_{st} \tag{73}$$

subject to the constraints

$$\sum_{s=1}^{S} \phi_{rst} x_{st} \leq 1, r = 1, 2, \ldots, R \text{ and } t = 1, 2, \ldots, T \tag{74}$$

$$x_{st} \geq 0, s = 1, 2, \ldots, S \text{ and } t = 1, 2, \ldots, T$$

Note that the water quality constraints may be changed from time to time. The plant efficiencies may increase to meet new flow conditions, but the increases will be irreversible.

7. BENEFITS OF STORAGE RESERVOIRS

The effect of an additional release from a storage reservoir during a normally low-flow period is a change in flow and storage conditions and an increase in the flows Q_{Ui} and Q_{Di} at all points i downstream of the reservoir. The augmented flows affect the optimal waste treatment constraint equations and hence the efficiencies and the total costs.

Suppose that we want to estimate the economic pollution abatement benefits of constructing a proposed set of dams and storage reservoirs in some specified sequence. Suppose that the individual dams start operating when construction is completed, at discrete points in time, $t = 0, 1, \ldots, T-1$, and that the operating release rules are specified. In this case we can compute the augmented stream flows at the various points in the system, and the different waste degradation times in the various reaches of the system. Then we can compute the optimum efficiencies under the long-run optimization objective described earlier. The optimum values of variables x_{it} and $x_{j(i)t}$ and the corresponding present value of total waste treatment costs will be different with and without the storage reservoirs. The saving in waste treatment costs realized on account of the reservoirs is the so-called 'pollution abatement benefit.' The benefits of a number of discrete alternative sets of reservoirs and operating release rules can be computed in the same way and compared.

It is important to note that the compiled benefits would occur only under the assumption that the decision-making authority would indeed adopt the computed optimal efficiencies as design parameters.

8. MATHEMATICAL PROGRAMMING ASPECTS

In this section we describe a differential algorithm which starts with an initial basic feasible solution, employs the simplex operation to move from one basic feasible solution to a better one, and finally employs the constrained gradient method to move from the best of the basic feasible solutions to the global optimum solution (which will occur on a bounding hyperplane or on the intersection of two or more bounding hyperplanes). The development follows that outlined by Wilde and Beightler (1967).

To simplify the notation, let us deal with the following standard problem format:

maximize

$$Z(\mathbf{x}) = \sum_{s=1}^{S} c_s \ln x_s, c_s > 0 \text{ for } s = 1, \ldots, S \tag{75}$$

subject to the constraints

$$\sum_{s=1}^{S} a_{rs} x_s \leq 1, r = 1, \ldots, R \tag{76}$$

$$x_s \geq 0, s = 1, \ldots, S \tag{77}$$

Some special properties of the objective function and the feasible solution region F defined by the constraints are given by the following lemmas.

Lemma 1. The objective function $Z(\mathbf{x})$ is strictly monotonic.

Proof. Given $Z(\mathbf{x}) = \sum_{s=1}^{S} c_s \ln x_s$, where $c_s > 0$, we obtain

$\partial Z(\mathbf{x})/\partial x_s = c_s/x_s > 0$ for $x_s \geq 0, s = 1, \ldots, S$. Thus the slopes are strictly positive for all $x_s \geq 0$, and $Z(\mathbf{x})$ is monotonically increasing.

Lemma 2. The objective function $Z(\mathbf{x})$ is strictly concave.

Proof. Consider any two points $\mathbf{x}^* > \mathbf{0}$ and $\mathbf{x}^{**} > \mathbf{0}$, and the linear interpolation $\mathbf{x} = \alpha\mathbf{x}^* + (1-\alpha)\,\mathbf{x}^{**}, 0 < \alpha < 1$. It is necessary to prove that the objective function is never over-estimated by a linear interpolation between any two points, viz.,

$$Z(\mathbf{x}) = Z(\alpha\mathbf{x}^* + (1-\alpha)\,\mathbf{x}^{**}) > \alpha\, Z(\mathbf{x}^*) + (1-\alpha)\, Z(\mathbf{x}^{**}).$$

Term by term, we want to prove that

$$\ln(\alpha x_s^* + (1-\alpha)\, x_s^{**}) > \alpha \ln x_s^* + (1-\alpha) \ln x_s^{**}.$$

Setting $x_s^*/x_s^{**} = u$ for convenience and rearranging, we want to prove that

$$\Psi(u) = \ln[\alpha u^{1-\alpha} + (1-\alpha)\, u^{-\alpha}] > 0$$

for $0 < u < 1$ and $1 < u \leq \infty$. We note first that $\Psi(1) = 0$ and

$$\frac{\partial\Psi(u)}{\partial u} = \frac{\alpha(u-1)}{u[u + \alpha/(1-\alpha)]}.$$

We observe that $\partial\Psi(u)/\partial u < 0$ when $0 < u < 1$, $\partial\Psi(u)/\partial u = 0$ when $u = 1$, and $\partial\Psi(u)/\partial u > 0$ when $1 < u \leq \infty$. Thus $\Psi(u)$ is U-shaped with minimum $\Psi(u) = 0$ at $u = 1$. Then $\Psi(u) > 0$ when $0 < u < 1$ and $1 < u \leq \infty$.

Lemma 3. There is only one local maximum and it is the global maximum.

Proof. Consider the global maximum point $\mathbf{x}^{**}$ and any other point $\mathbf{x}^* (\mathbf{x}^* \neq \mathbf{x}^{**})$ in the linear convex feasible solution region F (assuming that $\mathbf{x}^{**}$ exists, as when F is closed and bounded). Since F is convex, all points $\mathbf{x}$ on the line segment joining $\mathbf{x}^{**}$ and $\mathbf{x}^*$ are also in F. Thus $\mathbf{x} = \alpha\mathbf{x}^{**} + (1-\alpha)\,\mathbf{x}^*, 0 < \alpha < 1$. Moreover, the concavity of $Z(\mathbf{x})$ on this same line segment implies that

$$\begin{aligned} Z(\alpha\mathbf{x}^{**} + (1-\alpha)\,\mathbf{x}^*) &> \alpha\, Z(\mathbf{x}^{**}) + (1-\alpha)\, Z(\mathbf{x}^*) \\ &= Z(\mathbf{x}^*) + \alpha[Z(\mathbf{x}^{**}) - Z(\mathbf{x}^*)] \\ &> Z(\mathbf{x}^*). \end{aligned}$$

This last inequality follows because $\alpha > 0$ and $Z(\mathbf{x}^{**})$ is the global maximum. It follows that $Z(\mathbf{x}) > Z(\mathbf{x}^*)$ for all points $\mathbf{x}$ on the line segment, so that $\mathbf{x}^*$ cannot be a local maximum. Hence there cannot exist a local maximum point $\mathbf{x}^*$ distinctly different from the global maximum point $\mathbf{x}^{**}$.

Lemma 4. The global maximum point lies on the boundary of the feasible region F defined by the constraints (76), (77).

Proof. Let us make the contrary hypothesis that the global maximum $Z(\mathbf{x}^*)$ occurs at an interior point $\mathbf{x}^*$ of the feasible region F. Since $\mathbf{x}^*$ is an interior point, arbitrary perturbations $d\mathbf{x}^*$ are possible without violating the constraints. For $Z(\mathbf{x}^*)$ to be a maximum, it is necessary that $dZ(\mathbf{x}^*) \leq 0$ for all feasible perturbations $d\mathbf{x}^*$. To a first-order approximation, $dZ(\mathbf{x}^*)$ is given by

$$dZ = \frac{\partial Z}{\partial x_1} dx_1 + \ldots + \frac{\partial Z}{\partial x_S} dx_S.$$

From Lemma 1 we note that

$$\partial Z/\partial x_s = c_s/x_s > 0, s = 1, \ldots, S$$

If we select all $dx_s^* > 0$, then $dZ(\mathbf{x}^*) > 0$, which violates the necessary condition that $dZ(\mathbf{x}^*) \leq 0$ and contradicts the hypothesis that the global maximum occurs at an interior point.

8.1 Differential Algorithm

Introducing non-negative slack variables x_s, $s = S + 1, \ldots, S + R$ into the constraints (76), we reformulate the problem as follows:

maximize

$$Z(\mathbf{x}) = \sum_{s=1}^{S+R} c_s \ln x_s, \; c_s > 0 \text{ for } s = 1, \ldots, S$$
$$c_s = 0 \text{ for } s = S + 1, \ldots, S + R \tag{78}$$

subject to the constraints

$$f_r(\mathbf{x}) = \sum_{s=1}^{S} a_{rs} x_s + x_{S+r} - 1 = 0, r = 1, \ldots, R \tag{79}$$

$$x_s \geq 0, s = 1, \ldots, S + R \tag{80}$$

Now we have R equality constraints with R + S variables x_s. Without loss of generality, we can select R of the x_s and call them <u>state variables</u> u_m, $m = 1, \ldots, R$. The remaining x_s are called the <u>decision variables</u> v_n, $n = 1, \ldots, S$. For convenience we also relable the constraint coefficients, so that the problem can be reformulated as follows:

maximize

$$Z(\mathbf{u}, \mathbf{v}) = \sum_{m=1}^{R} c'_m \ln u_m + \sum_{n=1}^{S} c''_n \ln v_n \tag{81}$$

subject to the constraints

$$f_r(\mathbf{u}, \mathbf{v}) = \sum_{m=1}^{R} a_{rm} u_m + \sum_{n=1}^{S} b_{rn} v_n - 1 = 0,$$
$$r = 1, \ldots, R \tag{82}$$

$$u_m \geq 0, v_n \geq 0, m = 1, \ldots, R \text{ and}$$
$$n = 1, \ldots, S. \tag{83}$$

Here some of the coefficients c'_m and c''_n are positive and some are zero, and some of the coefficients a_{rm} and b_{rn} are zero. It is convenient to write the constraints (82) in matrix form:

$$\mathbf{Au} + \mathbf{Bv} - \mathbf{1} = \mathbf{0} \tag{84}$$

or $$\mathbf{u} = \mathbf{A}^{-1}\mathbf{1} - \mathbf{A}^{-1}\mathbf{Bv} \tag{85}$$

assuming that $\mathbf{A}$ is non-singular so that $\mathbf{A}^{-1}$ exists.

Now let us consider a feasible point $(\mathbf{u}, \mathbf{v}) \equiv \mathbf{x}$ and an infinitesimal perturbation $\mathbf{du}$ and $\mathbf{dv}$ to obtain a new point $(\mathbf{u} + \mathbf{du}, \mathbf{v} + \mathbf{dv})$. Near a feasible point there are both infeasible points where some $df_r \neq 0$ and feasible points where all $df_r = 0$. The perturbation of the decision variables, $\mathbf{dv}$, may be chosen arbitrarily, but the perturbation of the state variables, $\mathbf{du}$, must be constrained so that all $df_r = 0$ and the new point $(\mathbf{u} + \mathbf{du}, \mathbf{v} + \mathbf{dv})$ lies in the feasible region. Thus, to a first-order approximation,

$$df_r = \sum_{m=1}^{R} (\partial f_r/\partial u_m)\, du_m + \sum_{n=1}^{S} (\partial f_r/\partial v_n) dv_n = 0, r = 1, \ldots, R. \tag{86}$$

We note that $\partial f_r/\partial u_m = a_{rm}$ and $\partial f_r/\partial v_n = b_{rn}$, so that equations (86) can be written simply as

$$\sum_{m=1}^{R} a_{rm} du_m = -\sum_{n=1}^{S} b_{rn} dv_n, \; r = 1, \ldots, R \tag{87}$$

or, in matrix form,

$$\mathbf{du} = -\mathbf{A}^{-1}\mathbf{B}\mathbf{dv} \tag{88}$$

An element of $-\mathbf{A}^{-1}\mathbf{B}$ is called the constrained derivative $\delta u_m/\delta v_n$, where δ is used instead of ∂ to avoid confusing the constrained derivative with the unconstrained derivative $\partial u_m/\partial v_n$ (=0). We then write (88) as follows

$$du_m = \sum_{n=1}^{S} (\delta u_m/\delta v_n)\, dv_n \tag{89}$$

The objective function responds to the perturbation $\mathbf{dv}$ by an amount dZ, where

$$dZ = \sum_{m=1}^{R} (\partial Z/\partial u_m) du_m + \sum_{n=1}^{S} (\partial Z/\partial v_n) dv_n \tag{90}$$

Substituting from (89) into (90) we obtain

$$dZ = \sum_{n=1}^{S} [\partial Z/\partial v_n + \sum_{m=1}^{R} (\partial Z/\partial u_m)(\delta u_m/\delta v_n)]\, dv_n \tag{91}$$

We write the term in brackets as the constrained decision derivative:

$$\delta Z/\delta v_n = \partial Z/\partial v_n + \sum_{m=1}^{R} (\partial Z/\partial u_m)(\delta u_m/\delta v_n) \tag{92}$$

From (81) we obtain

$$\frac{\partial Z}{\partial u_m} = \begin{cases} c'_m/u_m, & \text{if } c'_m > 0 \\ 0, & \text{if } c'_m = 0 \end{cases} \tag{93}$$

$$\frac{\partial Z}{\partial v_n} = \begin{cases} c''_n/v_n, & \text{if } c''_n > 0 \\ 0, & \text{if } c''_n = 0 \end{cases} \tag{94}$$

It is also convenient to write $\delta u_m/\delta v_n = \alpha_{mn}$. Then we can write equations (89), (91) and (92) as follows:

$$du_m = \sum_{n=1}^{S} \alpha_{mn} dv_n \tag{95}$$

$$\delta Z/\delta v_n = c''_n/v_n + \sum_{m=1}^{R} (c'_m/u_m)\alpha_{mn} \tag{96}$$

$$dZ = \sum_{n=1}^{S} (\delta Z/\delta v_n) dv_n \tag{97}$$

In the algorithm we shall want to simplify the computations by perturbing only a single variable at a time. Suppose, for example, that we have a feasible solution, $\mathbf{u}^{(0)}$, $\mathbf{v}^{(0)}$ and $Z^{(0)}$, and that we change a single decision variable from $v_n^{(0)}$ to $v_n^{(n)}$. The state variables will then change from $u_m^{(0)}$ to $u_m^{(n)}$ and the objective function will change from $Z^{(0)}$ to $Z^{(n)}$. We have to insure that all the variables remain non-negative. Integrating (95) between $v_n^{(0)}$ and $v_n^{(n)}$, we obtain

$$u_m^{(n)} = u_m^{(0)} + \alpha_{mn}(v_n^{(n)} - v_n^{(0)}) \geqslant 0 \tag{98}$$

From the non-negativity condition of (98), $v_n^{(n)}$ must satisfy the following conditions for $m = 1, \ldots, R$.

$$\begin{aligned} &0 \leqslant v_n^{(n)} \leqslant v_n^{(0)} - u_m^{(0)}/\alpha_{mn}, \text{ if } \alpha_{mn} < 0 \\ &v_n^{(n)} \geqslant v_n^{(0)} - u_m^{(0)}/\alpha_{mn}, \text{ if } \alpha_{mn} > 0 \text{ and } v_n^{(0)} > u_m^{(0)}/\alpha_{mn} \end{aligned} \tag{99}$$

In any case, $v_n^{(n)} \geqslant 0$ and $u_n^{(n)} \geqslant 0$. Integrating (97), we obtain

$$Z^{(n)} = Z^{(0)} + \sum_{m=1}^{R} c'_m \ln (u_m^{(n)}/u_m^{(0)}) + c''_n \ln (v_n^{(n)}/v_n^{(0)}) \tag{100}$$

The strategy of the algorithm is to move from one basic feasible solution to a better one until the best of the basic feasible solutions has been reached. The global maximum may be bypassed as the last change of basis is made. However, the strategy is then to use a projected gradient search from the best basic feasible point to the global maximum point. The following steps are required.

Step 1. Find an initial basic feasible solution $\mathbf{u} = \mathbf{u}^{(0)}$, $\mathbf{v} = \mathbf{v}^{(0)}$, and $Z = Z^{(0)}$, by standard procedures. The constraints are satisfied by

$$\mathbf{u}^{(0)} = \mathbf{A}^{-1}1 \tag{101}$$

Step 2. Compute the decision derivatives $\delta Z/\delta v_n$ for $n = 1, \ldots, S$, at $\mathbf{u}^{(0)}, \mathbf{v}^{(0)}$. If a particular $\delta Z/\delta v_n \leqslant 0$ at $v_n^{(0)} = 0$, it is not feasible to choose the corresponding negative perturbation dv_n to generate a positive dZ. If all $\delta Z/\delta v_n \leqslant 0$, the global maximum has already been reached. However, if at least one $\delta Z/\delta v_n > 0$, a better solution exists. Corresponding to each positive $\delta Z/\delta v_n$ at $v_n^{(0)}$ compute

$$v_n^{(n)} = \underset{\alpha_{mn} < 0;\, m = 1, \ldots, S}{\text{minimum}} \left\{ \frac{-u_m^{(0)}}{\alpha_{mn}} \right\} \tag{102}$$

If a decision variable v_n were increased to this level $v_n^{(n)}$, the state variable u_k corresponding to the minimum ratio in (102) would be driven to zero ($u_k^{(n)} = 0$), while all the other state variables would remain non-negative ($u_m^{(n)} \geqslant 0$, all $m \neq k$). This is the standard simplex operation for changing a basis. (If all α_{mn} corresponding to a decision variable were positive, the decision variable could be increased without limit. Such is not possible in the context of the application described in this paper.)

Step 3. Corresponding to each $v_n^{(n)}$ obtained by equation (102) compute the change that would occur in the objective function:

$$Z^{(n)} - Z^{(0)} = \sum_{m=1}^{R} c'_m \ln (u_m^{(n)} - u_m^{(0)}) + c''_n \ln (v_n^{(n)}/v_n^{(0)}) \tag{103}$$

If all $Z^{(n)} - Z^{(0)} \leqslant 0$, then the maximum of the basic feasible solutions has already been reached. But if at least one $Z^{(n)} - Z^{(0)} > 0$, then at least one better basic feasible solution exists. In the latter case choose that decision value $v_n^{(n)}$ corresponding to the greatest immediate improvement (maximum $Z^{(n)} - Z^{(0)}$).

Step 4. Treating a new basic feasible solution as an initial solution (with the old decision variable becoming a state variable) repeat the steps until the best of the basic feasible solutions has been reached and no further improvement can be achieved by the simplex operation. Call the maximum basic feasible solution $\mathbf{u}^*, \mathbf{v}^* = \mathbf{O}$, and Z^*.

Step 5. Compute new decision derivatives $\delta Z/\delta v_n$ for $n = 1, \ldots, S$, at $\mathbf{u}^*, \mathbf{v}^*$. If all $\delta Z/\delta v_n \leqslant 0$, the global maximum coincides with the maximum basic feasible solution. If at least one $\delta Z/\delta v_n > 0$ then a better solution can be found by making a positive perturbation $dv_n > 0$, at $\mathbf{v}^* = \mathbf{O}$. Choose a finite step size Δv_n and write $v_n^{(n)} = v_n^* + \Delta v_n$. Then, from (98), compute the new state variables $u_m^{(n)}$, $m = 1, \ldots, R$, testing to make sure that all $u_m^{(n)} > 0$. Repeat the computation of $\delta Z/\delta v_n$ for the same decision variable at $\mathbf{v}^{(n)}$. If $v_n > 0$ and $\delta Z/\delta_n < 0$, further improvement can be realized by choosing a negative Δv_n at the next iteration. If $v_n \geqslant 0$ and $\delta Z/\delta v_n > 0$, further improvement can be realized by choosing a positive Δv_n at the next iteration. When $\delta Z/\delta v_n = 0$, $v_n > 0$, stop changing v_n and hold it fixed. (The solution is no longer a basic feasible solution, since $R + 1$ of the variables may be positive.)

Step 6. Holding the first decision variable v_n fixed at the value $v_n^{(n)}$, found in the preceding step, repeat the procedure for another decision variable. Repeat the procedure, one decision variable at a time, until the global maximum point $\tilde{\mathbf{u}}^{**}, \tilde{\mathbf{v}}^{**}$ is reached, at which point $\delta Z/\delta v_n = 0$ for all $v_n^{**} > 0$, and $\delta Z/\delta v_n \leqslant 0$ for all $v_n^{**} = 0$. (Because the iterations are finite, it may be necessary to terminate before the exact global maximum is reached.)

9. NUMERICAL EXAMPLE

Consider a system which has a main branch with defined points $i = 0, 1, \ldots, 5$, and a tributary entering at point $i = 2$ with defined points $j(i) = 0(2), 1(2), 2(2)$. Suppose that the following data are available.

Initial Point Data

Point	Flow million litres/day	B.O.D. milligrams/litre	D.O. milligrams/litre	Saturation D.O. milligrams/litre
0	600	1.66	9.55	10.00
0(2)	450	0.25	9.00	9.50

Waste Treatment Plant Data

Point	Waste Flow million litres/day	Raw B.O.D. Load milligrams/litre	Effluent B.O.D. milligrams/litre	Effluent D.O. milligrams/litre
1	40	150	x_1	1.0
3	150	150	x_2	1.0
4	75	150	x_3	1.0
1(2)	50	150	$x_{1(2)}$	1.0

Stream Data

Reach	Deoxygenation Rate Constant K	Reaeration Rate Constant L	Travel Time (days)
0	.35	1.00	0.24
1	.35	.65	1.05
2	.35	.65	0.05
3	.35	.65	1.10
4	.35	.65	0.08
0(1)	.35	.65	0.10
0(2)	.35	.65	0.30

Suppose that the D.O. saturation concentration is 9.5 milligrams per litre at points $i = 1, 2, \ldots, 5$ and points $j(i) = 1(2), 2(2)$. Suppose that the specified maximum allowable B.O.D. concentration immediately downstream of the points $i = 1, 2, \ldots, 5$ and $j(i) = 1(2), 2(2)$ is 4.0 milligrams per litre, and that the minimum allowable D.O. concentration on the oxygen sag curve is 6.16 milligrams per litre. It is easy to verify from equation (11) that the minimum allowable D.O. concentration downstream of each point $i = 1, 2, \ldots, 5$ and $j(i) = 1(2), 2(2)$ is 6.16 milligrams per litre.

The total annual costs of the waste treatment plants are given by the following equations (based on data presented by Frankel (1966), using 20-year amortization of capital costs, at 1963 price levels):

Plant Size Million litres/day	Total Annual Cost Dollars
40	480,000- 62,000 ln (x/z)
50	600,000- 75,000 ln (x/z)
75	855,000-107,000 ln (x/z)
150	1,460,000-190,000 ln (x/z)

Here z is the raw waste B.O.D. concentration and x is the effluent B.O.D. concentration. We suppose that the existing plants have 30 percent efficiencies.

From these data, one of the authors computed the necessary matrix elements, matrix inversions, and matrix multiplications for the constraints (65) and (66) in about three hours on a desk calculator. After rejecting five redundant constraints, the following problem formulation was obtained:

maximize $Z = 62 \ln x_1 + 75 \ln x_{1(2)} + 190 \ln x_3 + 107 \ln x_4$

Subject to the constraints

$$.024319\, x_1 \leqslant 1 \tag{C1}$$

$$.007218\, x_1 + .011733\, x_{1(2)} \leqslant 1 \tag{C2}$$

$$.006128\, x_1 + .009961\, x_{1(2)} + .033692\, x_3 \leqslant 1 \tag{C3}$$

$$.003725\, x_1 + .006055\, x_{1(2)} + .020480\, x_3 + .015102\, x_4 \leqslant 1 \tag{C4}$$

$$.000077\, x_1 - .006576\, x_{1(2)} \leqslant 1 \tag{C5}$$

$$.002147\, x_1 - .019395\, x_{1(2)} + .001313\, x_3 \leq 1 \qquad (C6)$$

$$.000496\, x_1 - .004243\, x_{1(2)} + .006267\, x_3 + .000343\, x_4 \leq 1 \qquad (C7)$$

$$.000583\, x_1 - .003223\, x_{1(2)} + .000613\, x_3 + .000778\, x_4 \leq 1 \qquad (C8)$$

$$.038650\, x_{1(2)} \leq 1 \qquad (C9)$$

Here the terms in the objective function are negative relevant costs in thousands of dollars.

Using the algorithm described earlier, one of the authors computed the global maximum solution in about three hours on a desk calculator. The solution is given as follows:

Point	Maximum Basic Feasible Solution B.O.D.	Efficiency	Global Maximum Solution B.O.D.	Efficiency
1	41.1	73%	30.9	79%
1(2)	25.9	83%	23.3	84%
3	14.6	90%	17.2	89%
4	25.8	83%	25.8	83%
Z	1331		1337	

Note that constraints C1, C3, C4 and C9 are 'tight' at the maximum basic feasible solution. Constraints C3 and C4 remain 'tight' at the global maximum solution. The global maximum occurs on the extreme edge defined by the intersection of the two bounding hyperplanes corresponding to constraints C3 and C4. For comparison, it is worth noting that if all the plants are designed for 90 percent efficiency (effluent B.O.D. = 14.6), the objective function takes the value z = 1136.

The optimal affluent B.O.D. concentrations (x's) can be substituted into the total annual cost equations to obtain the minimum cost.

10. ACCURACY, PRECISION, SENSITIVITY AND RELEVANCE

The authors have applied the above analysis to one river basin in Ontario* and are attempting to assess both the accuracy of pollution measurements and local costs and the statistical precision of available data for streams in Ontario. Once the accuracy and precision of data are estimated, the sensitivity of solutions to variations in the data will be tested. The authors are also examining the imputed marginal values of tightening or relaxing the water quality constraints (the Lagrange multiplier values $\lambda_r = \delta Z/\delta f_r$). Finally, the authors are soliciting opinions regarding the relevance of the model, and welcome any discussion of this paper.

ACKNOWLEDGEMENTS

The research leading to this paper was carried out under grants to Professor Clough from the National Research Council of Canada. The work was carried out in the Department of Industrial Engineering, University of Toronto, prior to the appointment of Clough as Professor of Management and Systems Engineering, University of

* The results are embodied in a confidential engineering report to the Government of Ontario concerning benefits of proposed engineering works.

Waterloo, and the appointment of Bayer as Assistant Professor in the School of Management, University of Alberta (Calgary). The authors are grateful to the N.R.C. and to the University of Toronto.

REFERENCES

CAMP, T. R., Water and Its Impurities, Reinhold, New York, 1963.

CLOUGH, DONALD J., Concepts in Management Science, Prentice-Hall, Inc., Englewood Cliffs, New Jersey, 1963.

DEAN, JOEL, Managerial Economics, Prentice-Hall, Inc., Englewood Cliffs, New Jersey, 1952.

DEININGER, R. W., 'Water Quality Management: The Planning of Economically Optimal Pollution Control Systems,' Ph.D. Thesis, Northwestern University, 1965.

DOBBINS, W. E. 'BOD and Oxygen Relationships in Streams', Proc. ASCE, **90** (SA3), June, 1964.

Federal Inter-Agency River Basin Committee on Water Resources, Proposed Practices for Economic Analysis of River Basin Projects, Washington, D.C., 1950.

FRANKEL, RICHARD JOEL, Water Quality Management: An Engineering-Economic Model for Domestic Waste Disposal, (University of California, Berkeley, Ph.D., 1965), University Microfilms, Inc., Xerox Corporation, Ann Arbor, Michigan, 1966. (Contains extensive references.)

LI, W. H., 'Unsteady Dissolved Oxygen Sag in a Stream,' Proc. ASCE, 88 (SA3), May, 1962

LOUCKS, DANIEL P., CHARLES S. REVELLE and WALTER R. LYNN 'Linear Programming Models for Water Pollution Control', Management Science, Vol. 14, No. 4, Dec. 1967.

O'CONNER, D. J., 'Oxygen Balance of An Estuary,' Trans. ASCE, **125**, 1960.

PHELPS, E. B., Stream Sanitation, John Wiley and Son, New York, 1944.

REVELLE, CHARLES S., DANIEL P. LOUCKS, WALTER R. LYNN, 'Linear Programming Applied to Water Quality Management,' Water Resources Research, Vol. 4, No. 1, Feb., 1968.

STREETER, H. W., and E. B. PHELPS, 'A Study of the Pollution and Natural Purification of the Ohio River,' United States Public Health Services, Bulletin, **146**, **1925**.

THOMAS, H. A., JR., 'Pollution Load Capacity of Streams,' Water and Sewage Wks, **95**, 1948.

WILDE, DOUGLASS J., and CHARLES S. BEIGHTLER, Foundations of Optimization, Prentice-Hall, Inc., New Jersey, 1967.

DISCUSSION

J. M. Lodal

My comment deals not with the technical aspects of Prof. Clough's programming model, but with the interpretation he makes of his results. He says that the cost difference between a control system with and without a particular reservoir is a measure of the net economic benefit one can assign to the reservoir project. Prof. Clough has simply found the least-cost way of achieving a specified objective and has not addressed the question of whether or not the objective was proper. So, for example, Prof. Clough's model would attribute 'benefits' to a reservoir project which improved the quality of water in a stream more cheaply than a system of treatment plants which gave the same improvement in quality, regardless of whether or not higher quality water was of any use.

Prof. D. G. Clough

Mr. Lodal argues that we have 'simply found the least-cost way of achieving a specified objective and have not addressed the question of whether or not the objective was proper'. As a matter of fact, we have found the least-cost way of meeting certain constraints imposed by various agencies according to their own unspecified objective functions which involve public health risks and other factors. In the real world there seems to be no hope of achieving a multi-agency consensus about measures of effectiveness and the formulation of some joint global objective function. In the real world we are faced with the problems of constitutionally delegated powers which are split among various levels of government and legal powers which are split among various departments of government and professional associations (e.g. Professional Engineering design standards, Medical Association health standard, Water Resources Commission pollution standards). Mr. Lodal's remark about whether or not higher water quality is of any use is irrelevant in the context of immovable constraints.

An Application to the Social Sciences of Mathemathical Programming

FRANCES E. HOBSON
University of Manitoba, Canada

SUMMARY

This transportation model is intended for use in allocating children to foster family care. The objective is to maximize benefit to the community as measured by expected total taxes paid over a lifetime by a child who has received foster care service for six years or more. For the U.S. 1960 this represents some 50, 000 (Canada 5, 000) as distinct from the 109, 000 who stay less than six years in foster homes and the 82, 000 placed in child welfare institutions for neglected, dependent, and emotionally disturbed children[8] (U.S. Children's Bureau (1966)).

It is assumed that a child from a particular occupation (or socio-economic) class family placed in a foster home of the same class is subject to the same social mobility probabilities as if he had remained with his family, where social mobility probabilities are defined as the transition probabilities of a Markoff process of order one: from state i (father's class) to state j (son's class).

Solutions are presented for different weights assigned for genetic and environmental effects applied to upwards and downwards change from source family to destination (i.e. foster) family. Linear constraints on this transportation problem are the number of children. The solution is presented for shortages of different destination classes. The use of the shadow prices (imputed costs) leads to the determination of an optimum living allowance for a child.

1. DATA REQUIRED

(1) P_{ij} = the social mobility probability of moving from state i (father's occupation or socio-economic class) to state j (son's class)
= transition probability of a markoff process of order one

(2) T_j = the estimated mean of the distribution of total taxes (income, succession duty, property, sales, excise) paid by a class j member

(3) a_i = the estimated number of class i children available to be placed in foster family care for at least six years (which is considered to be long enough to be 'in danger of growing up in foster care'[7] (U.S. Children's Bureau (1963)). For the U.S. 1960 this represents some 50, 000 (Canada 5, 000) as distinct from the 109, 000 who stay less than six years in foster homes and the 82, 000 placed in child welfare institutions for neglected, dependent, and emotionally disturbed children; by 1975 these estimates will be 302, 000 in foster care and 62, 000 in institutions 'assuming that the rate will continue to change in the direction and at the pace at which it was changing during the 1962-65 period'[8] (U.S. Children's Bureau (1966)).

(4) b_j = the estimated number of class j foster homes available

(5) w_k = the estimated weighting factor for genetic effect, where $(1 - w_k)$ estimates the environmental effect, for $k = |i - j|$, i.e. the size of the upwards or downwards change from source family to destination (foster) family.

2. THE TRANSPORTATION PROBLEM

To maximize

$$Z = \sum_{i=1}^{m} \sum_{j=1}^{n} c_{ij}x_{ij}$$

subject to:

$$\sum_{j=1}^{n} x_{ij} = a_i;\ a_i \geqslant 0;\ i = 1, 2, \ldots, m$$

$$\sum_{i=1}^{m} x_{ij} \leqslant b_j;\ b_j \geqslant 0;\ j = 1, 2, \ldots, n$$

$$\sum_{j=1}^{n} b_j \geqslant \sum_{i=1}^{m} a_i$$

$$x_{ij} \geqslant 0;\ i = 1, 2, \ldots, m;\ j = 1, 2, \ldots, n$$

$$0 \leqslant w_k \leqslant 1$$

where

$$c_{ij} = w_k c_{ii} + (1 - w_k)c_{jj}$$

$$c_{ii} = \sum_{j=1}^{n} p_{ij}T_j$$

i.e. it is assumed that a child from a particular class placed in a foster home of the same class is subject to the same social mobility probabilities as if he had remained with his family.

3. DATA USED

(1) p_{ij} = social mobility probabilities

			son's class 1	2	3	4	sum
	professional, business executive, self-employed	1	.38	.20	.41	.01	1.00
father's	clerical and sales	2	.23	.28	.48	.01	1.00
class	manual	3	.14	.12	.73	.01	1.00
	farm labour	4	.32	.12	.52	.04	1.00

These values are from[3] Lipset and Bendix (1966) who used as sources Rogoff (1953) for 9,890 respondents in Indianapolis, U.S.A., 1940 and Geiger (1951) for 26,607 respondents in Aarhus, Denmark, 1949 showing that they were in substantial agreement.

As at this date I do not agree with the current literature[4] (Mosteller (1968)) with his reference to Levine (1967) Ph.D. thesis, Harvard, that a social mobility matrix could be considered to be doubly stochastic—this implies that mobility does not depend on the relative sizes of the classes; however, by taking the work of Prais[6] (1955) and its source data of counts of people by seven classes of Glass[1] (1954) for 3,500 male respondents 18 years of age and over in England and Wales 1949 and converting it by an iterative procedure to a doubly stochastic matrix 5 × 5, he shows it to be in substantial agreement with a study in Denmark made by Svalastoga; I assume that the two Denmark studies referred to in this section are in agreement, hence that the data I am using is appropriate for my model (which simply considers moving from one generation to the next.)

To complete my review of the pertinent literature on this subject: Parzen[5] (1962) shows for the Prais[6] (1955) paper considering the data as a 3 × 3 row stochastic matrix, that the markoff chain is finite, irreducible, and aperiodic and, after many generations would yield a society with 6.7% in the upper class, 62.4% in the middle class, and 30.9% in the lower class.

(2) T_j = taxes: T_1 = \$3986.0; T_2 = \$1616.2; T_3 = \$745.7; T_4 = \$172.7 annual total taxes Canada 1957[2] (Goffman 1962) paid by a taxpaying unit which is approximately the same as a family unit—Goffman refers to Goldberg and Podoluk for this equivalence.

(3) a_i = supply of children: a_1 = 2,000; a_2 = 5,000; a_3 = 40,000; a_4 = 3,000. The computer runs were made on this U.S. estimated total of 50,000 rather than on the 5,000 Canadian estimate—only a scaling factor of 1/10 is needed to reduce these substantially impressive estimates which are summarized below: 1961 data show that for U.S. children in foster family care, 20% stay 6 years but less than 12 years and 9% stay 12 years and over[7] (U.S. Children's Bureau (1963)) and that the incidence rate per 1000 children under 18 years of age was 2.5[8] (U.S. Children's Bureau (1966)). Applying these to the 1961 Canadian population of 7,095,536 under 18 yields about 5,000. The distribution by social class was made on the basis of the best thinking available, in the absence of data.

(4) b_j = supply of foster homes. Let system I have b_1 = the number of Canadian taxpaying units paying taxes on income \$7,000 and over; b_2 on \$5,000-\$6,999; b_3 on \$1,000-\$4,999; b_4 on less than \$1000 (Goffman[2] (1962)). Let systems II, III, and IV be realistic supplies, to examine allocation.

System	b_1	b_2	b_3	b_4
I:	611,500	795,100	2,951,800	525,000
II:	2,001	5,001	40,001	3,001
III:	500	1,000	50,000	500
IV:	10	500	50,000	500

(5) w_k = weighting system for genetic effect:

System A: $w_0 = 1.0$; $w_1 = 0.1$; $w_2 = 0.3$; $w_3 = 0.5$: high weight for genetic effect

B: $w_0 = 1.0$; $w_1 = 0.1$; $w_2 = 0.2$; $w_3 = 0.3$: medium weight for genetic effect

C: $w_0 = 1.0$; $w_1 = 0.0$; $w_2 = 0.0$; $w_3 = 0.0$: no weight for genetic effect

4. SOLUTION TO THE TRANSPORTATION PROBLEM

The conclusions are for Z = annual taxes paid by 50,000 individuals. (The marriage rate between foster children is considered to be negligible so that the '50,000 individual tax payers' can be considered to be the same of '50,000 taxpaying units').

1 (a) For supply of foster homes system I all allocations are to class 1, with Z increasing as the weights change from system A to B to C, i.e. as the genetic weights decrease thus letting the environment of the class 1 foster home have more influence.

(b) For other supply systems with short supplies of class 1 foster homes, Z is smaller and decreases as the genetic weights decrease thus letting environment of the non-class-1 foster homes have more influence. The table presents Z.

		Genetic Weighting System		
		A: high	B: medium	C: none
supply of foster homes system	I	96, 471, 381	100, 029, 378	107, 269, 200
	II	70, 872, 630	70, 622, 076	70, 452, 797
	III	66, 848, 127	66, 651, 634	66, 041, 197
	IV	66, 267, 038	66, 084, 061	65, 410, 560

2 Outputs from our IBM 360-65 are quoted on pages 126-131

The shadow prices (imputed costs) show the cost per year of allocating a child to other than the optimal allocation.

5. CONCLUSION

The current living allowance for a child placed in a foster home is of the general order of $60 a month, varying slightly with the age of the child, but totalling to well under $1000 per child per annum. My proposal is that the optimal living allowance for a child placed in a foster home be computed from the shadow prices (imputed costs), multiplied by 50 years of tax paying, divided by 10 years of foster care.

Because of the limitations of the data currently available to me, I am forced to assume that a grown-up child instantly becomes a tax-paying member of a certain socio-economic class, with a certain probability, and pays the average taxes (as estimated in 1957) of that class for fifty years. Mr. M. G. Shaw has drawn my attention to the expected return on capital, say 8%, and he is, of course, justified. However, economic forecasting of income by socio-economic class and the total taxes paid thereon, interest rates, etc. would properly form an extension of this paper.

To continue with the MP, the maximum shadow price is where it is expected to be: in Table IC, page 10, as $847.31 as the cost of assigning a child to a class 3 foster home, assuming no genetic effect, i.e. that the environment of the foster home (for six years or more) is all-important. This $\$847.31 \times 50 \div 10 \doteqdot \$4,000$ as the first approximation to the maximum annual cost for allocating a child to other than the optimal allocation of class 1 (when a class 1 foster home is available).

Any additional living amounts up to this figure of $4, 000 per child per annum are economically feasible as an 'optimum living allowance for a child placed in a foster home for six years or more' under the assumptions mentioned and subject to the limitations of the data, e.g. the 36, 487 respondents as 'son's class' to the social mobility questionnaire had not all completed their careers and their mobility.

The implementation of this I leave to the imagination of the authorities concerned. It could be argued that institutions simulating class 1 homes are the solution but, not being a social scientist, I do not have the temerity to suggest it.

Note: I have not specifically mentioned:

(1) negative taxes of some $2500 per annum for individuals sentenced to jail, simply taking the $0 contribution as reported by taxes

I A $Z = 96,471,381.49999$

	1	2	3	4	
1	2,000	0	0	0	2,000
2	5,000	0	0	0	5,000
3	40,000	0	0	0	40,000
4	3,000	0	0	0	3,000
	611,500	795,000	2,951,800	525,000	

II A $Z = 70,872,629.8546$

	1	2	3	4
1	0	0	2,000	0
2	2,001	2,001	998	0
3	0	0	36,999	3,001
4	0	3,000	0	0
	2,001	5,001	40,001	3,001

III A $Z = 66,848,126.7999$

	1	2	3	4
1	0	0	2,000	0
2	500	0	4,500	0
3	0	0	40,000	0
4	0	1,000	1,500	500
	500	1,000	50,000	500

IV A $Z = 66,267,038.3579$

	1	2	3	4
1	0	0	2,000	0
2	10	0	4,990	0
3	0	0	40,000	0
4	0	500	2,000	500
	10	500	50,000	500

shadow prices

0.	374.76450	593.11840	140.62400
0.	374.76450	762.58080	280.15460
0.	205.30210	593.11840	83.66080
0.	235.23390	650.08160	140.62400

169.46240	169.46240	0.	56.96320
0.	0.	0.	27.03140
169.46240	0.	0.	0.
139.53060	0.	27.03140	27.03140

169.46240	196.49380	0.	56.96320
0.	27.03140	0.	27.03140
169.46240	27.03140	0.	0.
112.49920	0.	0.	0.

169.46240	196.49380	0.	56.96320
0.	27.03140	0.	27.03140
169.46240	27.03140	0.	0.
112.49920	0.	0.	0.

I B $Z = 100,029,378.3000$

	1	2	3	4
1	2,000	0	0	0
2	5,000	0	0	0
3	40,000	0	0	0
4	3,000	0	0	0
	611,500	795,100	2,951,800	525,000

II B $Z = 70,622,075.9541$

	1	2	3	4
1	0	0	2,000	0
2	2,001	2,999	0	0
3	0	2,002	37,997	1
4	0	0	0	3,000

III B $Z = 66,651,633.6999$

	1	2	3	4
1	0	0	2,000	0
2	500	1,000	3,500	0
3	0	0	40,000	0
4	0	0	2,500	500

IV B $Z = 66,084,060.7079$

	1	2	3	4
1	0	0	2,000	0
2	10	500	4,490	0
3	0	0	40,000	0
4	0	0	2,500	500

shadow prices

0	374.76450	677.84960	593.11840
0	374.76450	762.58080	266.63890
0	290.03330	677.84960	168.39200
0	477.89710	706.33120	196.87320

84.73070	84.73120	0	424.72640
0	0	.00050	13.51620
84.73070	0	0	0
56.24950	159.38260	.00040	0

84.73120	84.73170	0	424.72680
0	0	0	13.51610
84.73120	.00050	0	.00040
56.24960	159.38270	0	0

84.73120	84.73170	0	424.72680
0	0	0	13.51610
84.73120	.00050	0	.00040
56.24960	159.38270	0	0

I C $Z = 107,269,200$

	1	2	3	4
1	2,000	0	0	0
2	5,000	0	0	0
3	40,000	0	0	0
4	3,000	0	0	0
	611,500	795,100	2,951,800	525,000

→

II C $Z = 70,452,797,0829$

	1	2	3	4
1	0	0	2,000	0
2	0	5,000	0	0
3	2,001	1	37,997	1
4	0	0	0	3,000

→

III C $Z = 66,041,196.7998$

	1	2	3	4
1	0	0	2,000	0
2	500	1,000	3,500	0
3	0	0	40,000	0
4	0	0	2,500	500

→

IV C $Z = 65,410,560.4198$

	1	2	3	4
1	0	0	2,000	0
2	10	0	4,990	0
3	0	0	40,000	0
4	0	500	2,000	500

→

shadow prices

0	416.40500	847.31110	281.24800
0	416.40500	847.31200	281.24800
0	416.40500	847.31200	281.24800
0	416.40500	847.31200	281.24800

.00090	.00090	0	.00090
0	0	0	0
0	0	0	0
0	0	0	0

.00090	.00090	0	.00090
0	0	0	0
0	0	0	0
0	0	0	0

.00090	.00090	0	.00090
0	0	0	0
0	0	0	0
0	0	0	0

(2) additional allowances for foster children proceding to university

(3) using distributions of taxes by class, rather than averages

(4) using other social mobility probabilities

(5) the data used is for the population as a whole—children subjected to foster care for at least six years might be motivated differently as regards social mobility.

REFERENCES

1. Glass, D. V. et al (1954), Social Mobility in Britain, Routledge and Kegan Paul Ltd., London, p. 183.

2. Goffman, Irving Jay (1962), The Burden of Canadian Taxation, Canadian Tax Foundation, Toronto, Canada, p. 10 and p. 13.

3. Lipset, Seymour Martin and Bendix, Reinhard (1960), Social Mobility in Industrial Society, University of California Press, Berkeley and Los Angeles, p. 31.

4. Mosteller, Frederick (1968), Association and Estimation in Contingency Tables, Journal of the American Statistical Association, Volume 321, pp. 1-28.

5. Parzen, Emanuel (1962), Stochastic Processes, Holden-Day, Inc., San Francisco, pp. 257-258.

6. Prais, S. J. (1955), Measuring Social Mobility, Journal of the Royal Statistical Society Series A (General), Volume 118, pp. 56-66.

7. U.S. Department of Health, Education, and Welfare, Welfare Administration, Children's Bureau (1963), Children Problems and Services in Child Welfare Programs, p. 84.

8. U.S. Department of Health, Education, and Welfare, Welfare Administration, Children's Bureau (1966), Foster Care of Children: Major National Trends and Prospects, p. 8.

ACKNOWLEDGEMENT

To graduate student J. M. Wren for his suggestion of the measure of effectiveness of taxes (rather than total income earned over a lifetime).

SECTION 4 Matrix Generators and Output Analysers

A. J. Clark
L. Pessina
S. Spurkland
M. J. Dillon, P. M. Jenkins, Mary J. O'Brien

A System for the Specification and Generation of Matrices for Multi-period Production Scheduling Models

A. J. CLARK
Tate and Lyle Ltd., U.K.

There are three fairly clearly defined stages in the preparation of matrices for linear models, namely formulation, data collection, and generation. As shewn in Diagram 1 the output of each of these stages forms an important part of the input to the next. Thus the logical structure of the system being modelled is exposed during formulation. Augmented with numerical information obtained by data collection, it forms the basis of the model specification. Subsequently the generator processes this specification to produce the matrix.

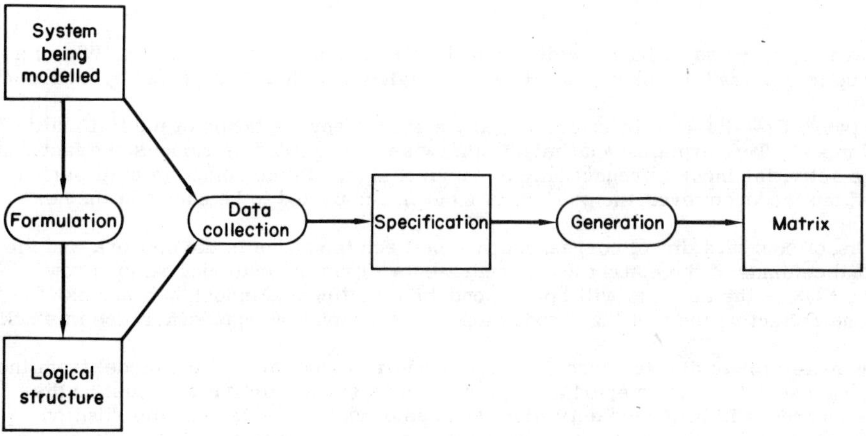

Diagram 1. The three stages in the preparation of the matrix.

1. FORMULATION

In the first stage—formulation—the system being modelled is stripped of its irrelevant detail, and reduced to a logical structure defining the activities, constraints, and their interactions. This logical structure, or model, can then be put in the unambiguous and visual form of a network flow diagram, with the activities and constraints shewn as nodes, and their interactions as connecting links.

A simple example of such a network is shewn in Diagram 2. Here the process being modelled is one in which two machines use two materials to produce two products, each of which may be stored. Overtime working is possible on the second machine.

If the network diagram of this system is examined it will be seen that the constraints on resources, such as the machine capacities, storage capacities, and material balance

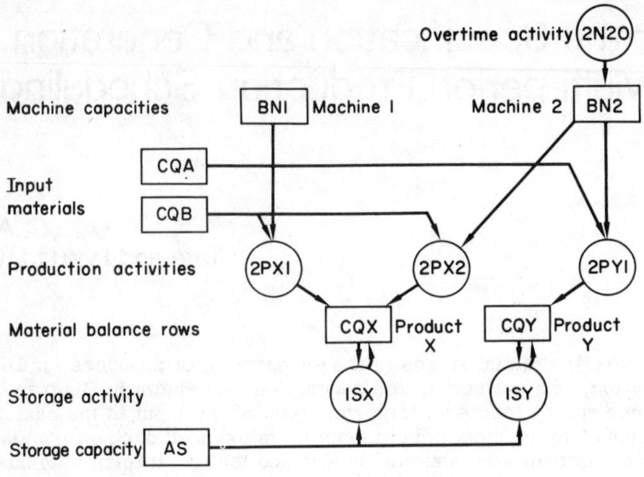

Diagram 2. Network flow-diagram of part of a production system.

rows, are represented by boxes, while the activities are enclosed in circles. Where a resource is required for, or produced by, an activity, then this is indicated by an arrow.

The network flow diagram is, of course, only a static representation of the full multi-period model. This explains what might otherwise be regarded as curious, the fact that storage activities input a product only to output it again. In the complete multiperiod form of the mode, however, the product is input in one period to be output in the next.

There is, of course, a direct correspondence between the nodes in the network, and the rows and columns of the matrix that it represents. Thus the material balance row labelled CQX in the diagram will correspond, in a multiperiod model, to a series of equations reflecting the fact that production less consumption equals increase in stock.

For the generator to derive correctly the full multi-period form of the model from the static representation, interperiod activities (such as storage activities) must be distinguished from within-period activities (such as production, sales and distribution activities). To make this distinction the first character of each activity's code name is used; the character 1 indicating an interperiod activity, and 2 a within-period activity.

In the same way three types of constraint must be distinguished. Constraints on interface resources (such as storage capacities) have code names beginning with A, within-period resources (production capacities) with B, and material balance rows with C.

It will be noticed that no numerical information is included in Diagram 2. However, though approximate numerical values may, and perhaps should, be inserted in these diagrams, if only as an aide-memoire, the real data, possibly retrieved from files prepared by other systems, will be inserted in the specification at the data collection stage.

Code names giving all the relevant attributes of each node must, however, be worked out during formulation and inserted, where possible, in the flow diagrams. Although it is not strictly necessary it is suggested that the specification of each type of attribute should use no more than one space within the code name, and that the names be so constructed that when sorted into alphanumeric sequence they fall into meaningful and useful groupings.

Bearing in mind the correspondence between the links of the network and the elements of the matrix, row nodes should, as a rule, be linked only to column nodes, and vice-versa. Dummy rows representing linear combinations of other rows may however be used, and these give rise to row-row links. References to dummy rows are then treated by the generator as indirect references to the linear combination of rows they represent. This device, and a similar device whereby activities are made indirectly to refer to rows, bounds, and cost of another activity, may be used to avoid repetitive specifications. An example, given later, shews the effects of the substitutions carried out by the generator in these cases.

2. DATA COLLECTION AND SPECIFICATION

The model having been formulated, and its logical structure exposed, we may begin to design data collection systems, the output of which is then inserted in the structure to form a complete specification of the model suitable for input to the generator.

In very general terms the following data items must be specified: for each node the upper bound, the lower bound, and the cost per unit; and for each link the resource required or produced per unit activity.

The use of bounds for row nodes may appear a little unusual; however as the sign convention whereby inputs to activities are positive (and outputs negative) is used throughout, upper bounds on rows correspond to the maximum amount of resource available, while lower bounds correspond to the minimum that must be used. This is simpler to specify and perhaps less confusing than the more usual right-hand-side with a greater than, less than, or equality condition, especially if the L.P. algorithm used allows both activities and constraints to be bounded above and below. The algebraic interpretation of a bounded row is shewn in Diagram 3.

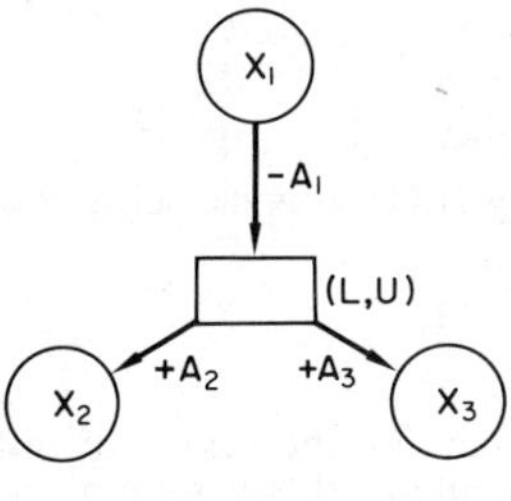

Diagram 3. Network representation of the bounded row

$$L \leqslant -A_1X_1 + A_2X_2 + A_3X_3 \leqslant U$$

In a multiperiod model some data will vary with time (for if all items were constant with time then a single period model would be used instead). These items must be specified as strings of time-value pairs shewing their variations with respect to an absolute scale of time. An example of such a time-value string, in a form acceptable to the generator, might be:

13. 2 (6840) 12. 0 (6841) 13. 2

This string could be used to specify a data item which falls from its normal value of 13. 2 to 12. 0 during week 6841.

The actual unit of time used in the specification will depend largely on the time span of the model and the degree of precision required. In the example given above the unit was one week, but it could equally well have been a year or an hour. Successive time units must be sequentially numbered; however they need not be consecutive—week 6901 might follow 6852 for example.

The units in which data items are measured should be consistent within themselves, and where appropriate, with the basic time unit chosen. Thus if a machine capacity is measured in hours, and a production activity in tons, then any connecting element must be in hours per ton, and the capacity of the machine quoted in hours per basic time unit.

To illustrate the way in which data items are specified for input to the generator consider once again the network in Diagram 2. Suppose that the production activity labelled 2PX1 requires 0.02 hours of machine capacity BN1 to produce one ton of product CQX. This would be specified as follows:

BN1 / 2PX1 = 0.02

CQX / 2PX1 = –1

Further, if the external costs associated with this activity were £1.50 per unit, then they would be specified as:

*C / 2PX1 = 1.50

where the node name '*C' is used to designate the cost row.

A number of special node names, all prefixed by an asterisk, are used to specify the cost row and bound vectors. They are:

Cost row	*C
Upper Bound	*U
Lower Bound	*L
Equality Bound	*E

Thus an upper limit of 164 hours/week of machine capacity BN1 (a week being the basic time unit) would be specified as:

BN1 / *U = 164

In a multiperiod model it is usually necessary to place bounds on the levels of interperiod activities at the opening and closing interfaces. Bounds of this type may be specified by attaching the letter O for opening or the letter C for closing to the appropriate bound vector name. Thus if in our example the opening value of the storage activity 1SX were 1200 tons, then this would be specified as *EO / 1SX = 1200.

If a group of elements all have the same value they may be specified together. Thus if activities 2PX1 and 2PX2 both have the same external cost then this could be specified as:

*C / (6 2PX1, 2PX2) = 1.50

or more compactly as

*C / 2PX (6 1, 2) = 1.50

The digit 6 which follows the opening bracket in the above statements defines an index to be used by the generator to control the string of codes within the brackets. The shorthand notation, of which the above statements are elementary examples, is used to specify blocks of similar names, which may be made to refer, through the system of indexing, to particular members of a corresponding block of data items. This facility, which is useful when there are structural repetitions in the model or when the same data item applies to a group of elements, will be explained more fully when the workings of the generator are discussed and more complex examples considered.

By using the algebraic operators + — * / data items may be combined together into expressions. This facility may be used to shew how an element value has been computed, or alternatively to apply a previously stored common variation or discount to a number of separate bounds or costs. Thus if the statement

S6 = 1.0 (6829) 0.5 (6830) 1.0 (6839) 0.5 (6840) 1.0

were used to store in S6 a vector of scaling factors reflecting statutory holidays and other seasonal factors affecting the hours worked per week, then these factors could subsequently be applied separately to a number of different items, as in the statements below:

BN1/ *U = 116(6833)164 * S6

BN2/ *U = 38 * S6

In the first of these statements the scaling factors stored in S6 are applied to a data item which itself varies with time; the time-value series that results is:

116 (6829) 58 (6830) 116 (6833) 164 (6839) 82 (6840) 164

Data items prepared by specially written data collection programs may be retrieved from intermediate storage devices for use in the matrix specification. References to such items begin with the letter X followed by the name of the external data item enclosed in brackets. For example the statement

S2 = X(SCALEA)

would retrieve the external data item named SCALEA, and store it internally in S2.

To refer to a group of external data items as a unit, the name of the first item must be specified together with the number of items in the group. Thus for example a statement such as

*E / 2V (1 L, M) (2 A, B, C, D) = X(DLA, 8)

could be used to retrieve eight consecutive data items, starting with the item labelled DLA, from an external file, and then to use them as fixed bounds for the activities 2VLA, 2VLB, . . . , 2VMD.

3. GENERATION

It will be noticed that the specification of the model is independent of the numbers and duration of the time periods for which a particular solution is to be obtained, all time-varying data having been specified relative to some absolute scale. Before attempting to obtain a particular solution, however, the specification must be completed by statements defining the actual model periods in terms of the same absolute scale.

Thus for example to specify a three period model with one time unit in the first period, and two in each the second and third, the following statements might be used:

T0 = 6849;

T1 = 6850;

T2 = 6851, 6852;

T3 = 6901, 6902;

The first statement, T0 = 6849, defines the opening interface, while the remaining statements simply list the basic time units to be included in each model period. The gap in sequence from 6853 to 6899 indicates that there are no basic time units identified by these numbers.

The table below sets out the model periods that would be generated by the set of statements above:

Unit No.	Date	Model Period
0	6849	—
1	6850	1
2	6851	2
3	6852	2
4	6901	3
5	6902	3

In addition to the model periods, the rate of interest to be used when discounting costs to present value must be defined in the preliminary specification. The statement

I = 0.0016

would for example be used to specify a rate of interest of 0.16% per time unit.

A number of statements defining the scaling factor series to be used in the main body of the specification will, as a rule, also be included at this point.

In processing the remainder of the specification the main functions of the generator are as follows:

1. To expand shorthand statements, evaluating expressions involving data items and associating the resulting series of element values with the appropriate row and column names.
2. To generate the multiperiod structure of the model, and to compute element values appropriate to each period.
3. To store matrix elements in core and, when they have all been generated, to output the matrix in the form required for the L.P. algorithm.

3.1 Expansion of shorthand statements

To illustrate the method used to expand shorthand statements and to establish correspondence between code names and data items, consider the following statements and their equivalent expansions.

C(1 L, M) / 2P = +1, −1

  CL / 2P = +1

  CM / 2P = +1

C(1 L, M) / 2P(1 X, Y) = +1, −1

  CL / 2PX = +1

  CM / 2PY = −1

C(1 L, M) / 2P(1 (6 X1, X2), Y) = +1, −1

  CL / 2PX1 = +1

  CL / 2PX2 = +1

  CM / 2PY = −1

C(1 L, M) / 2P(1 (2 X1, X2), Y) = +1, +2, −1

CL / 2PX1 = +1

CL / 2PX2 = +2

CM / 2PY = −1

Strings of codes, separated by commas and enclosed in brackets, are controlled by one of nine indices, the number of the index being the digit at the head of the string. Initially the values of all these indices are set equal to one, so that the first pair of code names generated uses the first items of coding in each string. The highest numbered index in use is then stepped through its range of values until all items of coding in the string have been used and the end of the string is reached. The next lowest index is then stepped by one and the process repeated until all items of coding in all strings have been used.

The first of the element values in the string following the equals sign is associated with each pair of code names generated until an index less than 6 is stepped. Thereafter, whenever an index numbered 5 or less is stepped, fresh element values are accessed. These items therefore must be arranged to correspond with the sequence index 5 within index 4 etc.

As each element value is accesssed, expressions involving data items (retrieved from external storage if necessary) are evaluated, and the result stored as a time series with one value at the opening interface, and one value for each time unit within the time span of the model. The computation of the actual element values in each period of the model is then based on this time series, the individual members of which are known as the unit values.

To illustrate the method used in preparing this vector of unit values consider the time-varying data item:

13(6835)20(6850)15 (6901) 12

With the model periods defined in the same way as before, the following vector of unit values would be generated

Date	Time Unit	Model Period	Unit Value
6849	0	—	20
6850	1	1	20
6851	2	2	15
6852	3	2	15
6901	4	3	15
6902	5	3	12

The four examples given above to illustrate the method used to expand shorthand statements are not intended as realistic examples of the use of this shorthand notation in practice—the expanded specification is in each case simpler than the shorthand form. To take a practical example, therefore, consider a model of a distribution system which supplies six depots (A, B, C, D, E, and F) from three sources (X, Y, and Z) with five products (A1, B1, B2, C1, and C2). Further suppose that source Z does not supply products C1 or C2; that the distribution costs associated with the product A1 differ from those of the other products; and that source X only supplies depots A, B, C, and D, while Y only supplies D, E, and F.

If we label the product rows with the initials CQ, and the transport activities with 2T, then the specification of the 53 distribution activities in this system would be:

```
: Product Flows:

    CQ(6 A1, B1, B2, C1, C2) (1 X, (7 A, B, C, D)) /
    2T(6 A1, B1, B2, C1, C2) X (7 A, B, C, D) = +1, –1;

    CQ(6 A1, B1, B2, C1, C2) (1 Y, (7 D, E, F)) /
    2T(6 A1, B1, B2, C1, C2) Y (7 D, E, F) = +1, –1;

    CQ(6 A1, B1, B2) (1 Z, (7 A, B, C, D, E, F)) /
    2T(6 A1, B1, B2) Z (7 A, B, C, D, E, F) = +1, –1;

: Distribution Costs:

    *C / 2T (1 A1, (6 B1, B2, C1, C2)) X (2 A, B, C, D) =

: From Source X       A     B     C     D:
: Product A1:         –,    –,    –,    –,
: Other Products:     –,    –,    –,    –;

    *C / 2T (1 A1, (6 B1, B2, C1, C2)) Y (2 D, E, F) =

: From Source Y       D     E     F:
: Product A1:         –,    –,    –,
: Other Products:     –,    –,    –;

    *C / 2T (1 A1, (6 B1, B2)) Z (2 A, B, C, D, E, F) =

: From Source Z       A     B     C     D     E     F:
: Product A1:         –,    –,    –,    –,    –,    –,
: Other Products:     –,    –,    –,    –,    –,    –;
```

It should perhaps be explained that the table headings, as they are enclosed in colons, are ignored by the generator. Dashes in the cost tables must, of course, be replaced by actual numerical values.

3.2 Computation of Element Values

To create the multiperiod form of the model the generator has to construct from an essentially static specification a series of elements each of which is appropriate to one model period or another.

To indicate the model period in which an activity takes place, or during which a constraint applies, the generator inserts an extra character at the end of each code name. For the opening (or zeroth) interface the character used is A. During the first period, and at the first interface, it is B. Subsequent periods and their corresponding interfaces are distinguished by the letters C, D, etc. Thus in a three period model a storage activity with the code 1S would exist in the four modified forms:

```
1S.....A 1S.....B 1S.....C 1S.....D
```

where the full stops are inserted to fill out the names to eight characters.

To determine element values appropriate to each model period the generator bases its computations on element value time series evaluated as data items are accessed

The individual members of this series are, it will be remembered, known as unit values.

A number of different methods are used to extract the individual element values from the element value time series. For example in some instances the actual unit at the interface between two model periods will be used, while in others the sum of the unit values over the model period will be taken.

There are in fact five essentially different methods of computation. To select the one that is appropriate in a particular instance, the generator relies on the first characters of the row and column code names which, it will be remembered, are used to distinguish between the various types of activity and constraint.

To illustrate these different methods of computation a number of examples, outlining the procedures used in all the possible cases, are given in the following paragraphs. Each of these examples is drawn from a specially fabricated model which is shewn in network form in Diagram 4.

It should be noted that, in order to avoid unnecessary complexity, the unit values used in all these examples are constant with time; however in each example the method that would be used with time-varying data is indicated. It should also be carefully noted that the model from which all these examples are drawn has three periods, the first of which contains one time unit, while the second and third have two each. Hence the element value time series on which the computations are based will have six unit values, one at the opening interface, and one for each of the five basic time units within the time span of the model.

<u>Interperiod Activities</u>

Interperiod activities are used to transfer stocks of resources subject to material balance constraints from one period to the next. Examples of such activities include

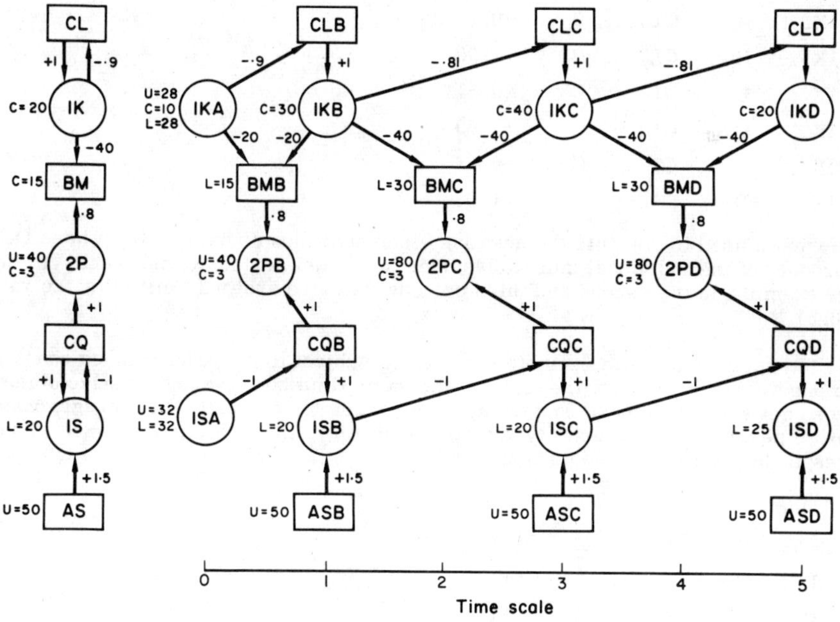

Diagram 4. Static network with corresponding multiperiod net.

the storage of materials, the retaining of labour forces, and the maintenance of stocks of plant.

These activities are identified by the initial character 1. Associated with each such activity there will as a rule be a material balance constraint, with initial letter C, for the resource being stored.

To model the storage of a resource two elements must be specified. For example the statements

CQ / 1S = +1

CQ / 1S = −1

would be used to define a storage activity which required one unit of a product CQ in one period to produce one unit of the same product in the next.

If the resource being stored depreciates with time, as for example a labour force depletes itself with time, then the quantity of resource produced by the storage activity will be less than the quantity input at an earlier time.

To model such a situation the statements

CL / 1K = +1

CL / 1K = −0.9

would be used, the element value −0.9 indicating a rate of depreciation of 10% per basic time unit.

The matrix elements generated in this case would be:

Column	Row	Value
1K.....A	CL.....B	−0.90
1K.....B	CL.....B	+1.00
1K.....B	CL.....C	−0.81
1K.....C	CL.....C	+1.00
1K.....C	CL.....D	−0.81
1K.....D	CL.....D	+1.00

The formula used to compute the negative element values above is $-|D|$, where D is the product of the individual unit values over the model period. Consequently, in the above example, in the second and third periods, as there are two time units, the value is −0.81.

While interperiod activities transfer resources subject to material balance constraints from one period to the next, they may at the same time both use and produce other resources not subject to these constraints. For example, to specify the storage capacity required per unit storage activity, the statement AS / 1S = 1.5 might be used. The corresponding matrix elements would be:

Column	Row	Value
1S.....B	AS.....B	+1.5
1S.....C	AS.....C	+1.5
1S.....D	AS.....D	+1.5

where in the general case of a time varying data item each element would take the unit value at the appropriate interface.

A quite different method of generation is used when an interperiod activity interacts with a within-period resource (initial letter B). To fix ideas consider the statement

BM / 1K = −40

which might be used for example, in modelling the retention of a labour force, to specify that the amount of work obtained from one man retained for a week is 40 man-hours. (Note that the sign of the element value is in accordance with our sign convention whereby the quantities of resources produced by activities have negative signs.)

To see how individual element values are computed from such a specification it should first of all be remembered that the interperiod activity 1K is treated by the generator as though it takes place instantaneously at the interfaces between model periods. Suppose then that at the start of a particular period 100 men are retained. If during the period 20 additional men are taken on then at the end of the period the labour force will comprise 120 men in all.

Assuming that the additional men are taken on uniformly over the period, the average size of the labour force during the period will be $\frac{1}{2}(100 + 120)$ men, and the number of man-hours work they will produce is therefore $\frac{1}{2}(100 + 120) \times 40$ man-hours per time unit.

To achieve this result two elements must be generated for each time period in the model. In the particular case that we are considering these elements would be

Column	Row	Value
1K. . . . . A	BM. . . . . B	−20
1K. . . . . B	BM. . . . . B	−20
1K. . . . . B	BM. C	−40
1K. C	BM. C	−40
1K. C	BM. . . . . D	−40
1K. . . . . D	BM. . . . . D	−40

The first element in each pair is, in the general case of a time-varying element value, computed as the sum of the unit values over the first half of the time period, while for the second member of each pair the sum over the second half of each period is used. Thus in the above example, as the first model period has but one time unit, its unit value has been split between the two elements.

Essentially the same method of computation is used when a cost is put on interperiod activity. Thus if a labour cost of £20 per man per week were specified as

*C / 1K = 20

then, assuming that there is no discounting to present value, the following cost row elements would be generated:

Column	Cost
1K. . . . . A	10
1K. . . . . B	30
1K. C	40
1K. . . . . D	20

where for example the cost on the acitivity 1K. . . . . B, which takes place at the interface between the first time period and the second, is the sum of the unit values over the second half of the first time period (£10) and the first half of the second time period (£20).

Bounds placed on interperiod activities, such as minimum stock levels for storage activities, apply at all interfaces other than the opening and closing interfaces. Thus the statement

*L / 1S = 20

which specifies a lower bound of 20 for the storage activity 1S, would in a three period model only yield two elements:

Column	Lower Bound
1S.....B	20
1S.....C	20

where the element values are the same as the unit values at the appropriate interfaces.

Bounds on interperiod activities at the opening and closing interfaces, i.e. the opening and closing stocks for the model as a whole, must be stated explicitly. For example to specify a fixed opening stock of 32 for the storage activity 1S, the statement

*EO / 1S = 32

must be used. Likewise the statement

*LC / 1S = 25

would be used to specify a minimum closing stock of 25.

Bounds may also be applied to interperiod constraints (first letter A). For example to place an upper limit of 50 on a storage capacity with code name AS the following statement would be used:

AS / *U = 50

The elements that would then be generated are

Row	Upper Bound
AS.....B	50
AS.....C	50
AS.....D	50

As with bounds on interperiod activities, each of the above elements values is the same as the corresponding unit value at the appropriate interface. The method of generation is however slightly different in as much as an element is generated for the closing interface.

As with all element values, the bounds on an interperiod constraint may vary with time. For example, if it were known that the storage capacity AS above was due to be reduced from 50 units to 30 units from time unit 6852 onwards (where 6852 was the last time unit in the second model period) then the above specification would become:

AS / *U = 50 (6851) 30

The six unit values in the element value time series would then be:

Date	Time Unit	Model Period	Unit Value
6849	0	—	50
6850	1	1	50
6851	2	2	50
6852	3	2	30
6901	4	3	30
6902	5	3	30

and the elements that would now be generated are:

Row	Upper Bound
AS.....B	50
AS.....C	30
AS.....D	30

Within-period Activities

Production, sales and distribution activities are all typical examples of the within-period activities (initial character 2) that occur in production scheduling models. Apart from the bounds, all the elements associated with these activities are generated in the same way, whether they be cost row elements, or elements associated with within-period rows (B) or material balance rows (C). (Note that within-period activities do not interact with interperiod resources.)

To take one example of this type of element consider the statement

BM / 2P = 0.8

which might represent the number of man-hours work required per unit of a production activity.

The matrix elements that this statement would cause to be generated are

Column	Row	Value
2P.....B	BM.....B	+0.8
2P.....C	BM.....C	+0.8
2P.....D	BM.....D	+0.8

where each element value is computed as the average of the unit values over the corresponding period.

Bounds on within-period activities are, on the other hand, computed not as the average but as the sum of the unit values over the corresponding model periods. Thus if the statement

*U / 2P = 40

were used to specify an upper limit of 40 tons per week on the production activity 2P, then the following matrix elements would be generated:

Column	Upper Bound
2P.....B	40
2P.....C	80
2P.....D	80

As would be expected, in the second and third model periods, as there are two time units, the upper bound is 80.

Bounds on within-period constraints (first letter B) and material balance rows (first letter C) are generated in exactly the same way, with element values computed as the sum of the unit values over the appropriate model periods.

Bounds do not necessarily have to be specified for each and every row and column. As a general rule those that are left unspecified are assumed to be non-constraining. There are however some exceptions. Thus if neither row bound is specified, then both are assumed to be zero; also if a column's lower bound is not specified, then it is assumed to be zero unless the corresponding upper bound is negative.

To bring together all the individual examples given in the foregoing paragraphs, consider the following specification of a hypothetical model:

<u>Storage</u>	
Storage Capacity Available	AS / *U = 50
Storage Capacity/ton stocked	AS / 1S = 1.5
Stock in	CQ / 1S = +1
Stock out	CQ / 1S = −1
Minimum stock	*L / 1S = 20
Opening Stock	*EO / 1S = 32
Closing Stock	*LC / 1S = 25
<u>Labour</u>	
Man-hours work/man-week	BM / 1K = −40
Labour force in	CL / 1K = +1
Labour force out	CL / 1K = −0.9
Cost/man-week	*C / 1K = 20
Opening Labour Force	*EO / 1K = 28
<u>Production</u>	
Fixed work in man-hours/week	BM / *L = 15
Man-hours/ton produced	BM / 2P = 0.8
Stock required/ton produced	CQ / 2P = +1
Maximum Production/week	*U / 2P = 40
Production Cost/ton	*C / 2P = 3

When processed by the generator to form a three period model with one time unit in the first period, and two time units in each of the second and third periods, the matrix on the facing page is produced. Static and multiperiod networks for this model are shewn in Diagram 4.

3.3 Matrix Storage and Output

As successive elements are generated using the methods described in the foregoing paragraphs, they are stored in core, each chained to the last element, if any, in the same column (or dummy row, in the case of row-row elements). At the same time lists of variable names, costs and bounds are prepared.

When the entire specification has been successfully processed the row and column

Matrix

	1KA	1KB	1KC	1KD	1SA	1SB	1SC	1SD	2PB	2PC	2PD	*L	*U
ASB	.	.	.	.	.	+1.5	.	.	.	.	.	none	50
ASC	.	.	.	.	.	.	+1.5	.	.	.	.	none	50
ASD	.	.	.	.	.	.	.	+1.5	.	.	.	none	50
BMB	−20	−20	.	.	.	.	.	.	+.8	.	.	15	none
BMC	.	−40	−40	.	.	.	.	.	.	+.8	.	30	none
BMD	.	.	−40	−40	.	.	.	.	.	.	+.8	30	none
CLB	−.90	+1	.	.	.	.	.	.	.	.	.	0	0
CLC	.	−.81	+1	.	.	.	.	.	.	.	.	0	0
CLD	.	.	−.81	+1	.	.	.	.	.	.	.	0	0
CQB	.	.	.	.	−1	+1	.	.	+1	.	.	0	0
CQC	.	.	.	.	.	−1	+1	.	.	+1	.	0	0
CQD	.	.	.	.	.	.	−1	+1	.	.	+1	0	0
*L	28	0	0	0	32	20	20	25	0	0	0		
*U	28	none	none	none	32	none	none	none	40	80	80		
*C	2	6	8	4	0	0	0	0	3	3	3		

Note: The last letter of each code name indicates the time period or interface, thus:

A Opening Interface

B First Period or Interface

C Second Period or Interface

D Third Period or Closing Interface.

Lower bounds are designated by *L, upper bounds by *U, and the cost row by *C.

names are sorted into the alpha-numeric sequence in which they will be output. The default bounds described earlier are then applied where none have been specified.

At this stage the matrix may contain a number of references to dummy rows, or references by one activity to another. In Diagram 5 the two networks shewn illustrate the method used by the generator to eliminate such dummy references.

Dummy rows, such as the row labelled CO in Diagram 5, are used in the specification of the model in much the same way as one might use a list of components for a sub-assembly; once defined the sub-assembly can be referred to as a unit so that the list of components of which it is made need not be repeated.

In our example the dummy row CO is specified as

BR / CO = 2

CQ1 / CO = 1

All references to this dummy row will be eliminated by substitution. Thus the statement

CO / 2V = 2

becomes

BR / 2V = 4

CQ1 / 2V = 2

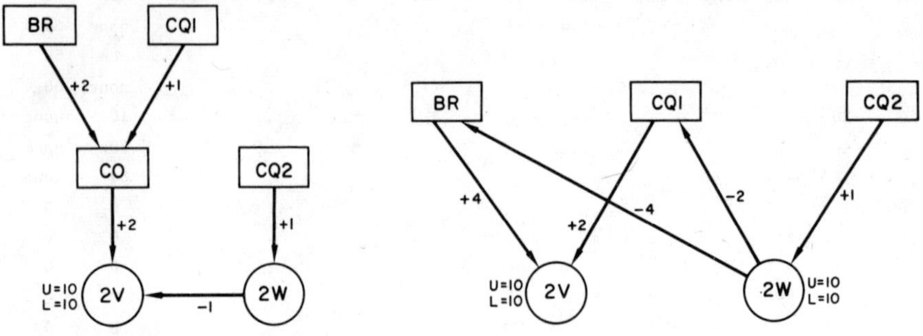

<u>Diagram 5.</u> Static networks before and after the elimination of row-row and column-column elements.

References by one activity to another, such as the reference by activity 2W to activity 2V in Diagram 5, are processed by the generator in much the same way as references to dummy rows. Thus the activity 2V, now defined as

BR / 2V = 4
CQ1 / 2V = 2

is eliminated from the statement

2V / 2W = −1

to yield

BR / 2W = −4
CQ1 / 2W = −2

Dummy references of this kind are used exclusively to model a particular situation in which a fixed demand, primarily satisfied by the one activity (2V), can be switched to another (2W), thereby releasing the resources required by the primary activity. In the example shewn in Diagram 5, the primary activity 2V is used to satisfy a fixed demand of 10 units/week. This demand can also be met by the activity 2W. To model this situation correctly, without introducing an additional demand constraint, the activity 2W must be amended so that, up to a limit of 10 units/week, it releases the resources required by the activity 2V. If the second network in Diagram 5 is examined it will be seen that this is indeed how it has been modelled.

Resolving all the dummy references as they are encountered, the generator finally proceeds to write out the entire matrix, with rows and columns in alpha-numeric sequence, onto an external file. The format in which the matrix is written is that required by the optimization program, in this case the I-B-M Mathematical Programming System.

Quelques Méthodes et Programmes Auxiliaires pour la Gestion de Modèles Mathématiques Linéaires

L. PESSINA
Montecatini Edison, Italy

SUMMARY

This paper describes a procedure, which has been implemented, to facilitate the use of linear programming models. It is particularly intended for models relating to short-term production problems, which are used frequently and need a rapid response.

The procedure uses the IBM MPS/360 Mathematical Programming System for the simplex calculations. It is supplemented by a set of FORTRAN programs that have been written primarily

(a) to provide ways of shortening and simplifying the task of assembling models, and particularly of adapting them successively to changes in the situation.

(b) to reduce the dimensions of the problem, by taking equations defining raw materials or products with no quantity limitations and using them to eliminate these variables from the rest of the problem, and

(c) to provide reports that can be consulted quickly and easily either about the input data or about the results of the simplex calculations.

Furthermore, it has been found necessary to consider problems with a substantial number of continuous variables together with a very limited number of zero-one variables. A routine has therefore been provided that uses MPS/360 to solve such problems in a reasonable amount of time by setting up a suite of alternative linear programming problems.

Editor's Note

Dr. Pessina's contribution to the NATO Conference on Applications of Mathematical Programming Techniques included 2 appendices: the first giving the output from a numerical example, and the second a listing of the FORTRAN programs of the system. Unfortunately these have had to be omitted from the present volume for reasons of space.

1. INTRODUCTION

L'objet de la présente est d'illustrer une procédure qui a été mise au point, en vue de faciliter l'utilisation des modèles de programmation linéaire. En particulier, des modèles qui, tout en se rapportant aux problèmes de gestion à court terme, sont employés fréquemment et nécessitent une réponse rapide.

La procédure exploite le 'software' IBM (MPS/360—Mathematical Programming System) en ce qui concerne la solution du simplexe et est complétée par une série de programmes en langage Fortran, qui ont été établis principalement afin de:

— réaliser des mesures qui permettent d'abréger et faciliter la phase de mise au point des modèles et surtout leur adaptation successive selon les changements de situation.

— réduire les dimensions des problèmes au moyen d'une opération d'élimination/remplacement des équations relatives aux ressources et produits ayant des contraintes économiques mais sans contraintes de quantité.

— obtenir des reports qui peuvent être consultés rapidement et aisément soit pour les données d'entrée que pour les résultats de la solution du simplexe.

De plus, en raison de la nécessité de faire face aux problèmes présentant un considérable nombre de variables 'continues' par rapport à un nombre très limité de variables 'bivalentes' (du type 0-1) on a établi une routine, dans l'attente de programmes appropriés et efficaces, laquelle exploite d'une façon convenable le code MPS permettant ainsi de résoudre le problème, au moyen de l'algorithme de la programmation linéaire, dans un délai de temps raisonnable.

2. DESCRIPTION DE LA PROCÉDURE

La procédure PROLIN est articulée en trois phases distinctes: La première consiste dans:

— Préparation et codage des données d'entrée du modèle (2. 1)

— PROLIN 1 — PROLIN 2 — elaboration et impression des données d'entrée (2. 2 et 2. 3)

La seconde phase consiste dans l'exécution des calculs d'optimisation et de post-optimisation, au moyen du Code 'MPS' (2. 4)

La troisième phase consiste dans:

— PROLIN 4 — PROLIN 5 — PROLIN 6 — elaboration et impression des résultats. (2. 5 et 2. 6)

Le déroulement de la procédure est explicité par l'organigramme de la figure 1.

2.1 Préparation et codage des données d'entrée du modèle.

Les données d'entrée qu'il faut préparer sont les suivantes:

— Code et déscription alphabétique des lignes de la matrice (contraintes, fonctions économiques)

— Code et déscription alphabétique des colonnes de la matrice (variables, vecteurs des membres, vecteurs 'range' des seconds membres)

— Coefficients de la matrice avec leurs codes de colonne et de ligne.

Les 'formats' pour la préparation des données dont on vient de parler sont explicités par la figure 2.

Le codage des colonnes et des lignes de la matrice doit être réalisée suivant les normes spécifiées dans les tableaux de la figure 3.

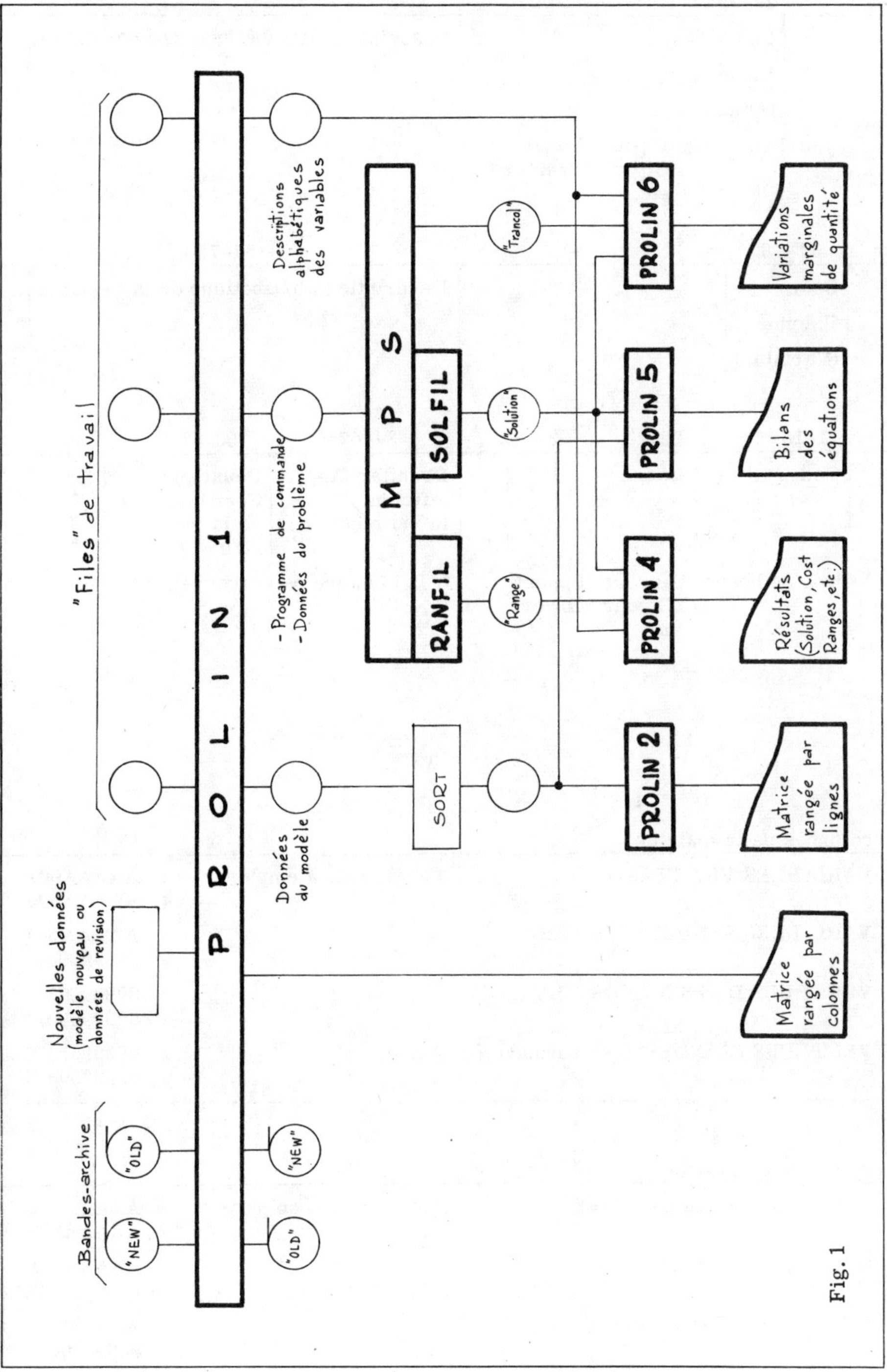
"Files" de travail
P R O L I N 1
Bandes-archive
"NEW"
"OLD"
"OLD"
"NEW"
Nouvelles données (modèle nouveau ou données de revision)
Données du modèle
- Programme de commande
- Données du problème
Descriptions alphabétiques des variables
SORT
M P S
RANFIL
SOLFIL
"Range"
"Solution"
"Trancol"
PROLIN 2
PROLIN 4
PROLIN 5
PROLIN 6
Matrice rangée par colonnes
Matrice rangée par lignes
Résultats (Solution, Cost Ranges, etc.)
Bilans des équations
Variations marginales de quantité

Fig. 1

	22-29		31-32		34-73
	Code Ligne (Equation)		*		Description alphabétique de l'équation

* Type d'équation: E pour 'Equal'
G pour 'Greater'
etc.

	11-18		34-73
	Code Colonne (Variable)		Description alphabétique de la variable

	11-18		22-29		31-32		34-47		50-63		66
	Code		Code		* *		Premier Coefficient de la matrice		Deuxième Coefficient de la matrice		Jk * * *

* Type de borne: UP pour 'Upper'
LO pour 'Lower'
etc.

* Jk = indice de regroupement
*
*

Fig. 2

colonnes de la matrice

VARIABLES VIRTUELLES	code à employer	⩾ AAAAA001 ⩽ AAAAA500
VARIABLES STRUCTURELLES	" " "	⩾ AAAAA501 ⩽ 99999000
VECTEURS DES SECONDS MEMBRES	" " "	⩾ 99999001 ⩽ 99999500
VECTEURS 'RANGES' (des seconds membres)	" " "	⩾ 99999501 ⩽ 99999998

lignes de la matrice

FONCTIONS ÉCONOMIQUES	code à employer	⩾ AAAAA001 ⩽ AAAAA100
BORNES DES VARIABLES	" " "	⩾ AAAAA501 ⩽ AAAAA999
EQUATIONS (Contraintes)	" " "	⩾ AAAAB ⩽ 99999000

Fig. 3

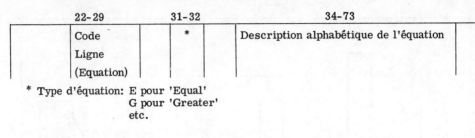

2.2 Prolin 1

Le programme Proline 1 réalise les opérations suivantes:

- — Gestion des bandes-archive des modèles (2.2.1)
  - — Enregistrement d'un nouveau modèle (2.2.1.1)
  - — Mise à jour d'un modèle déjà enregistré (2.2.1.2)
- — Contrôle du modèle—Détection des erreurs (2.2.2)
- — Etablissement du 'Programme de commande' pour 'MPS' (2.2.3)
- — Etablissement du 'file' des données pour 'MPS' (2.2.4)
  - — Changement de la valeur des coefficients (2.2.4.1)
  - — Elimination des équations virtuelles (2.2.4.2)
  - — Etablissement des données pour problèmes ayant des 'charges fixes' (2.2.4.3)
- — Impression du report—Etablissement des 'files' pour les programmes successifs. (2.2.5)

2.2.1 Gestion des bandes-archive des modèles

Pour faciliter la gestion de plusieurs modèles, nous avons pensé de constituer des bandes-archive. Sur chaque bande on a prévu d'enregistrer 10 différents modèles, disposés en séquence et identifiés par un mot à 8 caractères.

La gestion des bandes-archive est realisée de la manière suivante:

Enregistrement d'un nouveau modèle

Les données d'entrée doivent être préparées sur des cartes perforées ou tout autre support, d'après les normes indiquées au par. 2.1. Ce 'stream' de données ne sera plus utilisé après enregistrement.

A l'intérieur du 'stream' les données peuvent être disposées dans n'importe quel ordre; au moyen d'une routine de sélection, le programme range ces mêmes données en ordre croissant sur les codes de colonne/ligne et il les enregistre ensuite dans cet ordre sur la bande-archive.

Mise à jour d'un modèle déjà enregistré

Seules les données de mise à jour doivent être préparées sur des cartes. Il n'y a aucune limite au nombre de données qui peuvent être modifiées dans une seule élaboration.

Le 'format' pour ces données est celui déjà spécifié (v.par. 2.1). Il faut toutefois rajouter un indice de mise à jour à la colonne 9 de la carte.

Quatre indices de mise à jour sont prévus:

(a) — Indice 'E': pour éliminer totalement une variable (description + coefficients)

(b) — Indice 'A': pour annuler une donnée (description variable ou description équation ou coefficient).

(c) — Indice 'S': pour remplacer une donnée (idem)

(d) — Indice 'I': pour insérer une nouvelle donnée (idem)

Dans ce cas aussi, les données peuvent être disposées dans n'importe quel ordre; la même routine de sélection prévue à propos du premier enregistrement permet de les ranger en ordre croissant sur les codes colonne/ligne.

A ce moment, la mise à jour des données enregistrées dans la bande-archive est realisée au moyen d'une routine de 'merge'. Quelques contrôles sur les codes des données et sur les indices de mise à jour ont été insérés dans la routine de 'merge' pour détecter, dans la limite du possible, les instructions de mise à jour incohérentes. Une liste des mises à jour qui ont eu lieu effectivement dans le modèle va être imprimée pendant l'élaboration; les mises à jour incohérentes, qui n'ont pas eu lieu, vont être signalées par des messages d'erreur.

2.2.2 Contrôle des données—Détection des erreurs

Pendant cette phase, le 'data-set' relatif au modèle en élaboration est soumis à des contrôles qui permettent de mettre en évidence des omissions, des erreurs de codification, et des incohérences, comme par exemple:

- Colonnes ou lignes codées, mais dépourvues de coefficients.
- Colonnes ou lignes dépourvues de description ou avec deux descriptions.
- Exploitation de codes non conformes aux normes établies, etc.

2.2.3 Etablissement du 'Programme de commande' pour 'MPS'

Le code 'MPS' est composé d'une série de routines et de procédures. L'exécution coordonnée d'une série de ces procédures se réalise par l'intervention d'un programme de commande composé par des 'statements' où l'on précise le nom du modèle, de la fonction économique, de la ligne des bornes, etc., ainsi que les noms des procédures que l'on désire: 'Solution', 'Range', Trancol', etc. Pour éviter chaque fois la préparation manuelle de ce programme de commande on a prévu une routine qui, d'après des simples indications données par quelques cartes de contrôle insérées au début de l'élaboration, réalise automatiquement les 'statements' nécessaires au fur et à mesure.

Le programme établi de cette façon est 'écrit' sur un 'file' sur lequel seront enregistrées ensuite les données pour la résolution du problème.

2.2.4 Etablissement du 'file' des données pour 'MPS'

Le 'file' des données pour le code 'MPS' doit être subdivisé dans les sections suivantes:

- 'Rows' — Codes des lignes de la matrice
- 'Columns' — Données des colonnes de la matrice
- 'RHS' — Vecteurs des seconds membres
- 'Ranges' — Vecteurs 'ranges' des seconds membres
- 'Bounds' — Bornes des variables

Le Prolin 1 tire de la bande-archive les données nécessaires aux différentes sections et constitue le 'file' dans l'ordre requis. La codification adoptée (v. par. 2.1) et l'ordre de rangement utilisé pour la bande-archive (v. par. 2.2.1) permettent d'effectuer aisement et rapidement la préparation du 'file'.

Au cours de la préparation de ce 'file' le programme réalise les élaborations indiquées aux paragraphes suivants.

Changement de la valeur des coefficients

Les procédures de mise à jour des matrices du type 'Revise' du code MPS ou du type décrit ci-dessus permettent d'effectuer sur cette même matrice toutes les modifications qui se rendent nécessaires. Il ne faut pas oublier, en tout cas, que pour la mise en oeuvre de ces procédures, il est nécessaire de préparer manuellement toutes les données de modification; tout cela provoque évidemment des pertes de temps et, éventuellement, des imprécisions.

Il faut donc recourir à ces procédures seulement dans les cas indispensables, comme p.e.: correction d'erreurs, modifications imprévisibles.

Par conséquent, dans tous les cas où il s'agit de modifier une matrice selon des critères préétablis, dans le but de la prédisposer pour une nouvelle solution, ou, en général, pour plusieurs solutions différentes entre elles, (qui peuvent être demandées le même jour ou dans des periodes de temps successives), il faut pouvoir disposer de procédures et de routines qui réalisent automatiquement, par un nombre limité d'instructions, les transformations au fur et à mesure que celles-ci se rendent nécessaires.

Le code 'MPS' prévoit un certain nombre de procédures paramétriques ('Paraobj', 'Pararhs') et une routine de 'Scale' qui permettent de réaliser une partie des dites transformations.

Pour compléter ce qui est prévu par le 'MPS', une routine généralisée a été insérée dans le Prolin 1 telle qu'elle puisse permettre d'effectuer le changement de la valeur d'une partie, ou bien, le cas échéant, de tous les éléments de la matrice. Le fonctionnement de cette routine est le suivant:

Pour chaque élément de la matrice (y compris les éléments des fonctions économiques, des bornes, des vecteurs des seconds membres et des vecteurs 'ranges' des seconds membres) une possibilité a été prévue de mémoriser sur la bande-archive deux valeurs numériques et, à côté de celles-ci, un indice de regroupement (v. par. 2.1 — fig. 2).

A l'état actuel le programme prévoit un nombre maximum de sept indices différents (numérotés de 1 à 7) car ce nombre parfaît suffisant pour le moment; il est toutefois possible de les augmenter, si cela se rend nécessaire, par des simples modifications du programme.

A l'aide de ces indices, il est donc possible de réunir n'importe quel élément de la matrice dans un nombre de groupes séparés, variable entre 1 et 7.

Il faut préciser que, en conséquence de ce qu'on vient de dire, au autre groupe va se former, et précisément celui relatif aux éléments dépourvus d'indice; ces derniers n'entrent toutefois pas dans la routine et pour ceux-là on utilise, pour la résolution du problème, la valeur du premier coefficient. Pour les éléments munis d'indice, la valeur du coefficient qu'on utilise pour la résolution du problème est calculée par la relation suivante:

$$C_{ij} = C1_{ij} + (C2_{ij} - C1_{ij}) \cdot X_k$$

dans laquelle:

— $C1_{ij}$ et $C2_{ij}$ sont les valeurs, respectivement du premier et du deuxième coefficient, mémorisées dans la bande-archive.

— X_k est le paramètre qu'on définit au fur et à mesure, suivant les exigences, pour chacun des différents groupes. Les sept valeurs de X_k sont introduites au début de l'élaboration par une seule carte de contrôle.

Il est évident qu'en attribuant à X_k des valeurs comprises entre 0 et 1 on obtiendra des valeurs de C_{ij} comprises entre $C1_{ij}$ et $C2_{ij}$; par des valeurs de X_k extérieures à l'intervalle $0 \div 1$ on obtiendra une extrapolation.

En attribuant à tous les coefficients $C1_{ij}$ (ou $C2_{ij}$) d'une groupe la valeur de zéro, il est possible d'obtenir en même temps et par une seule instruction le 'Scale' de tous les éléments du groupe.

Une application pratique de cette routine a été effectuée en vue d'adapter rapidement aux exigences les modèles qui comprenaient des éléments variables suivant les saisons.

Elimination des 'équations virtuelles'

Dans la plupart des modèles de programmation linéaire il y a un certain nombre d'équations concernant des ressources dont la disponibilité n'est pas limitée.

En général, dans les modèles relatifs à des procédés d'industries chimiques, les services auxiliaires comme par exemple l'énergie électrique, la vapeur, l'eau, le fuel-oil, etc. sont des ressources non limitées; plusieurs matières subsidiaires sont disponibles en quantités illimitées.

Ce même raisonnement peut être parfois valable non suelement pour les facteurs productifs, mais aussi pour quelques produits ou sous-produits qu'on peur écouler en quantité bien supérieure à la production.

Nous avons pu constater que dans nos modèles le nombre d'équations dont nous venons de parler représente en moyenne 10 ÷ 15% du nombre total.

Pour les équations de ce genre le 'shadow price' coincide avec le coefficient économique de la correspondante variable d'entrée ou de sortie (prix de la ressource, ou revenue du produit). Ce 'shadow price' reste constant pour n'importe quelle solution du modèle.

Il est donc évident que toutes ces équations peuvent être remplacées, en ce qui concerne la solution du simplexe, par une seule équation dont les éléments soient calculés en faisant la somme des produits des coefficients des équations précédentes par les 'shadow prices' qui leur correspondent.

L'exemple que nous allons citer pourra mieux expliquer ce que nous venons d'exposer.

Considérons la matrice A (v. Figure 4) dans laquelle:

- EV_1, EV_2, EV_3 représentent les équations relatives aux ressources non limitées, que nous allons appeler par la suite 'équations virtuelles'.
- VV_1, VV_2, VV_3 représentent les variables d'entrée (ou de sortie) relatives à ces équations, et C_1, C_2, C_3 les coefficients économiques correspondants. Nous appellerons ces variables 'variables virtuelles'.
- $EN_1, EN_2, \ldots EN_n$ représentent les équations 'normales' du problème.
- $VN_1, VN_2, \ldots VN_n$ sont les variables 'normales' du problème, qui ont des coefficients sur les équations virtuelles.
- FE est la fonction économique.

Nous transformons la matrice A dans la matrice A', dans laquelle:

les variables et les équations virtuelles ont été éliminées et remplacées par l'équation ERV dont les coefficients $\Sigma_1, \Sigma_2, \Sigma_3, \Sigma_4, \ldots \Sigma_n$ ont été calculés:

$$\Sigma_1 = a_1 \cdot C_1 + a_2 \cdot C_2 + a_3 \cdot C_3$$

$$\Sigma_2 = b_1 \cdot C_1 + b_3 \cdot C_3$$

$$\Sigma_3 = d_2 \cdot C_2 + d_3 \cdot C_3$$

$$\Sigma_4 = 0$$

.

.

.

$$\Sigma_n = \ldots\ldots\ldots\ldots\ldots$$

- VS est la variable d'écart de l'équation ERV

Du point de vue théorique on pourrait se passer de l'équation ERV et inclure les valeurs Σ_i directement dans la fonction économique, en les additionnant aux CE_1,

MATRICE A

variab. virtuelles			variables normales											
VV_1	VV_2	VV_3	VN_1	VN_2	VN_3	VN_4	...	...	...	...	VN_n			
C_1	C_2	C_3	CE_1	CE_2	CE_3	CE_4	...	...	...	...	CE_n		FE	fonct. économ.
-1	–	–	a_1	b_1	–	–							EV_1	equations virtuelles
–	-1	–	a_2	–	d_2	–							EV_2	
–	–	-1	a_3	b_3	d_3	–							EV_3	
			X		X	X							EN_1	equations normales
			X	X	X								EN_2	
													...	
			X			X							...	
				X		X							EN_n	

MATRICE A'

variab. solde	variables normales											
VS	VN_1	VN_2	VN_3	VN_4	...	...	...	...	VN_n			
-1	CE_1	CE_2	CE_3	CE_4	...	...	...	...	CE_n		FE	fonct. économ.
-1	Σ_1	Σ_2	Σ_3	Σ_4	...	...	...	...	Σ_n		ERV	eq. remplacant les eq. virtuelles
	X		X	X							EN_1	equations normales
	X	X	X								EN_2	
											...	
	X			X							...	
		X		X							EN_n	

Fig. 4

$CE_2, \ldots CE_n$ qui leur correspondent. Du point de vue pratique, nous préférons maintenir ces deux lignes séparées.

En tout cas, en ce qui concerne la solution du problème, la matrice transformée A' correspond parfaitement à la matrice A originaire.

Par cette artifice nous pouvons réduire d'une manière assez concrète les temps de solution du simplexe.

Dans le tableau suivant nous indiquons quelques données de comparaison.

Ordinateur IBM 360/50 — 250 000 bytes — code MPS					
				temps employé pour l'optimisation	
n° équations normales	n° équations virtuelles	n° variables	n° coefficients	du modèle tel quel	du modèle transformé
915	180	2722	11086	205'	115'

L'opération qui nous permet d'éliminer les équations virtuelles est effectuée automatiquement par le Prolin 1 pendant la phase de préparation du 'file' des données pour MPS. L'identification de ces équations est obtenue au moyen du code des variables virtuelles qui leur correspondent (v. par. 2. 1 'Préparation et codage des données').

Etablissement des données pour la solution de problèmes ayant des 'charges fixes'.

Pour quelques-uns de nos modèles de programmation de la production, nous nous sommes trouvés dans la nécessité de tenir compte aussi, en ce qui concerne l'optimisation, des charges fixes se rapportant aux activités de production. Des problèmes de ce genre sont compris dans la catégorie des 'Programmes mixtes à variables bivalentes'.

A l'état actuel de nos travaux nous ne disposons pas de programmes efficients pour résoudre de tels problèmes.

Pour les cas indispensables nous avons donc recouru à l'algorithme de la P.L. en analysant les 2^n solutions possibles ('n' représente le nombre des charges fixes prises en considération).

Il est évident que, voulant réduire à des limites raisonnables les temps d'élaboration, nous avons dû aussi nous limiter à l'examen d'un nombre assez réduit de composantes fixes.

Cela dit, la méthode adoptée est la suivante:

Considérons un problème avec un nombre quelconque de variables et supposons qu'une seulement de ces variables ait des coûts fixes; appelons X_i cette variable (par exemple une installation de production) et K_i son coût fixe; supposons que la variable ait aussi un coût proportionnel S_i.

La fonction de coût de cette variable doit donc assumer la valeur:

$$C_i = S_i \cdot X_i + K_i \qquad \text{si } X_i > 0$$

et la valeur $C_i = 0$ si $X_i = 0$

Oublions pour le moment la constante K_i et introduisons uniquement S_i dans la fonction économique du modèle; resolvons avec l'algorithme de PL le modèle dans les deux conditions suivantes:

1e solution: X_i est libre d'assumer une valeur quelconque $\geqslant 0$
2e solution: X_i est fixé à la valeur de zéro.

Des deux solutions ainsi obtenues nous déduisons les deux valeurs Z_1 et Z_2 de la fonction économique; nous corrigeons à ce moment-là la valeur de Z_1 en soustrayant la valeur de K_i:

$$Z'_1 = Z_1 - K_i$$

Ensuite en faisant une comparaison entre Z'_1 et Z_2 nous choisirons la meilleure des deux solutions*.

En généralisant le critère que nous venons d'exposer, nous pouvons conclure que pour un nombre 'n' de variables affectées de charges fixes nous devrons comparer entre elles les valeurs des 2^n fonctions économiques:

$$Z'_j = Z_j - \sum_{i=1}^{n} K_i, \text{ pour toutes } X_i \neq 0, \forall\ j \in \{1, 2, \ldots 2^n\}$$

---

* Il est évident que si X restait spontanément à zero même dans la première solution, les deux alternatives seraient coincidentes indépendamment de la valeur de K_i, c'est à dire $Z_1 = Z_2$.

en choisissant comme solution optimale celle qui donne lieu à la valeur maximum de Z'_j.

Pour l'exécution pratique de ce que nous venons d'exposer on a inséré dans le Prolin 1 une routine qui sert à:

— préparer les données nécessaires pour réaliser le blocage à zéro des variables X_i qui, d'une solution à l'autre, doivent assumer telle condition (cela est obtenu en attribuant aux X_i un coefficient économique fortement négatif).

— calculer, pour chaque solution, la somme des charges fixes K_i qui lui corresponde.

— préparer un 'programme de commande' pour 'MPS', tel qu'on puisse faire exécuter en séquence les 2^n optimisation, le calcul des valeurs Z'_j, et, enfin, la recherche de la meilleure entre elles.

Les données d'entrée à livrer pour cette routine se limitent aux codes des variables X_i et aux valeurs des charges fixes K_i qui leur correspondent.

Nous avons utilisé cette procédure en plusieurs occasions sur un modèle à 130 lignes, 700 colonnes, 1800 coefficients, 8 variables X_i; le temps de 256 solutions a été d'environ 50 minutes.

2.2.5 <u>Impression du report. Etablissement des 'files' pour les programmes successifs.</u>

Au cours des élaborations illustrées aux paragraphes précédents, le Prolin 1 fournit un report sur lequel tous les éléments du modèle sont reproduits en ordre de colonne (v. Annexe A—page 3). Chaque variable est munie de sa description alphabétique et pour chaque coefficient de la variable on indique le code et la description alphabétique de l'équation qui lui correspond.

Pour tous les éléments du modèle on trouve les deux valeurs mémorisées du coefficient, l'indice du groupe d'appartenance, ainsi que la valeur du 'coefficient calculé' qui est utilisé pour la résolution du problème.

Dans ce report on trouve aussi tous les coefficients concernants les équation 'virtuelles' (qui, en effet, ne font pas partie des données pour la résolution du simplexe) ainsi que la valeur Σ_i qui les remplace.

Au cours des élaborations dont nous venons de parler, le Prolin 1 prépare aussi sur un 'file' toutes les données nécessaires à obtenir le report du modèle rangé par ligne (v. Prolin 2), ainsi que le report des bilans des équations (v. Prolin 5).

Sur un dernier 'file' on prépare aussi les descriptions alphabétiques des variables, qui serviront pour les programmes Prolin 4 et Prolin 6.

2.3 Prolin 2

Le Prolin 2 est un programme qui sert à obtenir un report avec tous les éléments du modèle rangés par ligne. Chaque équation est munie de sa description alphabétique et pour chaque coefficient de l'équation on indique le code et la description alphabétique de la variable qui lui correspond.

Le report comprend aussi les équations 'virtuelles'.

Quant aux coefficients, le report indique seulement la valeur du 'coefficient calculé' qui est utilisé pour la resolution du problème.

Le Prolin 2 est un programme très simple qui utilise le 'file' des données prédisposées par le Prolin 1, après une sélection ('Sort') sur les codes ligne/colonne. Le temps nécessaire pour son exécution est déterminé uniquement par la vitesse des unités d'entrée/sortie. Emploi de ce report s'est révélé très utile pendant la phase de mise au point de nouveaux modèles et à l'occasion des mises à jour qui comprenaient un nombre élevé de modifications.

Dans le cas de modifications peu nombreuses cette phase de la procédure peut être omise, parce qu'il suffit de disposer du report du modèle rangé par colonnes fourni par le Prolin 1.

2.4 Calculs d'optimisation et de post-optimisation

Cette phase est réalisée au moyen du code 'MPS'.

Les sorties des procédures 'Solution', 'Range', et 'Trancol' ne sont pas imprimées directement mais, après conversion pour qu'elles soient accessibles au langage Fortran, elles sont enregistrées sur 'file'; cela pour permettre que les résultats de ces procédures puissent se présenter dans une forme claire et complète, et pour permettre de les utiliser en partie pour d'autres élaborations (Prolin 5).

Pour cette opération d'enregistrement on utilise les subroutines du 'READCOMM' ('Read Communication Format') qui sont rappelées et opportunément rassemblées au moyen des programmes auxiliaires 'Solfil' et 'Ranfil'.

2.5 Prolin 5

Le Prolin 5 est un programme qui sert pour obtenir un report sur lequel sont indiqués, par rapport aux valeurs de la solution optimale, les bilans à quantité de chaque équation du modèle, y compris les équations virtuelles. Comme pour les autres reports, ce dernier comprend les descriptions alphabétiques. Pour réaliser ce report, le Prolin 5 exploite le même 'file' que l'on utilise pour le Prolin 2, ainsi que le 'file' des valeurs de la solution optimale préparé pendant la phase 'MPS' ('File Solution' v.par. 2.4).

La disponibilité de ce report s'est révélé très utile surtout au cours de l'examen des résultats du modèle avec les responsables des directions de production.

2.6 Prolin 4—Prolin 6

Les instruments les plus efficaces pour qu'on puisse se faire une opinion claire sur la validité et la stabilité de la solution optimal d'un modèle sont obtenus, comme l'on sait, par les procédures d'analyse marginale de post-optimisation.

Les procédures du code 'MPS' qui effectuent ces analyses sont les procédures 'Range' et 'Trancol'.

Pour pouvoir consulter les résultats d'une manière pratique et rapide, nous avons envisagé l'opportunité de les ranger et de les compléter avec leurs descriptions alphabétiques. Cela se réalise par les programmes Prolin 4 et Prolin 6.

Dans le report fourni par le Prolin 4 toutes les variables du modèle sont disposées dans le même ordre d'entrée. Pour chaque variable on reproduit les données suivantes:

- Code et description alphabétique
- Condition (dans la base, à la borne supérieure ou inférieure, etc.)
- Quantité optimale
- Borne inférieure et supérieure d'entrée.

— Variations quantitatives (en diminution ou en augmentation par rapport à la valeur optimale) qui correspondent aux valeurs extrêmes du 'Cost Range'.
— Coefficient économique d'entrée
— Valeurs (minimum et maximum) du coefficient économique où les quantités de la base optimale restent inchangées ('Cost Range')
— Codes des variables entrant et sortant de la base optimale, en correspondance des valeurs du coefficient économique hors du 'Cost Range', ainsi que le sens de variation qui leur correspond (A = augmentation, D = diminution).

Le report donné par le Prolin 6 indique les variations marginales de niveau des variables 'de base' par rapport à l'introduction des variables 'hors base'.

3. CONCLUSION

La procédure illustrée résulte de l'expérience acquise à la suite de l'utilisation positive de la programmation linéaire pendant plusieurs années au sein de notre Société.

En effet, cette expérience nous permet de considérer la rapidité de préparation, la possibilité de réduire les erreurs ainsi que la facilité de constater celles qui se sont glissées inévitablement et enfin la rapidité d'interpréter les résultats, comme instruments indispensables pour l'utilisation profitable de la programmation mathématique dans la gestion de l'enterprise.

DISCUSSION ON THE PAPERS BY A. CLARK AND L. PESSINA

D. Edmonds

In response to Mr. M. Arbabi, I would like to mention our experience with Crash in MPS/360.

First its usefulness is very model dependent and it has been known to increase running time slightly.

Second when we use Crash we use Phase B only.

Third we only use Crash in models where Phase B completes as opposed to terminating with Etas ex core.

M. G. Shaw

I should like to ask both Mr. Clark and Dr. Pessina three questions:

1. How much programming and analytical effort was required to set up their systems?
2. How long did their systems take to get on to a production basis?
3. Did their generators automatically set up a starting basis?

A. Clark

1.	Model Building	6 man months
	Programming	4 man months
	Testing & Debugging	2 man months
2.	20 months	
3.	No	

Mr. Edmonds's comment was noted.

L. Pessina

1. Le temps pour réaliser la procédure à été de 6 personnes-mois.

2. Le temps machine necessaire à preparer le file pour MPS depend du nombre des elements de la matrice, suivant une loi de variation à peu près linéaire.

   A titre d'exemple pour un modèle à 10, 000 coefficients le temps sur ordinateur 360/50 est environ de 10 minutes, sans impression du 'listing'.

3. Cette procédure est déjà comprise dans le code MPS.

The Problem Formulation Language of the NCC Linear Programming System

SVERRE SPURKLAND
Norwegian Computing Centre, Norway

1. INTRODUCTION

Input to the programme is two-phased.

Any number of matrices, vectors and name-lists may be read from cards onto magnetic tape. This part of input is envisaged as rather stable.

We refer to the matrices and vectors on magnetic tape as input matrices.

The second part of input is the problem formulating language.

MS-statements operate upon input matrices, the one by one unit matrix which is called E, and matrices previously defined by MS-statements.

The most characteristic feature of the matrix generator is that the LP-problem matrix is not established within the computer.

A basic feasible optimal solution for one problem may be used as initial solution for another problem. The two problems may be completely different. A big problem may for example be solved stepwise in that new variables and constraints are added step by step, the final solution at one step being used as intial solution for the next.

Problems are identified and connected by LP-statements.

The general LP-problem considered is rather too general in most cases, the form of any particular problem is, however, easily stated by a MAX-MIN-statement

By the use of PRINT, FORM and WRITE-statements, output may be made highly selective regarding both the conditions under which output is to be done and as regards how and what is to be printed, or written on magnetic tape.

The conditional IF-statements may be useful in that any ad hoc features may be added to the programme.

The information pertaining to a problem starts with a LP-statement and ends with the next LP-statement (or with an END card). To each LP-statement there must be one and only one MAX-MIN-statement, any number of paired PRINT and FORM-statements, and any number of WRITE and IF-statements.

The MS statements are global.

The programme has been in operation for about one year.

2. MAX–MIN STATEMENTS

The general linear programming problem considered is:

maximize $z = (c + \alpha d)X$

subject to $\alpha \underline{w} + \underline{u} \leqslant v + BX \leqslant \bar{u} + \alpha \bar{w}$

and $\underline{x} \leqslant X \leqslant \bar{x}$

α is a parameter and a solution to the problem is wanted for all values of α in the interval $(\underline{\alpha}, \bar{\alpha})$.

c, d, $\underline{w}$, $\underline{u}$, v, $\bar{u}$, $\bar{w}$ $\underline{x}$ and $\bar{x}$ are given vectors, ($1 \times n$ or $n \times 1$ matrices) and B is a given matrix.

The problem stated above is given to the programme in the following form:

$$\text{MAX } ((c + \alpha \,.\, d) \,.\, X/\alpha \,.\, \underline{w} + \underline{u} < v + B \,.\, X < \bar{u} + \alpha \,.\, \bar{w}, \underline{x} < x < \bar{x}) \ :$$

The vector X must always be denoted by the single letter X.

The other vector and α and B may be given any name.

A vector may also be denoted by a number. All the elements are then taken to be equal to that number.

If $\underline{w} = \bar{w}$ and $\underline{u} = \bar{u}$ we may write:

$$\text{MAX } ((c + \alpha \,.\, d) \,.\, X/v + B \,.\, X = \bar{u} + \alpha \,.\, \bar{w}, \underline{x} < X < \bar{x}) \ :$$

As a rule only some of the vectors are non trivial, and the programme accepts

$$\text{MAX } ((c + \alpha \,.\, d) \,.\, X/v + B \,.\, X < \bar{u} + \alpha \,.\, \bar{w}, \underline{x} < X < \bar{x}) \ :$$

and

$$\text{MAX } ((c + \alpha \,.\, d) \,.\, X/\alpha \,.\, \underline{w} + \underline{u} < v + B \,.\, X, \underline{x} < X < \bar{x}) \ :$$

The programme will also accept a number of forms obtained from the forms stated above by the following rules:

$(c + \alpha \,.\, d)$ may be substituted by $(\alpha \,.\, d)$ or c

$\alpha \,.\, \underline{w} + \underline{u}$ may be substituted by $\alpha \,.\, \underline{w}$ or $\underline{u}$

$\bar{u} + \alpha \,.\, \bar{w}$ may be substututed by $\alpha \,.\, \bar{w}$ or $\bar{u}$

$v + B \,.\, X$ may be substituted by $B \,.\, X$

, $\underline{x} < X < \bar{x}$ may be substituted by , $\underline{x} < X$ or

, $x < \bar{x}$ or may be deleted.

MAX may be substituted by MIN.

Altogether the number of acceptable MAX-MIN statements is 864.

3. LP–STATEMENTS

More than one problem may be solved in a sequence.

A basic solution of one problem may be used as the initial solution for another problem. Problem name, initial-solution–directive and parameter range are given in the LP-statement.

The general form of this is:

LP nl (α), USE n2 (Y), $\underline{\alpha} < \alpha < \overline{\alpha}$:

n1 is the name given to the problem.

α is the parameter name that is used in the MIN-MAX statement $\underline{\alpha}$ and $\overline{\alpha}$ are numbers giving the range of α.

n2 is the name of another parametric problem, the number y is within the range of that parameter.

If available, the solution of problem n2 with parameter value y is used as initial solution for problem n1 with parameter value $\overline{\alpha}$.

When the solution of n1 with parameter value $\overline{\alpha}$ has been obtained normal parametric steps are taken until α reaches the value $\underline{\alpha}$.

There is only one constraint on the use of the USE-directive. Let A and B be the matrices of problems n1 and n2. Then the dimensions of A must be at least as great as that of B, and the left—uppermost part of A must be identical with B. The programme does not check that this rule is followed, and trouble is very likely to arise if the rule is violated.

There are five other forms of the LP-statement.

LP n1 (α), USE n2, $\underline{\alpha} < \alpha < \overline{\alpha}$:

The problem n2 is not parametric.

LP n1, USE n2 (y) :

The problem n1 is not parametric.

LP n1, USE n2 :

Neither n1 nor n2 is parametric.

LP n1 (α), $\underline{\alpha} < \alpha < \overline{\alpha}$:

Standard initial solution is used, that is the problem n1 with parameter value $\overline{\alpha}$ is solved from scratch.

LP n1 :

The problem is not parametric and standard initial solution is used.

4. MATRIX—STATEMENTS

Any number of input matrices and vectors may be operated upon by matrix-statements.

The letter E is the name of the one by one unit matrix. The form and meaning of matrix-statements are most easily clarified by some examples.

The matrix, UNIT, defined by the statement

MS UNIT = E(100)(1, 1) :

is a 100 × 100 unit matrix.

The meaning of the statement is that the matrix E is to be repeated 100 times, the displacement between each location of E being (1, 1). That is the indices of E is incremented by (1, 1) for each repetition.

The matrix, ROW, defined by

MS ROW = E(100) (0, 1) :

is a 1 × 100 matrix, all elements of which are equal to one.

The matrix, FROMTO, defined by

MS FROMTO = UNIT (20) (0, 100) + :

MS + (100, 0) ROW (20) (1, 100) :

is a transport problem matrix with 100 origins and 20 destinations.

The meaning of the second term is that the first of the twenty occurrences of ROW is located from (100, 0). The displacement between each location of ROW is (1, 100).

The most general term accepted in a matrix statement is (h, k) Z . A(n) (s, p) the sum of any number of such terms is accepted.

The location pair (h, k) may be deleted if it is (0, 0).

The multiplicand Z is a number which is multiplied into every element of the matrix A.

The displacement pair (s, p) and the repetition number n may be deleted if n is 1.

s and p may have negative values.

One special matrix statement is accepted. The matrix B defined by

MS B = (TRANSP) FROMTO

is the transpose of FROMTO.

The content of a general matrix statement is given by

$$A = \sum_q Q_q B_q + \sum_t (\text{TRANSP})\ B_t$$

where B_q and B_t are known matrices and

$$Q_q B_q = (h_q, k_q) Z_q \cdot B_q(n_q)(s_q, p_q)$$

The operators Q_p are distributive

$$Q_q(B_1 + B_2) = Q_q B_1 + Q_q B_2$$

and commutative

$$Q_1 Q_2 = Q_2 Q_1$$

(TRANSP) and Q_q do not commute however, in fact

(TRANSP) $Q_q B_q$ Q'_q (TRANSP) B_q

where

$$Q'_q = (k_q, h_q) Z_q (n_q)(p_q, s_q)$$

It is seen that

$$\prod_q Qq = S \prod R_q$$

where

$$S = \left(\sum_q h_q, \sum_q k_q\right) \prod_q Z_q$$

and

$$R_q = (n_q)(s_q, p_q)$$

5. PRINT AND FORM-STATEMENTS

Output on the high speed printer is controlled by PRINT and FORM-statements.

PRINT-statements state when printing is wanted, FORM-statements. how and what should be printed.

A PRINT-statement must always be followed by a FORM-statement.

Six PRINT-statements are available:

PRINT ITRE (n1, n2, n3) :

n1, n2, n3 are integers.

Printing in accordance with the FORM-card following the PRINT-card is to be done every n2 iterations, from iteration n1 until iteration n3.

PRINT TIME (n1, n2, n3) :

Printing is to be done every n2 seconds from time n1 to time n3. The first iteration starts at time zero.

PRINT STEP (a1, a2, a3) :

a1, a2, a3 are fixpoint numbers.

Printing is to be done when the parameter α in the problem is a1, a1 + a2, a1 + 2 . a2, etc. until a3.

```
PRINT    COND (C1)          :
PRINT    COND (C1, C2)      :
PRINT    COND (C1, C2, C3)  :
```

C1, C2, C3 are the names of possible conditions of the problem. For the time being only FINITE, INFIN and NONSOL are recognized.

For example: PRINT COND (INFIN, NONSOL) means that printing according to the FORM-card following should be done if the solution is either non-existent or infinite.

The two most important forms of FORM-statements are exemplified by:

FORM (8F8.4) T(R) TLIST :

and

FORM (6E10.2) T(R . SET) TR000 :

Here are (8F8. 4) and (6E10. 2) formats interpreted as in FORTRAN. T is the name used for the dual variables and R is the set of critical constraints, and so T(R) is the non-zero dual variables. TLIST is the name of a name-list, the elements of which are to be used as names for the elements of T. This name list should be read from cards onto tape together with the input matrices and vectors. If the last character of the namelist-name is zero, names will be generated by the programme as exemplified by TR000, TR001, ... , TR999. SET is a set of indices, and R. SET is the product of the two sets R and SET. If for example R = (0, 3, 8) and SET = (0, 2, 8, 10) only the element T_8 of T which, in this example will be called TR008 will be printed.

The non-default values of a vector define an index set, for example the set SET may be defined by

MS SET = E(4)(0, 2) :

The non-zero primal variables, slack variables and reduced cost are denoted by X(N), U(Q) and K(M). Subvectors of these may be denoted X(N . SET) etc. The printing of these and a number of other vectors may be directed in the same way as shown in the examples above.

The features described so far are completely implemented.

6. WRITE-STATEMENTS

Output on magnetic tape is controlled by WRITE-statements, exemplified by

```
WRITE  TIME (900, 900, 90000)     X(N . SET)   :
WRITE  STEP (5, 13, 700)          T(R)         :
WRITE  COND(FINITE)               U(Q)         :
WRITE  ITRE (7, 11, 450)          A(N)         :
```

7. IF-STATEMENTS

Three forms are accepted:

```
IF   C1   THEN   GO   TO   A1   ELSE   GO     TO   A2 :
IF   C1   THEN   GO   TO   A1 :
IF   C1   THEN   GO   TO   A1   ELSE   STOP   :
```

A1 and A2 are user-programmed subroutines.

An Approach to Matrix Generation and Report Writing for a Class of Large-Scale Linear Programming Models

MARTIN J. DILLON, P. M. JENKINS, MARY J. O'BRIEN
Research Analysis Corporation, U.S.A.

SUMMARY

This paper presents an approach to matrix generation and report writing for a class of large-scale linear programming models. The approach differs from some commercially available general systems in that a combinatorial logic based on particular mnemonics is discarded in favor of a generative approach, which operates on sequential integers by implicit rules. The basis for a general matrix-generator language is described in terms of the logical operation required to produce vector and row names and to output the matrix coefficients. These operations require models to be conceived as a group of 'vector sets.' A vector set is defined as a collection of vectors and their associated constraints, which can be generated by consistent logic. The greater part of the paper is concerned with the definition of the logic required within a vector set and methods of partitioning a model into a number of vector sets. A general report writer has been developed, based on the sequential integer mnemonic system. Finally, the present status and possible applications of this matrix-generation approach is discussed.

1. INTRODUCTION

Linear programming has been used by the US Department of Defense (DOD) as an important tool for the annual consideration of strategic mobility and resource allocation requirements.* The size of the matrices increased from a 300-constraint model in 1964-1965 to a 3000-constraint version in 1966-1967. By this time, the generation by hand or even by semiautomated methods had become a very onerous task. Furthermore, DOD requirements have changed with increased experience. No longer is it sufficient to describe a single deployment scenario with analysis restricted to postoptimal or minor structural alterations. Rather the DOD prefers to consider sets of possible scenarios and activity options that necessitate separate models to describe. Since a deployment activity is described in a consistent manner as the removal, transportation, and delivery of a commodity, the DOD requirement is for a generator based on a consistent description of the deployment activities but capable of grouping these activities to represent a variety of assumptions and scenarios. One of the prime characteristics of a generator system should therefore be a means of achieving this kind of structural change within the matrix description.

The 1968 model became the first of the series to be generated entirely automatically. The generator was written in FORTRAN for the IBM 7040/44 to run under LPIII linear programming package and also under MPS/360, the IBM mathematical programming system on the IBM 360 series of computers. The matrix generator so produced contained characteristics such that general applications of the approach could be envisaged, certainly for use by the military and Department of Transportation user.

---

* A more detailed description of the history is given in 'A Large-Scale Mathematical Programming Model of the US Department of Defense Mobility Resources Acquisition and Allocation Problem,' Jenkins, P.M., Mary J. O'Brien, and Justin C. Whiton, also presented at this conference.

This paper discusses an approach and presents a language and mnemonic system that, it is thought, considerably reduce the onus on model builders who require a flexible matrix generator and a report-writing facility. Although matrix generation can be useful to model builders both as a means of reducing the quantity of input required and as a means of increasing the flexibility of the system, the builder must pay the price of considering his model in the light of the matrix-generator formulation.

A typical model builder first conceives a matrix in terms of the constraints and the characteristics of the vectors that enter into them, this being the only satisfactory method of describing the problem space for optimization. On the other hand, definition of the elements tends to force consideration of the matrix from the viewpoint of vector definition. This is a result of the nature of vectors (in general, a description of many of the problem variables) and the nature of constraints (usually a summation over some of the variables making up a vector). This approach considers a matrix as a group of vector sets; a vector set is defined as a group of vectors that may be generated by consistent logic and which enter into constraints in an identical or at least very similar manner. Each vector set is generated independently and contains its inherent logic. The original matrix generators were written in FORTRAN, but it was found that the logical statements necessary to define vector sets were limited in the sense that from vector set to vector set the same statements were required in differing combinations. This led to the possibility of building a general system of matrix generation, the logical operations being defined in a library and accessed via key phrases.

2. APPROACH

The approach, unlike some general matrix generators, uses a generative rather than a combinatorial logic. With the combinatorial approach a variable* is assigned a mnemonic and combined with other named variables. Matrix coefficients are described by a defined position on the basis of the mnemonic name. This seems too restrictive since each variable has to be named independently; because the names have no inherent structure, operations must be defined explicitly. A simple example may better explain. Assume two variables named 'toffee' and 'nougat' which enter into the vector names, and assume they enter into the same type of constraint. There is no implicit way of proceeding from 'toffee' to 'nougat' either in the vectors or constraints, or to access data arrays. However, to preempt the generative approach, if

$$\text{toffee} + 1 = \text{nougat}$$

most of these difficulties vanish. Whenever possible the generative approach assigns sequential integer values for the variables that enter into the vectors. Each variable type is also given a unique position in the vector name. The construction of names for vectors and constraints is generative, using integer values as a code in the sense that, given the logic to produce one vector name, many more may be generated by selectively and successively incrementing the values of the positions. This is possible primarily because integer values, unlike character strings (that are used as mnemonics in the combinatorial approach), can be used to access data arrays conveniently.

It should be noted that the ultimate form of the vector names, depending on the linear programming system being used, need not necessarily be a string of numbers like '178893.' If enough characters are available for vector and row names in the linear programming system, these same integers may be used to access data arrays that

---

* During the succeeding discussion, the term 'variable' is used to denote a quality that enters into vectors and constraints. The terms column and vector are interchangeable, as are row and constraint.

contain alphabetic mnemonic codes for vector name construction. Thus in an inter-city transportation problem, for example, San Francisco (SF) = 1, New York City (NY) = 4, London (LO) = 3. When the combination of a movement from San Francisco via New York to London is accessed via 1, 4, 3, the mnemonic becomes SF.NY.LO. This method in fact defines the logic of the model quite independently from the mnemonics since vector names may be contracted, expanded, or otherwise altered merely by changing the mnemonic data arrays.

It seems advisable at this stage to define a language and the tools with which to discuss the implementation of the generative approach to matrix building. Examples are used liberally to illustrate certain points in the discussion.

3. LOGIC WITHIN VECTOR SETS

For the purposes of discussion it is assumed that vector names may be defined in terms of six indexes: i, j, k, l, m, n. A vector ijk$l$mn will be named, IJKLMN (i.e., integer values are assigned to the indexes and define the name). A typical constraint

$$\sum_{i=e}^{f} \sum_{k=c}^{d} \sum_{l=a}^{b} x_{ijklmn}$$

will be named OJOOMN. The indexes summed over will be assigned a zero in the constraint, and the remainder form a set of rows defined by the allowable combinations of J, M, and N. These two rules form the basis of the generative system. Each problem variable represented in the vector is assigned a unique position in the vector name (i to n) and retains that position whenever possible. Any variable index summed over becomes zero or some other indicative value.

3.1 Vector Generation

With the system described, the types of operations required to produce vectors are comparatively limited. A typical vector set has some fixed positions indicative of the nature of the set. The other positions take values between defined limits. All combinations of the values are generated, some being discarded as unallowable or not required.

Example 1 will clarify:

In our system IJKLMN

Assume M = aircraft and fixed

N = commodity and fixed

these values defining the vector set.

Let I be the origin

L be the destination

J be the time period

Let K in this example be an unassigned position

I may have r values 1, . . . , r

L may have s values r + 1, . . . , s + r + 1

J may have t values 1, . . . , t

Vector generation is accomplished by fixing M, N and K, then incrementing the values of I from 1,r and L from r + 1 to s + r + 1, and J from 1, t. However, it may be that not all origins may serve all destinations, in which case a data array must be accessed to determine whether IL form an acceptable combination. The accessing may be accomplished through the indexes r and s. For example, if the position in the array ORIDST (r, s) is not zero then the pair is acceptable.

Another type of test may arise with time period J. The commodity N may not be required at every time period, and the requirement may end prior to the 't' th period. An array NREQ (L, J) stores the time periods at which there is a requirement for N at destination L. Since deliveries are possible at any time period until no later requirement exists, the test must involve accessing the array with all J values from the present 'j' to the 't' th with any nonzero value providing an acceptable L, J combination.

The types of operation required to produce a vector within a set may be defined as: *

(1) Fixed variable value.

(2) Increment for all values between given limits. (Let this operation be abbreviated to INC.)

(3) INC with simple yes/no test to an array.

(4) INC with test for any positive response from a series of stored values.

The general matrix-generator language for example 1 would be:

```
VECTORS
FIX COL5 = M, COL6 = N, COL3 = 0
INC COL1 (1, r)
INC COL4 (r + 1, s + r + 1)
TEST ORIDST (COL1, COL4 − r)
INC COL2 (1, t)
TESTF NREQ (COL4 − r, COL2)
```

The arrays ORIDST and NREQ are zero/one arrays, one indicating an allowable combination. When a combination is allowed, generation proceeds; when a combination is not allowed, a transfer is made to the nearest preceding INC. The FORTRAN matrix generator produced by these statements is listed in App B. TEST is the code word for a one-shot test in an array. TESTF indicates that tests in the NREQ array must be made from the present value of the variable in the nearest preceding INC to the maximum value. Any positive response is acceptable.

3.2 Constraint Generation

Constraint definition is, in the simplest case, that indicated earlier. The indexes summed over become zero. In our earlier example, if a constraint was required summing the total quantity of commodity N delivered during all time periods to the destination L, all vectors of IJOLMN would enter a constraint OOOLON. Another type of operation is one that requires a series of constraints to be generated.

An example would be the necessity of time phasing the cumulative requirements at destination L of commodity N, in which case a series of constraints OJOLON is required for every J value from the present 'j' th to the final 't' th. This may incorporate a test for acceptable L, J combinations. Occasionally a series of constraints must be generated from a given combination of vector variables. An example would be a vector describing an operation that requires five inputs. Thus, from the vector IJOLMN, constraints of the type OOKLOO, for example, would be generated, where K would have values 1 through 5. The types of operation required to produce constraints

---

* This is not necessarily an exclusive list of possible operations. For an explanation of the elements of the general language see App A.

by operation on the vector variables can be broadly classified as follows: *

(1) Fix mnemonic value.

(2) Simple transfer of value from the vector to the row.

(3) Simple transfer with a test to decide whether the constraint is required.

(4) Transfer with incrementation to produce a series of constraints with or without tests to determine their desirability.

(5) Access to a routine that will generate the required constraint (s). These may be generated using the above rules, but a separate routine may be advisable owing either to the complexity of the logic or to the general nature of the constraints (i.e., they are used in many vector sets). †

The matrix-generator formulation of constraints for some of the examples would be:

```
BLOCK
FIX ROW1 = 0, ROW2 = 0, ROW3 = 0, ROW5 = 0
ROW 4 = COL4, ROW6 = COL6
COVAL = 1.
OUT GE
END
BLOCK
FIX ROW1 = 0, ROW3 = 0, ROW5 = 0
ROW4 = COL4, ROW6 = COL6
INC ROW2 FROM COL2 TO T
TESTF NREQ (COL4, ROW2)
COVAL = 1.
OUT GE
END
```

Each block produces one or a set of constraints. Any operations within a block are restricted to that block, thus INC (incrementation) of ROW2 from COL2, extends only to the END. As in vector generation, TEST and TESTF determine whether a particular combination is to be allowed. COVAL is the matrix coefficient and will be discussed later. The command 'OUT' produces a matrix element in the format required by the linear programming package to be used for solving the problem. Generally a string of characters representing the vector name, the constraint name, and the coefficient is output. When generation of elements is finished, it is necessary to list all constraints along with indicators of their types, i.e., objective, equal, greater or equal, and less or equal. In the generator, the constraint type is given in the 'OUT' command, OUT 'ind,' where ind is either EQ, GE, LE, or OB. All constraints generated are stored along with their types to provide the linear programming package with the requisite list.

The matrix coefficient can be obtained in any of three ways:

(1) Fixed value

(2) From a data array

(3) From an expression

---

* This is not an exclusive list.

† In practice, anything that is difficult to define by a generative relation is usually difficult to define by combinatorial methods. It is always possible using the system given here to form the vectors and constraints combinatorially in separate vector sets.

The accessing of the data arrays is accomplished through the integer values of the variables making up the vector or constraint.

In summary, all vectors and row names are defined by positive integer values that have positional uniqueness. Vectors are generated by incrementation of positional values. Acceptable vectors are chosen by logical rules that are limited in number. Row names are derived from vector names by transference, again employing only a limited library of logical statements.

4. VECTOR SETS

A vector set is one of those irritating things that is easy to define and describe but can be difficult to apply in particular situations. A vector set is a group of vectors that may be generated by incrementation and with consistent logic and that enter into constraints in an identical or similar manner. However, the model builder is frequently presented with the problem of partitioning a model into vector sets in a manner that often cuts across his method of conception. For example, a transportation problem may be conceived as operations between cities, each city being described as input/output of commodities and their transportation means. If each city has similar characteristics but transportation means differ widely, the model may best be broken up by transportation-type vector sets rather than city vector sets. This will be particularly true if there is no logical reason for going from one transportation method to another (sea and air means, for example).

In order to effectively use vector sets as a means of matrix generation, the model builder must attempt to break up the model in various ways to determine which sets produce the most consistent and simple logical sets. Again, an example may serve to indicate the kind of considerations required:

Example 2.

I = destinations of some type, of which there are 'r'

J = origins and let there be 's'

K = intermediate positions that may or may not be included in I, J, = 'X'

M = methods of transportation including air and ground means = 'M'

N = commodities required to be moved = 'c'

L = discrete time periods = 't'

The model must represent the movement of the commodities from destinations to origins either directly or via some intermediate position. Objective functions can be such things as least-cost or shortest time of delivery. The model builder's problem is to partition the model into consistent sets. Time periods will probably always be contained within the vector sets. This is usually true of time-dependent models unless the mode of operation varies with time. The method of transportation may restrict the type of commodity carried. If this is true, these may form separate vector sets. If all commodities (or almost all) may be carried with no significant difference in the constraints required, it may not be necessary to form many vector sets. It will probably be essential to treat air and ground delivery means separately. It would also be wise in most circumstances to have different vector sets for direct origin-destination routes from those via an intermediate site, as constraints will probably be necessary for the intermediate sites. If some operations involve changing commodities at intermediate sites, this may mean more vector sets. Thus a possible break-up of the model might be:

Two main categories.

(a) Air delivery systems

(b) Ground delivery systems

Within each category operations would be split up.

(1) Direct origin to destination

(2) Origin via intermediate to destination

(3) Origin to delivery at intermediate to destination

This will produce six vector sets.

There is almost inevitably a great deal of logic that is consistent across vector sets, e.g., in the previous example, a tonnage constraint for the quantity of commodity delivered at a destination. Vector-set mnemonics should be structured so that advantage may be taken of this consistency. Any general system should allow access to a special series of statements defined by arguments that generate the constraint. In the above example for the tonnage constraint, perhaps all vector sets produce the constraint IOOONO, which is independent of origin, time period, or transportation means. The arguments would then be COL1 (I) and COL5 (N). Within a vector set the constraint would be defined

```
BLOCK
ACCESS COMDEST (COL1, COL5)
END
```

where COMDEST is a series of statements operating from the argument values and independent of their origin. In general it seems more desirable to have many vector sets with comparatively simple statements than to have a few sets that contain involved logic. The ability to access groups of statements outside the program flow makes the former even more attractive.

5. DATA

One of the characteristics of most large-scale linear programming models is the vast quantity of data required. At present it is thought that the general system should have little capability to carry out arithmetical operations on data. It is preferable that large quantities of data generated from some smaller quantity of basic data should be handled by a preprocessor outside the matrix generator. Unlike many industrial matrices, DOD matrices do not tend to operate from a large number of comparatively small tables of data. Rather, a small quantity of basic data must be processed into the form required from the coefficients in the matrix. The processing, which results in a data expansion of more than two orders of magnitude, is best handled outside the general program flow and made available to the general program in data arrays.

6. REPORT WRITING

Report writing in linear programming refers to the processing of solutions from linear programming runs for presentation in a convenient format. Operations, in general, consist of searching for vectors in a solution, grouping them, summing or adjusting their activity levels, and constructing tables from them with appropriate titles, column headings, and row names.

The advantages of sequential integer codes to any form of report writing are obvious when the number of vectors that may appear in a solution is very large. To name vectors that one would like to represent in a table individually is onerous. Since each position in the vector name is assigned uniquely to a variable type in the problem and the specific variables are assigned integer values by incrementation, describing characteristics of a vector set for presentation in a table is never difficult. The system in current use has a language designed to exploit these features of the generator and combines the specification of sets of vectors with table descriptors. The two in conjunction determine which individual vectors are to be sought in a solution and how they are to appear in tabular form.

The set specification is comprised of positional descriptors containing a combination of specific values and range of values equivalent to those in the generator code. A position may be defined in one of three ways: by an integer value; by an asterisk; or by an 'OR' expression. The first specifies that the position must have a unique value. The asterisk indicates either that the value in this position is unconstrained or that the position values are defined in the table descriptors. The 'OR' expression is used to define a subset of acceptable values for that position. Thus:

SET A . B . C . (X . Y . Z). * . *

defines: Positions 1 through 3 to be A, B, C, respectively
Position 4 to be X.OR.Y.OR.Z.
Positions 5 and 6 to be either unconstrained or defined in the table descriptors.

The table descriptors take the form:

COLUMNS: COLn = r-j

ROWS: COLn = k-l

where 'n' determines the position in the vector name. In COLUMNS, both the shape of the table and the vectors to be included in each column are defined ('1-9' would direct a table of nine columns to be constructed). Similarly, ROWS determines implicitly the number of rows a table is to have and fixes the nth position in the vector name for each row. In conjunction with the SET specification, the table descriptors determine that in the ith column and the jth row, vectors with the ith value from the starting column count and the jth value from the starting row count will be considered for the table.

The report writer output is primarily tabular with the ability to include text descriptive of the type of run or the information contained in tables.

7. DISCUSSION

7.1 Present Status

At present no general system exists for the conversion of the language described into a matrix generator tied to the particular user. The approach is being examined for its applicability to transportation and nontransportation models. It is also applied to various classes of model formulations to examine its aptitude.

For production models, the language is being used as an intermediate stage between a model builder, unskilled in programming, and a FORTRAN programmer. The programmer requires only a library of FORTRAN statements equivalent to a phrase in the general language. Even at this stage the time required to generate a model has been decreased by a factor of three or more.

The foundation of a general system is in existence. Utility routines dealing with the mechanics of matrix generation have been developed for the IBM 7040/44 and are easily converted into the IBM 360-series format. The utility routines affect the naming of columns and rows, the production of matrix elements, the storing of row names, and the eventual listing of the matrix in the required linear programming format.

Preliminary work on a general system has been started, especially in regard to the input of data in a convenient manner. (This is always a problem in a general system.)

On the other hand, the report writer is in an advanced stage of development. It is a general system not necessarily tied to any mnemonic code, which has a language and syntax in use in production runs. It can be added to the end of a linear programming run and will operate on LPIII or MPS/360 solutions.

7.2 Applications

The type of problem for which this approach seems appropriate are those models that have a large number of variables defining the vectors (or constraints) where most of the permutations of the variables are possible vector candidates. These characteristics are frequently found in large problems containing a large number of related vectors and constraints. Thus, it has been of use in the DOD models, and its use for the transportation models has been suggested.

Use of this technique for large industrial matrices, refinery models, blending models, and woodcutting models has not been examined owing to the lack of availability of large models rather than the scarcity of examples of the type of model. In the smaller examples, because of the ordered nature of the matrices, it can be applied with ease. App C describes a simple transshipment problem and the matrix generator associated with it.

It may seem that a great deal of sophistication in conception by the modeler is required in order to apply the ideas of vector sets and logic within a vector set. This may be true, but it is difficult in general practice to conceive of a large model without partitioning the vectors and constraints into groups dependent upon the type of operation involved. The alternative is surely a conception of unrelated vectors and constraints. With large matrices a conceptualizer is forced to consider the model in the light of what is expedient for the general matrix-generator systems. The advantages to be accrued by automated matrix generation and report writing far outweigh the disinclinations of the builder to learn and apply a general system, whether this one or some other.

APPENDIX A: GENERAL MATRIX GENERATOR LANGUAGE

The following gives a brief description of a subset of the General Language, which deals with the definition of a vector set and its associated restraints:

The terms in the language:

Delimiters	Operators	Macros
	FIX	MACRO
VECTORS	INC	ACCESS
BLOCK	TEST	
END	TESTF	
	OUT	

Explanation and Use

I. Delimiters:

VECTORS: initiates the definition of a vector set.

VECTORS Statement

VECTORS 'M'

where 'M' is an integer value that names the set relative to other sets.

BLOCK: heads a sequence of statements for constraint generation that are to be treated as a unit and whose execution is to be completed before moving on to the next BLOCK.

END: determines the end of a constraint BLOCK.

Example:

```
VECTORS 1
    .
    .    (sequence of statements defining vectors)
    .
    .
BLOCK
    .
    .    (sequence of statements defining constraints)
    .
    .
END
BLOCK
    .
    .
    .
    .
END
VECTORS 2
    .
    .    (vector definition and sequence of blocks for next vector set)
    .
    .
END
```

II. Operators:

FIX: fixes the values of one or more of the variable names used in the language to refer to the positions in the vector and row name mnemonic system.

The variables are COL n and ROW n where n ranges from 1 through the maximum number of positions used.

FIX Statement:

FIX COLn = i, COLn' = l', . . .

Where 'i' is an integer, or variable with an integer value. It may be used as an index to an array containing mnemonics for the COLn position.

INC: determines an iterative procedure whose range extends:

(a) when it occurs within the VECTORS section, to the last END statement in the last BLOCK of the set.

(b) when it occurs within a BLOCK, to the END which terminates the BLOCK.

The number of iterations is determined by parameters in the INC statement.

INC Statement:

INC 'variable name,' FROM exp_1 TO exp_2 where 'variable name' is either COLn or ROWn or another variable name local to the set and 'exp_1', for our purposes here can be considered an integer value.

Within the range of the INC, the variable takes on a value from exp_1 to exp_2, increasing by one for every iteration until exp_2 is reached and the INC is terminated.

For those familiar with ALGOL, FORTRAN, or PL/1, INC statements act similarly to DO loops, with a similar ability to be nested.

Example:

INC COL4, From 1 to 10

INC COL5, From 1 to 6

TEST: causes an 0, 1 decision array to be accessed. If the test is 1, execution proceeds to the next statement, if it is 0, execution transfers to the previous INC.

TEST Statement:

'ARRAYNAME'

(exp. 1, exp. 2, exp. 3)

where 'arrayname' is the name of an array either previously defined in the GENERAL DATA area of the program or is local to the current VECTORS set, and 'exp' are expressions whose values are integers pointing to a position in the array, either 1, 2, or 3 expressions may be used.

TESTF: differs from TEST in two critical respects:

(a) Its use requires that one 'exp' in the array pointers uses the incrementing variable in the preceding INC.

(b) Where TEST tests only one position in the named array, TESTF tests all positions forward from the current value of INC variable until it either reaches the maximum size of the INC variable or finds a position with one in it. Otherwise, TESTF acts as TEST:

when a 1 is encountered in the array being tested, execution proceeds. If no 1 is found, execution transfers to the previous INC.

TESTF Statement:

TESTF 'Arrayname' (exp 1, exp 2, exp 3)

OUT: produces one matrix element each time it is executed.

OUT Statement:

OUT 'ind'

where ind is either EQ, LE, GE, or OB.

One additional statement is necessary:

COVAL = exp

This statement assigns the value of 'exp' to all elements generated between it and the next COVAL assignment.

III. Macros:

MACRO: heads a series of statements that are to be executed as a unit.

MACRO Statement:
MACRO 'macro name' (arg list) where 'macro name' is the name used in the ACCESS statement when the MACRO is to be executed, 'arg list' are variable names used in the MACRO as dummies, whose values are to be obtained during execution from the ACCESS statement.

Example:
```
MACRO    VECSET (I, J)
  .
  .      (statements to be executed)
  .
END
```

ACCESS: causes a MACRO to be executed.

ACCESS Statement:

ACCESS 'mainname' (arg list) where mainname refers to a previously defined MACRO and arg list gives the variables whose values are to be used in place of the dummy variable in the MACRO.

Example: ACCESS VECSET (COL1, COL2)

APPENDIX B: VECTOR GENERATION, A FORTRAN EXAMPLE

```
      SUBROUTINE    AIRCOM

      Declarations

      COL5 = M

      COL6 = N

      COL3 = O

      COL1 = O

1     COL1 = COL1 + 1
      IF (COL1.GT.R) RETURN
      COL4 = O

2     COL4 = COL4 + 1
      IF (COL4.GT.S) GO TO 1
      IF (ORIDST (COL1, COL4) EQ.O) GO TO 2
      COL2 = O

3     COL2 = COL2 + 1
      IF (COL2.GT.T) GO TO 2
      I1 = COL2

4     IF (NREQ (COL4, I1).EQ.1) GO TO 5
      I1 = I1 + 1
      IF (I1.GT.T) GO TO 2
      GO TO $

5     COLNAM = NAME (1)

C     NAME IS A FUNCTION THAT FORMS THE VECTOR NAME FROM THE
      ELEMENTS COL1-6

C     THIS COMPLETES THE VECTOR NAME
```

APPENDIX C: A SHORTEST ROUTINE PROBLEM

A single quantity of commodity is required to be trans-shipped from an origin through intermediate sites to a destination. The network of possible routes is shown on the facing page.

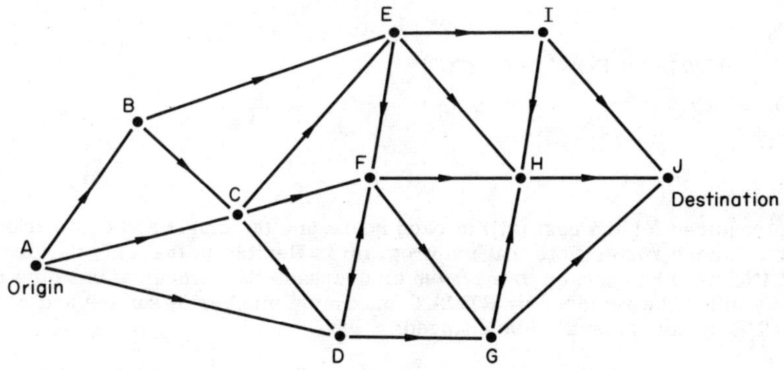

A Transhipment Network

Each route from origin 'i' to destination 'j' has a cost or value associated with it 'aij.' There are in the above example 22 possible origin and destination combinations.

The matrix requires each vector, an origin-destination pair, to have three elements; a cost row, an outgoing row, and an incoming row (the latter two have matrix coefficients of 1. and -1. respectively).

The general matrix generator program to produce this would be:

```
COLUMNS
INC     ROUT FROM 1 TO 22
FIX     COL1 = ORIG (ROUT)
FIX     COL2 = DEST (ROUT)
BLOCK
FIX     ROW 1 = 0, ROW 2 = 0
```

Comment: This will be the cost row

```
COVAL = COST (ROUT)
OUT OB
END
BLOCK
FIX     ROW1 = COL1, ROW2 = 0
COVAL = 1
```

Comment: This is the constraint for the origin

```
OUT EQ
END
BLOCK
FIX       ROW1 = 0, ROW2 = COL2
COVAL = -1
OUT EQ
END
```

Arrays are required for the cost (aij) of each route, and the origin and destination associated with each route. Note that the program is flexible in that only the limits in the first INC need be changed to increase or decrease the number of possible routes The '22' may in fact be replaced by RTMAX (maximum number of routes) and read in as data. The arrays must also be changed.

Although the problem shown belongs to the small, easily formulated class of problems, the power of a matrix generator can be seen. As soon as the clear, underlying logic of the matrix can be identified and expressed simply in the matrix generator, the problem reduces itself to listing the costs, the origins, and the destinations of routes, in any order and with no limitation on the number.

SECTION 5 Project Scheduling

E. Balas

Project Scheduling with Resource Constraints*

EGON BALAS
Carnegie-Mellon University, U.S.A.

1. INTRODUCTION

Critical path and network flow techniques can solve project scheduling problems of a realistic size when there are no resource constraints; the introduction of the latter, however, changes the problem into one in which the number of variables and constraints depends on the number of time periods, and which existing methods can solve only for a very small size. The purpose of this paper is to reformulate the scheduling problem with resource constraints in a way that should eliminate the dependence of problem size on the number of time periods and should make this problem amenable to critical path—and network flow techniques. Thus, scheduling with resource constraints is shown to be equivalent to the problem of finding an optimal selection of arcs in a disjunctive graph with stability conditions. 'Simple' machine sequencing and machine sequencing with sets of identical machines are shown to be special cases of this model. The scheduling problem with resource constraints can thus be solved by generating a sequence of PERT networks which satisfy certain stability conditions.

2. 'PURE' SCHEDULING PROBLEMS

Scheduling or sequencing is perhaps the most frequently occurring class of real-world problems for which we have no optimization technique sufficiently powerful to cope with them for anything like a realistic size. Not that these problems are mathematically very complicated. On the contrary, they can usually be translated by utterly simple mathematical models for which methods of solution are readily available. But due to the combinatorial nature of these problems, the methods can usually solve, even on the fastest computers, only problems of a very small size.

The advent of PERT (Malcolm, Roseboom, Clark and Fazar, 1959) and CPM (Kelley, 1961) was a breakthrough in this area, which made it possible for a 'pure' scheduling problem to be numerically solved for practically any size that might occur. By a 'pure' scheduling problem we mean that of finding real numbers $t_i \geq 0, i \in N = \{1, \ldots, n\}$, and

$$\min \{t_n \mid t_j - t_i \geq d_{ij}, (i, j) \in H\} \tag{1}$$

where the d_{ij} are given nonnegative integers and H is a specified subset of the Cartesian product $N \times N$. The variable t_i designates the time when activity i is started, d_{ij} is the minimum time lapse between the starting of activities i and j, and H is the set of pairs of activities for which a precedence relationship is given.

Scheduling problems of this type, in which one wants to minimize the completion time of a project, given the (fixed) completion times of the activities of the project and the precedence relationships between these activities, can now be solved for projects involving more than 100, 000 activities. A refinement of this problem, in which the completion times of some activities are variable rather than fixed, with completion costs varying in inverse proportion to completion times, and the objective is either the mini-

---

* The research underlying this paper was partially supported by the New York Scientific Center, IBM, and by the U.S. Office of Naval Research.

mization of total completion time subject to a constraint on total cost, or the minimization of total cost subject to a constraint on total completion time, can also be solved for sizes comparable to the ones mentioned above.

3. THE IMPACT OF RESOURCE CONSTRAINTS

The reason that the above problems can be solved efficiently is that they can be stated as longest path problems or network flow problems. However, for most real-world situations these models are oversimplified, because they assume unlimited resources. To make them more realistic, one has to introduce various types of resource constraints. But as soon as this is done, the problem ceases to be a longest path—or a network flow problem, and the size of projects for which an optimal solution can be found, shrinks from about 100, 000 to about 10* activities! Indeed, if resource i is used by activities $j \in J_i$, a constraint set of the type

$$\sum_{j \in J_i} a_{ij} x_j^k \leq b_i \qquad i = 1, \ldots, q$$

$$x_j^k = \begin{cases} 1 & \text{if } t_j < k \leq t_j + c_j \\ 0 & \text{otherwise} \end{cases} \qquad j = 1, \ldots, n \tag{2}$$

must hold for each time period k. Here b_i is the availability of resource i, a_{ij} is the amount of resource i required by the (indivisible) activity j, t_j is as in (1), while c_j is the (fixed) completion time of activity j. The constraints (2) then express the condition that the amount of resource i required in any given time period k by all the activities being deployed in that time period, should not exceed the availability of resource i.

Thus, if there are 10 activities and 15 time periods, there will be 150 zero-one variables x_j^k, which is at the very limit of what existing integer programming codes can handle. Furthermore, additional 0-1 variables have to be introduced in order to explicitate the relationship between the x_j^k and the t_j variables, or, alternatively, to recast the model only in terms of 0-1 variables. The very efficient techniques mentioned above are of no avail here.

The problem arises then to find another formulation of scheduling under resource constraints, which should permit the use of the critical path—or network flow techniques, or some generalization of these methods, and should somehow avoid the need to consider as many constraints (and variables) for each resource, as there are time periods.

4. A SPECIAL CASE: 'SIMPLE' MACHINE SEQUENCING

As a starting point, we shall use the 'simple' machine sequencing problem, which can easily be seen to be a scheduling problem under resource constraints, and which can be reduced to a sequence of critical path problems.

By 'simple' machine sequencing we mean the problem of finding an optimal sequence for processing m items (or lots of items) on q machines, when each item has to be processed on a specified sequence of machines, i.e.

(a) a given operation on a given item has to be carried out on a specified machine;

---

* We are talking about algorithms for finding an optimal solution. There are, however, various heuristic techniques for finding a feasible solution for larger problems.

(b) the operations to be carried out on a given item are ordered by a set of precedence relationships of the type $t_j - t_i \geqslant d_{ij}$;

(c) there is freedom of choice as to the sequence of operations to be carried out on a given machine;

(d) one is looking for a sequence minimizing t_n, the total time needed for processing all items.

The objective (d) subject to the conditions (b) constitutes a problem of type (1), i.e. a 'pure' scheduling problem. Indeed, if we associate

(α) a node j with each operation, including two dummy operations: 'start' and 'end';

(β) an arc (i, j) with each pair of operations pertinent to the same item (lot of items) and adjacent in the sequence defined by the precedence relationships (b);

(γ) a length (real number) d_{ij} with each arc (i, j), expressing the time that must lapse between the starting of operations i and j;

then we obtain a PERT network of the type shown in fig. 1a and 1b, where operations $n_{h-1} + 1, \ldots, n_h$ are the ones pertinent to item h.

However, we have to introduce the additional condition, that there should be no time-overlap between operations to be carried out on the same machine.

In other words, for each time period k, a constraint set of the form

$$\sum_{j \in J_i} x_j^k \leqslant 1 \qquad i = 1, \ldots, q$$

$$x_j^k = \begin{cases} 1 \text{ if } t_j < k \leqslant t_j + c_j \\ 0 \quad \text{otherwise} \end{cases} \qquad j = 1, \ldots, n \tag{2'}$$

must hold, where the index set J_i designates the operations to be carried out on machine i, while c_j is the completion time of operation j.

Obviously then, the 'simple' machine sequencing problem is a scheduling problem with special resource constraints, in which $b_i = 1, a_{ij} = 1, \forall$ i, j.

However, this special scheduling problem with resource constraints can be reduced to a sequence of critical path problems, when formulated in terms of disjunctive graphs. Moreover, the time dimension of the problem is being 'done away with' in this formulation, in the sense that the variables and constraints of the problem are not time-dependent (are not related to any particular time period).

5. DISJUNCTIVE GRAPHS

Two arcs of a graph D are said to form a disjunctive pair, if any path in D is allowed to meet at most one member of the pair. A graph D containing disjunctive arcs is called a disjunctive graph. We shall denote a disjunctive graph by

$$D = (N; A, B) \tag{3}$$

where N is the set of nodes, A the set of conjunctive arcs, and B the set of (pairwise) disjunctive arcs.

Consider now the (conjunctive) graph defined by α, β, γ above, pictured in fig. 1, and denote it by G = (X, Z), with X as its set of nodes and Z as its set of arcs. Let $X_i \subset X$ be the subset of nodes associated with operations to be carried out on machine i, for i = 1, ..., q, where the sets X_i are disjoint. Consider then the disjunctive graph

$$D = (X; Z, W) \tag{4a}$$

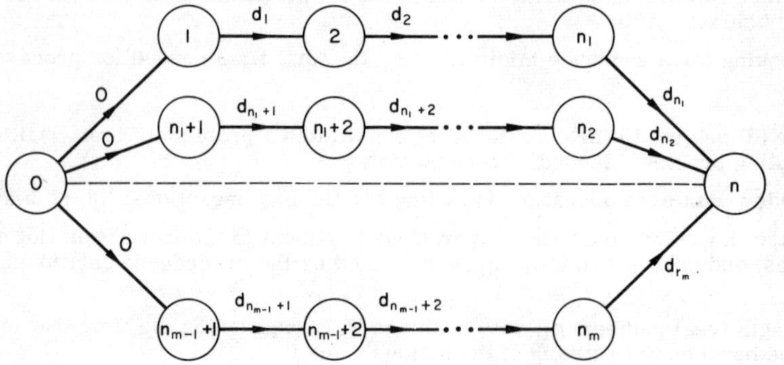

Fig. 1a

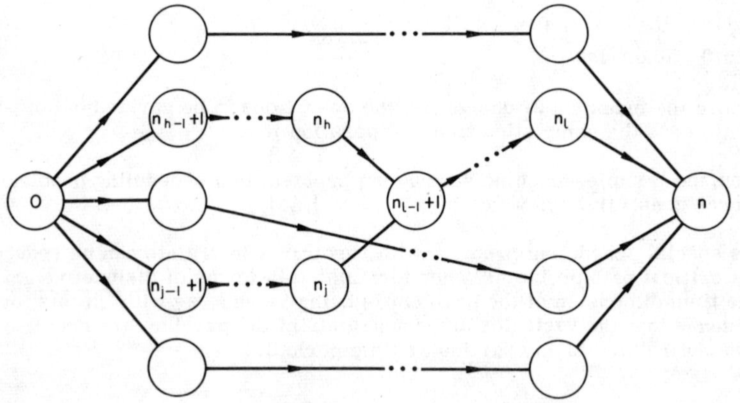

Fig. 1b

obtained from G by introducing the set W of (pairwise) disjunctive arcs, defined by (4b) and (4c):

$$W = \bigcup_{i=1}^{q} \{(j, k) \in X_i \times X_i \mid j \neq k, \text{ and there is no path in G from j to k}\} \tag{4b}$$

$$\left\{\begin{matrix}(j, k) \text{ and } (r, s) \text{ form} \\ \text{a disjunctive pair}\end{matrix}\right\} \Leftrightarrow \left\{\begin{matrix}(j, k) \in W, (r, s) \in W, \\ \text{and } r = k, s = j\end{matrix}\right\} \tag{4c}$$

Each disjunctive arc (j, k) is assigned a length $d_{jk}(=c_j$, the completion time of operation j). Obviously, $d_{jk}(=c_j)$ and $d_{kj}(=c_k)$ are generally not equal. Then the set W of (pairwise) disjunctive arcs in D translates the condition that no time-overlap is allowed between any two operations that have to be performed on the same machine.

The pair of disjunctive arcs can be pictured like in fig. 2. Two examples of disjunctive graphs are shown in fig. 3 and 4. Fig. 3 represents a 'simple' machine sequencing problem with 3 items and 3 machines, whereas fig. 4 pictures one with 5 items and 4 machines.

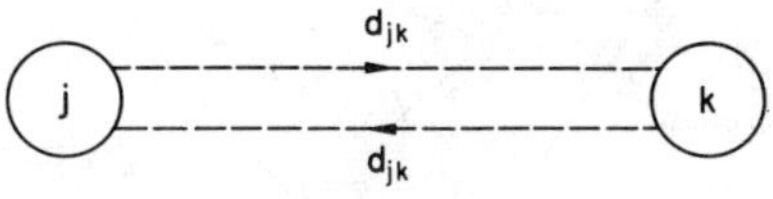

Fig. 2

If we drop the first and the last node, then each subset of nodes connected by conjunctive arcs is associated with operations to be performed on the same item, and each subset of nodes connected by disjunctive arcs is associated with operations to be carried out on the same machine.

A subset of W containing at most one arc of each disjunctive pair, is called a selection (of disjunctive arcs). A selection $C_h \subset W$ containing exactly one arc of each disjunctive pair is termed complete. Let

$$C = \{C_1, \ldots, C_p\} \tag{5}$$

be the set of complete selections. Obviously, if $\sigma = \frac{1}{2} \mid W \mid$ is the number of disjunctive pairs of arcs, then the number of distinct complete selections is $p = 2^{\sigma}$.

Each complete selection generates a (conjunctive) graph of the form

$$G_h = (X, Z \cup C_h) = (X, Z_h) \tag{6}$$

Some graphs G_h of the type (6) possibly contain circuits (i.e. closed paths); others do not. Let

$$C' = \{C_h \in C \mid G_h = (X, Z \cup C_h) \text{ has no circuits}\}. \tag{7}$$

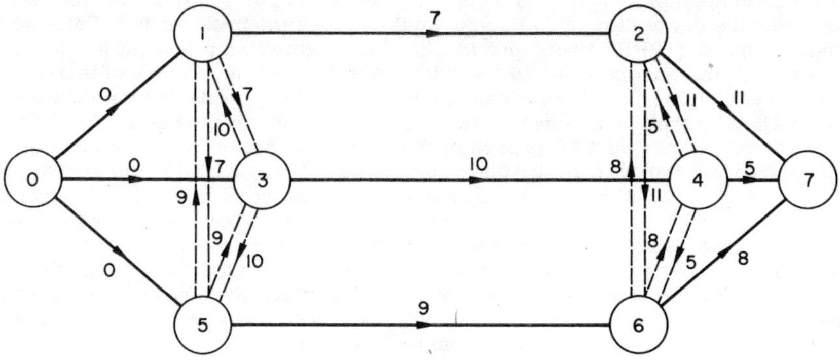

Fig. 3.

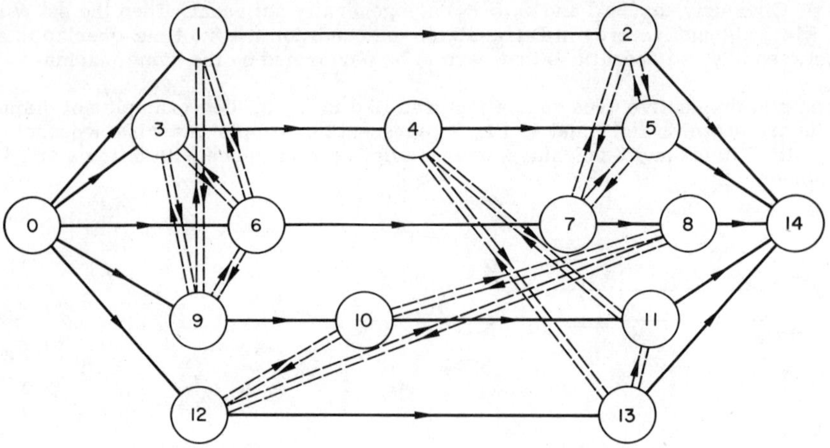

Fig. 4.

6. (2-STATE) MINIMAXIMAL PATHS IN D

A critical path in a graph G_h generated by $C_h \in C'$ is defined as a longest path from source to sink (i.e. connecting the two nodes associated with the dummy operations 'start' and 'end'), the length of a path being defined as the sum of the lengths of the arcs met by the path. Let v_h denote the length of a critical path in G_h.

A critical path in a graph $G_p = (X, Z \cup C_p)$ is a <u>(2-state) minimaximal path in D</u>, if

$$v_p = \min_{C_h \in C'} v_h \tag{8}$$

The term '2-state' will be explained later. The 'simple' machine sequencing problem is equivalent to the problem of finding a <u>(2-state) minimaximal path in a disjunctive graph D</u>. Methods for solving this latter problem have been given by Roy and Sussmann (1964), Balas (1966, 1968) and Raymond (1968).

The second method proposed by Balas (1968) consists of an implicit enumeration procedure over the set of disjunctive arcs, which reduces the problem to a finite sequence of slightly amended PERT problems, in which one has to find a longest and a second longest path from certain nodes to the source and to the sink. The results are used to derive an evaluation of the improvement one can obtain by replacing various disjunctive arcs (j, k) by their complements (k, j). At any given stage, some of the disjunctive arcs are fixed (for the current branch of the search-tree), while others are free (to be considered for replacement by their complements); but the only candidates which actually need to be considered for replacement are the free arcs on the critical paths of the current PERT network. The evaluations mentioned above are then used to direct the search so as to go from any given PERT network to the 'best' adjacent one (adjacent = differing by exactly one disjunctive arc). At each stage the shortest critical path found so far is used as a current upper bound for the search, whereas a critical path in the PERT network containing only the fixed arcs yields a current lower bound, which is then tested against the current upper bound.

While computational experience is not yet available for determining the size of problems that this method can solve in a reasonable time, the following features seem to be worth mentioning:

(a) it can start with any feasible sequence, for instance the one actually used in production, or generated by some heuristic procedure;

(b) it finds at an early stage a local optimum (a sequence that cannot be improved by replacing any one disjunctive arc by its complement); thus one can stop at any moment with a reasonably 'good' feasible sequence;

(c) storage requirements are limited to a set of binary vectors whose cardinality cannot exceed at any given stage the number of disjunctive pairs of arcs.

7. MACHINE SEQUENCING WITH SETS OF IDENTICAL MACHINES

A generalization of the 'simple' machine sequencing problem consists in allowing for sets of identical machines: each operation has to be performed not necessarily on a specified machine, but on any machine in a specified set (of identical machines). In this case the PERT graph of fig. 1 representing the sequence of operations for each item is the same as before, but the resource constraints to be superimposed are different: they have to express the condition that for each set of identical machines, the number of operations (to be performed on machines in that set) for which there is a time-overlap, should not exceed the number of machines in the set. In other words, for each time period k, a constraint set of the form

$$\sum_{j \in J_i} x_j^k \leqslant b_i \qquad i = 1, \ldots, q$$

$$x_j^k = \begin{cases} 1 \text{ if } t_j < k \leqslant t_j + c_j \\ 0 \quad \text{otherwise} \end{cases} \qquad j = 1, \ldots, n \tag{2''}$$

must hold, where the index set J_i designates the operations to be performed on machines in set i, b_i stands for the number of machines in set i, while c_j, as before, is the completion time of operation j.

To cover this case, we shall extend the concept of a minimaximal path to the case when the minimum is sought over the set of graphs G_h generated by all, not necessarily complete, selections of disjunctive arcs, satisfying certain stability conditions. Since considering only complete selections allows for only 2 possible states for each disjunctive pair of arcs (one or the other member of the pair is in the selection), whereas considering not necessarily complete selections allows for a 3rd state (no member of the pair is in the selection), this generalized concept will be termed a 3-state minimaximal path as opposed to the 2-state one considered earlier.

On the other hand, it turns out that with an appropriate definition of the stability conditions mentioned above, this model can be made sufficiently comprehensive to translate the general scheduling problem with resource constraints (1), (2). Hence we shall formulate the model for this latter case, and subsequently specialize it for machine sequencing with sets of identical machines.

8. 3-STATE MINIMAXIMAL PATHS WITH STABILITY CONDITIONS

A necessary condition for problem (1), (2) to have a solution is that the graph G defined by the precedence constraints of (1) has no circuits. We shall henceforth make this assumption.

Let G = (X, Z) be the graph (PERT network) defined by (1), let $X_i \subset X$ be the subset of nodes associated with the set J_i in (2), i.e. with the i-th resource constraint (with the set of activities using the i-th resource), and let i = 1, . . . , q. We then want to consider a disjunctive graph D = (X; Z, W) defined as in (4a, b, c), with the additional feature that for $i = 1, \ldots, q$, we associate with each node $j \in X_i$, the coefficient a_{ij} of x_j^k in (2). Since the sets X_i are not nessarily disjoint, each node j will have as many coefficients a_{ij} associated with it, as there are sets X_i containing j, i.e. a coefficient a_{ij} is associated with each pair $(X_i, j \in X_i)$.

Fig. 5a pictures the PERT network of a 'pure' scheduling problem with 23 activities (Muth and Thompson, 1963), while fig. 5b shows the disjunctive graph we would associate with this problem if the following set of resource constraints were superimposed:

$$4x_3^k + x_7^k + 3x_9^k \leqslant 4$$

$$5x_6^k + 3x_7^k + 7x_8^k + 2x_9^k \leqslant 10$$

$$x_{15}^k + 4x_{19}^k + 2x_{22}^k \leqslant 4$$

$$3x_{15}^k + 5x_{16}^k \quad x_{22}^k \leqslant 7$$

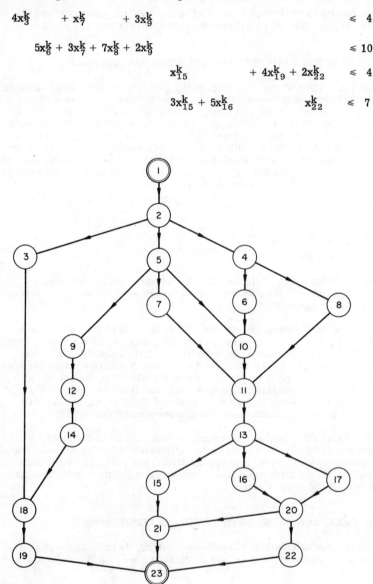

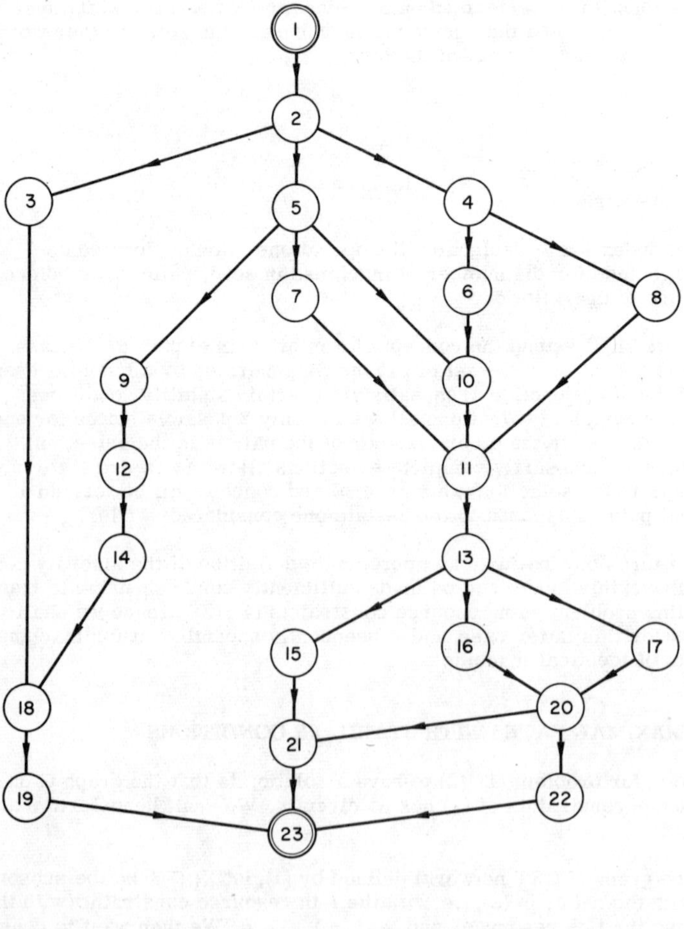

Fig. 5a.

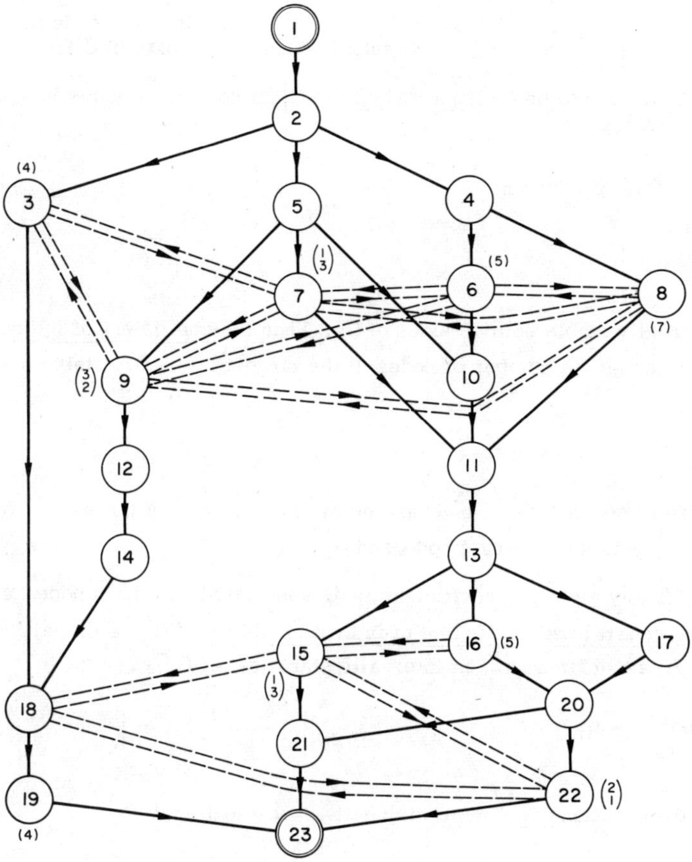

Fig. 5b.

With the disjunctive graph D defined as above, let $S_h \subset W$ be a (not necessarily complete) selection of disjunctive arcs.

Let us denote

$$\overline{Z} = \{(j, k) \in X \times X \mid \text{there is a path in G from j to k}\}. \quad (9)$$

Now consider the (conjunctive) graph

$$G_h = (X, Z \cup S_h) = (X, Z_h) \quad (10)$$

and the associated graphs of the form $G_h^i = (X_i, Z_h^i)$, $i = 1, \ldots, q$, where

$$Z_h^i = \{X_i \times X_i\} \cap \{S_h \cup \overline{Z}\}. \tag{11}$$

In other words, the arcs of G_h^i are all arcs in the selection S_h having both ends in X_i, plus an arc (j, k) for all pairs $j \in X_i$, $k \in X_i$ such that there is a path in G from j to k.

A set of nodes $Y \subset X_i$ is said to be internally stable in G_h^i, if no pair of nodes in Y is connected by an arc of G_h^i, i.e. if

$$j \in Y, k \in Y \Rightarrow (j, k) \notin Z_h^i, (k, j) \notin Z_h^i \tag{12}$$

Let

$$Y_h^i = \{Y_1, \ldots, Y_t\} \tag{13}$$

be the family of internally stable sets of nodes of G_h^i. Then the coefficient of internal stability of G_h^i is defined as the number of nodes in the largest internally stable set, i.e.

$$\alpha_h^i = \alpha(G_h^i) = \max_{Y \in Y_h^i} |Y| \tag{14}$$

Obviously, if all pairs of nodes in G_h^i are connected by an arc, then the largest set of non-connected nodes consists of a single node, and $\alpha_h^i = 1$.

We now recall that, for any given i, a coefficient a_{ij} is associated with each node j of G_h^i, and we define the generalized coefficient of internal stability of G_h^i as the largest sum of coefficients a_{ij} associated with an internally stable set of G_h^i, i.e.

$$\beta_h^i = \beta(G_h^i) = \max_{Y \in Y_h^i} \sum_{j \in Y} a_{ij} \tag{15}$$

Let S be the set of all selections $S_h \subset W$ of disjunctive arcs such that

$$\beta_h^i = \beta(G_h^i) \leqslant b_i, i = 1, \ldots, q, \tag{16}$$

and let

$$S' = \{S_h \in S \mid G_h = (X, Z \cup S_h) \text{ has no circuits}\}. \tag{17}$$

The elements of S′ will be called feasible selections, and v_h will denote the length of a critical path in G_h.

A critical path in a graph $G_p = (X, Z \cup S_p)$ is a 3-state minimaximal path in D with stability conditions (16), if

$$v_p = \min_{S_h \in S'} v_h. \tag{18}$$

The feasible selection S_p generating G_p is then called optimal.

We shall denote by P_1 the problem (1), (2) with the added assumption on (1) that it defines a graph G = (X, Z) without circuits; and by P_2 the problem of finding an optimal selection (with an associated minimaximal path) in a disjunctive graph D = (X; Z, W) defined by (4), with stability conditions (16).

9. EQUIVALENCE OF P_1 AND P_2

Let us denote by $(\bar{t}, \bar{x})$ a solution to P_1, and by S_p a solution to P_2 (an optimal selection in D, with v_p as the length of a minimaximal path).

The following theorems establish the equivalence of P_1 and P_2.

Theorem 1. P_1 and P_2 are solvable if and only if

$$\max_{j \in J_i} a_{ij} \leq b_i, \; i = 1, \ldots, q. \tag{19}$$

Proof.

Necessity: Follows directly from (2) for P_1 and from (16) for P_2.

Sufficiency:

For P_2: If (19) holds, any complete selection S_h such that $G_h = (X, Z \cup S_h)$ has no circuits, is feasible.

For P_1: For each resource i, let $j_1, \ldots, j_i$ be a permutation of the elements of J_i, such that $j_k > j_h$ if G contains a path from j_h to j_k. Let $\overline{P}_1$ be the problem consisting of (1) and a set of constraints of the type

$$t_{j_2} - t_{j_1} \geq c_{j_1}, t_{j_3} - t_{j_2} \geq c_{j_2}, \ldots, t_{j_i} - t_{j_{i-1}} \leq c_{j_{i-1}} \tag{20}$$

for each $i = 1, \ldots, q$.

$\overline{P}_1$ is a critical path problem which always has a solution (the constraints (20) cannot give rise to circuits, since there is no path in G from j_k to j_h if $j_k > j_h$). It is easy to see that any such solution also satisfies (2).

We shall now introduce a lemma needed for the proof of the next theorem.

First, we shall observe that if the unit time period is taken sufficiently small for all d_{ij} (and c_j) to be integers when expressed in this unit, then P_1, if solvable, also has an integer solution. Indeed, let the time period index k in (2) range over $1, \ldots, K$, with K sufficiently large (for instance

$$K = \sum_{(i,j) \in Z} d_{ij},$$

with all d_{ij} integers), and let $(\hat{t}, \hat{x})$ be a vector satisfying the constraints of P_1. Now suppose some $\hat{t}_j$ are non-integer. Then, denoting by $[\hat{t}_j]$ the largest integer contained in $\hat{t}_j$, it is easy to see that $(\tilde{t}, \hat{x})$, where $\tilde{t}_j = [\hat{t}_j]$, also satisfies the constraints of P_1.

Now let S_h be a selection in D, such that $G_h = (X, Z \cup S_h)$ has no circuits. Let a variable t_j be associated with each node $j \in X (j = 1, \ldots, n$ where $n = |X|)$, and let $\hat{t} = (\hat{t}_j)$ be a vector with integer components satisfying the precedence constraints defined by G_h. Further let $\hat{x} = (\hat{x}_j^k)$ be defined by

$$\hat{x}_j^k = \begin{cases} 1 \text{ if } \hat{t}_j < k_j \leq \hat{t}_j + c_j & j = 1, \ldots, n; \\ 0 \quad \text{otherwise} & k = 1, \ldots, K \end{cases} \tag{21}$$

and let

$$Q^k = \{j \in X | \hat{x}_j^k = 1\}, k = 1, \ldots, K. \tag{22}$$

Lemma. For each $k = 1, \ldots, K$ and each $i = 1, \ldots, q$, the set $Q^k \cap X_i$ either is empty, or is an internally stable set of G_h^i.

Proof. Suppose the lemma is false. Then there exists $r \in Q^k \cap X_i$ and $s \in Q^k \cap X_i$ such that either $(r, s) \in Z_h^i$ or $(s, r) \in Z_h^i$. Suppose $(r, s) \in Z_h^i$ (an analogous reasoning holds for the opposite case). Then

$$\hat{t}_s - \hat{t}_r \geqslant d_{rs} = c_r. \tag{23}$$

But $r \in Q^k$, $s \in Q^k$ implies $\hat{x}_r^k = \hat{x}_s^k = 1$, and from (21) we have

$$\hat{t}_s < k \leqslant \hat{t}_r + d_{rs} \tag{24}$$

which contradicts (23).

Let us recall now that $G = (X, Z)$ is the graph (without circuits) defined by the constraints of (1), $D = (X; Z, W)$ is the disjunctive graph defined from G by (4) with stability conditions (16), and $\overline{Z}$ is defined by (9).

Theorem 2. If $(\overline{t}, \overline{x})$ is a solution to P_1, the set S_h defined by

$$(r, s) \in S_h \Leftrightarrow \begin{Bmatrix} r \in J_i,\ s \in J_i \text{ for some } i,\ (r, s) \notin \overline{Z}, \\ \overline{t}_s - \overline{t}_r \geqslant d_{rs} = c_r \end{Bmatrix} \tag{25}$$

is a solution to P_2 (an optimal selection in D), such that $v_h = \overline{t}_n$.

Conversely, if S_p is a solution to P_2 and $\hat{t}$ solves the problem of minimizing t_n subject to the precedence constraints defined by $G_p = (X, Z \cup S_p)$, then $(\hat{t}, \hat{x})$, where $\hat{x}$ is defined by $\hat{t}$ via (21), solves P_1 and $\hat{t}_n = v_p$.

Proof.

(a) From (25) it follows that

(i) S_h is a selection in D;

(ii) $G_h = (X, Z \cup S_h)$ has no circuits;

(iii) $\overline{t}$ satisfies the precedence constraints defined by G_h.

To show that S_h is a feasible selection, i.e. that it satisfies the stability conditions (16), we shall suppose that it does not, i.e. $\beta_h^i < b_i$ for some i. Let Y_i be the internally stable set of G_h^i which defines β_h^i, i.e.

$$\sum_{j \in Y_i} a_{ij} = \max_{Y \in Y_h^i} \sum_{j \in Y} a_{ij} = \beta_h^i > b_i \tag{26}$$

Next we shall show that there is at least one k for which $j \notin Y_i \Rightarrow \overline{x}_j^k = 1$. From the internal stability of Y_i and from (25) it follows that

$$r \in Y_i,\ s \in Y_i \Rightarrow r \in J_i,\ s \in J_i,\ (r, s) \notin S_h,\ (s, r) \notin S_h$$

$$\Rightarrow -c_s < \overline{t}_s - \overline{t}_r < c_r. \tag{27}$$

I has been shown above, that if P_1 is solvable, it has an integer solution. From this and from (25) it is clear that the first part of the theorem holds if and only if it holds when $\overline{t}$ is required to be an integer in each component. Now let us assume that $\overline{t}$ satisfies this requirement, and let $\overline{t}_* = \max_{r \in Y_i} \overline{t}_r$.

Then for $k = \bar{t}_* + 1$ we have from (27)

$$\bar{t}_j < k \leq \bar{t}_j + c_j, \qquad \forall j \in Y_i \tag{28}$$

and hence $\bar{x}_j^k = 1$ for $k = \bar{t}_* + 1$ and all $j \in Y_i$.

Accordingly, for each i such that $\beta_h^i > b_i$, there is at least one k for which

$$\sum_{j \in Y_i} a_{ij} \bar{x}_j^k > b_i \tag{29}$$

But this contradicts the assumption that $(\bar{t}, \bar{x})$ is a solution to P_1. Hence S_h is a feasible selection. Further, from (iii) above it follows that $v_h \leq \bar{t}_n$, where v_h is the length of a critical path in G_h. This together with the second part of the theorem to be proved below, shows that $v_h. = \bar{t}_n$ and S_h is optimal.

(b) $\hat{t}$ obviously satisfies the precedence constraints of (1), since $Z \subset Z_p$. According to the Lemma, for any given i and k, the set of indices $\{j \in J_i | \hat{x}_j^k = 1\}$ is associated with an internally stable set Y of nodes of G_p^i. But this implies, in view of (16),

$$\sum_{j \in J_i} a_{ij} \bar{x}_j^k \leq \max_{Y \in Y_p^i} \sum_{j \in Y} a_{ij} = \beta_p^i \leq b_i, \qquad \forall i, k \tag{30}$$

which means that $(\hat{t}, \hat{x})$ also satisfies (2).

Since the definition of $\hat{t}$ implies $\hat{t}_n = v_p$, where v_p is the length of a critical path in $G_p = (X, Z \cup S_p)$, we have $\tilde{t}_n \leq v_p$ for any vector $(\tilde{t}, \tilde{x})$ that solves P_1. If, however, there exists $(\tilde{t}, \tilde{x})$ such that $\tilde{t}_n < v_p$, then $\tilde{t}$ defines via (25) a selection S_h in D which has been shown under (a) to be feasible and such that $\tilde{t}$ satisfies the precedence constraints of $G_h = (X, Z \cup S_h)$. But then $\tilde{t}_n = v_h < v_p$, which contradicts the assumed optimality of S_p. Hence $(\tilde{t}, \tilde{x})$ with the above property cannot exist, and $(\hat{t}, \hat{x})$ with $\hat{t}_n = v_p$, solves P_1.

<u>Remark.</u> In the case of machine sequencing with sets of identical machines, when (2) takes the special form of (2″), the stability conditions (16) become

$$\alpha_h^{i} = \alpha(G_h^i) = \max_{Y \in Y_h^i} |Y| \leq b_i, \quad i = 1, \ldots, q \tag{16'}$$

i.e. the generalized coefficient of internal stability β_h^i reduces to the usual coefficient of internal stability α_h^i. Further, the sets X_i are disjoint:

$$\bigcap_{i=1}^{q} X_i = \phi. \tag{31}$$

10. CONCLUSION

The scheduling problem with resource constraints has been shown to be equivalent to the problem of finding an optimal selection (and a related 3-state minimaximal path) in a disjunctive graph with stability conditions. In this model, the number of variables and constraints is independent of the number of time periods. Also, this formulation makes our problem amenable to an implicit enumeration technique of the type discussed above for the simple machine sequencing problem (Balas, 1968), which solves it by generating a sequence of PERT networks satisfying the stability conditions; but the description of this technique is left to another paper.

REFERENCES

1. BALAS, E. (1966), 'Finding a Minimaximal Path in a Disjunctive PERT Network.' Theory of Graphs, International Symposium, Rome. Dunod, Paris; Gordon and Breach, New York.

2. BALAS, E. (1968), 'Machine Sequencing via Disjunctive Graphs: an Implicit Enumeration Algorithm.' Management Sciences Research Report No. 125, Carnegie-Mellon University.

3. KELLEY, J. E., Jr. (1961), 'Critical Path Planning and Scheduling: Mathematical Basis,' Operations Research, 9, p. 296-320.

4. MALCOLM, D. G , ROSEBOOM, J. H., CLARK, C. E. and FAZAR, W. (1959), 'Application of a Technique for Research and Development Program Evaluation.' Operations Research, 7, p. 646-669.

5. MUTH, J. R. and THOMPSON, G. L., ed. (1963), Industrial Scheduling, Prentice Hall.

6. RAIMOND, J. F. (1968), 'An Algorithm for the Exact Solution of the Machine Scheduling Problem.' IBM New York Scientific Center Report No. 320-2930.

7. ROY, B. and SUSSMANN, B. (1964), 'Les problèmes d'ordonnancement avec contraintes disjonctives.' SEMA, Note D.S. 9 bis.

SECTION 6 Non-Convex Programming Methods

G. Zoutendijk
K. Spielberg
P.H. Randolph, G.E. Swinson, M.E. Walker
P. Rech, L.G. Barton
P. Gray

Mixed Integer Programming and the Warehouse Allocation Problem

G. ZOUTENDIJK
University of Leiden, The Netherlands

1. INTRODUCTION

In this paper an attempt will be made to give a short introduction to the field of mixed integer programming (part 2) and to discuss, in part 3, four papers in these proceedings which are concerned with warehouse allocation. Some final remarks will then be made in part 4. Maybe the reason that I have been asked to act as 'rapporteur' is that about 13 years ago I started my work in mathematical programming by studying a simple warehouse allocation problem. A number of depots, say m, had to be built to supply a known demand b_j, $j = 1, \ldots, n$ of a certain homogeneous product to n customers, customer j situated at a location with coordinates (x_j, y_j). Transportation cost was supposed to be proportional to geographical distance. The capacity a_i of each depot was given and

$$\sum_{i=1}^{m} a_i \geq \sum_{j=1}^{n} b_j$$

was supposed to hold. The way we tried to solve this problem was the following:

1. Choose some initial allocation of the m warehouses.
2. Calculate the distances and solve a transportation problem to determine the amount x_{ij} of the demand b_j at destination j to be supplied by depot i.
3. Given the x_{ij} from 2, solve for each depot i a weighted Torricellian problem

   $$\min_{x,y} \sum_{j \in J_i} x_{ij} d_{ij}, \quad \text{where}$$

   $$J_i = \{j \mid x_{ij} > 0\}, \; d_{ij} = \sqrt{(x - x_j)^2 + (y - y_j)^2}$$

   A special computational method for this convex minimization problem is easily developed.
4. Repeat 2-3 until no further improvement can be made.

The algorithm worked nicely and, as could be expected, led to a local minimum which sometimes had to be built on top of a mountain or in the middle of a lake.

It is not the latter difficulty but the problem of local optima and the necessity to avoid these local optima, that I would like to emphasize first. The situation in warehouse allocation is considerably worse than in many continuous nonlinear programming models coming from practical problems. In the latter case the present way of doing things is usually a good starting point for an algorithm; it is not likely that the optimum we obtain from this starting point is far off in value from the global optimum. Moreover the model would probably not be valid in that completely different situation. Not so in warehouse allocation and other integer programming problems. Any algorithm which just tries to obtain marginal improvements starting from a good guess will therefore be highly unreliable.

2. METHODS FOR THE MIXED INTEGER PROGRAMMING PROBLEM

We will restrict ourselves to the so-called 0-1 mixed integer programming problem:

$$\text{Max } \{p^T x + q^T w \mid Ax + Bw \leqslant b, x \geqslant 0, w_j = 0 \text{ or } 1\}, \tag{1}$$

where

p and x are vectors of length n_1, q and w of length n_2 and b of length m; A and B are m by n_1 and m by n_2 matrices, respectively.

The general mixed integer linear programming problem can be formulated as a 0-1 problem, although this is not always recommendable.

We can distinguish between three classes of methods:

1. Branch and bound methods which start from the linear programming solution
2. Dual decomposition methods
3. Direct enumeration methods.

2.1 Branch and bound methods [10], [5], [9]

These methods apply a tree search starting from the solution of the linear program obtained by replacing the integer requirements $w_k = 0$ or 1 by $0 \leqslant w_k \leqslant 1$. This will give an upper bound for the value of the solution of the mixed integer problem. Next a tree is being developed by taking one of the integer variables with a fractional value in the solution, by assigning the values 0 and 1, respectively, to it and by solving the two resulting linear programs. After one step the situation may be like

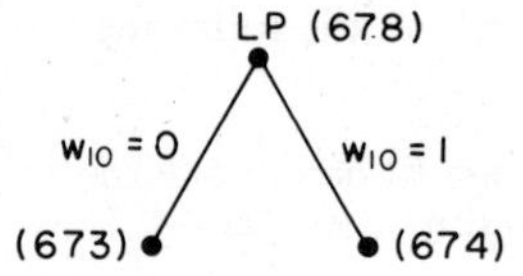

(between brackets the value of the objective function)

Next we continue the tree search from one of the nodes, e.g. the one with value 674. Again an integer variable with fractional value is given the values 0 or 1. After two more steps the situation might be:

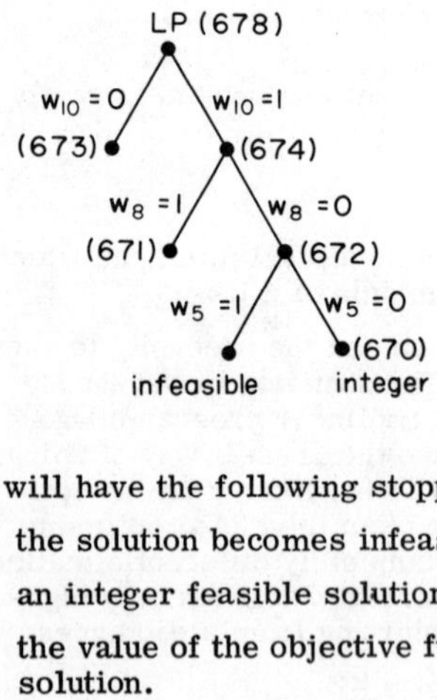

We will have the following stopping rules:

1. the solution becomes infeasible;
2. an integer feasible solution is obtained;
3. the value of the objective function becomes less than the value of a known integer solution.

When this has happened the node will be closed and the search will be continued from another node which is still open.

Several questions arise in this respect:

(1) What is the proper tree. It is not necessary, for instance, that we branch on the basis of one variable being set to 0 or 1; we can take more variables and do the branching on the basis of relations like $w_3 = w_4 = w_7 = 0$ and $w_3 + w_4 + w_7 \geqslant 1$.

The latter relation has then to be added to the linear programming tableau.

If, for instance, for a given intermediate linear programming solution, the set $k = \{1, \ldots, n_2\}$ is partitioned into $K_0 = \{k \mid w_k = 0$ has been assigned$\}$, $K_1 = \{k \mid w_k = 1$ has been assigned$\}$ and $K_2 = K - (K_0 + K_1) = \{k \mid w_k$ not yet assigned$\}$, and if, by accident, within K_2 some w_k variables are 0 or 1, say for $k \in K_{20}$ and K_{21}, respectively, then we could continue our tree in the following way:

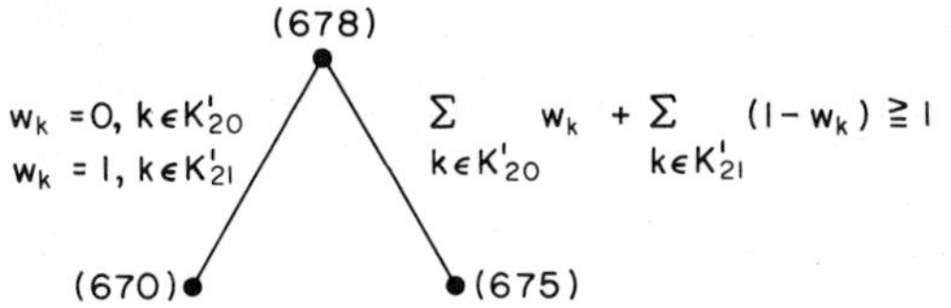

with $K'_{20} \subset K_{20}$ and $K'_{21} \subset K_{21}$ (in the sets K' we could for instance include those eligible k for which the dual variable belonging to w_k or $1 - w_k$ is at least $\alpha\bar{u}$ with $\alpha > 0$ and $\bar{u}$ being the maximum of the dual values to be considered.

The node at the left-hand side would not entail any new calculations. We can continue with this node and choose a variable with $k \in K_2 - (K'_{20} + K'_{21})$ to which the values 0 and 1 will then be assigned. The same idea can be applied to any node.

(2) Which is the proper variable to be chosen. Several proposals have been made and tried out (see [5] and [9]) like:

- take the one with value closest to $\frac{1}{2}$ in the final LP tableau,
- take the one with value closest to 0 or 1 in the final LP tableau,
- take the one for which $|q_k|$ is largest,
- calculate for each $k \in K_2$ with fractional value the decrease in the objective function value if w_k is set to 0 or 1 and one dual simplex pivotal step would be made; select the one for which the difference of the two decreases is largest in absolute value.

The latter criterion which looks attractive at first sight might be hard to apply computationally since some rows of the complete final LP tableau would be required. These, however, are not immediately available in most large LP computer programs. Nevertheless it might be worth the effort.

(3) Which tree discipline has to be used. Possibilities:

- always continue from the node with highest value; the amount of tree-searching would then be minimized; however, storage requirements could become prohibitive, while a considerable amount of searching and data-transfer would be required; moreover it will take a long time before a first integer feasible solution will be found which might be a disadvantage if computer time is limited and an approximate solution is preferred to no solution at all.
- keep to the right as long as possible (assuming that for each dichotomy the best one is written at the right-hand side), then go back to the node with the highest value;

— keep to the right as long as possible, then go back to the one which is closest in terms of treedistance.

Several compromises might be possible. For instance, let h denote the depth of the tree (i.e. the number of assignments made), let v_1, h_1 denote value and depth of the current (or closest) node, let v_2, h_2 denote value and depth of the open node with highest value ($v_2 \geqslant v_1, h_1 > h_2$) let

$$\Delta v = \frac{v_2 - v_1}{h_1 - h_2}$$

then calculate for each open node with $h \geqslant h_2$, $z = v_2 - (h - h_2)\Delta v$ and choose as next node the one for which $v - z$ is largest.

The answer as to which method is best will be dependent on the type of computer used, as well as on the way of programming. It might even be problem dependent, although this is less obvious.

A survey of branch and bound methods in general can be found in [11].

2.2 Dual decomposition [4]

In this method a sequence of linear programming problems, obtained by substituting certain integer values for the w_k is solved, as well as a sequence of pure integer programming problems (except for one variable), each one differing from the previous one by having at least one additional constraint. Therefore, the underlying assumption of the dual decomposition method is that it is essentially simpler to solve pure integer programming problems than to solve mixed problems. This assumption is probably true. For the pure integer problem there are two different approaches towards a solution:

— cutting plane methods [8], in which the associated LP problem ($0 \leqslant w_k \leqslant 1$ instead of $w_k = 0$ or 1) is first solved; from the final LP tableau a new relation is then derived which the current solution does not but the optimum integer solution has to satisfy and this process is continued until an integer solution is obtained which will then be the optimum one. This method, although theoretically sound, often behaves in an erratic way in practice; the number of cuts to apply is not predictable and often excessively large. A good method to find these cuts has still to be invented.

— direct enumeration methods for which the reader is referred to [3], [1], [7] and [14]. These are also tree search methods. However, for the 0-1 pure integer problem many exclusion rules can be formulated which speed up the treesearch to a considerable extent and make the solution of rather large problems possible. These direct enumeration methods do not start from the final solution of an associated linear programming problem but they use the original relations. They work with so-called partial solutions in which some of the variables have already been given the value 0 or 1, while the other ones are still unassigned. During the start all variables are unassigned. The branching process in the tree consists of assigning 0 or 1 to one or more variables, selected according to some rules. Compulsory choices might speed up the search.

— branch and bound methods of the type described in 2.1 which are of course also applicable to pure integer problems.

In order to understand the dual decomposition method we rewrite the original problem (1) in the following way:

$$\hat{w}_0 = \underset{w}{\text{Max}} \{w_0 \mid w_0 \leqslant q^T w + \underset{x}{\max} \{p^T x \mid Ax \leqslant b - Bw, x \geqslant 0\}; w_k = 0 \text{ or } 1\}$$

Given an integer solution w^h at step h we first solve the linear programming problem

$$\text{Max}\ \{p^T x \mid Ax \leqslant b - Bw^h, x \geqslant 0\},$$

using the dual simplex algorithm.

There are two possibilities:

(1) The problem is infeasible. We will then find a dual extreme ray r (those elements of the matrix row which shows the infeasibility that belongs to the dual variables u_i). Obviously $r^T(b - Bw^h) < 0$ holds since the dual problem has an infinite solution. Assuming that for the optimal choice of the integer solution there will be feasibility we will therefore add the requirement $r^T(b - Bw) \geqslant 0$, i.e. $r^T Bw \leqslant r^T b$ to the next pure integer sub-problem. We can then continue the solution of the LP problem neglecting the infeasibility row as potential pivot row. Again we will arrive in situation (1) or (2), etc.

(2) The problem is feasible, so that we obtain the optimum solution x^h and its dual u^h. The value for the whole problem will then be

$$p^T x^h + q^T w^h = (u^h)^T (b - Bw^h) + q^T w^h \leqslant \hat{w}_0$$

If in the LP sub-problem w^h would be replaced by the (unknown) optimum integer solution (this would entail a change of the right-hand side, hence of the objective vector in the dual problem), then u^h would still be a feasible dual solution in this new LP problem and we will have ($\hat{u}$ being the optimum dual solution of this problem):

$$(u^h)^T (b - B\hat{w}) + q^T \hat{w} \geqslant (\hat{u})^T (b - B\hat{w}) + q^T \hat{w} = \hat{w}_0$$

We will therefore require in the next pure integer sub-problem:

$$(u^h)^T (b - Bw) + q^T w \geqslant w_0, \text{ i.e. } \{(u^h)^T B - q^T\} w + w_0 \leqslant (u^h)^T b$$

The previous solution of the integer sub-problem, (w^h, w_0^h) will certainly not satisfy this new relation.

We are now in a position to describe the dual decomposition method:

(a) Start with an integer solution $w^0 = 0$

(b) At step h solve the LP problem

$$\text{Max}\ \{p^T x \mid Ax \leqslant b - Bw^h, x \geqslant 0\} \quad \text{and}$$

obtain the relations $r^T Bw \leqslant r^T b$, if any, and

$$\{(u^h)^T B - q^T\} w + w_0 \leqslant (u^h)^T b$$

(c) Solve the 'pure' integer programming problem

$$w_0^{h+1} = \text{Max}\ \{w_0 \mid \text{all relations obtained from previous LP sub-problems}\}$$

With the solution w^{h+1} step b will now be repeated.

At each step we will have the following useful inequality:

$$\underset{l \leqslant h}{\text{Max}}\ \{p^T x^l + q^T w^l\} \leqslant \Lambda_0 \leqslant w_0^h$$

(only those values of l should be considered for which the LP problem is feasible).

As soon as there is equality the overall optimum solution $(\hat{x}, \hat{w})$ has been reached. It is not difficult to prove that this will take place after a finite number of steps.

Several questions immediately arise in connection with this method. The computational efficiency and therefore the practical usefulness of the method may very well be dependent on a satisfactory answer to these questions:

(1) Like in any other decomposition method starting with a good integer solution hardly helps to speed up calculations. The first integer subproblem would only have one relation and the next integer solution would probably be far from the optimum one. Can this be avoided, for instance by adding many relations of the type $r^T(b - Bw) \geqslant 0$ to the integer sub-problem, also for those rows which did not lead to infeasibility in the LP problem (hence $r^T(b - Bw^0) \geqslant 0$ holds but we nevertheless require it for w', etc.). An efficient way of extracting as much useful information as possible from the LP problem will be of utmost importance since the number of steps will be reduced.

(2) How long should old relations be retained in the integer sub-problem. At one extreme we have the possibility of retaining them all, at the other extreme the possibility of omitting a relation as soon as it is no longer binding.

(3) Wouldn't it be better in the integer sub-problem to stop at the first feasible integer solution which in most enumerative algorithms can be found rapidly.

(4) It is possible in the integer sub-problem to make an efficient use of information left from the previous integer sub-problem, for instance by recalculating the value for each open node and for each node which has been closed as a consequence of rules 2 and 3 of section 2.1. In a direct enumeration algorithm like the one described in [14] this would mean that the alternative of a compulsory choice should also be stored (unless it is certain that it will be infeasible in all other integer sub-problems).

2.3 Direct enumeration

Direct enumeration could also be applied to the mixed integer problem in a straightforward way. LP problems of the type Max $\{p^T x \mid Ax \leqslant b - B\overline{w}, x \geqslant 0\}$ would then have to be solved for each possible $\overline{w}$. A tree could be developed for this. Stopping rules would be rather weak, however, so that such a procedure would not be recommendable except in special problems.

We will instead outline a composite approach:

Let us assume that a 'reasonable' approximation of the integer solution, say $\overline{w}$, is known (if not, any w can be taken as such). Let us now modify the problem in such a way that $w = 0$ is a reasonable starting solution. Hence, replace b by $b - B\overline{w}$; $B_{.k}$ (k-th column of B) by $-B_{.k}$ and q_k by $-q_k$ for those k for which $\overline{w}_k = 1$. We will assume that this has been done, so that $\overline{w} = 0$ is a good starting solution.

We now consider (first for $h = 0$, $b^0 = b$ and $K_0^h = K_1^h = \phi$):

$$P_1 : v_1^h = \text{Max}\,\{p^T x \mid Ax \leqslant b^h, x \geqslant 0\} + \sum_{k \in K_1^h} q_k;$$

$$P_2 : v_2^h = \text{Max}\,\{p^T x + q^T w \mid Ax + Bw \leqslant b, x \geqslant 0, 0 \leqslant w \leqslant 1; \; w_k = 0 \text{ for } k \in K_0^h, w_k = 1 \text{ for } k \in K_1^h\}$$

We define

$$b^h = b - \sum_{k \in K_1^h} B_{.k}$$

so that $v_1^h \leqslant v_2^h$ will hold for all h.

From P_2 we will select a w_k variable to which the values 0 and 1 will be assigned (using one of the criteria discussed in section 2.1). This will lead to a new problem P_1 and two new problems P_2. Originally all variables are unassigned ($K_0^h = K_1^h = \phi$).

We will obtain a tree like (dots denoting unassigned variables):

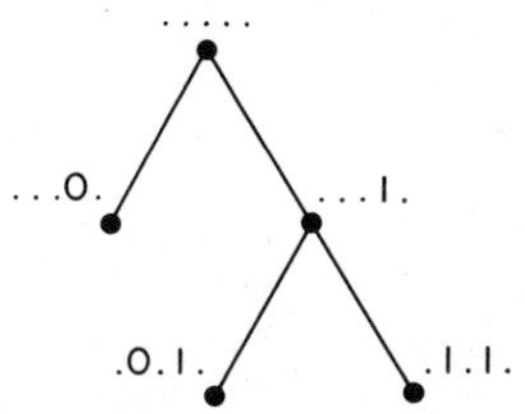

Each node has associated with it sets K_0 and K_1 as well as the two values v_1 and v_2, $v_1 \leqslant v_2$. Note that v_1 does not change when a 0 assignment is made. A node will be closed:

(a) if P_2 becomes infeasible;

(b) if $v_2 < v_1^*$ where v_1^* is the best value obtained for P_1 up to now;

(c) if $v_1 + \Sigma\{q_k \mid q_k > 0$ and k unassigned$\} < v_1^*$;

(d) if $v_1 = v_2$

We will then continue the search starting from another node which is still open. This treesearch is hardly different from the one applied in the branch and bound method. Rule c is the only additional stopping rule. However, it will now be possible to derive additional constraints from the problems P_1 in the same way as is being done in the dual decomposition method. These constraints could be added to the problems P_2: As a result the treesearch will probably be speeded up since condition a or d will be reached sooner. Just as in section 2.1 other trees are possible and might have advantages, e.g. assignment of more than one variable at a time, (see note (1) in 2.1). It is hardly possible to discuss the relative advantages of the various methods. Computational experience is only scarcely available. However, it seems to be a safe statement by now that practical mixed integer programming problems with up to 100 0-1 variables can be solved and that the solution of much larger problems is often possible. For further information, especially on applications, the reader is referred to [2].

3. FOUR PAPERS ON WAREHOUSE ALLOCATION AND RELATED PROBLEMS

In this part, four more or less related papers on warehouse allocation will be briefly discussed. Further details, examples and other information can be found in the papers themselves which all appear in these proceedings. We will use the same notation as in part 2, so that our notation will differ from the one used in the paper under discussion.

3.1 K. Spielberg, On Solving Plant Location Problems

This paper deals with the problem of selecting a number of plant locations from m possible sites. At site i there will be a fixed charge q_i and a capacity a_i. Demands b_j are given (j = 1, ... , n) as well as the unit shipment costs c_{ij}. The mixed integer programming formulation of this problem is as follows:

Minimize

$$\sum_{i=1}^{m} \sum_{j=1}^{n} c_{ij} x_{ij} + \sum_{i=1}^{m} q_i w_i$$

subject to

$$\sum_{i=1}^{m} x_{ij} \geq b_j \; (j = 1, \ldots, n)$$

$$\sum_{j=1}^{n} x_{ij} \leq a_i w_i \; (i = 1, \ldots, m)$$

$$w_i = 0 \text{ or } 1,$$

where the Boolean (0-1) variables w_i make sure that there will be either no warehouse at site i or, if there is one, that the fixed charge q_i is taken account of.

Let us first notice that for any choice of the w_i the problem reduces to a transportation problem for which efficient special algorithms exist.

It seems obvious therefore to use a dual decomposition method in this case. According to Spielberg's experience, however, this approach has been disappointing. He has therefore designed a method of the type, described in 2.3, i.e. a mixture of branch and bound, enumeration and dual decomposition.

Several algorithms and some computational results are given.

3.2 P.H. Randolph, G.E. Swinson and M.E. Walker, A Nonlinear Programming Warehouse Allocation Problem

In this paper customers have to be assigned to warehouses. There are m warehouses with capacities a_i, $i = 1, \ldots, m$ and n customers. Each warehouse has a pre-determined geographical service area which is fixed.

It is therefore possible to introduce sets

$$J_i = \{j \in \{1, \ldots, n\} \mid j \text{ can be serviced by warehouse } i\};$$

$$I_j = \{i \in \{1, \ldots, m\} \mid i \text{ can serve } j\}$$

The product to be supplied can only be transported in integral quantities. The demand x_j at customer j is variable and unlimited.

The pay-off function at j will be denoted by $p_j(x_j)$ and will be nonlinear and non-concave. In this particular problem transportation cost could be neglected. Therefore, if x_{ij} is transported from warehouse i to customer j, the problem formulation is:

Maximize

$$\sum_{j=1}^{n} p_j(x_j)$$

subject to

$$\sum_{j \in J_i} x_{ij} \leq a_i$$

$$\sum_{i \in I_j} x_{ij} = x_j, \qquad x_{ij} \text{ integer.}$$

To solve this problem the authors make use of dynamic programming.

They first partition the set of all customers into 'sectors', customer j' belonging to the sector consisting of all j for which $I_j = I_{j'}$ (see example).

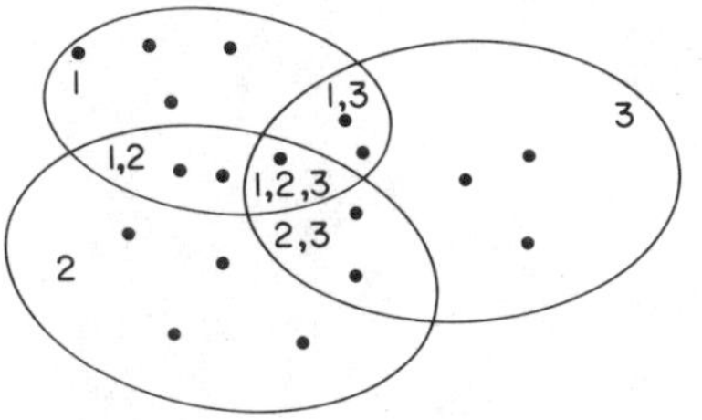

3 service area; 8 sectors.
Dots represent customers.

The method consists of three steps:

(a) Determine an initial allocation from each warehouse to each sector.

(b) Knowing what is available in the sector, determine in each sector an optimal allocation to customers.

(c) Make a re-allocation to the sectors for one warehouse and repeat this for other warehouses, one at a time, until no improvement in the total pay-off can be obtained any more.

In step (a) as good an estimate as possible is made. The problems in step (b) can be solved using dynamic programming. If the total amount available in a sector equals x and if the number of customers in the sector equals k, then writing

$$f_k(x) = \max_{x_1,\dots,x_k} \left\{ \sum_{j=1}^{k} p_j(x_j) \mid \sum_{j=1}^{k} x_j = x \right\}$$

we have the functional equation:

$$f_k(x) = \max_{0 \leq x_k \leq x} [P_k(x_k) + f_{k-1}(x - x_k)],\ f_0(x) = 0.$$

from which $f_k(x)$ can be found through enumeration.

In step (c) one warehouse is singled out and the allocation of its items to the sectors is changed in such a way that total pay-off will be maximized (note that the allocations of the other warehouses to the sectors are not changed). This again can be formulated as a dynamic programming problem with one constraint and will be solved in the same way as the within-sector allocation problem above. Step (c) is then carried out for another warehouse and this process is repeated until an entire cycle (all m warehouses one after the other) passes without a change in pay-off. We have then obtained a local maximum that might be far from the global maximum. Which local maximum we arrive at is of course dependent on the initial allocation to warehouses, as well as on the order in which the warehouses are chosen in step (c). The authors have also succeeded in establishing an upper bound for the solution, again by using dynamic programming. Whether this upper bound will help much in practical situations remains to be seen. The computational efficiency of this dynamic programming approach, compared to tree search methods, is subject to doubt.

There are at least two other methods of solution for this problem:

(1) Formulation as a mixed integer programming problem:

Maximize

$$\sum_{j=1}^{n} \sum_{k=1}^{b_j} q_{jk} w_{jk}$$

subject to

$$\sum_{j \in J_i} x_{ij} \leq a_i$$

$$\sum_{i \in I_j} x_{ij} = \sum_{k=1}^{b_j} k w_{jk}$$

$$\sum_{k=1}^{b_j} w_{jk} \leq 0,\ j = 1, \ldots, n;\ w_{jk} = 0 \text{ or } 1;$$

in which $q_{jk} = p_j(k)$, while b_j is an upperbound for x_j.

The integer requirements for x_{ij} can be omitted since transportation theory tells us that they will be satisfied automatically for any integer choice of the w_{jk}. In view of the special structure for fixed w_{jk} the dual decomposition method of section 2.2 is the obvious method to apply.

(2) The problem is a special case of the problem considered by Rech and Barton which will be the subject of the next section.

3.3 P. Rech and L. G. Barton, A Nonconvex Transportation Algorithm

This paper is not restricted to warehouse allocation but discusses the general problem of minimizing a non-convex separable piecewise linear function, the variables of which are quantities to be transported and satisfy the usual transportation type constraints (limited availability at sources, limited requirements at destinations, balance equations at transhipment points, capacity restrictions on routes). A typical example of such a problem is the warehouse allocation problem under economies of scale [13].

For convex transportation programming (minimization of a convex separable function with transportation type constraints) the out-of-kilter algorithm has been developed by Fulkerson [6]. This algorithm is being used as a subroutine by the authors. Their method is a branch-and-bound technique which is clearly described in their paper, so that we can restrict ourselves to the main points. First the non-convex functions occurring in the objective function are approximated by their convex envelopes (see figure 3 of the paper). Then the resulting convex minimization problem is solved resulting in a solution with value v_1^0. This solution is feasible in the original problem (but in general not optimal). It will there have a value v_2^0. Denoting the optimal value of the original problem by $\hat{v}$ the inequality $v_1^0 \leq \hat{v} \leq v_0^2$ will hold as is proved in the paper. Next the error resulting from the convex approximation is calculated for each non-convex function by substituting the solution obtained and the variable x_j is selected for which this error is largest.

The corresponding cost function is now partitioned according to $x_j \leq \mu_j$ and $x_j \geq \mu_j$. For both parts a new (closer) convex approximation is calculated and two convex minimization problems are solved. This is the branching part of the method on the basis of which a tree is developed. The search will be continued with the node for which v_1 is lowest. If infeasibility is encountered the node will be closed. As upper-bound we have the minimum of all v_2 values calculated, as lower bound v_1. The two bounds will grow together when the search is continued. For μ_j we can take the first breakpoint to the left (or to the right) of the current value of x_j.

This paper can be considered as an important contribution in the field of mathematical programming. It includes as special cases the warehouse allocation problems discussed in 3.1, 3.2 and 3.4. For the problem considered by Spielberg it would not be necessary to introduce the Boolean variables w_k but a direct search could be initiated using the method just described. It would be surprising, however, if this would give a computational advantage.

3.4 P. Gray, Exact Solution of the Site Selection Problem by Mixed Integer Programming

The problem considered by the author is the fixed charge problem:

Minimize

$$c^T x + q^T w$$

subject to

$$Ax \leqslant b, x \geqslant 0,$$

$$w_k = 1 \text{ if } \sum_{j \in J_k} x_j > 0,$$

$$= 0 \text{ otherwise}$$

where the q_k are fixed charges, while $J_1, J_2, \ldots, J - \Sigma J_k$ form a partitioning of $\{1, \ldots, n\}$. The plant allocation problem studied by Steinberg is a special case.

By introducing upper bounds d_k for the sums

$$\sum_{j \in J_k} x_j$$

we obtain the following mixed integer programming problem:

Minimize

$$c^T x + q^T w$$

subject to

$$Ax \leqslant b, x \geqslant 0,$$

$$\sum_{j \in J_k} x_j \leqslant d_k w_k$$

$$w_k = 0 \text{ or } 1$$

The problem is solved by using a direct enumeration algorithm. First the associated linear program is solved. Rounding each w_k with fractional value to 1 gives a feasible solution of the mixed integer problem with value $\bar{v}$. Next all w_k are set to 1 and the resulting LP is solved. If $q_k > 0$ for all k—which is assumed—the variable cost $c^T x$ is minimized in this way; this minimum will be denoted by v_1. We clearly have $q^T w \leqslant \bar{v} - v_1$. This additional constraint is used to reduce the search of integer solutions w. A lower bound for $q^T w$ can also be obtained, for instance by replacing c by 0 and by then solving the associated LP problem. If a value v_2 is so obtained for $q^T w$, then

$q^T w \geq v_2$ will have to hold. The upperbound can be gradually improved since $\bar{v}$ can be adjusted as soon as a better feasible solution has been found. A feasibility test can be included: if for a w which satisfies upper and lower bounds we set $x_j = 0$ if $j \in J_k$ and $w_k = 0$ has been assigned and if the resulting system of inequalities is inconsistent, then the node will be closed. Extension of the method to non-linear separable objective functions is also discussed.

Comparing this method with the one suggested by Steinberg it can be concluded that the latter will need a smaller tree (more powerful exclusion rules) but more work per node. This might be an advantage, especially for larger problems.

4. FURTHER REMARKS

All enumeration algorithms have their limitations; in spite of the most sophisticated exclusion rules they remain combinatorial. If a formulation as a mixed integer programming problem is possible, but at the cost of introducing many additional integer variables, then any direct enumeration method which does not need this formulation is nearly always to be preferred. This has been clearly demonstrated for the travellin salesman problem, where a direct branch and bound method has been very successful [12]. This means that we can expect more special methods for special problems. If still larger problems have to be solved—and there are many large ones of considerable practical importance—heuristic decision rules will be needed to direct and limit the treesearch. There should then be more possibilities for human judgment and experience to be applied during the calculation process. Direct man-machine interaction will probably be of much importance in the future for the solution of the nonconvex programming problems which we have discussed.

REFERENCES

1. BALAS, E., An Additive Algorithm for Solving Linear Programs with Zero-One Variables, Operations Research **13** (1965), 517-546.
2. BALINSKI, M. L., Integer Programming: Methods, Uses, Computation, Mgt. Science **12** (1965) 253-313.
3. BENDERS, J. F. et al., Discrete Variables Optimization Problems, paper presented at the Mathematical Programming Symposium, Santa Monica, March 1959.
4. BENDERS, J. F., Partitioning Procedures for Solving Mixed Variables Programming Problems, Num. Math. **4** (1962) 238-252.
5. DAKIN, R. J., A Tree-search Algorithm for Mixed Integer Programming Problem Computer J. **8** (1965) 250-255.
6. FULKERSON, D. A., An Out-of-Kilter Method for Minimal-Cost Flow Problems, J. SIAM **9** (1961) 18-27.
7. GLOVER, F., A Multiphase-Dual Algorithm for the Zero-One Integer Programming Problem, Operations Research **13** (1965) 879-919.
8. GOMORY, R. E., An Algorithm for Integer Solutions to Linear Programs, pp. 269-302 of R. L. Greves and P. Wolfe (eds.), Recent Advances in Mathematical Programming, McGraw-Hill, 1963.
9. HERVE, P., Résolution des Programmes Linéaires à Variable Mixtes par la Procédure S.E.P., Metra **6** (1967) 77-91.
10. LAND, A. H. and A. G. DOIG, An Automatic Method of Solving Discrete Programming Problems, Econometrica **28** (1960) 497-520.
11. LAWLER, E. L. and D. E. WOOD, Branch and Bound Methods: A Survey, Operation Research **14** (1966) 699-719.

12. LITTLE, J. D. C., K. G. MURTY, D. W. SWEENEY and C. KAREL, An Algorithm for the Travelling Salesman Problem, Operations Research 11 (1963) 972-989.

13. MANNE, A. S., Plant Location under Economies of Scale—Decentralization and Computation, Mgt. Science 11 (1964) 213-235.

14. ZOUTENDIJK, G., Enumeration Algorithms for the Pure and Mixed Integer Programming Problem, to appear in: Proceedings of the 1967 Mathematical Programming Symposium, Princeton.

On Solving Plant Location Problems

KURT SPIELBERG
I.B.M. New York Scientific Center, U.S.A.

1. INTRODUCTION

It is well known that certain types of plant location problems can be formulated as mixed-integer programming problems and be solved by enumerative search algorithms. Among recent efforts in the area of 'simple plant location problems' (see later for definition) we may cite the references [1, 2, 3]. The 'capacitated' problem is considered in [4]. See also [5] for a brief summary.

While it is easy to propose an algorithm for solving such problems, e.g. by specializing some existent general approach to mixed-integer programming, it is not easy at all to construct an algorithm which promises to be efficient. Problems of practical size can be expected to be large, involving more than say 50 possible plant sites, and, therefore, can be expected to pose substantial, sometimes insurmountable, computational difficulties.

The present paper draws upon earlier experience of the author [2, 3] which suggests the great desirability of 'adapting' the algorithm to the data at hand. It proposes an enumerative algorithm which can, in some sense, be adapted dynamically. This should serve to make the basic devices of the algorithm, such as systematic use of 'associated' Benders inequalities and solution of auxiliary subproblems, maximally effective.

Some of the basic notions of this paper have been developed for the general mixed integer problem in collaboration with M. Guignard and will be presented in the more general context elsewhere. We believe that the resulting search structures and algorithms are, for [0, 1] mixed integer problems, more general and flexible than, e.g., those of Land and Doig, Dakin, Balas, without appearing to be unduly complex. One may hope then that these algorithms will be effective for wider classes of problems and data than other algorithms in current use.

2. SEARCH PROCEDURE, SUBPROBLEMS

Consider the plant location problem in the form:

minimize

$$z = \Sigma c_{ij}x_{ij} + \Sigma f_i\eta_i$$

subject to

$$\Sigma x_{ij} \geqslant r_j,\ j = 1, 2, \ldots, n \qquad (2.1)$$

$$\Sigma x_{ij} \leqslant a_i\eta_i,\ i = 1, 2, \ldots, m$$

There are m plants or possible plant sites with capacities a_i and n customers with requirements r_j. There are unit shipment costs c_{ij} and fixed charges f_i. The charge f_i is incurred if and only if there is any shipment from i, that is if

$$\sum_j x_{ij} > 0.$$

The problem is to find a pattern of open plants $i(\eta_i = 1)$ and a corresponding pattern of shipments x_{ij} which minimize the total cost.

In addition to problem (2. 1), the 'capacitated plant location problem', one may consider the 'simple plant location problem' which arises from (2. 1) as the bounds a_i are removed. This problem can be viewed as having the same form as (2. 1), with $r_j = 1$ (all j), the x_{ij} reinterpreted as the fraction of the demand at j supplied by plant i and c_{ij} the cost of shipping an amount of the commodity equal to the total requirement at j from plant i.

Evidently, the problem can be resolved by enumeration. One considers a sequence $\{\eta^s\}$ of m-vectors η which exhaust all possible patterns of open and closed plants. For a fixed η, problem (2. 1) becomes $T(\eta)$, a linear programming problem in x of very special structure. For 'Simple Plant Location' this problem is resolved by determining, for each j, the route (i(j), j) with minimal shipping cost c(i(j), j) over those i for which $\eta_i = 1$, and setting $x(i(j), j) = 1$. For 'Capacitated Plant Location', the linear program becomes a standard Transportation Problem with equality constraints, if a fictitious demand point (an additional column, indexed by, say, l) with requirement

$$r_l = \sum_{i \in I} a_i - \sum_{j=1}^{n} r_j \qquad (2.2)$$

(I. . . index set of open plants)

is adjoined to the problem. The problem $T(\eta)$ is considered infeasible when $r_l < 0$, and also whenever costs of unreasonable size, representing prohibited routes (i, j), must be incurred if the requirements are to be satisfied.

Consider now the manner in which the sequence $\{\eta^s\}$ is to be realized, i.e., the search procedure to be used. There are several well known procedures, which can be classified, according to the geometric appearance of the search tree with which the graph of integer points is scanned, as Single Branch and Multi-Branch procedures (see the survey [5]). The former type of search has been advocated by E. Balas (e.g., [6]) and others for zero-one integer programming, the latter by Land and Doig [7] for mixed integer programming.

In the following we describe a procedure which is of the Single Branch type, but is much more flexible than the original Balas scheme, in that it is adapted for mixed integer programming (primarily with (0, 1) integer variables) and in that adjacent vectors $y^{\nu-}, y^{\nu}, y^{\nu+}$ (see Fig. 1) may be viewed as differing from each other, for certain well-defined purposes, in more than one component. In the Balas algorithm the search origin ϕ corresponds to a 'natural' value for the zero-one vector η, say to $\eta = 0$, or possibly to $\eta = e = (1, 1, \ldots, 1)$, i.e., $\bar{\eta} = e - \eta = \phi$. A 'forward step' in the search, say from node ν to ν^+, is determined by selecting a 'link i*', i.e., setting

$$\eta_{i*} = 1 \text{ (or } \bar{\eta}_{i*} = 1).$$

At level k one has k components of η fixed at 1, possibly also some others, say i_e, 'cancelled' (fixed at 0), if it was determined at a level $k' \leq k$ that no improved solution could be obtained with $y_{i_e} = 1$. The number next to a link in Fig. 1 would identify the components fixed at one.

More generally, of course, one may identify a forward step with either fixing a component at one or at zero, $i* > 0$ or $i* < 0$ in Fig. 1 (Compare, e.g., ref [10]). This requires, however, certain associated conventions as far as the computational activity at node ν is concerned. No work in this direction appears to have been published.

In the following, we generalize the search by associating with a general node ν (and corresponding level k) a partitioning of index set I into four subsets:

Ω (open), C (closed), F1 (free, type 1) and F2 (free, type 2).

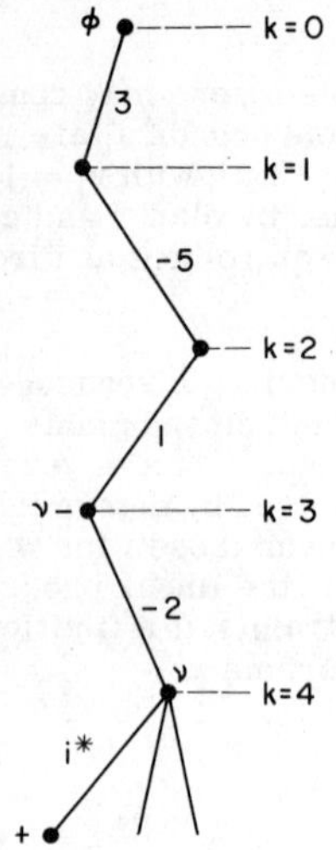

Fig. 1.

In ref. [3] this partitioning of F was used only once, at a restart of the search procedure. Free Plants of type 1 were considered as tentatively open (free to be closed on a forward step), those of type 2 tentatively closed (free to be opened on a forward step). This partitioning had a drastic effect on the search procedure. When exploited completely it sometimes led to almost immediate resolution of the problem (see section on computational results), but also, if this was not the case, to a considerable increase in computational requirement per iteration.

The procedure described in the sequel is a natural generalization, in that it provides for a 'dynamic' repartitioning at any node. The particular algorithmic details suggested are, moreover, believed to allow the choice of procedures which benefit from the partitioning without incurring heavy computational penalties.

Consider, then, at node ν the partitioning

$$I^{\nu} = \Omega^{\nu} + C^{\nu} + F1^{\nu} + F2^{\nu} \tag{2.3}$$

and the resolution of three problems:

TP^{ν}: the original problem (2.1) with the fixed values $\eta_i = 1 (i \in \Omega^{\nu})$

and $\eta_i = 0 (i \in C^{\nu})$ substituted.

$$\eta_i \ldots 0 \text{ \underline{or} } 1 \text{ for } i \in F^{\nu} \tag{2.4}$$

TR^{ν}: the 'relaxed' problem. Same as TP^{ν}, except for:

$$0 \leq \eta_i \leq 1, i \in F^{\nu} \tag{2.5}$$

TZ^{ν}: the 'constrained' problem. Same as TP^{ν}, except for:

$$\begin{aligned} \eta_i &= 1 \text{ for } i \in F1^{\nu} \\ \eta_i &= 0 \text{ for } i \in F2^{\nu} \end{aligned} \tag{2.6}$$

Denote the objective functions by Z, Z_r and ζ, respectively. One evidently has:

$$Z_r \leqslant Z \leqslant \zeta \tag{2.7}$$

It is convenient, at node ν, to introduce new variables y, in accordance with

$$\begin{aligned} y_i &= \eta_i \text{ for } i \in C + F2 \\ y_i &= 1 - \eta_i \text{ for } i \in \Omega + F1 \end{aligned} \tag{2.8}$$

In terms of the variables y_i, the subproblems at ν are treated by the method of Balas, a forward step being associated with setting some y_i, $i \in F^\nu$, equal to one. The condition (2.6), characterizing TZ^ν, becomes:

$$y_i = 0 \text{ for } i \in F^\nu \tag{2.9}$$

Of course, the definition of the y_i depends on the state (specified the partitioning (2.3)) at ν.

The search procedure terminates when node ν is at level k = 0 and $F^\nu = \phi$.

3. THE STATE AT NODE

The fundamental device for directing and curtailing the search is an inequality on the y_i, $i \in F$, associated with node ν through auxiliary problem TR^ν. It is possible to use that device alone (algorithm 1), but possibly also once or several times supplemented by a complete restart at some good solution point, say $\hat{\eta}$.

On such a restart one defines the state at level k = 0, node ν (ν simply counts all nodes) by setting

$$\begin{aligned} y_i &= \eta_i \quad \text{for } i: \hat{\eta}_i = 0 \\ y_i &= 1 - \eta_i \text{ for } i: \hat{\eta}_i = 1 \end{aligned} \tag{3.1}$$

and maintains this state for at least some higher levels of k. Effectively, one thus examines the 'vicinity' of the good solution point $\hat{\eta}$, and hopes that the various exclusion devices of the search are particularly effective in this vicinity.

Experience shows that it may be essential to couple this procedure with a strategy which gives priority to the examination of candidates which are 'unlikely' candidates for improvement (see section on 'strategy' and computational experience).

More generally, consider as second fundamental device the solution of the relaxed problem TR^ν.

The solution of TR^ν furnishes several, immediate tests which may lead to curtailment of the search.

(i) If TR^ν is feasible, evidently the more constrained problem TP^ν is also infeasible. Hence, node ν is abandoned and the search reverts to ν^- (with $\eta_{i_{\nu-\nu}}$ changing to $1 - \eta_{i_{\nu-\nu}}$, $i_{\nu-\nu}$ being the link which led from ν^- to ν).

(ii) If Z_r^ν attains or exceeds z*, an upper bound on z (usually the value of the best objective function found so far), the search may revert to ν^- on account of (2.7).

(iii) If the solution to TR^ν happens to be integral in y, it also solves TP^ν, and the search may again revert to ν^- after recording of the solution.

In the algorithms to be presented, however, we are primarily interested in using the information contained in the solution ${}^r y^\nu$ (or ${}^r\eta^\nu$) to determine the search pattern for problems TP^ν and TZ^ν, that is to determine the state at ν. In passing it may be noted that this is similar in intent to the procedures of Gomory which involve solving an 'asymptotic' problem, after the solution to a linear programming problem is at hand (see [5]). It is quite conceivable that such procedures may find a natural place in an enumerative method.

For the present we are content with the following simple rounding procedure. Suppose that TR has been solved at ν (of course, in general, one may decide to solve TR only at a subset of the iterations, e.g. at preselected levels k, and usually only when k is reached on a forward step from a lower level). Preselect a rounding parameter $0 \leqslant \delta \leqslant 1$, which may perhaps best be set to something like $\frac{1}{2}$ initially and closer to 0 later, i.e., in particular, after any restart.

Then one may define the state at ν (for purposes of solving TZ^ν and TP^ν, by partitioning F in accordance with:

$$\begin{aligned} &i \to F1 \text{ if } {}^r\eta_i{}^\nu \geqslant \delta \\ &i \to F2 \text{ if } {}^r\eta_i{}^\nu < \delta \end{aligned} \tag{3.2}$$

The variables y_i for TZ^ν are then defined by (2.8).

One disadvantage of (3.2) might be that it can easily lead to infeasibility of TZ^ν, e.g. by virtue of $r_l < 0$ (see 2.2, with $I = \Omega + F1$). It may be desirable, though certainly not necessary, to avoid such infeasibilities when possible. One way of doing this might be to check r_l whenever an index i is to be transferred to index set F2 by virtue of (3.2), and to avoid the transfer if it would render r_l negative. Another way of doing it would be to make δ small. Certainly, $\delta = 0$ would guarantee that TZ^ν is feasible whenever TR^ν is.

After a restart it is evidently undesirable to use (3.2) at low levels k (except with $\delta = 0$), since this would negate the effect of the restart. It is easy to make allowances for such considerations in a computer program.

4. THE ASSOCIATED INEQUALITY

Consider problem TP^ν:

$$\begin{aligned} \min \Big\{ z' = z - &\sum_{\Omega^\nu + F1^\nu} f_i = cx + \sum_{F2^\nu} f_i y_i - \sum_{F1^\nu} f_i y_i \\ \Sigma x_{ij} \quad &\leqslant a_i, \qquad i \in \Omega^\nu \\ \Sigma x_{ij} + a_i y_i \quad &\leqslant a_i, \qquad i \in F1^\nu \\ \Sigma x_{ij} - a_i y_i \quad &\leqslant 0, \qquad i \in F2^\nu \\ \Sigma x_{ij} \quad &= 0, \qquad i \in C^\nu \\ -\Sigma x_{ij} \quad &\leqslant -r_j, \qquad j \in J = \{1, 2, \ldots, n\} \\ x_{ij} \geqslant 0, y_i \ldots (0,1) &\text{ for } i \in F^\nu, y_i = 0 \ldots i \in \Omega^\nu + C^\nu \Big\} \end{aligned} \tag{4.1}$$

It is of the form

$$\min \quad \{z' = cx + qy \mid Dx + Qy \leqslant b, x \geqslant 0, y \in S\} \tag{4.2}$$

considered by Benders in [8].

For any fixed y, say $y = y_*^\nu$, one may attempt to solve the resulting linear programming problem and its dual. When the primal problem is feasible, one obtains a vector u^ν consisting of the optimal dual variables: $u^\nu = (u_i^\nu, v_j^\nu)$. The m variables u_i^ν correspond to the first m constraints, the n variables v_j^ν to the last n constraints of (4. 1). When the primal problem is infeasible, there exists an extremal half-ray in dual space, of the form $u^\nu + \lambda\bar{u}^\nu$, along which the dual objective function goes to infinity. Corresponding to u^ν and $\bar{u}^\nu$ one has the inequalities

$$(q + u^\nu Q)y - u^\nu b \leq z^{*\prime} \tag{4. 3}$$

and $\bar{u}^\nu(b - Qy) \geq 0$ (4. 4)

as necessary conditions on y, which must be satisfied for the existence of a solution to (1) with $z' \leq z^{*\prime}$. ($z^{*\prime}$ stands for some known upper bound on z').

For the purposes of the algorithms of this paper we shall assume that a vector u^ν is computed whenever the level k of the search is increased and is saved to be used again when the search returns to level k from a higher level. The same could be done with $\bar{u}^\nu$, but we assume that infeasibility of the primal problem is relatively unlikely, so that $\bar{u}^\nu$ may be recomputed if necessary, without undue loss of time.

In the case of problem (4. 1) one has

$$Q = \begin{bmatrix} \delta_1 a_1 & & & \\ & \delta_2 a_2 & & \\ & & \ddots & \\ & & & \delta_m a_m \\ & \phi & & \end{bmatrix}, \; b = \begin{bmatrix} \\ a_i\tilde{\delta}_i \\ \\ \\ -r_j \end{bmatrix}, \quad \begin{array}{l} \delta_i = 1, i \in F1 \\ \delta_i = -1, i \in F2 \\ \delta_i = 0, i \in \Omega + C \\ \tilde{\delta}_i = 1, i \in F1 + \Omega \\ \tilde{\delta}_i = 0, i \in F2 + C \end{array} \tag{4. 5}$$

so that (4. 3) becomes:

$$\sum_{i \in F1^\nu} (-f_i + u_i^\nu a_i)y_i + \sum_{i \in F2^\nu} (f_i - u_i^\nu a_i)y_i - \sum_{F1^\nu + \Omega^\nu} u_i^\nu a_i + \sum_J r_j v_j^\nu \leq z^{*\prime} \tag{4. 6}$$

It is clear that problem TP can be resolved by a consideration of TZ^ν at all nodes ν of interest. Problem TZ^ν is TP^ν with y fixed at ϕ. Its dual then is the transportation problem

$$\max \{-ub \mid c_{ij} + u_i - v_j \geq 0, u_i \geq 0, v_j \geq 0\} \tag{4. 7}$$

with the objective function

$$t_d = - \sum_{i \in F1^\nu + \Omega^\nu} u_i a_i + \sum_{j \in J} r_j v_j \tag{4. 8}$$

We shall use t_d^ν to denote t_d evaluated at u^ν. Inequality (4. 6) then is of the form

$$\sum_{F1^\nu} (-f_i + u_i^\nu a_i)y_i + \sum_{F2^\nu} (f_i - u_i^\nu a_i)y_i \leq z^* - \zeta^\nu, \tag{4. 9}$$

with $$\zeta^\nu = \sum_{\Omega^\nu + F1^\nu} f_i + t_d^\nu \tag{4. 10}$$

the optimal objective function of TZ^ν, if TZ^ν is feasible. (z^* is any available upper bound on z, the o.f. of (4. 1)).

The case $z^* > \zeta^\nu$ can be neglected.

When it arises for feasible TZ^ν, then ζ^ν is an improvement over z^* and replaces the latter. When it arises for infeasible TZ^ν, we shall use (4.4) for purposes of the search and u^ν only as starting value for $TZ^{\nu+}$.

Hence, the case of interest is the one with $z^* - \zeta^\nu \leqslant 0$. It is then possible to obtain information from inequality (4.9), by treating it much in the same manner as one treats 'infeasible' inequalities in zero-one integer programming of the single Branch (Balas) type (compare also with the notions: 'complete reduction', 'preferred variables' in [9, 5]). Write (4.9) schematically in the form

$$\sum_{I1} \alpha_i y_i - \sum_{I2} \beta_i y_i \leqslant -\gamma \tag{4.11}$$

with $(\alpha_i, \beta_i) \geqslant 0, \gamma \geqslant 0$.

Define $t = -\Sigma\beta_i$. Then it is clear that

$$\text{condition 1: } -\gamma - t < 0 \tag{4.12}$$

implies that inequality (4.9) can not be satisfied at successor nodes to ν, and

$$\text{condition 2: } -\gamma - t - \alpha_i{}^* < 0,\ i^* \in F^\nu \text{ implies the same if } y_i{}^* \text{ is set to 1.} \tag{4.13}$$

Hence, condition 1 justifies reverting to the predecessor node ν^-, and condition 2 justifies cancellation of i^*, that is fixing $y_i{}^*$ at 0 for problem TP^ν.

Another equivalent way of interpreting (4.9) is to write it as:

$$\zeta^\nu - \sum_{i \in F^\nu} g_i{}^\nu y_i \leqslant z^* \tag{4.14}$$

and view $\sum_{i \in F^\nu} |g_i|$ as the maximal decrease in ζ^ν over the successor nodes μ to ν. The $g_i{}^\nu$ may then be interpreted as 'gain functions' and interpreted as the maximal 'gain' (reduction in ζ^ν) attainable from fixing y_i at 1.

This notion is useful, if

(i) more careful physical consideration of a specific problem allows the construction of tighter (smaller) gain functions, and therefore more powerful tests of types (4.12) and (4.13), or if

(ii) it is possible to derive, for a specific problem, relations such as:

$$g_i{}^\mu \leqslant g_i{}^\nu \tag{4.15}$$

(μ ranging over all successor nodes to ν). Relation (4.15) warrants the cancellation of i at ν, whenever $g_i{}^\nu \leqslant 0$.

In terms of (4.11), property (4.15) guarantees that the coefficients of a given $y_i{}^*$ in (4.11) in associated inequalities at successor nodes to ν can not be smaller than at ν. If a coefficient, say of i^*, is positive, it will remain so. Since the order of fixing components at 1 is arbitrary, one will usually choose those with largest negative coefficient (largest gain). It is clear, then, that components with positive coefficients need never be chosen, in the sense that conditions (4.12) or (4.13) will become effective before a choice of i^* is necessary.

Now, consider the case of infeasibility, leading to inequality (4.4). Infeasibility with $y = y_*^\nu = \phi$ implies that

$$\bar{u}^\nu b - \sum_{i \in \Omega^\nu} (\bar{u}^\nu Q)_i < 0.$$ The condition (4.4), interpreted at ν

with $y_i = 0 (i \in C^\nu)$, $y_i = 1 (i \in \Omega^\nu)$ and $y_i = y_i (i \in F^\nu)$, leads to the inequality

$$\bar{u}^\nu b - \sum_{i \in \Omega^\nu} (\bar{u}^\nu Q)_i - \sum_{i \in F^\nu} (\bar{u}^\nu Q)_i y_i \geqslant 0,$$

$$\text{or:} \quad \sum_{i \in F^\nu} (\bar{u}^\nu \,.\, Q)_i y_i \leqslant \bar{u}^\nu \,.\, b - \sum_{i \in \Omega^\nu} (\bar{u}^\nu \,.\, Q)_i < 0 \tag{4.16}$$

This is again an inequality of form (4.11), and can therefore be used in the same manner.

We shall not dwell on the computation of $\bar{u}$. When the infeasibility of TZ^ν is of the type $r_l < 0$, the heuristic device of opening plant $i^* \in F2^\nu$, such that a_{i^*} is maximal over a_i, $i \in F2^\nu$, may be as good as any. When the infeasibility arises from the necessity of including a route (i, j) with infinite cost, it is also easy to think of heuristic devices of plants to be opened. A formal method of obtaining $\bar{u}$ (see [11]) based upon the dual transportation Problem Algorithm of Ford and Fulkerson (see [12]), utilizes the fact that the necessity of including a large cost implies a dual variable change at the point $u = u^\nu$, in which one of the dual variables goes to infinity.

5. SOLVING THE SUBPROBLEMS TZ^ν, TR^ν

Computationally, the most important step is the resolution of subproblem TZ^ν. The relaxed problem TR^ν certainly need not be resolved at every node, but rather at some preselected nodes only, e.g. those at given levels. Actually, problem TZ^ν is replaced by the related problem $\widetilde{TZ}^\nu$, the problem with the additional column l, having the requirement

$$r_l = \sum_{i \in \Omega^\nu + F1^\nu} a_i - \sum_{j \in J} r_j,$$

and costs $c_{il} = 0$, $i \in \Omega^\nu + F1^\nu$. Note, that TZ^ν is a problem with m rows and inequality constraints, whereas $\widetilde{TZ}^\nu$ is a problem with $|\Omega^\nu + F1^\nu|$ rows and equality constraints.

The solution to TZ^ν is determined from that for $\widetilde{TZ}^\nu$ in the obvious manner

$$\zeta^\nu = \tilde{\zeta}^\nu, \qquad x^\nu = \tilde{x}^\nu \tag{5.1}$$

except for the dual variables $u = (u_i, v_j)$, for which it is necessary to augment the obvious relations

$$u_i^\nu = \tilde{u}_i^\nu, i \in \Omega^\nu + F1^\nu, \qquad v_j^\nu = \tilde{v}_j^\nu, j \in J, \tag{5.2}$$

by a determination of the additional variables $u_i^\nu (i \in F2 + C)$ in accordance with the duality constraints for TZ^ν:

$$c_{ij} + u_i^\nu - v_j^\nu \geqslant 0 \qquad i \in F2 + C, j \in J \tag{5.3}$$

The optimal dual objective function for the transportation problem

$$t_d^\nu = - \sum_{i \in \Omega + F1} a_i u_i^\nu + \sum_{j \in J} r_j v_j^\nu \tag{5.4}$$

must, in accordance with (5.1) and (5.2) be same as $\tilde{t}_d^\nu = t_d^\nu + r_l \tilde{v}_l^\nu$, so that $\tilde{v}_l^\nu$ must be zero. (The trivial case $r_l = 0$ can be disregarded). Since $u_i^\nu (i \in F2 + C)$ does not appear in t_d^ν, it may be choosen so as to make the inequalities (4. 9) most effective i.e., so as to minimize the gain functions $g_i (i \in F2^\nu)$, or maximize the $(f_i - u_i^\nu a_i)$ in inequality (4. 9).

$$u_i^\nu = \max_j (0, v_j^\nu - c_{ij}) \tag{5. 5}$$

The fact that $\tilde{v}_l^\nu$ must be zero can also be shown from the complementary slackness relations:

$$\begin{aligned} &x_{ij}^\nu (c_{ij} + u_i^\nu - v_j^\nu) = 0 \qquad i \in \Omega^\nu + F1^\nu, j \in J + \{l\} \\ &u_i^\nu (a_i - \sum_{j \in J} x_{ij}^\nu) = 0, \quad v_j^\nu (\sum_{i \in \Omega^\nu + F1^\nu} x_{ij}^\nu - r_j) = 0 \end{aligned} \tag{5. 6}$$

Suppose all $u_i^\nu > 0 (i \in \Omega + F1)$. This would imply

$$a_i = \sum_{j \in J} x_{ij}^\nu (i \in \Omega + F1), \text{ which in turn signifies}$$

$$r_l \sum_{i \in \Omega + F1} a_i - \sum_{j \in J} r_j = 0.$$

Hence we may assume the existence of an index $i_* \in \Omega + F1$, such that $u_{i*}^\nu = 0 = \tilde{u}_{i*}^\nu$. But then the fact that $c_{i*l} = 0$, in conjunction with the constraint $c_{ij} + \tilde{u}_{i*}^\nu - \tilde{v}_l^\nu \geq 0$ for $\tilde{T}Z^\nu$, implies $\tilde{v}_l^\nu = 0$.

These observations are important because they show the need for distinguishing between the labels α_i and β_j used in the Ford-Fulkerson Flow algorithm and the dual variables u_i, v_j. The α_i, β_j must eventually be normalized so that $\beta_l = 0$. In particular, it is, therefore, not justified to assume that the u_i, v_j non-decrease monotonically in the dual solution procedure, as do the α_i, β_j. Consider now the computationally important problem of updating a solution to TZ^ν to one at $TZ^{\nu+}$. Since state $\nu+$ may be closely related to state ν, one will wish to start the solution procedure (by which we shall always mean the dual algorithm of Ford-Fulkerson) with a starting set of feasible labels $(\alpha_i, \beta_j)_0$ which make the dual objective function t_d (to be maximized) for the transportation problem of TZ^ν as large as possible.

Let, in general, the partitioning at node ν be given by $\{C^\nu, \Omega^\nu, F1^\nu, F2^\nu\}$. Suppose that the procedures advocated in this paper lead to a partitioning

$$\{C^\nu + \Delta C, \Omega^\nu + \Delta\Omega, F1^\nu - \Delta F1, F2^\nu - \Delta F2\} \text{ at}$$

successor node $\nu+$. Of course, one has

$$\Delta C + \Delta\Omega = \Delta F1 + \Delta F2 \tag{5. 7}$$

Plants $i \in \Delta F1$, which were treated as open at ν, are now closed, plants $i \in \Delta F2$ are now open. The fictitious demand r_l for $\tilde{T}Z^{\nu+}$ becomes

$$r_l^{\nu+} = r_l^\nu + \sum_{i \in \Delta F2} a_i - \sum_{i \in \Delta F1} a_i \tag{5. 8}$$

Consider the optimal dual solution to TZ^ν. By virtue of (5. 3), the (u_i^ν, v_j^ν) are dually feasible for problem $\tilde{T}Z^{\nu+}$.

They can therefore be used as a starting set of labels for the dual algorithm applied to the transportation problem of $\tilde{T}Z^{\nu+}$.

$$(\alpha_i^{\nu+}, \beta_j^{\nu+})_0 = (u_i^\nu, v_j^\nu), \quad i \in F1^{\nu+} + \Omega^{\nu+}, j \in J + \{l\} \tag{5.9}$$

The shipments in rows which are removed ($i \in \Delta F1$) are returned to the corresponding row supplies and column demands and the flow algorithm is restarted.

A simple example:

$F1^{\nu} = \{1, 2, 3\}$, $F2^{\nu} = \{4\}$

$r_l = 13 - 8 = 5$

Solution:

f_i	a_i \ r_j	3	4	1	r_l^{ν} 5
4	4	2	4	1	0
1	6	1	3	2	0
1	3	1	1	2	0
1	5	1	1	2	0
		c_{ij}			l

u_i^{ν} \ v_j^{ν}	1	3	1	0
0	1	1	①	③
0	③	①	1	②
2	2	③	3	2
2	2	0	3	2

The encircled entries are shipments, the other entries are the quantities $e_{ij} \equiv c_{ij} + u_i - v_j$. The dual variable u_4^{ν} is computed from (5). The dual objective function is:

$$t_d^{\nu} = -\sum_{F1^{\nu}+\Omega^{\nu}} a_i u_i^{\nu} + \sum r_j v_j^{\nu} = -6 + 16 = 10,$$

which, of course, equals the shipping cost. The objective function for TZ^{ν} is $\zeta^{\nu} = 6 + 10 = 16$. Now let the search explore whether the node $\gamma+$, with the partitioning $\Omega^{\nu+} = \{4\}$, $F1^{\nu+} = \{1, 2, 3\}$ gives a better solution. One starts with the labels $\alpha_i = (0, 0, 2, 2)$ and $\beta_j = (1, 3, 1, 0)$ and with a new requirement $b_l = 5$.

s_i	α_i	d_j / β_j	0 / 1	0 / 3	0 / 1	5 / 0	
0	0		1	1	①	③	
0	0		③	①	1	②	(2,1)
0	2		2	③	3	2	(2,3)
5	2		2	0	3	2	(4,5)
			(2,1)	(4,5)		(2,1)	Labels

(s_i, d_j) denote the units to be shipped. The labeling procedure allows shipment of one more unit along the path indicated, and then comes to a halt with rows $I = \{3, 4\}$ and columns $J = \{2\}$ labeled. ($e_{22} \rightarrow 0$. Hence no labeling of row 2). The dual labels $\alpha_i : i \notin I$ and $\beta_j : j \notin J$ are then augmented by $\delta = \min_{I\bar{J}} e_{ij} = 2$. ($\bar{J} = \{j | j \notin J\}$). .

s_i	α_i	d_j / β_j	0 / 3	0 / 3	0 / 3	4 / 2
	2		1	3	①	③
	2		③	2	1	③
	2		0	③	1	0
4	2		0	①	1	0

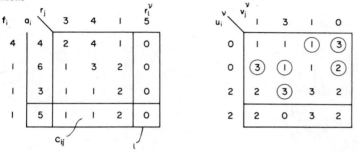

It is now clear that the remaining shipment of 4 units is assigned to route $(4, 4)$ at zero cost.

The dual variables, optimal for $\nu+$, are $u_i = (0, 0, 0, 0)$, $v_j = (1, 1, 1, 0)$ after normalization (subtraction of 2 from (α_j, β_j)). $t_d^{\nu+} = -0 + 8$, $\zeta^{\nu+} = 7 + 8 = 15 < \zeta^{\nu}$.

Suppose, now, that we explore the closing of plant 1:

$$\nu++: \ \Omega = \{4\}, \ F1 = \{2, 3\}, \ C = \{1\}$$

d_j	0	0	1	0	
β_j	1	1	1	0	
s_i α_i					
1 0	③	2	1	②	(4,1)
0 0	0	③	1	0	(2,1)
0 0	0	①	1	④	(4,1)
	(2,1)	(4,1)	(2,1)		

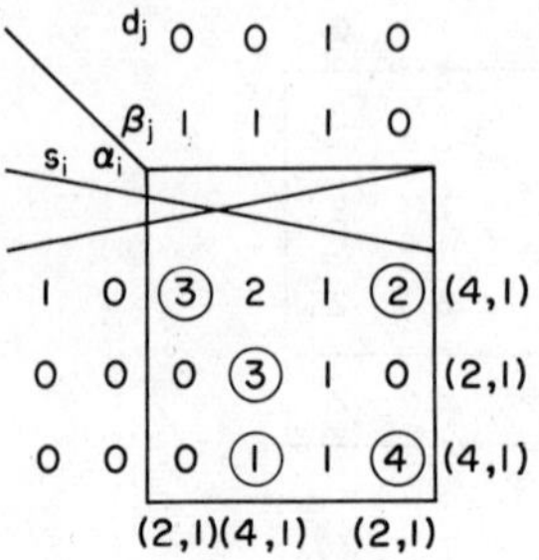

r_l now becomes $14 - 8 = 6$, so that in addition to shipment $x_{1,4}$, one unit of either $x_{2,4}$ or $x_{4,4}$ should be returned. One will, usually, prefer to return the shipment for $i \notin \Omega$.

$$I = \{2, 3, 4\}, \ J = \{1, 2, 4\} \ \delta = 1$$

d_j	0	0	1	0
β_j	1	1	2	0
s_i α_i				
1 0	③	2	0	②
0 0	0	③	0	0
0 0	0	①	0	④

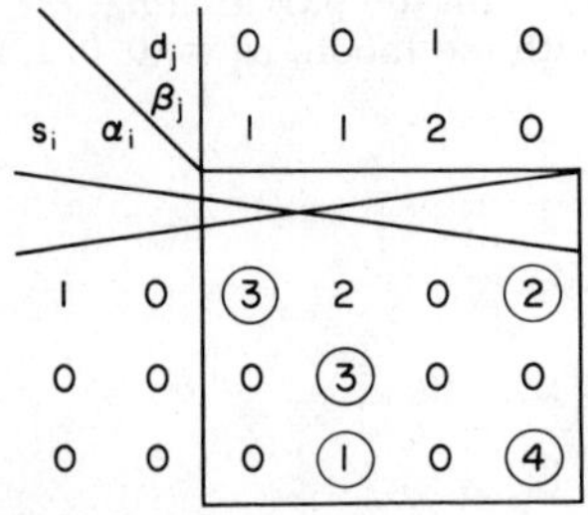

Evidently the shipment is assigned to route $(2, 3)$

$$t_p = 3 + 2 + 3 + 1 = 9$$
$$t_d = -0 + (3 + 4 + 2) = 9$$
$$\zeta = 9 + 3 = 12$$

It should be noted that, on <u>return</u> to a node ν of the search tree from a successor node, the transportation problem is entirely unchanged.

An index i_* has been moved from F1 to Ω, or from F2 to C. Therefore, it is advocated that the dual variables (u_i, v_j) associated with a node at level k are saved, at relatively modest memory requirement, when level k is left for a higher level. On return to k the e_{ij} can be recomputed and the primal solution re-established at little cost.

Similar things could be said about problem TR^{ν}. We shall restrict ourselves to pointing out the well-known fact that TR^{ν} can be replaced by an associated standard transportation problem $\widehat{TR}^{\nu}$, if a fictitious demand r_l is introduced as above and if the costs c_{ij} are replaced by

$$c_{ij}^{\nu} = c_{ij} + f_i/a_i \text{ for } i \in F1 \qquad (5.10)$$

6. THREE TYPES OF ALGORITHMS

It is possible, even though somewhat arbitrary, to conceive of three types of algorithms of increasing complexity, say type 1, type 2, type 3.

6.1 Algorithms with Natural Search Origin. Simple Search

Within this category we may, perhaps, distinguish two cases. First, there are plant location problems for which it is pretty clear what the final solution should look like, e.g., high fixed charges and a dense matrix (few prohibited routes) may make it apparent that only few plants will be chosen to be open. Or low fixed charges and a sparse matrix (many prohibited routes) may suggest that almost all plants will be open in the final solution. In these two cases it seems reasonable to start the search procedures at the origins $\eta = 0$ or $\eta = e = (1, 1, \ldots, 1)$, respectively, and to forego the use of 'dynamic' or 'adaptive' devices during the search.

Secondly, a particular search origin may be 'natural' in the sense that the convenient property (Equation 4.15) can be shown to hold with this origin. E.g., it can be shown that the origins $\eta = 0$ and $\eta = e$ are, in this sense, natural for the Simple Plant Location Problem, but that only $\eta = e$ is natural for the Capacitated Plant Location Problem (see section 5).

This property may then be sufficiently important to warrant remaining with a simple search scheme, rather than using (Equation 3.2) to make dynamic redefinitions of the search.

It is possible, though not likely, that one may wish to use a simple search procedure, in order to avoid the increase in computing time required by the dynamic relocation procedure. In this connection, the requirement of efficient 'updating' of solutions to subproblems, or other pertinent data, as the search progresses from node to node, becomes important. Such updating can be done most easily with type 1 algorithms, because of the fact that 'adjacent' vectors y^s in the search differ in one component only.

For specific algorithms of type 1, applied to the simple plant location problem, see reference [2]. They are given for both natural origins, $\eta = 0$ and $\eta = e$. The computational results reveal the enormous dependence on the structure of the data, and consitute convincing evidence for the need of more flexible, 'adaptive', algorithms.

In section 7, nevertheless, we give one algorithm of the simple type for the <u>capacitated</u> problem.

6.2 Dynamic, or Adaptive, Algorithms

The purpose of the current paper is essentially to introduce the notion of a dynamically redefinable, or adaptive, 'state' of the search (section 3) and to combine it with efficient use of the associated inequality (section 4). It is hoped that this results in a search which is as well adapted to the data structure as can be expected from the still moderate computational effort.

The second algorithm of section 7, designed for the capacitated problem, but easily altered to fit the uncapacitated problem, is only one way in which these notions can be used. The reader will have little difficulty in altering it so as to, perhaps, make it more suitable for his particular problems.

6.3 Complex Algorithms for Problems having Special Structure

Certain problems may allow the introduction of devices more complicated than the ones advocated here. E.g., in [3] it is shown that the notion of 'gain function' can be generalized for the Simple Plant Location Problem. In addition to the 'global' gain functions defined by equation 4.14 one may define 'local' gain functions which can be used in cancellation tests of considerable power. To compute and update the various gain functions, the algorithm carries along four sets of minimal cost entries (one for each j), the minimization being executed over various subsets of F.

For certain problems, the cancellation tests are powerful enough to ensure termination of the search within a few iterations after attainment of the optimal solution and an appropriate restart. With other problems the tests seem still to be powerful, but the computational effort required per iteration proves to be excessive.

One of the purposes of developing the algorithms (of type 2) of this paper is to provide procedures which have some of the power of the above method, without (hopefully) the corresponding excessive computational requirements.

7. NEW ALGORITHMS FOR THE CAPACITATED AND SIMPLE PLANT LOCATION PROBLEM

In the following we give a type 1 algorithm for the Capacitated Plant Location Problem and a type 2 algorithm for the same problem. The latter, however, is also intended to be used, with obvious modifications, of which some will be pointed out, for the Simple Plant Location Problem.

Because of the ample treatment given in the earlier sections, the algorithms can be stated in brief outline form.

7.1 A Type 1 Algorithm for the Capacitated Problem

The algorithms are simplest when the search starts at the point $\eta = e$, i.e., with all plants tentatively open, to be closed on forward steps. As a consequence of this choice, infeasibility of problem TZ^{ν} implies infeasibility of TP^{ν} and, therefore, permits a 'backup' (reversion to the predecessor node $\nu-$).

Furthermore, one can show that property (4.15) is valid. (It is not valid in general). The case of a search with 'fixed state' defined by search origin $\eta = e$ is, of course, a special case of the general search described in B and C. The various relations remain valid, with F2 always $= \phi$. The gain functions g_i are, therefore, given by

$$g_i^{\nu} = f_i - u_i^{\nu} \cdot a_i, \; i \in F^{\nu} \tag{7.1}$$

A successor node μ to node ν is reached by the process of closing plants and possibly fixing plants open. The transportation problems of $TZ^{\nu}, TZ^{\nu+1}, \ldots, TZ^{\mu}$ differ from each other only in a non-decreasing number of closed plants. We may suppose that at node ν there is at least one nonzero shipment in column l, so that $v_l = 0$. In going from ν to ν^+ one must return some shipment to a plant, say to $i*$. But then $e_{i*l} = u_{i*} - v_l = 0$ and $u_{i*} = v_l = 0$, and row $i*$ and column l will get labeled until the shipment has been reassigned. This implies that v_l remains zero in every dual variable change, i.e. remains zero until the flow problem has been resolved.

In a computer program the partitioning $(F1^{\nu}, F2^{\nu}, \Omega^{\nu}, C^{\nu})$ can be recorded conveniently in two 'state' vectors of size m + n, which are updated algorithmically.

The instructions are assumed to be followed consecutively, unless stated otherwise.

1. Input of data: a, r, c, f, tolerance τ, etc.

   Initialization: $k = \nu = 0$, $z^* = \infty$, $F1 = I$, etc.

2. Solve 'projected' problem $TZ^{\nu+}$.

   (a) If $TZ^{\nu+}$ is infeasible, go to 3b.

   (b) If $z^* > \zeta^{\nu+}$, replace z^* by $\zeta^{\nu+}$ and record solution $(x, u_i, v_j)^{\nu+}$.

Compute projected gains $g_i^{\nu+}, i \in F^{\nu+}$.

Set $\Sigma^{\nu+} = \Sigma(g_i^{\nu+})_+$ (sum over positive gain)

3. (a) If $(\zeta^{\nu+} - \Sigma^{\nu+} - z^* + \tau) \geq 0$, go to 4.

Otherwise, cancel $i \in F^{\nu+}$ if $g_i^{\nu+} \leq 0$ ('projected cancellation')
(Cancellation: fix plant open, or transfer i from F to Ω)
If $F^{\nu+} \neq 0$, go to 5.
If $F^{\nu+} = 0$, go to 4.
(b) If $\nu > 0$, go to 4b.

4. (a) If k = 0, terminate.
(b) Do not complete forward step on i_*.
(Closing of plant i_*)
Transfer i_* from C to Ω.
Reverse projected cancellations

If $(\zeta^{\nu} - \Sigma^{\nu} - z^* + \tau) \geq 0$, go to 7

Retrieve u_i^{ν}, v_j^{ν} and compute e_{ij}^{ν}
Go to 6b (resume computation at node ν)

5. (a) Transfer projected data from erasable to permanent storage
Save $u_i^{\nu+}, v_j^{\nu+}, \zeta^{\nu+}$.
If k = 0, go to 6.
(b) Record i_*.

6. (a) Replace k by k + 1, ν by $\nu + 1$.
(b) After a forward step, go to e.
(c) Recompute g_i^{ν}, Σ^{ν}. Retrieve ζ^{ν} from memory
Augment ζ^{ν} by f_{i_*} if $\sum_J x_{i_*j} = 0$.

If $(\zeta^{\nu} - \Sigma^{\nu} - z^* + \tau) \geq 0$, go to 7.
(d) Recover the solution to TZ^{ν} by retrieving (u_i^{ν}, v_j^{ν}) from memory, recomputing e_{ij}^{ν} and using a flow algorithm to assign the shipments.
(e) Select i^*, the next link (the next plant to be closed), such that
$g_{i^*} = \max g_i$ over $i \in F^{\nu}$.
(f) Reduce Σ^{ν} by g_{i^*}.
Transfer i^* from F^{ν} to C^{ν}.
Recompute r_l. If $r_l \geq 0$, go to g.
If $r_l < 0$ and $F^{\nu} = \phi$, go to 7.
Otherwise, go to e.
(g) Initialize the projected problem $TZ^{\nu+}$, by returning shipments in row i^* to the requirements and corresponding shipments in column l to the sources. Go to 2.

7. Backup: if k = 0, terminate.

Retrieve i_*, the link from $\nu-$ to ν.
Transfer indices cancelled at ν from C^{ν} to F^{ν}.
Transfer i_* from C^{ν} to Ω^{ν}.
Set k = k − 2; go to 6.

7.2 A Type 2 Algorithm for the Capacited Problem

To keep the description simple, we assume that TR^{ν} is solved at every node and do not describe the imposition of an arbitrary origin and the restart procedure. Variations in strategy are also omitted.

1. Input of data a, r, c, f, etc.

   Tolerance τ, rounding parameter δ.
   Initialization: $k = \nu = 0$, etc.
   $z^* = \infty$, $F1 = I$

2. Solve projected problem $TR^{\nu+}$.

   (a) If infeasible: $\nu = 0$, go to 7 (backup), $\nu \neq 0$, go to b.
   If $z_r \geqslant z^*$, go to b.
   If ${}^r y^{\nu}$ is integral, replace z^* by z_r, record the solution and go to b.
   If $\nu = 0$, go to 3b. Otherwise to 3a.

   (b) Cancel i^*. Go to 6.

3. (a) Record i^*.

   (b) Replace k by $k + 1$, ν by $\nu + 1$.
   After a forward step, go to 4.
   If $F^{\nu} = \phi$, go to 7.
   Retrieve (u_i^{ν}, v_j^{ν}) from memory.
   Compute e_{ij}^{ν}.
   Go to 5 ('State' is present in memory).

4. Compute new state:

   (i) Compute $\sum = \sum_{i \in \Omega^{\nu} + F1^{\nu}} a_i$

   If $i \in F2^{\nu}$ and ${}^r y_i^{\nu} > \delta$, transfer i from $F2^{\nu}$ to $F1^{\nu}$ and increase Σ by a_i.

   (ii) If $i \in F1^{\nu}$ and ${}^r y_i^{\nu} \leqslant \delta$

   and if $\sum - \sum_J r_j - a_i \geqslant 0$,

   transfer i from $F1^{\nu}$ to $F2^{\nu}$ and decrease Σ by a_i.

5. Solve TZ^{ν}

   Store u^{ν} in memory.
   Compute Benders inequality (4.11) from u^{ν} (or $\bar{u}^{\nu}$) if TZ^{ν} is feasible (infeasible).
   If (4.11) can not be satisfied at successor nodes with improvement in z larger than τ, go to 7.
   Otherwise, cancel any $i_* \in F^{\nu}$ which satisfy (4.13).

6. Selection of next link i^*:

   (a) If F^{ν} is empty, go to 7.
   Select i^* such that $g_{i*} = \max g_i$ over $i \in F^{\nu}$ and

   $\Sigma - \Sigma r_j - a_{i*} \geqslant 0$

   for $i_* \in F1^{\nu}$.

   If there exists no such i^* (which implies that $F2^{\nu}$ is empty), go to 7.
   Set up projected problem $TR^{\nu+}$.
   Go to 2.

7. Backup

If $k = 0$, terminate.

Retrieve i_*, the link from μ to ν.

Transfer indices cancelled at γ from C^ν or Ω^ν to F^ν.

Transfer i_* from C^ν to Ω^ν, or from Ω^ν to C^ν.

Set $k = k - 2$. Go to 3b.

8. COMPUTATIONAL RESULTS, STRATEGY, OUTLOOK

The algorithms of section 7 have been programmed and used to solve small problems. Reliable results for larger problems are, unfortunately not yet available.

It may be of interest, at any rate, to reproduce from [3] the results which demonstrate the difference between type 1 and type 3 algorithms for the Simple Plant Location Problem.

Consider the results exhibited in the table:

The programs SPLT1, PL13N are based upon type 1 algorithms with $\eta = e$, the program PLON upon a type 1 algorithm with $\eta = 0$. The comparison program, with 'generalized' origin, is a rather complicated type 3 program.

SYMBOLS:

it2 ...	iteration number at which last solution obtained
it3 ...	iteration number after which optimality ascertained
it4 ...	a count which measures how often a successor node was explored and immediately abandoned
of1 (of2) ...	first (last) value of the objective function
t, t′ ...	computation times in minutes. Normal entry ... 360/50, ... 7094, ... 360/40
ρ ...	restart iteration
σ ...	strategy after restart
nop ...	no. of components η_i equal to 1 (no. of open plants)
— ...	entry not obtained or recorded
~ ...	approximate or estimated entry

For a description of the data and an extensive discussion, see ref. [3]. Here it suffices to observe that the comparisons above the horizontal line are favorable for the type 3 algorithm, while those below are unfavorable.

What is most important, we believe, is the very drastic improvement exhibited by runs no. 1, 2, 11-13. These computer runs demonstrate that an insight into the data structure and a corresponding adaptation of algorithm to problem can serve to render the problem trivially easy. They appear encouraging enough to justify further work into the direction of program-data adaptation.

By strategy we mean primarily the manner in which the next link is chosen at a node, and secondarily such matters as choice of tolerance restart iteration, etc. Here it need only be pointed out that the success of some of the runs recorded in the table was heavily dependent upon the strategy. The fundamental pattern was that before restart it was desirable to improve the 'current' solution maximally (locally) during the next step, whereas after restart it was almost mandatory to do the opposite, i.e. to explore locally unfavorable continuations. Evidently the latter was successful because it led to quick establishment of the infeasibility of such attempts, with resulting simplifications in the remaining problem.

Comparative Results for Programs with

					RIGID ORIGIN					GENERALIZED ORIGIN							
#	P	m	n	τ	Program	it2	of2	it3	t′	σ	ρ	it2	of2	nop	it3	it4	t
1	L1	52	52	—	PLON	915	50754.	947	11.2	C	50	49	50754.	3	67	102	2.75
2	M1	52	52	—	SPLT1	33	1457.3	4863	12.7	C	50	33	1457.4	19	51	86	1.17
3	F3	90	100	—	PL13N	510	5339.1	3429	37	C	100	251	5339.8	24	259	824	27
4	G4	100	150	—	—	—	—	—	—	C	200	165	2288.6	3	682	2511	(159)
5	H3	100	150	—	PL13N	147	9212.9	993	30	C	150	277	9212.7	38	283	586	20
6	D1	60	80	—	SPLT1	44	1427.3	139	1.12	C	50	44	1427.3	34	51	57	1.5
7	D50	60	80	100	SPLT1	40	2475.9	24200	(90)	C	100	41	2475.9	19	219	401	6.9
8	D50	60	80	50	—	—	—	—	—	C	100	41	2475.9	19	647	1278	19.4
9	D25	60	80	50	SPLT1	32	1881.9	2189	5.3	C	100	33	1881.9	27	105	149	2.0
10	D25	60	80	—	SPLT1	124	1868.7	10263	33.3	C	100	170	1868.7	25	~175	—	2.8
11	D25	60	80	—	SPLT1	124	1868.7	10263	33.3	D	100	134	1868.7	25	137	202	2.5
12	C1	60	80	—	SPLT1	37	1093.1	11055	55.2	C	50	37	1093.1	23	83	139	1.7
13	C1	60	80	—	SPLT1	37	1093.1	11055	55.2	D	50	37	1093.1	23	97	153	1.8
14	C25	60	80	100	SPLT1	48	1363.5	16000	(45.3)	C	100	48	1363.5	12	243	559	10.17
15	C100000	60	80	100	SPLT1	57	31628.	10145	16.1	C	100	59	31646.	3?	1501	6513	(78)
16	B100	30	80	—	SPLT1	25	1899.9	1245	3.44	C	100	25	1900	5	805	1903	13.7
17	G2	100	150	—	PLON	56	4459.9	229	70	C	100	434	4460.1	2	628	2909	167
18	G2	100	150	—	PLON	56	4459.9	229	70	C	100	384	4460.1	2	460	2909	126
19	C50	60	80	100	SPLT1	67	1569.2	41261	128	C	100	73	1569.2	7	1181	4155	(80)

The above is oversimplified. It was necessary to make additional minor adjustments which were quite reasonable a posteriori, i.e. after some experience with a program, but difficult to foresee a priori (see again ref. [3]). Corresponding decisions on strategy for the algorithms of this paper will have to await some fairly substantial numerical experimentation.

Finally, some words should be said about the probable success of enumerative methods as discussed here. No matter what is done, problems with a large number of plants become intractable on account of the inherent combinatorial difficulties. At some point the need must arise for concentration upon heuristic rather than finite schemes. Among heuristic procedures one might consider the use of approximative and rapid solution schemes for the transportation problem (see e.g. [13]). Or one might invoke 'probabilistic' search schemes, see e.g. [14, 15, 16], which permit curtailment of the search by a discarding (cancellation) of continuations which can not be proven to be unnecessary, but only to be choices which are unlikely to lead to improved solutions.

It seems, however, that such devices can advantageously be fitted into the structure of algorithms as proposed here, as devices for increasing the rate of convergence (with a corresponding decrease in reliability) of the enumerative procedure.

REFERENCES

1. Efroymson, M. A., Ray, T. L., 'A Branch-Bound Algorithm for Plant Location'. Operations Research, Vol. 14, 1966, pp. 361-368.

2. Spielberg, K., 'An Algorithm for the Simple Plant Location Problem with some Side Conditions', Report # 2900, IBM Data Processing Division, New-York Scientific Center, May, 1967.

3. Spielberg, K., 'Plant Location with Generalized Search Origin', Report #320-2927, IBM Data Processing Division, New-York Scientific Center, 1968.

4. Gray, P., 'Mixed Integer Programming Algorithms for Site Selection and other Fixed Charge Problems having Capacity Constraints', Technical Report # 101, Dept. of OR and Dept. of Stat., Stanford Univ., Stanford, 1967.

5. Balinski, M. L., Spielberg, K., 'Methods for Integer Programming: Algebraic, Combinatorial and Enumerative', to appear in 'Progress in Operations Research,' Vol. # 3, editor: J. Aronofsky, J. Wiley, 1968.

6. Balas, E., 'An Additive Algorithm for Solving Linear Programs with Zero-One Variables', Operations Research, Vol. 13, 1965, pp. 517-546.

7. Land, A. H., Doig, A. G., 'An Automatic Method of Solving Discrete Programming Problems', Econometrica, Vol. 28, 1960, pp. 497-520.

8. Benders, J. F., 'Partitioning Procedures for Solving Mixed-Variables Programming Problems', Numerische Mathematik, Vol. 4, 1962, pp. 238-252.

9. Lemke, C. E., Spielberg, K., 'Direct Search Zero-One and Mixed Integer Programming', Operations Research, Vol. 15, 1967, pp. 892-914.

10. Geoffrion, A. M., 'Integer Programming by Implicit Enumeration and Balas Method', Siam Review, Vol. 9, 1967, pp. 178-190.

11. Spielberg, K., 'On the Fixed Charged Transportation Problem', Proceedings of the 19th National Conference, A. C. M., 1964, pp. A1. 1-1 to A1. 1-13.

12. Ford, L. R. Jr, Fulkerson, D. R., 'Flows in Networks', Princeton University Press, 1962.

13. Duby, J.-J., 'Traitement d'un problème de transport de grandes dimensions sur un ordinateur de petite puissance', Etude no 25, IBM France-Développement Scientifique, 1966.

14. Reiter, S., Rice, D. B., 'Discrete Optimizing Solution Procedures for Linear and Nonlinear Integer Programming Problems', Institute Paper # 109, Inst. for Res. in the Behavioral, Economic and Management Science, Purdue Univ., 1965.

15. Graves, G. W., Whinston A. B., 'A New Approach to Discrete Mathematical Programming', Working Paper # 117, Western Management Science Inst., 1967.

16. Glover, F., 'A Note on Some Stochastic Approaches to Nonstochastic Discrete Programming', Man. Sc. Res. Report # 111, Carnegic Inst. of Tech., Grad. School of Ind. Adm., 1967.

A Non-Linear Programming Warehouse Allocation Problem

P. H. RANDOLPH
New Mexico State University, U.S.A.
G. E. SWINSON
Braddock, Dunn and McDonald, Inc., U.S.A.
M. E. WALKER
SENSEA, U.S.A.

SUMMARY

A nonlinear programming method is developed for the optimum allocation of a fixed stockpile of resources. The method lends itself to problems whose phsyical characteristics are completely general in nature. The successive approximation procedure of dynamic programming serves as a format for the computational scheme used in the method. This leads to a relative maximum, and the particular relative maximum found depends on the initial policy space estimate. By choosing initial policies sequentially at random a set of relative maxima is obtained, and the problem becomes one of optimal stopping rules. However, the application of stopping rules requires the specification of an upper bound to the relative maxima. This upper bound may itself be a random variable. In the context of a warehouse allocation problem a procedure is presented in which optimal stopping rules are used to determine a sequential plan for sampling from the set of relative maxima and the set of upper bounds.

1. INTRODUCTION

The optimum distribution of available resources has been a problem of growing emphasis during the last decade. Problems whose physical characteristics include such mathematical niceties as linearity, convexity, etc., have been thoroughly massaged, and absolute optimum solution techniques are available in the present inventory of literature. A comparable collection of absolute optimum techniques does not exist at the present time for problems which do not contain mathematical pleasantries in their characteristics. For general problems of nonlinear programming there exist gradient methods, various modifications of linear programming, dynamic programming, and the sampling methods of Reiter and Sherman (1965). For the problem of this paper, a merger of the last two techniques, namely dynamic programming and sampling, along with the optimal stopping rule procedures of Randolph, Swinson and Walker (1968), seemed to be the most feasible approach.

The method presented in this paper evolved from the necessity to determine optimum allocations for problems which possess non-linear, separable objective functions and a multiplicity of inter-dependent constraints. In particular, it is presented in the context of evaluating the number of items to be shipped from a set of warehouses to a group of customers where the service areas of the warehouses overlap.

2. DESCRIPTION OF THE PROBLEM

Let us begin with a brief examination of the single warehouse problem. Suppose there are l customers serviced by one warehouse and that the total number of units available at the warehouse is b. Let y_k be the number of items to be supplied to customer k, $k = 1, 2, \ldots, l$. We will assume that the payoff $r_k(y_k)$ to the supplier for supplying y_k items to customer k is an arbitrary but known function. Then the problem is to find the number of units to supply each customer from the total fixed stockpile such that the total payoff is maximized; that is, we wish to find y_k satisfying

$$\sum_{k=1}^{l} y_k \leqslant b$$

y_k a nonnegative integer

and which maximizes

$$\sum_{k=1}^{l} r_k(y_k).$$

This is a well known problem in dynamic programming which can be solved readily, and thus the single warehouse problem presents no real difficulties.

When there is more than one warehouse, then the problem is usually more complex. Suppose a supplier has m warehouses scattered around the country to service l customers. Each warehouse is able to service a given set of customers, but, on the other hand each customer could possibly be supplied by more than one warehouse; that is, warehouses have overlapping geographical areas of service as illustrated by Figure 1 for three warehouses serving twenty customers.

As before let y_k be the total number of items to be supplied to the k-th customer, $k = 1, \ldots, l$, by all warehouses that are able to supply this customer and let the payoff to the supplier for supplying a total of y_k integral units to the k-th customer be an arbitrary but known function denoted by $r_k(y_k)$.

Let the total number of items each warehouse has available be denoted by b_i, $i = 1, \ldots, m$. Also, let the double subscripted variable y_{ik}, denote the number of items to be supplied by warehouse i to customer k. Then, if we define

$$K_i = \{k \mid \text{customer k in the area serviced by warehouse i}\}$$

$$L_K = \{i \mid \text{warehouse i is able to supply customer k}\}$$

the problem is to find y_k and y_{ik} which satisfy

$$\sum_{k \in K_i} y_{ik} \leqslant b_i \qquad i = 1, \ldots, m$$

$$\sum_{i \in L_k} y_{ik} = y_k \qquad k = 1, \ldots, l$$

y_{ik} a nonnegative integer for all i and k and which maximizes the total payoff to the supplier

$$\sum_{k=1}^{l} r_k(y_k)$$

If the warehouse service areas overlap, as implied by the model formulated above, then the dynamic programming procedure as suggested for the single warehouse problem will no longer apply. However, by decomposing the problem into a set of subproblems for areas we call sectors, it will be possible to use dynamic programming within each sector and then combine these results for the solution to the total problem.

3. THE OPTIMUM WITHIN-SECTOR ALLOCATION

Let a sector be defined as that area serviced by a particular set of warehouses (In Figure 1 we have seven sectors numbered 1, . . . , 7). Without loss of generality, the

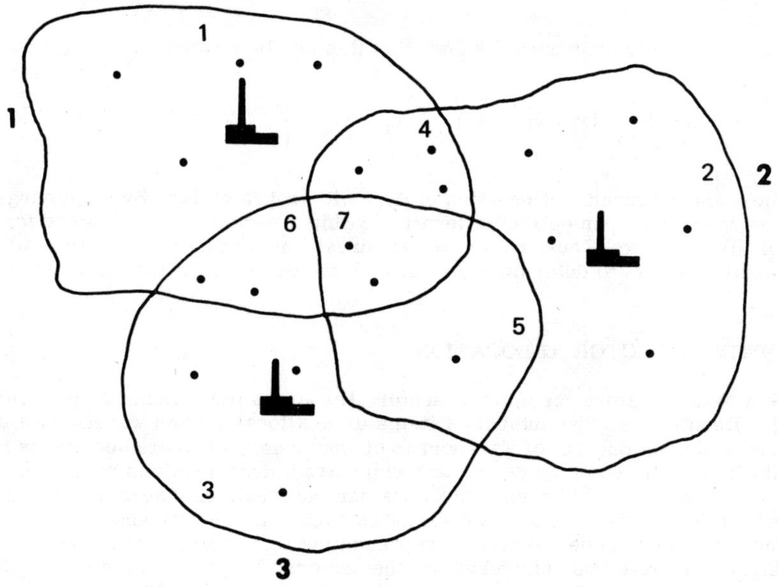

Fig. 1. Illustration of three warehouses serving twenty customers

customers in the j-th sector can be denoted by 1, 2, . . . k_j. For the k-th customer in this j-th sector, where k = 1, 2, . . . k_j, let us as before denote by y_k, a total number of items to be supplied this customer from the limited stockpile. These y_k items can all come from one warehouse, or they could be split among the several warehouses which are able to supply the sector. We can then state that the total payoff over all k_j points in sector j is given by

$$R_{k_j}(x_j) = R_{k_j}(y_1, \ldots, y_{k_j}) = \sum_{k=1}^{k_j} r_k(y_k)$$

where

$$x_j = \sum_{k=1}^{k_j} y_k \text{ is a total number of items allocated to sector j.}$$

Although the optimum value of x_j, the sector allocation, is not known, we are now in a position to determine the optimum allocation y_k within a sector for any sector allocation x_j.

Briefly, the dynamic programming scheme is as follows: Suppose we denote by $f_{k_j}(x_j)$ the total sector payoff when an optimal within-sector-allocation policy is employed over the k_j customers in the sector and when the sector has a total of x_j items allocated to the totality of customers in the sector. That is,

$$f_{k_j}(x_j) = \max_{y} [R_{k_j}(y_1, \ldots, y_{k_j})] = \max_{y} \sum_{k=1}^{k_j} r_k(y_k)$$

where

$$y = [y_1, \ldots, y_{k_j}]$$

and $\sum_{k=1}^{k_j} y_k = x_j$. However, it is well known that this can be written as

$$f_{k_j}(x_j) = \max_{0 \leqslant y_{k_j} \leqslant x_j} [r_{k_j}(y_{k_j}) + f_{k_j-1}(x_j - y_{k_j})]$$

which is the standard functional equation of dynamic programming. By employing this function recursively, an optimal allocation within a sector can be obtained for each sector allocation x_j. Thus, sector payoff curves can be generated quite readily, and the objective now is to determine the optimal values of the sector allocations, x_j.

4. THE OPTIMUM SECTOR ALLOCATION

The problem of determining the optimal sector allocations is formulated in the following manner. Denote by x_{ij} the number of items to be allocated from warehouse i to sector j. Let I_j denote that set of warehouses of whose service areas sector j is a subset. Likwise, let J_i denote those sectors which are a decomposition of the service area of warehouse i. As an example, consider the idealized diagram of three overlapping warehouse service areas depicted in Figure 2. The warehouses are denoted by the numbers outside the circles representing their service areas and the sector designations have been placed within the sectors. For this example, $I_5 = \{2, 3\}$ since sector 5 is a subset of the service areas of warehouses 2 and 3. Also, $J_2 = \{2, 4, 5, 7\}$ since the service area of warehouse 2 can be decomposed into the sectors numbered 2, 4, 5, and 7.

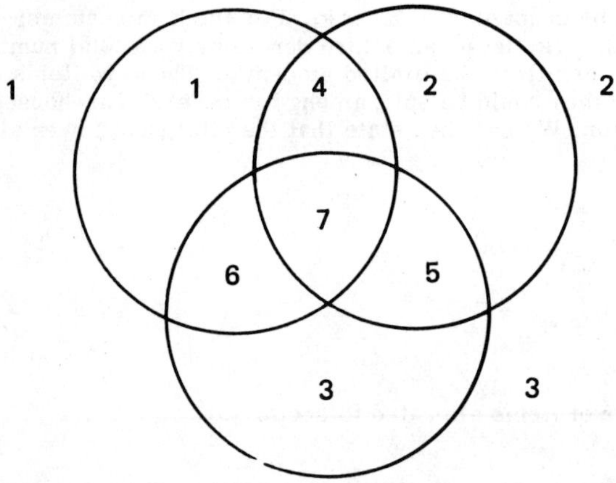

Fig. 2. Idealized service area diagram for three warehouses

Finally, let the total number of warehouses be denoted by m and the total number of sectors by n. For the above example, m = 3 and n = 7.

Using the above notation, we can state the following problem: Find x_{ij} which satisfy the following set of linear equations

$$\sum_{j \in J_i} x_{ij} = b_i, \qquad i = 1, \ldots, m \qquad \text{(warehouse constraints)}$$

$$\sum_{i \in I_j} x_{ij} = x_j, \qquad j = 1, \ldots, n \qquad \text{(sector constraints)}$$

x_{ij}, x_j non-negative integers

and which maximizes the total payoff function

$$F(x) = \sum_{j=1}^{n} f_j(x_j)$$

where x is the vector of the sector allocations (i.e., $x = [x_1, \ldots, x_n]$) and where $f_j(x_j)$ is the function denoting the optimal return over the k_j customers of the j-th sector for an allocation of x_j.

To determine the optimal sector allocations, it is tempting to continue the dynamic programming procedure. However, the number of constraints involved in this problem make it prohibitive to attempt a solution by the standard dynamic programming approach. Probably the most feasible technique for finding optimum sector allocations is the method of successive approximation.

The method of successive approximation is applied as follows. An initial allocation of the warehouse resources to sectors, x^0, which satisfies the set of warehouse and sector constraints is first obtained. This initial allocation can be described by the matrix

$$x^0 = \begin{bmatrix} x^0_{11} & x^0_{12} & \cdots & x^0_{1m} \\ x^0_{21} & x^0_{22} & \cdots & x^0_{2n} \\ \cdot & \cdot & & \cdot \\ \cdot & \cdot & & \cdot \\ \cdot & \cdot & & \cdot \\ x^0_{m1} & x^0_{m2} & \cdots & x^0_{mn} \end{bmatrix}$$

where $x^0_{ij} = 0$ if $j \in J_i$ and $x^0_{ij} \geqslant 0$ if $j \in J_i$. The row totals of x^0 correspond to the sizes of the warehouse stockpiles,

$$\sum_{j=1}^{n} x^0_{ij} = b_i \qquad i = 1, \ldots, m,$$

and the column totals correspond to the initial sector allocations,

$$\sum_{i=1}^{n} x^0_{ij} = x^0_j \qquad j = 1, \ldots, n.$$

The initial allocation matrix yields a total initial payoff of

$$F(x^0) = \sum_{j=1}^{n} f_j(x^0_j)$$

Now, focus on the sectors of one warehouse, such as warehouse number 1, and let the allocation of items from warehouse 1 to these sectors vary. This gives a payoff of

$$F(x) = \sum_{j\in J_1} f_j(x_{1j} + \sum_{\substack{i\in I_j \\ i\neq 1}} x^0_{ij}) + \sum_{j\notin J_1} f_j(x^0_j)$$

where, of course,

$$x_{1j} + \sum_{\substack{i \in I_j \\ i \neq 1}} x_{ij}^0 = x_j$$

is the total allocation to sector j in which

$$\sum_{\substack{i \in I_j \\ i \neq 1}} x_{ij}^0,$$

units are held fixed. Translating the origin to the point

$$\sum_{j \in I_j} x_{ij}^0,\ j \in J_1, i \neq 1,$$

yields a new sector allocation variable,

$$\bar{x}_j = x_j - \sum_{\substack{i \in I_j \\ i \neq 1}} x_{ij}^0,$$

and a new sector payoff function

$$\bar{f}_j(\bar{x}_j) = f_j\Big(x_{1j} + \sum_{\substack{i \in I_j \\ i \neq 1}} x_{ij}^0\Big).$$

This new sector variable, $\bar{x}_j$, is, of course, the variable x_{1j}. Thus $\bar{f}_j(\bar{x}_j)$ represents the total payoff in sector j if x_{1j} items are assigned to sector j from warehouse 1 and all other warehouse allocations to sector j are held constant. For example, $\bar{f}_j(0)$ would equal the total payoff in sector j due to items being allocated to that sector only from warehouses other than warehouse 1, and therefore

$$\bar{f}_j(0) = f_j\Big(\sum_{\substack{i \in I_j \\ i \neq 1}} x_{ij}\Big).$$

The total payoff, F(x), can be written in terms of the new sector allocation variables and payoff functions as

$$F(x) = \sum_{j \in J_1} \bar{f}_j(\bar{x}_j) + \sum_{j \notin J_1} f_j(x_j^0)$$

Now, the payoff function, F(x), is to be maximized over the set of $x_{1j}, j \in J_1$ subject to the single restriction that

$$\sum_{j \in J_1} x_{1j} = b_1$$

Since this is a problem with just one restraint, the standard dynamic programming procedure can be used to optimize the allocation of the items to the sectors lying within the service area of the first warehouse.

Denote by x^1 the solution found by optimizing over the sectors of the first warehouse.

The components x_{1j}^1 are the optimal first warehouse allocations found above and

$$x_{ij}^1 = x_{ij}^0, i \neq 1,$$

are the original allocations for all other warehouses. Using the new allocation, x^1, proceed to some warehouse other than the first, say the second, and optimally allocate its resource over the sectors it covers while keeping all other warehouse allocations fixed. This yields an allocation which we can denote by x^2. This procedure is repeated for all warehouses in the problem. Each new allocation yields a total payoff for the whole problem that is at least as great as the previous allocation. The entire process of allowing each warehouse to reallocate its resources is then repeated until an entire cycle passes within which no sector payoff changes from that of the previous circle. The final allocation is at least a relative optimum solution.

Thus, the method begins with an initial allocation and converges upon a relative maximum. By repeating the process with initial allocations x^0 sufficiently distant from one another, we can expect to determine a number of relative maxima in this way and, hopefully, the absolute maximum.

On the other hand, it can be shown that if a different order of warehouse optimization is used, it is also possible to generate a set of relative maxima. Thus we have two methods of generating relative maxima; namely, by varying the initial allocation or by varying the order of considering the warehouses. Either method can be used by itself or a combination of the two methods may be used.

The method of successive approximation can, therefore, be used to generate a set of relative maxima. The problem of determining if a given set of relative maxima includes the absolute maximum is one that plagues the optimization field. Optimal stopping rules can be utilized to terminate sampling from the space of relative maxima even when it is not certain that the absolute maximum has been found. However, an essential requirement for the application of stopping rules is the specification of an upper bound to the set of relative maxima.

5. AN UPPER BOUND ON THE OPTIMAL SOLUTION

To find an upper bound, dynamic programming is used to combine the sector payoff functions, but with the problem restriction modified. Denote by $G(s_n)$ the return from an n sector problem when s_n units of resources are available and an optimal sector allocation policy is employed. Then the functional equation is

$$G_n(s_n) = \max_{0 \leq x_j \leq h} [f_j(x_j) + G_{n-1}(s_n - x_j)] \quad j \epsilon \{1, \ldots, n\}$$

where $h = \min\left[\sum_{i \epsilon I_j} b_i, s_n\right]$ and $u_j \leq s_n \leq v_j$ are the limits for s_n and will depend on the order in which the sectors are combined. As an example, consider the simple two warehouse problem indicated in Figure 3. Suppose the sector payoff functions in the example are combined in the order 1, 3, 2. Then

$$u_1 = 0 \qquad u_3 = b_1$$

$$u_2 = v_2 = b_1 + b_2$$

$$v_1 = b_1 \qquad v_3 = b_1 + b_2$$

On the other hand, if the order of combination is 1, 2, 3, then

$$u_1 = 0 \qquad u_2 = 0$$

$$u_3 = v_3 = b_1 + b_2$$

$$v_1 = b_1 \qquad v_2 = b_1 + b_2$$

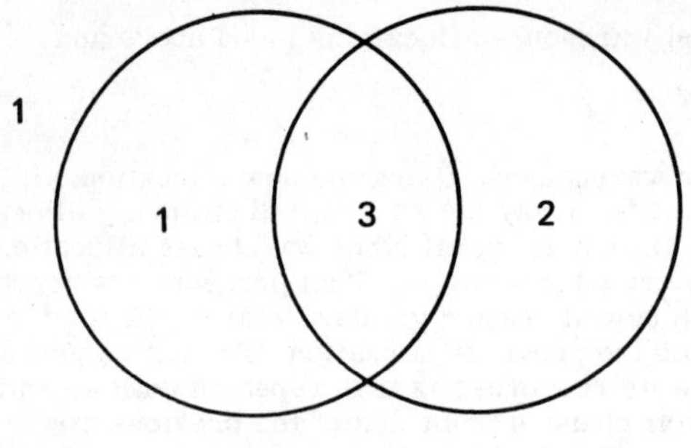

Fig. 3. A simple two warehouse problem to illustrate upper bound calculation

Note that there are no further restraints and thus the problem has been modified somewhat. Finally, after all n sectors are included, the final allocation results, call it x, which gives $G_n(x)$. This again is a local maximum and is also dependent on the order in which the sectors are combined. Furthermore, if per chance the original constraints happen to be satisfied, this is also an absolute maximum to the original problem.

If a relative optimal return for the original problem is denoted by $\phi(x)$, then it is obvious that

$$\phi(x) \leqslant G_n(x)$$

for all possible allocations. In particular,

$$\text{Max } \phi(x) \leqslant \min G_n(x)$$

Although experience has indicated that this relation is usually an equality, we have not been able to prove that an equality will exist in general. However, from a practical point of view this is not important. Instead, the smallest $G_n(x)$ that has been observed can be treated as an upper bound to $\phi(x)$. Then by optimal stopping rules, a procedure can be determined for stopping the computation even when the observed maximum of $\phi(x)$ is strictly less than the observed minimum of $G_n(x)$.

6. SEQUENTIAL SAMPLING AND OPTIMAL STOPPING RULES

The stopping rule discussed in this section has been detailed by Randolph (1968), and its application to maximization problems of the type presented in this paper has been described by Randolph, Swinson and Walker (1968). Therefore, the procedure will be merely sketched here.

Suppose we denote the first M relative payoff maxima by $\phi_1, \ldots \phi_M$ and the first N upper bound values by $g_1, \ldots, g_N$. Let $\phi(M) = \max\limits_{0 \leqslant \mu \leqslant M} \phi_\mu$ and $g(N) = \min\limits_{0 \leqslant \nu \leqslant N} g_\nu$. Then we can define $d(M, N) = g(N) - \phi(M)$ as the interval of uncertainty at stage (M, N).

The interval of uncertainty, d(M, N), is a non-negative random variable. If $d(M, N) = 0$, then the payoff allocation, x, corresponding to $\phi(M)$ is an absolute maximum. If $d(M, N) > 0$, the payoff allocation may be an absolute maximum or it may be only a relative maximum. If the payoff allocation corresponding to $\phi(M)$ is an absolute maximum, and if $d(M, N) > 0$, then the upper bound allocation corresponding to g(N) violates the problem constraints. On the other hand, if perchance the upper bound allocation corresponding to g(N) does not violate the constraints and if $d(M, N) > 0$, then $\phi(M)$ is not an absolute maximum. Thus, whenever $d(M, N) > 0$, there exists an uncertainty in the problem; either the absolute maximum payoff has not been found, or a feasible upper bound allocation has not been discovered, or both.

Suppose at stage (M, N) the non-negative real line is partitioned using the points

$$\phi(M) - (r-2)d, \ldots, \phi(M) - d, \phi(M), g(N), g(N) + d, \ldots, g(N) + (s-2)d$$

where the M and N indices on d have been dropped. We choose r such that there exists no payoff observation in the interval $(0, \phi(M) - (r-2)d]$ but there is at least one payoff observation in the interval $(\phi(M) - (r-2)d, \phi(M) - (r-3)d]$. Likewise, s is chosen such that there is no upper bound observation in $[g(N) + (s-2)d, \infty]$ but at least one upper bound observation lies in the interval $[g(N) + (s-3)d, g(N) + (s-2)d]$. Thus, we have constructed a total of $r + s - 1$ partitions of the real line where

$$r = [(\max_{\mu} \phi_{\mu} - \min_{\mu} \phi_{\mu})/d(M, N)] + 3$$

$$s = [(\max_{\nu} g_{\nu} - \min_{\nu} g_{\nu})/d(M, N)] + 3$$

where $[\cdot]$ is the usual greatest integer function.

If a Direchlet prior density function is specified for the parameters of a multinomial distribution of the relative payoff maxima and the initial probabilities are assumed uniform, then the net expected gain in $\phi(M)$ for making one more relative maximum observation beyond the M-th observation is given by

$$I_M(\phi) = \frac{d(M, N)}{2(r + M)} - c_{\phi}$$

where c_{ϕ} is the relative cost of calculating a relative payoff maximum. For similar assumptions on the distribution of upper bound values, the net expected improvement in g(N) for making one more upper bound observation beyond the N-th observation is given by

$$I_N(g) = \frac{d(M, N)}{2(s + N)} - c_g$$

where c_g is the relative cost of calculating an upper bound.

The optimum sequential plan for sampling requires that the value of $I_M(\phi)$ be determined at each stage of the problem solution. If at any stage $I_M(\phi) \leq 0$, then the cost of computing another relative maximum is at least as great as the expected improvement from this observation and all sampling can be stopped. On the other hand, if $I_M(\phi) > 0$, then the value of $I_N(g)$ must also be determined. If $I_N(g) > I_M(\phi) > 0$, then an upper bound should be computed in the next stage of the problem solution; if $I_M(\phi) \geq I_N(g)$ the next observation should come from the relative maxima space. Sampling is continued sequentially from the set of relative maxima and the set of upper bounds until $I_M(\phi) \leq 0$, at which time the sampling is stopped.

7. AN EXAMPLE—OPTIMIZING A SECTOR ALLOCATION

To illustrate the computational format used to optimize sector allocations, let us once again consider the idealized example of three overlapping warehouse service areas of Figure 2. Assume that each warehouse has a stockpile of 8 items; i.e., $b_1 = b_2 = b_3 = 8$. Also, let us assume that the following sector payoff curves apply.

SECTOR PAYOFF CURVES FOR THREE OVERLAPPING WAREHOUSE SERVICE AREAS

x_j	0	1	2	3	4	5	6	7	8	9	10	11	12	13	14	15	16	17	18	19	20	21	22	23	24
$f_1(x_1)$	0	3	4	5	6	7	7	7	7																
$f_2(x_2)$	0	2	3	4	5	5	5	5	5																
$f_3(x_3)$	0	2	4	6	7	8	8	8	8																
$f_4(x_4)$	0	0	2	3	4	6	8	9	10	11	11	11	11	11	11	11	11								
$f_5(x_5)$	0	6	8	9	9	9	9	9	9	9	9	9	9	9	9	9	9								
$f_6(x_6)$	0	0	0	0	0	0	0	0	0	0	0	1	2	4	7	13	30								
$f_7(x_7)$	0	1	2	3	4	5	6	7	8	9	10	11	12	12	12	12	12	12	12	12	12	12	12	12	12

To begin the optimization procedure, an initial allocation of the warehouses resources must be made. Let the following initial allocation matrix be chosen:

		S_1	S_2	S_3	S_4	S_5	S_6	S_7
	W_1	1	X	X	0	X	7	0
$x^0 =$	W_2	X	2	X	2	2	X	2
	W_3	X	X	1	X	1	5	1

where S_j denotes sector j, W_i denotes warehouse i, and $x_{ij}^0 = X$ implies sector j does not lie within the service area of warehouse i. This initial allocation matrix yields a total problem payoff of 24.

The first four steps of the sector allocation procedure outlined in this paper, are summarized by the tables appearing on the next two pages. In the first step, the W_2 and W_3 sector allocations are held fixed and the W_1 sector allocations are optimized. This yields the allocation matrix x^1 and a total problem payoff of 29. In steps 2 and 3, respectively, the W_2 and W_3 allocations are optimized while, in each case, the other warehouse allocations are held constant. These steps yield the intermediate allocations x^2 and x^3 and total problem payoffs of 31 and 38, respectively.

Step 4 begins the second cycle of the optimization procedure. W_1 is again allowed to reallocate its resources. Note that there was no change in the W_1 allocation. It is easily verified that, were W_2 allowed to reallocate its resources, no change would occur there either. Obviously, then, since the existing W_3 allocation is optimal with respect to the existing W_1 and W_2 allocations, the optimization procedure can be halted at this point with the conclusion that the total problem payoff of 38, corresponding to the allocation matrix x^1, represents at least a relative maximum solution to the allocation problem.

Let us also compute a value of the upper bound by combining the sectors in the three overlapping warehouse problem in the order 3, 6, 5, 7, 1, 4, 2. This order of combination implies

$$G_1(s_1) = f_3(s_1) \qquad 0 \leqslant s_1 \leqslant 8$$

$$G_2(s_2) = \max_{0 \leqslant x_6 \leqslant h} [f_6(x_6) + G_1(s_2 - x_6)] \qquad 0 \leqslant s_2 \leqslant 16, \quad h = \min[16, s_2]$$

$$\vdots$$

$$G_7(s_7) = \max_{0 \leqslant x_2 \leqslant h} [f_2(x_2) + G_6(s_7 - x_2)] \qquad 24 \leqslant s_7 \leqslant 24, \quad h = \min[8, s_7]$$

OPTIMIZING A SECTOR ALLOCATION

1. Fix $W_2 + W_3$ Allocations—Optimize w.r.t. W_1.

x_{1j}	0	1	2	3	4	5	6	7	8
$\bar{f}_1(x_{11})$	0	3	4	5	6	7	7	7	7
$\bar{f}_4(x_{14})$	2	3	4	6	8	9	10	11	11
$\bar{f}_6(x_{16})$	0	0	0	0	0	0	1	2	4
$\bar{f}_7(x_{17})$	3	4	5	6	7	8	9	10	11

Let $h_2(z) = \max\limits_{0 \leq x_{14} \leq z} [\bar{f}_4(x_{14}) + \bar{f}_1(z - x_{14})]$

z	0	1	2	3	4	5	6	7	8
$h_2(z)$	2	5	6	7	9	11	12	13	14
x_{14}	0	0	0	0	3	4	4	4	4

Let $h_3(z) = \max\limits_{0 \leq x_{16} \leq z} [\bar{f}_6(x_{16}) + h_2(z - x_{16})]$

z	0	1	2	3	4	5	6	7	8
$h_3(z)$	2	5	6	7	9	11	12	13	14
x_{16}	0	0	0	0	0	0	0	0	0

Let $h_4(8) = \max\limits_{0 \leq x_{17} \leq 8} [\bar{f}_7(x_{17}) + h_3(8 - x_{17})]$

$= 17; x_{17} = 0$

$x^1 =$

	S_1	S_2	S_3	S_4	S_5	S_6	S_7
W_1	4	X	X	4	X	0	0
W_2	X	2	X	2	2	X	2
W_3	X	X	1	X	1	5	1

Total Problem Payoff = 29

2. Fix $W_1 + W_3$ Allocations—Optimize w.r.t. W_2.

x_{2j}	0	1	2	3	4	5	6	7	8
$\bar{f}_2(x_{22})$	0	2	3	4	5	5	5	5	5
$\bar{f}_4(x_{24})$	4	6	8	9	10	11	11	11	11
$\bar{f}_5(x_{25})$	6	8	9	9	9	9	9	9	9
$\bar{f}_7(x_{27})$	1	2	3	4	5	6	7	8	9

Let $h_2(z) = \max\limits_{0 \leq x_{24} \leq z} [\bar{f}_4(x_{24}) + \bar{f}_2(z - x_{24})]$

z	0	1	2	3	4	5	6	7	8
$h_2(z)$	4	6	8	10	11	12	13	14	15
x_{24}	0	1	1	2	2	4	5	5	5

Let $h_3(z) = \max\limits_{0 \leq x_{25} \leq z} [\bar{f}_5(x_{25}) + h_2(z - x_{25})]$

z	0	1	2	3	4	5	6	7	8
$h_3(z)$	10	12	14	16	18	19	20	21	22
x_{25}	0	1	1	1	1	2	2	2	2

Let $h_4(8) = \max\limits_{0 \leq x_{27} \leq 8} [\bar{f}_7(x_{27}) + h_3(8 - x_{27})]$

$= 23; x_{27} = 0$

$x^2 =$

	S_1	S_2	S_3	S_4	S_5	S_6	S_7
W_1	4	X	X	4	X	0	0
W_2	X	1	X	5	2	X	0
W_3	X	X	1	X	1	5	1

Total Problem Payoff = 31

OPTIMIZING A SECTOR ALLOCATION

3. Fix $W_1 + W_2$ Allocations—Optimize w.r.t. W_3.

x_{3j}	0	1	2	3	4	5	6	7	8
$\bar{f}_3(x_{33})$	0	2	4	6	7	8	8	8	8
$\bar{f}_5(x_{35})$	8	9	9	9	9	9	9	9	9
$\bar{f}_6(x_{36})$	0	0	0	0	0	0	0	0	0
$\bar{f}_7(x_{37})$	0	1	2	3	4	5	6	7	8

Let $h_2(z) = \max\limits_{0 \le x_{35} \le z} [\bar{f}_5(x_{35}) + \bar{f}_3(z - x_{35})]$

z	0	1	2	3	4	5	6	7	8
$h_2(z)$	8	10	12	14	15	16	17	17	17
x_{35}	0	0	0	0	1	1	1	1	3

Let $h_3(z) = \max\limits_{0 \le x_{36} \le z} [\bar{f}_6(x_{36}) + h_2(z - x_{36})]$

z	0	1	2	3	4	5	6	7	8
$h_3(z)$	8	10	12	14	15	16	17	17	17
x_{36}	0	0	0	0	0	0	0	0	0

Let $h_4(8) = \max\limits_{0 \le x_{37} \le 8} [\bar{f}_7(x_{37}) + h_3(z - x_{37})]$

$= 19; x_{37} = 2$

$x^3 =$

	S_1	S_2	S_3	S_4	S_5	S_6	S_7
W_1	4	X	X	4	X	0	0
W_2	X	1	X	5	2	X	0
W_3	X	X	5	X	1	0	2

Total Problem Payoff = 38

4. Fix $W_2 + W_3$ Allocations—Optimize w.r.t. W_1.

x_{1j}	0	1	2	3	4	5	6	7	8
$\bar{f}_1(x_{11})$	0	3	4	5	6	7	7	7	7
$\bar{f}_4(x_{14})$	6	8	9	10	11	11	11	11	11
$\bar{f}_6(x_{16})$	0	0	0	0	0	0	0	0	0
$\bar{f}_7(x_{17})$	2	3	4	5	6	7	8	9	10

Let $h_2(z) = \max\limits_{0 \le x_{14} \le z} [\bar{f}_4(x_{14}) + \bar{f}_1(Z - x_{14})]$

z	0	1	2	3	4	5	6	7	8
$h_2(z)$	6	9	11	12	13	14	15	16	17
x_{14}	0	0	1	2	3	4	4	4	4

Let $h_3(z) = \max\limits_{0 \le x_{16} \le z} [\bar{f}_6(x_{16}) + h_2(z - x_{16})]$

z	0	1	2	3	4	5	6	7	8
$h_3(z)$	6	9	11	12	13	14	15	16	17
x_{16}	0	0	0	0	0	0	0	0	0

Let $h_4(8) = \max\limits_{0 \le x_{17} \le 8} [\bar{f}_7(x_{17}) + h_3(z - x_{17})]$

$= 19; x_{17} = 0$

$x^1 =$

	S_1	S_2	S_3	S_4	S_5	S_6	S_7
W_1	4	X	X	4	X	0	0
W_2	X	1	X	5	2	X	0
W_3	X	X	5	X	1	0	2

Total Problem Payoff = 38

A summary of the computation of $G_7(24)$, the upper bound, appears in the table on the following page. Note that an upper bound of 47 results.

To apply optimal stopping rules we note that $\phi(1) = \phi_1 = 38$ and $g(1) = g_1 = 47$. This yields $d(1, 1) = 9$ and $r = s = 3$. If $c_\phi = .03$, then we find $I_1(\phi) = 1.095$ as the net expected improvement in the relative maximum through one more observation on the set of relative maxima. Since $I_1(\phi) > 0$, we compare $I_1(\phi)$ with $I_1(g)$. If $c_g = .05$, then $I_1(g) = 1.075$. Since $I_1(\phi) > I_1(g)$, we should calculate another value of the relative maximum in the next stage of the problem solution.

Suppose the second initial allocation was chosen to be

$x^0 =$

	S_1	S_2	S_3	S_4	S_5	S_6	S_7
W_1	0	X	X	0	X	8	0
W_2	X	2	X	2	2	X	2
W_3	X	X	0	X	0	8	0

Then, the successive approximation procedure would lead to the following final allocation matrix (x^f),

$x^f =$

	S_1	S_2	S_3	S_4	S_5	S_6	S_7
W_1	0	X	X	0	X	8	0
W_2	X	1	X	6	1	X	0
W_3	X	X	0	X	0	8	0

which yields a total problem payoff of 46. Since this total payoff is higher than that obtained previously, our first effort evidently produced only a local optimum solution. The problem still remains of determining if this new payoff represents an absolute maximum solution.

If we apply optimal stopping rules, we see for $\phi(2) = \phi_2 = 46$ that $d(2, 1) = 1$ and $r = 11$ and, thus, $I_2(\phi) = .00846$ which again is positive. For this new value of d, we also obtain $s = 3$ and $I_1(g) = .0750$. Therefore, at stage $(M, N) = (2, 1)$ we have $I_2(\phi) < I_1(g)$ and we should now make a second observation on the upper bound.

Suppose the order of combining the sectors for the second upper bound computation was chosen to be 1, 4, 6, 7, 3, 5, 2. It can be shown that this order of combining sectors yields an upper bound value of 50. Therefore, the minimum upper bound, g(2), is still equal to 47. Since $d(2, 2) = d(2, 1)$, $I_2(\phi)$ has not changed and thus it is still profitable to continue sampling. The value of s has changed which implies we must compute the new expected improvement on the upper bound. This computation yields $I_2(g) = .0125$. Therefore, the results of stage (2, 2) indicate $I_1(\phi) < I_2(g)$ which implies we should again compute another upper bound.

Let the order of sector combination for the upper bound calculation corresponding to stage (2, 3) be chosen to be 2, 5, 4, 7, 3, 6, 1. An upper bound of 46 results from this order of combination which implies $I_2(\phi) \doteq -c_\phi = -.03$ since $d(2, 3) = g(3) - \phi(2) = 0$. Therefore, the stopping rule would terminate all further sampling. Furthermore, since $\max_\mu \phi_\mu = \min_\nu g_\nu$, we know that the payoff of 46 is an absolute maximum to the warehouse allocation problem, and the corresponding final allocation matrix is the maximizing solution.

AN UPPER BOUND COMPUTATION

SECTOR COMBINATION ORDER: 3, 6, 5, 7, 1, 4, 2

s_j	0	1	2	3	4	5	6	7	8	9	10	11	12	13	14	15	16	17	18	19	20	21	22	23	24
$G_1(s_1)$	0	2	4	6	7	8	8	8	8																
x_3	0*	1	2	3	4	5	6	7	8																
$G_2(s_2)$	0	2	4	6	7	8	8	8	8	8	8	8	8	8	8	13	30								
x_6	0	0	0	0	0	0	0	0	0	0	0	0	0	0	0	15	16*								
$G_3(s_3)$	0	6	8	10	12	14	15	16	17	17	17	17	17	17	17	17	30	36	38	39	39	39	39	39	39
x_5	0	1	1	1	1	2	2	2	3	3	3	3	3	3	3	3	0	1	2*	3	4	5	6	7	8
$G_4(s_4)$	0	6	8	10	12	14	15	16	17	18	19	20	21	22	23	24	30	36	38	39	40	41	42	43	44
x_7	0	0	0	0	0	0	0	0	0	1	2	3	4	5	6	7	0	0	0*	0	1	2	3	4	5
$G_5(s_5)$									17	18	21	22	23	24	25	26	30	36	39	41	42	43	44	45	46
x_1									0	1	1	1	1	1	1	1	0	0	1	1*	1	1	1	1	1
$G_6(s_6)$									17	18	21	22	23	24	25	27	30	36	39	41	42	43	44	45	47
x_4									0	0	0	0	0	0	0	5	0	0	0	0	0	0	0	0	5*
$G_7(s_7)$																	30	36	39	41	43	44	45	46	47†
x_2																	0	0	0	1	1	1	1	1	0*

* Sector allocation yielding upper bound payoff of 47. Note that this allocation violates the problem constraints $(\sum_{j \in J_2} x_j < b_2)$.

† Value of the upper bound.

REFERENCES

1. RANDOLPH, PAUL H., (1968) 'An Optimal Stopping Rule for Multinomial Observations,' Metrika (to appear).
2. RANDOLPH, P. H., G. E. SWINSON and M. E. WALKER, (1968) 'Nonlinear Integer Programming through Optimal Stopping Rules,' Statistical Laboratory Technical Report Series No. 5, New Mexico State University, Las Cruces, New Mexico.
3. REITER, S. and G. SHERMAN, 'Discrete Optimizing,' (1965) Journal of the Society of Industrial and Applied Mathematics, 13 : 864-889.

A Non-Convex Transportation Algorithm

P. RECH, L. G. BARTON
Shell Development Co., U.S.A.

1. INTRODUCTION

This paper deals with a solution method for solving nonconvex transportation problems. Such a problem consists of determining a minimal cost transportation program for shipping a single commodity, available at m sources, to n destinations with known demands. These destinations are connected to the supply points by a transportation network which can have features such as multiple modes of transportation, transhipment points, and restrictions on the amount which can be shipped from one point to another (an example of this is the capacity of a pipeline). The problem is nonlinear in the sense that some cost functions $C(X_{ij})$ can be nonconvex piecewise linear cost functions similar to the function illustrated in Figure 1, where X_{ij} represents the amount of the commodity which is shipped from point i to point j of the transportation network. A particular case of such a cost function is the concave piecewise linear cost function which arises whenever economies of scale are present, i.e., wherever the unit shipping costs are decreasing.

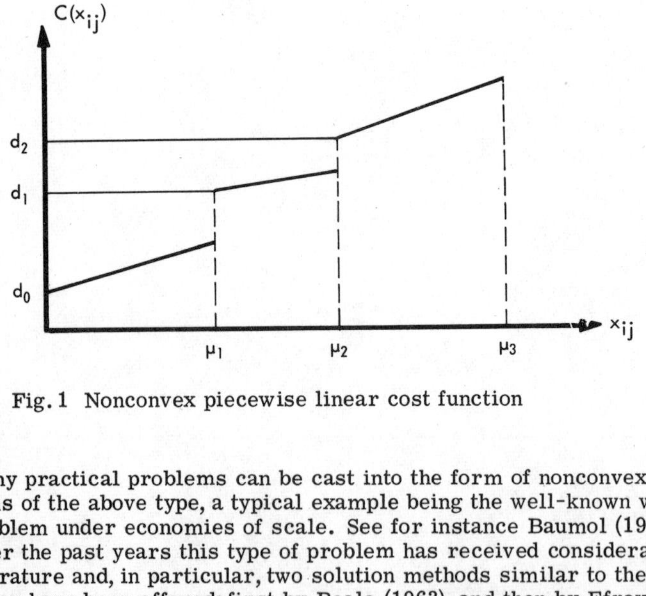

Fig. 1 Nonconvex piecewise linear cost function

Many practical problems can be cast into the form of nonconvex transportation problems of the above type, a typical example being the well-known warehouse location problem under economies of scale. See for instance Baumol (1958) and Manne (1964). Over the past years this type of problem has received considerable attention in the literature and, in particular, two solution methods similar to the one discussed in this paper have been offered; first by Beale (1963), and then by Efroymson and Ray (1966). We believe that the contribution of our approach lies in the fact that we view the problem as a minimal cost network flow problem. This allows us to overcome quite easily a certain number of difficulties inherent in the general method. This, combined with the practical attractiveness of the whole approach, justifies this exposition and further illustration.

Any distribution problem involving only one commodity can be cast into a minimal cost network flow problem; see for instance Fulkerson (1962). The example illustrated in Figure 2 is a straightforward transportation problem involving the two supply points 1 and 2 and the two demand points 4 and 5. Availabilities at nodes 1 and 2 are represented as the upper bounds a_1 and a_2, on the flows X_{01} and X_{02}, whereas the demands at nodes 4 and 5 are represented as lower bounds on the flows X_{46} and X_{56}. Associated with each arc (i, j) of the network is a cost function $C(X_{ij})$ which may not be linear. For instance, the cost $C(X_{14})$ might exclusively represent the linear transportation cost from node 1 to node 4, whereas the cost function $C(X_{01})$ could include the procurement cost of the commodity at node 1 (which might exhibit economies of scale) and even the handling cost at this node. If we call respectively the artificial nodes 0 and 6, source and sink, then the minimal cost network flow problem

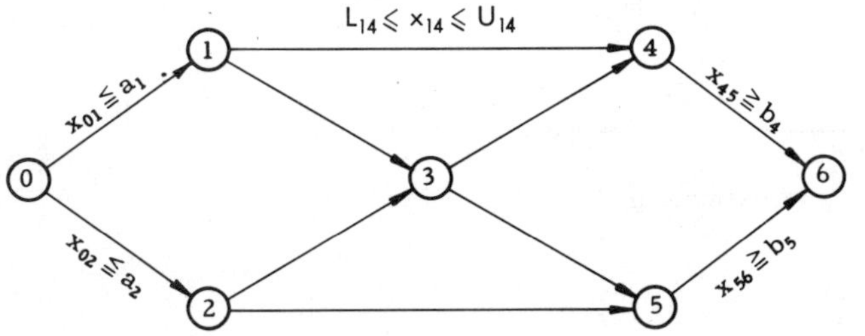

Fig. 2 Directed distribution network

(a_1 and a_2 are supplies at nodes 1 and 2, whereas b_4 and b_5 are demands at 4 and 5)

consists of finding a feasible flow from the source to the sink which minimizes the overall cost. When all the cost functions are convex piecewise linear functions, then such a problem can be solved straightforwardly by the out-of-kilter algorithm developed by Fulkerson (1962). However, when some cost functions are nonconvex, then some sort of iterative procedure is required, and it appears that a judicious application of the branch and bound method could very well be the most satisfactory answer to this need.

In this paper the branch and bound method, see for instance Lawler and Wood (1966), is applied to decompose the original problem into a series of convex piecewise linear suboptimization problems. To solve each one of the latter we use the out-of-kilter algorithm which is well suited for this purpose. (This is essentially due to the fact that it can be initiated with any infeasible flow.) Computationally the method appears to be quite efficient, and presents a certain number of advantages which have, in our opinion, important practical implications. In fact, we believe that it epitomizes an approach towards real life optimization problems that will probably come to age with the advent of the time-sharing computers. This approach lies somewhere between the rigorous optimization methods and the heuristic ones and will be characterized by the emphasis on means of allowing intuitive knowledge and experience to bear on the solution procedures. Obviously, great learning advantages could accrue from such 'controlled' optimization methods.

2. BRANCH AND BOUND SOLUTION METHOD

Let us consider a minimal cost network flow problem involving mostly linear cost functions, but also including some nonconvex piecewise linear ones. We denote the latter by $C_k(X_k), k \in M$, where M is the set of indices corresponding to nonconvex cost functions.

To solve this problem we apply the branch-and bound procedure as follows. We first approximate all nonconvex cost functions $C_k(X_k)$ by their convex envelopes $\underline{C}_k(X_k)$ as illustrated in Figure 3.

Replacing the cost functions $C_k(X_k)$ by their convex piecewise linear approximations $\underline{C}_k(X_k)$ transforms the original problem into an approximate convex problem which

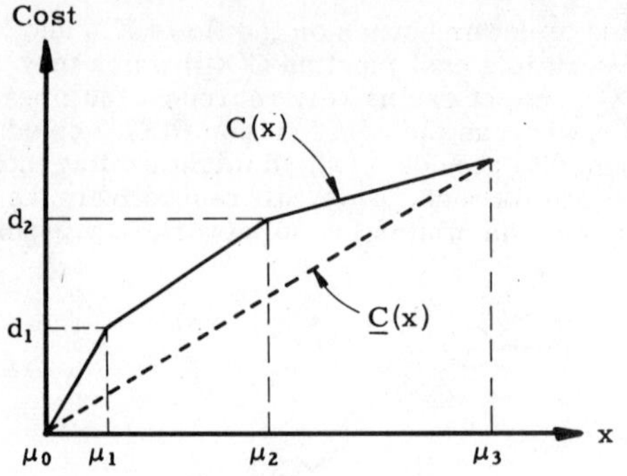

(a) Linear approximation

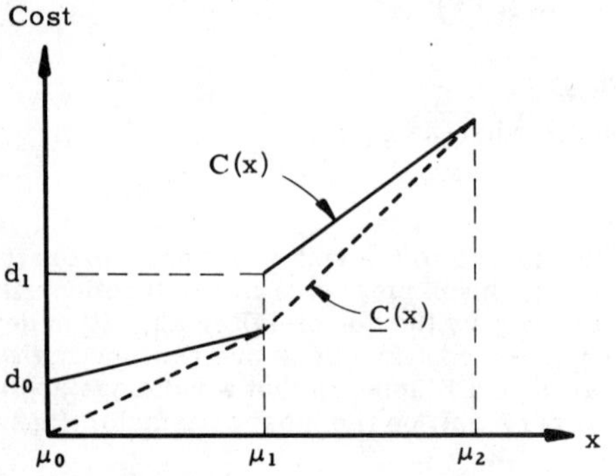

(b) Convex approximation

Fig. 3 Convex envelopes of cost functions

can be solved by the out-of-kilter algorithm. Let us denote by X(1) the solution vector of this problem and by $\underline{Z}(1)$ the optimal value of the objective function corresponding to it. Since X(1) must also be a feasible solution to the original problem we can compute the true value of the objective function corresponding to this solution (as opposed to its approximate value $\underline{Z}(1)$), and we denote it by $\overline{Z}(1)$. At this point, the following crucial observation must be made.

Lemma 1: $\underline{Z}(1)$ and $\overline{Z}(1)$ are respectively lower and upper bounds to the optimal value, say Z^*, of the original problem.

Proof: Since X(1) is a feasible solution to the original problem, we must have $\overline{Z}(1) \geqslant Z^*$. Furthermore, if X^* is an optimal solution to the original problem, then it is also a feasible solution to the approximate problem. Hence, the corresponding value of the objective function, say $\underline{Z}^*$, must satisfy $\underline{Z}(1) \leqslant \underline{Z}^*$. But the convex envelopes have the property that $\underline{C}_k(X_k) \leqslant C_k(X_k)$ for all $X_k \in D_k$, where D_k is the domain of definition of these functions. Hence, $\underline{Z}^* \leqslant Z^*$ and, therefore,

$$\underline{Z}(1) \leqslant \underline{Z}^* \leqslant Z^* \leqslant \overline{Z}(1). \text{ Q.E.D.} \qquad (1)$$

The solution of this first approximate problem constitutes the first bounding step.

The branching is done as follows. For each nonconvex cost function we determine the error introduced by the convex approximation, i.e., we compute the value

$$\Delta_k = C_k[X_k(1)] - \underline{C}_k[X_k(1)] \text{ for all } k \in M. \tag{2}$$

If Δ_s is the largest of all these values, then we leave all the approximate functions $\underline{C}_k(X_k)$ unchanged except $\underline{C}_s(X_s)$ which we replace by a better approximate function derived as follows. We partition the cost function $C_s(X_s)$ into two parts in a manner to be discussed later and approximate these as was done previously with the whole function, by their convex piecewise linear envelopes, say $L(X_s)$ and $R(X_s)$. This partitioning is illustrated in Figure 4. This new approximation of $C_s(X_s)$ is an improvement over the previous approximation $\underline{C}_s(X_s)$; however, it is no longer convex and, therefore, the new approximate problem can no longer be solved by the out-of-kilter algorithm. To overcome this difficulty we set up two new approximate problems.

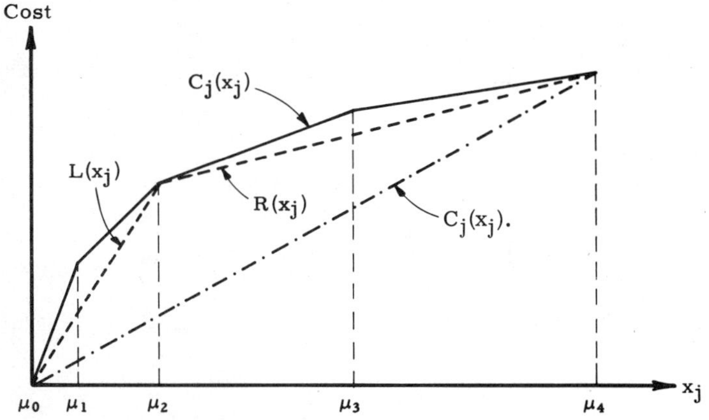

Fig. 4 Improved approximation of $C_j(x_j)$

In the first one, say problem 2, $C_s(X_s)$ is replaced by $L(X_s)$ and the additional restriction that $X_s \leq \mu_2$ (for the case illustrated in Figure 4). The second one, say problem 3, is based on $R(X_s)$ and the additional restriction that $X_s \geq \mu_2$. By construction, these problems are now convex piecewise linear and therefore we can solve them with the out-of-kilter algorithm. This can be done very efficiently by starting from the previous solution X(1), replacing only $C_s(X_s)$ by either $L(X_s)$ or $R(X_s)$ and adding the proper bound on the flow X_s. If X(2) and X(3) are the solutions to these two approximate problems (both do not necessarily exist) and if, as before, $\underline{Z}(2)$, $\underline{Z}(3)$ and $\overline{Z}(2)$, $\overline{Z}(3)$ are the corresponding lower and upper bounds, then it can be easily shown that we must have

$$\underline{Z}(1) < \text{Min}[\underline{Z}(2), \underline{Z}(3)] \leq Z^* \leq \text{Min}[\overline{Z}(1), \overline{Z}(2), \overline{Z}(3)] \tag{3}$$

If optimality has not been reached, i.e., if the upper bound is not equal the lower bound, then the above branching procedure can be repeated starting from the approximate problem which gives the minimum value of all the $\underline{Z}(t)$, where $\underline{Z}(t)$ is the optimal value of the objective function of the tth approximating problem. Using an inductive proof, it can be shown easily that the following generalization of (1) and (3) must hold.

Lemma 2: Let $\underline{Z}$ and $\overline{Z}$ be respectively the lower and upper bounds of Z* obtained by solving T approximate problems. If $\underline{Z} = \underline{Z}(t)$ then, branching out from the tth problem, two new approximate problems, say T + 1 and T + 2, can be solved which will give the improved bounds $\underline{\underline{Z}}$ and $\overline{\overline{Z}}$ satisfying the relation

$$\underline{Z} < \underline{\underline{Z}} \leq Z^* \leq \overline{\overline{Z}} \leq \overline{Z} \tag{4}$$

To summarize, we illustrate in Figure 5 the progress of the branch and bound procedure towards optimality. Node i of the tree corresponds to the ith approximate problem. The following relations must hold:

(a) $\underline{Z}(2) = \mathrm{Min}[\underline{Z}(2), \underline{Z}(3)] > \underline{Z}(1)$

(b) $\underline{Z}(4) = \mathrm{Min}[\underline{Z}(3), \underline{Z}(4)]$ (assuming that problem 5 had no feasible solution)

(c) $\underline{Z}(3) = \mathrm{Min}[\underline{Z}(3), \underline{Z}(6), \underline{Z}(7)]$

(d) $\underline{Z}(9) = \mathrm{Min}[\underline{Z}(6), \underline{Z}(7), \underline{Z}(8), \underline{Z}(9)]$

(e) $\underline{Z}(9) = \overline{Z}(9)$ which, with (d), indicates that optimality has been reached.

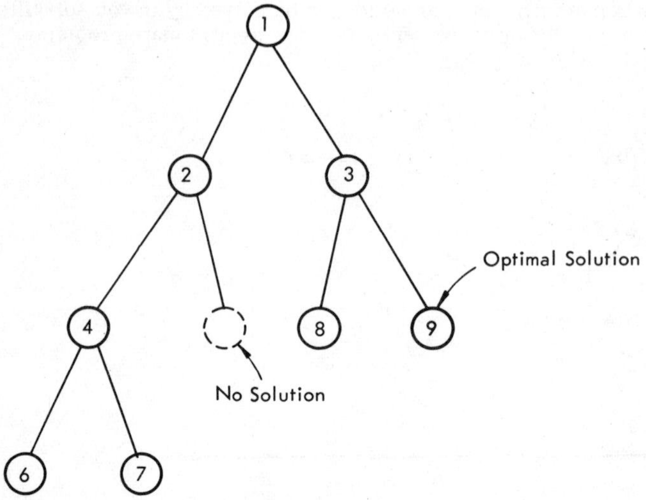

Fig. 5 Solution tree of branch and bound procedure

3. COMPUTATIONAL PROCEDURES

A few computational remarks are now in order. First, let us recall some of the essential characteristics of the out-of-kilter procedure for solving minimal cost network flow problems (Fulkerson (1962)).

(1) The directed network can have multiple arcs between any two nodes; thus, convex piecewise linear cost functions can be handled by replacing one arc by several arcs with increasing unit costs.

(2) For each arc a lower bound on the flow as well as an upper bound can be specified. This feature is essential for the handling of the flows corresponding to our approximate cost functions L(X) and R(X).

(3) The unit costs can be either positive or negative provided, of course, that there exist no cycle in the network along which an infinite solution could be obtained. This feature allows the introduction for instance, of sales prices, and thus profit maximization problems can be solved.

(4) The method can be initiated with any flow (satisfying the conservation equation at each node), feasible or not, i.e., it can violate upper or lower bound restrictions. Similarly, any set of node numbers (see (5) below) can be given. This feature is essential for the branch and bound method because the various suboptimization problems to be solved vary from each other only by a few cost coefficients and possibly a few

upper and lower bounds. Hence, the previous primal and dual solutions can be used to initiate the solution procedure of a new suboptimization problem. This solution is 'out-of-kilter' but computational experience indicates that the optimization procedure brings it back into kilter in relatively few iterations.

(5) Associated with each optimal solution is a set of node numbers which are optimal dual variables. Furthermore, the relative costs (marginal values) corresponding to all the bounds are also given. Besides their useful economic significance, the latter could also possibly be used to improve the branching procedure of the method developed in this paper. However, we have made no attempt in this direction.

(6) Finally, it must be mentioned that the out-of-kilter algorithm is developed for integral costs only. Hence, some numerical errors are introduced in the present method which are due to the necessary truncations of the unit costs associated with the segments of the cost functions considered. Although this introduces some programming complications if a computer code is to be developed, these errors do not present (for practical purposes) any real limitation, because they can be adequately controlled by proper scaling. However, this scaling is not trivial if all possibilities of cycling are to be avoided in a computer program.

Let us now go back to the branching procedure. We recall that we suggested to branch out from the problem which gives the lowest value $\underline{Z}$ of the approximated objective functions.* Associated with its optimal solution, say $X(k)$, are the errors

$$\Delta_j = C_j(X_j(k)) - \underline{C}_j(X_j(k)) \text{ for all } j \in M.$$

If Δ_s is the minimal value of all Δ_j's, then the domain of definition, say D_s, of the approximate cost function $\underline{C}_s(X_s)$ considered (which is not necessarily all the domain of definition of $C_s(X_s)$) can be partitioned at the first break point at the left of the value $X_s(k)$. This partitioning, which is illustrated in Figure 6, has given us satisfactory results. However, in some circumstances a special case can occur where the strict application of the preceding rule leads to cycling. Such a case is illustrated in Figure 7. When this occurs the difficulty can be overcome by partitioning the domain D_s at the first break point on the right of $X_s(k)$.

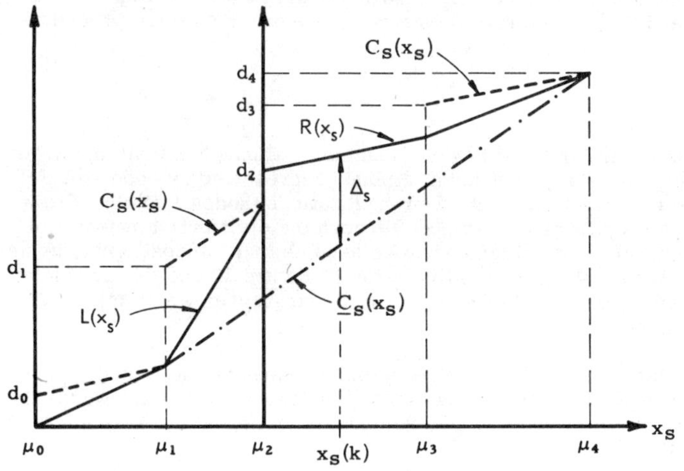

Fig. 6 Convex approximate functions $L(x_s)$ and $R(x_s)$ when $\mu_2 < x_s(k) \leqslant \mu_3$

---

* However, this is not necessary, for alternative methods see Lawler and Wood (1966).

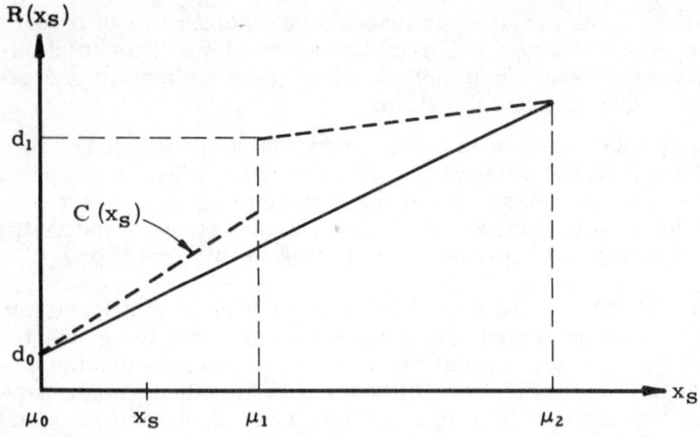

Fig. 7 Special situation for which cycling can occur

Finally, some remarks about the bounding procedures are in order. We recall that at each iteration we obtain a lower bound $\underline{Z}$ and an upper bound $\overline{Z}$ on the optimal value of the objective function of the original problem. More precisely, Lemma 2 shows that at each iteration the value $\overline{Z} - \underline{Z}$ is strictly decreasing and that, therefore, the method must converge to the optimal solution. Although, by construction of the approximate problems, this convergence must occur in a finite number of steps, it is conceivable that many iterations could be required to reach optimality. In such cases it might be desirable to stop the computations whenever the relative value $(\overline{Z} - \underline{Z})/\underline{Z}$ falls below a specified (or computed) tolerence ϵ. Obviously, for practical problems this procedure is perfectly acceptable and even recommended because it saves significant time. Corresponding to the example given hereafter, Figure 8 illustrates the convergence of the bounds. One last remark which is of importance is that it is possible to use judgement or experience with the system being studied to generate at the beginning of the process some reasonable solutions which will give a better upper bound. It is also conceivable that such information could be used to improve the branching criteria and thus accelerate the convergence of the solution procedure.

4. AN EXAMPLE

As an example, let us consider the following imaginary, although plausible, warehouse location problem. A product is produced at a plant (represented by node s of Figure 9) and is distributed to ten warehouses at different locations (nodes 1 to 10). From these warehouses the product can be shipped through the distribution network of Figure 9 to customers with known demands. We assume that the cost functions associated with the first ten arcs are those illustrated in Figure 10 and that all the other costs are linear. The problem is to choose a set of warehouses which minimizes the overall distribution costs.

This particular problem was easily solved with our computer code, based on the method of this paper, in 27 iterations. The history of the convergence towards the optimal solution is illustrated in Figure 8. Some practical problems of similar structures were solved as easily.

ACKNOWLEDGEMENTS

The authors acknowledge with pleasure the helpful assistance received in this study from their colleagues R. Tibrewala of Shell Oil Company, New York, and Mrs. C. B. Witze of Shell Development Company, Emeryville.

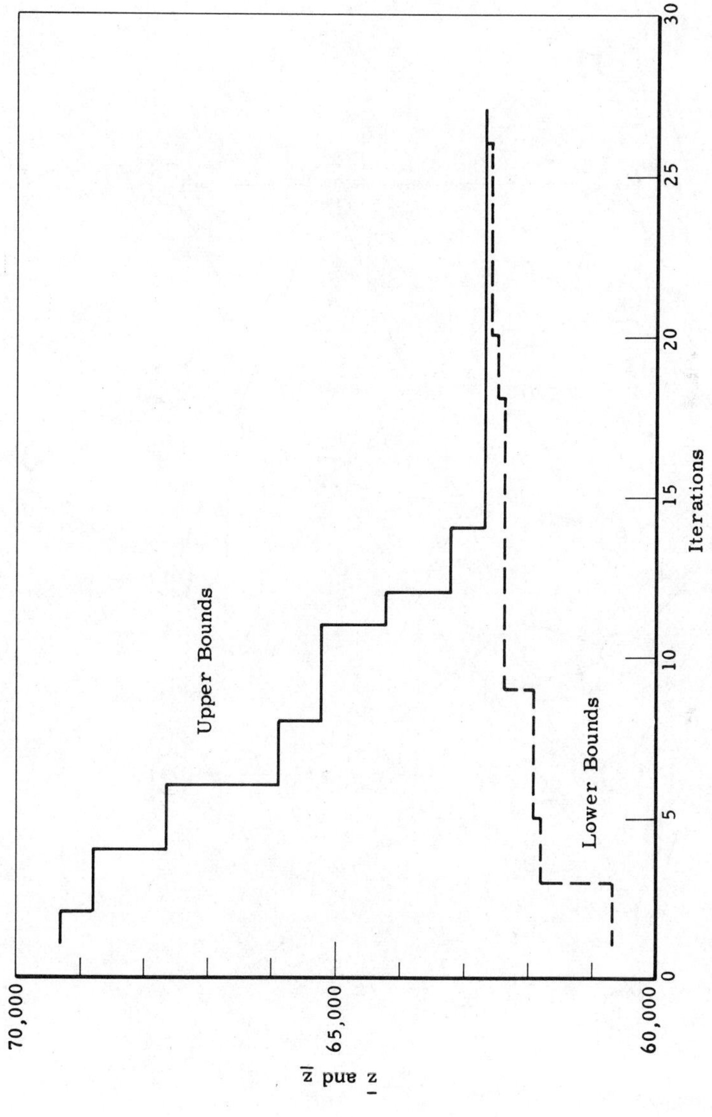

Fig. 8 Examples of convergence of the bounds

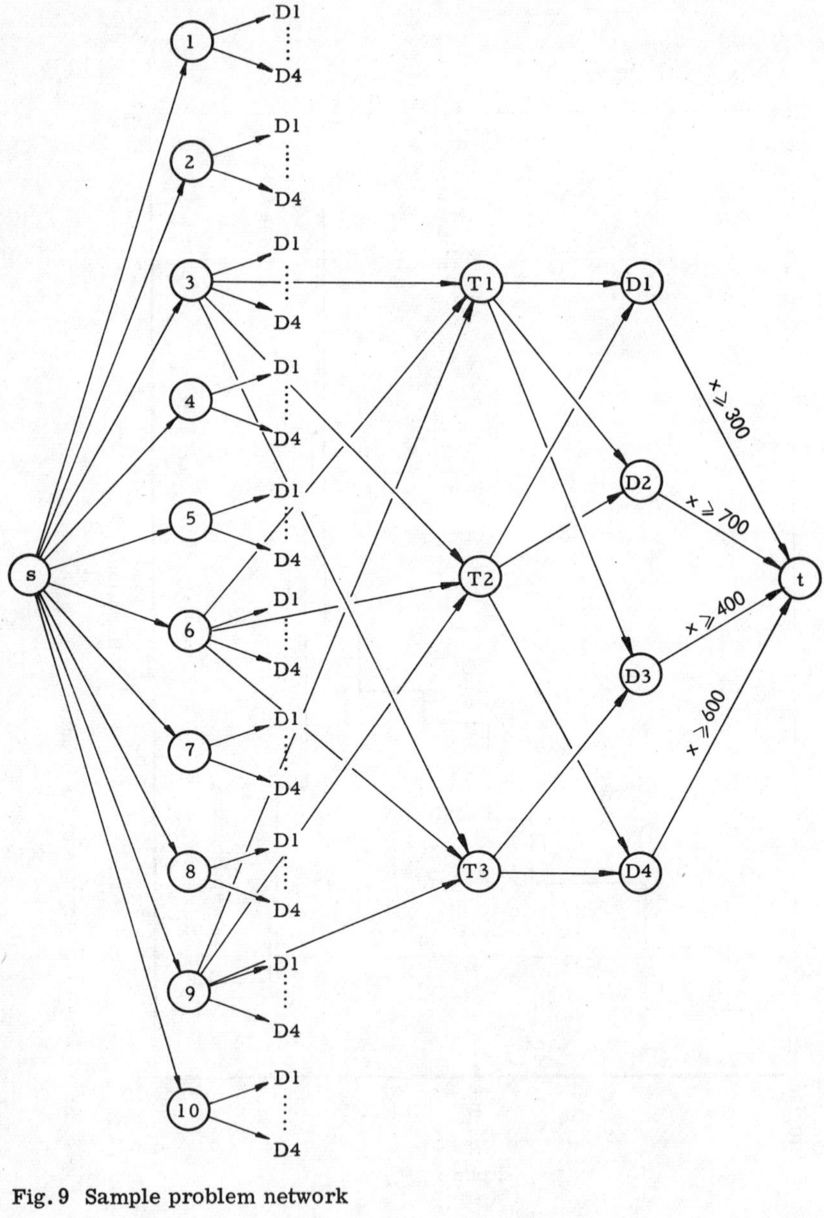

Fig. 9 Sample problem network

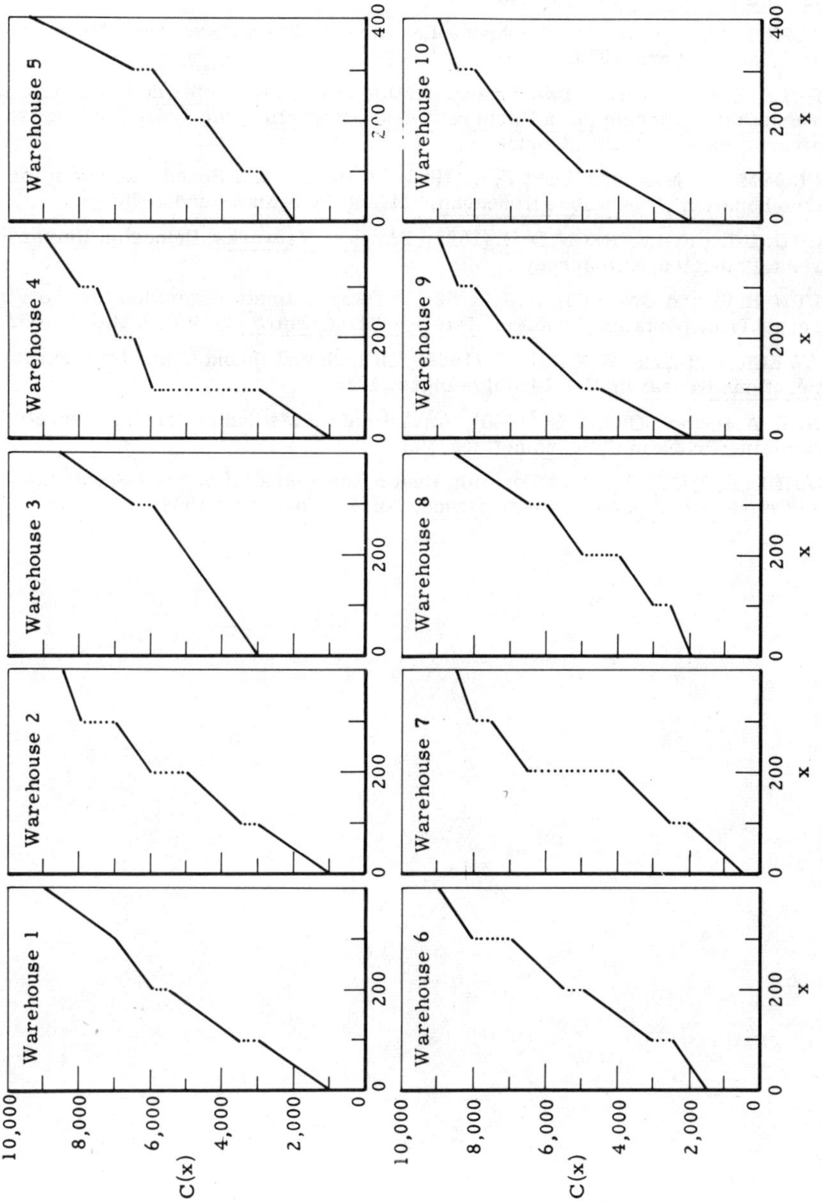

Fig. 10 Cost function for test problem

REFERENCES

(1) BALINSKI, M. L. (1965), 'Integer Programming: Methods, Uses, Computation', Management Science, Vol. 12, No. 3.

(2) BAUMOL, W. J. (1958), 'A Warehouse Location Problem', Operations Research, Vol. 6, March-April 1958.

(3) BEALE, E. M. L. (1963), 'Two Transportation Problems', Actes de la 3ième Conférence Internationale de la Recherche Opérationnelle, Oslo 1963, The English Universities Press Ltd., London.

(4) EFROYMSON, M. A. and RAY, T. L. (1966), 'A Branch and Bound Algorithm for Plant Location', Operations Research, Vol. 14, No. 3, May-June 1966.

(5) FORD, L. R., FULKERSON, D. R. (1962), Flows in Networks, Princeton University Press, Princeton, New Jersey.

(6) KUHN, H. W. and BAUMOL, W. J. (1962), 'An Approximate Algorithm for the Fixed Charge Transportation Problem', Nav. Res. Log. Quarterly, Vol. 9, March 1962.

(7) LAWLER, E. L., and WOOD, D. E. (1966), 'Branch and Bound Methods: A Survey', Operations Research, Vol. 14, July-August 1966.

(8) LAND, A. H. and DOIG, A. G. (1960), 'An Automatic Method of Solving Discrete Programming Problems', Econometrica, Vol. 28.

(9) MANNE, A. S. (1964), 'Plant Location Under Economies of Scale—Decentralization and Computation', Management Science, Vol. 11, November 1964.

Exact Solution of the Site Selection Problem by Mixed Integer Programming

P. GRAY
Stanford Research Institute, U.S.A.

SUMMARY

This paper presents algorithms for solving site-selection and similar fixed-charge problems with upper bound constraints. The basic approach to obtaining algorithms for exact solutions is to formulate the problems as mixed integer programs and to solve these programs by decomposing them into a master problem (which is an integer program) and a series of subproblems which are linear programs. As in all decomposition schemes, heavy use is made of the fact that large portions of the constraint matrix contain only zero elements. In addition, the fact that an optimal solution must occur at an extreme point plays a central role.

To achieve computationally feasible algorithms, the number of vertices that have to be examined must be reduced to manageable proportions, while still guaranteeing that the optimal solution will be found. This is accomplished by adapting the bound-and-scan integer programming algorithm by F. S. Hillier to the fixed-charge problem. The decomposition idea is applied in this paper to two classes of problems and detailed algorithms are presented for them:

- A fixed-charge problem with linear variable costs in which a fixed charge is associated with each continuous variable that appears at non-zero level.
- The same problem with separable concave or convex costs rather than linear costs.

1. INTRODUCTION

1.1 Statement of the Problem

The purpose of this paper is to present algorithms for solving the site-selection and similar fixed-charge problems. These problems typically arise in the following way: a number, possibly a large number, of sites are known to be available at which facilities can be established for performing a given service. The set of facilities must provide service at a specified level of effectiveness. Fixed costs are associated with each site if a facility is opened there, and variable costs are associated with operations. Upper bound constraints on the allowable size of a facility at each site must be taken into account. The problem to be solved by the operations researcher is to select a subset of the available sites that meets the constraints and that minimizes the total cost.

This problem which arises in many contexts, can be formulated as a mixed-integer programming problem—that is, a programming problem in which some of the variables take on values 1 and 0 (corresponding to a site being opened or not) and the rest are continuous variables. If there are no upper bounds on the size of the facilities, then the branch-and-bound algorithm presented by Efroymson and Ray (1966) is applicable. However, imposing upper bounds on facility size invalidates the Efroymson-Ray method. It is this upper-bounded problem to which this paper is primarily addressed.

More generally, the fixed-charge problem considered may be stated as:

$$\text{Minimize} \sum_{i=1}^{m} c_i(x_i) + \sum_{j=1}^{p} f_j y_j,$$

subject to

$$\mathbf{A\,x} \geqslant \mathbf{b}$$

$$\mathbf{x} \leqslant \mathbf{m}$$

$$y_j = \begin{cases} 1 \text{ if } \sum_{i \in S_j} x_i > 0 \\ \\ 0 \text{ otherwise} \end{cases} \qquad \text{for } j = 1, 2, \ldots, p$$

$$\mathbf{x} \geqslant 0$$

where **A** is an m × n matrix, **x** and **m** are n-vectors, **b** is an m-vector, c(·) are functions of a single variable, the f_j are positive constants, p ⩽ m, and the S_j are subsets of { 1, 2, . . . , m}.

This general structure encompasses each of the special cases for which computational algorithms have been developed. The **A** matrix is considered either as a general matrix or as a matrix having special structure, such as all elements non-negative or a transportation-problem matrix. Both the case in which a fixed charge is associated with each component of x(p = m and the sets S_j each contain one element) and the case where a fixed charge is associated with disjoint groups of components of x are investigated. Both linear and separable non-linear variable cost structures are considered.

The fixed-charge problem has appeared in the operations research literature at least since 1954, when Hirsch and Dantzig formulated the problem and showed that if the objective function is linear, an optimal solution would occur at an extreme point of the constraint set, **A x** ⩾ **b**. This important result is used in each of the algorithms presented here.

1.2 Decomposition Approach

The basic approach to solving these mixed integer programs is to decompose them. Benders (1962) presented a decomposition algorithm in which he reduced the mixed problem to the solution of a series of integer programs. The present work differs from Benders' in that the approach is to decompose the problem into a master problem and a series of linear programs. As in all decomposition schemes, heavy use is made of the fact that large portions of the constraint matrix contain only zero elements. In the fixed-charge problem, the constraint matrix typically contains a large number of constraints that define the requirements on the continuous variables. In addition, there are integer constraints related to the fixed charges and, finally, there are constraints that relate the fixed and variable costs. The last reflect the conditions under which the fixed charges are incurred. They can be thought of as providing a weak coupling between the discrete and continuous parts of the problem.

It is the weak coupling that leads to decomposition. Specifying the values of the integer variables so that the integer constraints are satisfied produces a sub-problem which involves only continuous variables and hence can be solved by the now classical simplex method. Once the variable cost has been minimized by the simplex method and the fixed cost specified through the integer variables, the total cost is known. A systematic search through the allowable integer values, then, is all that is required to find an optimal solution.

The key question to be resolved for computability in any search scheme is how to reduce the size of the search to manageable proportions, while guaranteeing that the optimal solution will be found. Branch-and-bound methods provide the necessary technique for examining the integer parts of the problem. Of the various such algorithms currently available, the bound-and scan algorithm by F. S. Hillier (1966) was selected for use in many of the numerical calculations. This algorithm is particularly efficient if a good sub-optimal solution is available. Some attention is therefore paid to finding initial feasible solutions.

2. FORMULATION

2.1 General Decomposition

Consider the mixed-integer linear programming problem in which all variables have finite upper bounds:*

$$\left.\begin{aligned} &\text{Minimize } \mathbf{c}_1\mathbf{x} + \mathbf{c}_2\mathbf{y}, \\ &\text{subject to } \mathbf{A}_1\mathbf{x} + \mathbf{A}_2\mathbf{y} \geqslant \mathbf{b} \\ &\qquad \mathbf{x} \leqslant \mathbf{m}_1 \\ &\qquad \mathbf{y} \leqslant \mathbf{m}_2 \\ &\qquad \mathbf{x}, \mathbf{y} \geqslant 0 \\ &\qquad \mathbf{y} \text{ integer.} \end{aligned}\right\} \text{Problem I}$$

If at least one vector $(\mathbf{x}, \mathbf{y})$ exists which satisfies the constraints, an optimal solution to the problem must exist since all components of $\mathbf{x}$ and $\mathbf{y}$ are bounded above and below. Furthermore, since each component y_i of $\mathbf{y}$ is bounded between 0 and m_i, and $\mathbf{y}$ is defined only at integer lattice points, there are only a finite number of feasible values of $\mathbf{y}$.

Let $\mathbf{y}_1$ be a particular vector that satisfies the constraints, $0 \leqslant \mathbf{y} \leqslant \mathbf{m}_2$ and $\mathbf{y}$ integer. For this fixed value of $\mathbf{y}$, the original problem reduces to:

$$\left.\begin{aligned} &\text{Minimize } \mathbf{c}_1\mathbf{x} \\ &\text{subject to } \mathbf{A}_1\mathbf{x} \geqslant \mathbf{b} - \mathbf{A}_2\mathbf{y}_1 \\ &\qquad \mathbf{x} \leqslant \mathbf{m}_1 \\ &\qquad \mathbf{x} \geqslant 0 \end{aligned}\right\} \text{Problem II}$$

and Problem II is a linear program that can be solved by the simplex method. If Problem II does not have a feasible solution for this particular $\mathbf{y}_1$ then Problem I also does not have a feasible solution for $\mathbf{y}_1$. If, however, an optimal solution $\mathbf{x}_1$ is found for Problem II, then the value of the objective function for Problem I becomes $\mathbf{c}_1\mathbf{x}_1 + \mathbf{c}_2\mathbf{y}_1$.

The definitions of Problem I and II immediately suggest an algorithm for solving the mixed-integer programming problem:

(1) Find a feasible solution to Problem I.†

(2) Enumerate the finite set of vectors $\mathbf{y}_1$ that satisfy the constraints $0 \leqslant \mathbf{y} \leqslant \mathbf{m}_2$, $\mathbf{y}$ integer.

(3) For each $\mathbf{y}_1$ found in Step 2, solve Problem II. If Problem II is infeasible for this value $\mathbf{y}_1$, so is Problem I. If an optimal solution $\mathbf{x}_1$ to Problem II is found, evaluate $\mathbf{c}_1\mathbf{x}_1 + \mathbf{c}_2\mathbf{y}_1$.

(4) The optimal solution is

$$\min_{\text{all } y_1} \{\mathbf{c}_1\mathbf{x}_1 + \mathbf{c}_2\mathbf{y}_1\}$$

---

* $\mathbf{c}_1, \mathbf{m}_1$ and $\mathbf{x}$ are n_1-vectors; $\mathbf{c}_2, \mathbf{m}_2, \mathbf{y}$ are n_2-vectors; $\mathbf{A}_1$ is an $(m \times n_1)$ matrix, and $\mathbf{A}_2$ is an $(m \times n_2)$ matrix.

† This step could be dispensed with. If the exhaustive search of Step 2 does not turn up a feasible solution to Problem II, no feasible solution exists.

Although this decomposition procedure is guaranteed to produce an optimal solution to Problem I, it does not, on the face of it, appear particularly attractive computationally for problems of even moderate size. For example, a problem containing ten integer variables each with range 0-1 would require attempting to solve 1024 linear programs; for twenty such variables the number of linear programs to be solved jumps to 10^6.

However, the decomposition procedure can be made workable for a wide range of mixed-integer programming problems, particularly fixed-charge problems, by making use of the detailed structure of the constraints to reduce drastically the number of vectors $\mathbf{y}_1$ (and hence the number of linear programs) examined.

2.2 A Linear Fixed-Charge Problem

The linear fixed-charge problem is a special case of Problem I in which each component of y can take on only the values 0 or 1 and the value 1 is assumed only if some combination of components of x is non-zero. If a fixed charge, f_i, is incurred for each activity engaged in, the problem can be stated in the form:*

Minimize

$$\mathbf{cx} + \mathbf{fy}$$

subject to

$$\mathbf{Ax} \geqslant \mathbf{b}$$

$$\mathbf{x} \leqslant \mathbf{m}$$

$$y_i = \begin{cases} 0 \text{ if } x_i = 0 \\ 1 \text{ if } x_i > 0 \end{cases}$$

$$\mathbf{x} \geqslant 0$$

so that the activities must satisfy certain constraints and cannot exceed upper bounds. A typical problem of this sort is the warehouse location problem with stockpile costs. The vector **x** corresponds to the supplies to be located at potential warehouses; **b** to the demands that must be met, and **m** to the maximal capacity of the warehouses. The objective is to select a set of warehouses that satisfy the demands, stay within the size constraints on individual sites, and minimize total cost. The total cost has two components: the variable-charge component, resulting from the allocation of stockpile to the warehouses; and the fixed-charge component resulting from the costs associated with opening up each warehouse selected.

For present purposes, this fixed-charge problem can be written in the form:†

$$\left.\begin{array}{lr} \text{Minimize} & \mathbf{fy} + \mathbf{cx} \\ \text{subject to} & \mathbf{Ax} \geqslant \mathbf{b} \\ & \mathbf{My} - \mathbf{x} \geqslant \mathbf{0} \\ & \mathbf{y} \leqslant 1 \\ & \mathbf{x}, \mathbf{y} \leqslant 0 \\ & \mathbf{y} \text{ integer.} \end{array}\right\} \text{Problem III}$$

---

* This formulation assumes $n_1 = n_2$. In general, n_1 can be greater than n_2, with y_i taking on a value 1 if any one of several components of x is non-zero.

† In Problem III, **M** denotes a diagonal matrix with value m_j in the jth position on the diagonal corresponding to the upper bound on x_j.

This statement of the problem presupposes that all components of f and c are strictly positive (no free goods). In this case, the constraints $\mathbf{My} - \mathbf{x} \geqslant \mathbf{0}$ and $\mathbf{y} \leqslant 1$ assure that, in the optimal solution, $\mathbf{x}$ will be bounded by m and that y_i will be 0 if x_i is 0. This results from the fact that if x_i is 0 in the optimal solution, then $y_i = 0$ satisfies the constraints and minimizes the objective function. On the other hand, if x_i is positive, it cannot exceed m_i because the maximum feasible value of y_i is 1.

If the integrality requirement on **y** were dropped, Problem III would become a conventional linear program and could be solved by the simplex method. The linear program solution* will typically call for fractional values of some or all components of y. Although this is not a feasible solution to Problem III it can be made into one by simply rounding each fractional y_i up to 1. Thus, a feasible solution to this mixed-integer problem can be found readily. Furthermore, the value of the objective function for this initial solution (call it $\mathbf{L}_0$), also provides a bound on the total fixed charge **fy**.

This bound is found by solving a second linear program. Consider the case in which all fixed charges are incurred (e.g., all warehouses are opened). For this case, Problem III reduces to the linear program:

Minimize

$$\mathbf{cx}$$

subject to

$$\mathbf{Ax} \geqslant \mathbf{b} \qquad \text{Problem IV}$$

$$\mathbf{0} \leqslant \mathbf{x} \leqslant \mathbf{m}.$$

If Problem III has a solution when considered as a linear program, then so will Problem IV, since it has the same constraint set and objective function in **x** and all the y_i components are fixed at their upper bound of 1.

The variable cost $\mathbf{cx}_0$ found in this way is the minimum variable cost, since it corresponds to the case in which all activities are available. If we now subtract $\mathbf{cx}_0$ from $\mathbf{L}_0$ we obtain at once an upper bound, **FMAX**, on the fixed cost **fy** in the optimal solution of Problem III, where

$$\mathbf{FMAX} = \mathbf{L}_0 - \mathbf{cx}_0$$

Thus, we can adjoin an additional constraint to Problem III, namely, $\mathbf{fy} \leqslant \mathbf{FMAX}$. This constraint enables us to reject at once any value of the vector **y** for which $\mathbf{fy} > \mathbf{FMAX}$. Similarly, a lower bound on the value of **fy**, in the optimal solution, call it **FMIN**, can be established. In terms of the warehouse location problem, if it is possible to supply to all demands from any site then **FMIN** is equal to the fixed cost of the cheapest warehouse and **FMIN** provides no new information. However, if this situation does not obtain, the constraint $\mathbf{fy} \geqslant \mathbf{FMIN}$ can be adjoined to Problem III. A loose **FMIN** bound is found by solving Problem III as a linear program with the variable-costs vector set to zero.

Although no method has been found for improving FMIN as calculations proceed, it is possible to improve **FMAX**. Each time a value of the objective function, $\mathbf{L}_i$, of Problem III is found which is better than any found thus far, **FMAX** can be reduced to $\mathbf{L}_i - \mathbf{cx}_0$, and the constraint $\mathbf{fy} \leqslant \mathbf{FMAX}$ tightened.

---

* If no feasible solution to the linear program exists, there is no feasible solution to Problem III.

Summarizing in terms of the algorithm presented in Part A, for the fixed-charge problem the procedure is

(1) Find a feasible solution to Problem III. This can be done by solving Problem III as a linear program and rounding all fractional values of y_i up to 1.

(2) Find values for **FMAX** and **FMIN** and adjoin the constraints $\mathbf{fy} \geq \mathbf{FMIN}, \mathbf{fy} \leq \mathbf{FMAX}$.

(3) Enumerate the finite set of vectors $\mathbf{y}_1$ that satisfy the constraints $\mathbf{0} \leq \mathbf{y} \leq 1$, $\mathbf{y}$ integer, $\mathbf{fy} \leq \mathbf{FMAX}, \mathbf{fy} \geq \mathbf{FMIN}$.

(4) For each $\mathbf{y}_1$ found in Step 3, solve the problem:

Minimize

$$\mathbf{cx},$$

subject to

$$\mathbf{Ax} \geq \mathbf{b}$$

$$\mathbf{x} \leq \mathbf{my}_1$$

$$\mathbf{x} \leq \mathbf{0}$$

If no feasible solution exists, find a new $\mathbf{y}_1$. If an optimal solution $\mathbf{x}_1$ is found, evaluate $\mathbf{c}_1\mathbf{x}_1 + \mathbf{f}_1\mathbf{y}_1$. Determine if **FMAX** can be reduced and, if it can, tighten the **FMAX** bound.

(5) The optimal solution is

$$\underset{\text{all } \mathbf{y}_1}{\text{minimum}} \{\mathbf{cx}_1 + \mathbf{fy}_1\}$$

These steps are the essence of the computational method presented in Gray (1968). Further refinements are introduced there to speed the computations, particularly in reducing the number of vectors $\mathbf{y}_1$ for which linear programs must be solved.

2.3 Non-Linear Variable Costs

The discussions thus far have been restricted to linear variable costs. Concave and convex costs which can be approximated by piecewise linear costs can also be treated. The only restriction imposed is that the objective function be separable.

The concave cost case is of particular importance because it is often encountered in warehouse location problems.* The concave cost function in Figure 1 can be replaced by three separate linear cost functions as shown in Figure 2. Note that the lower envelope of the three cost functions is the original concave cost. Replacing the variable x with three variables (x_1, x_2, and x_3) allows the concave cost function to be replaced by three linear segments, each having a different associated fixed cost (f_1, f_2, and f_3 in Figure 2). Thus, the problem has been expanded to six variables, three of them fixed-charge variables, to remove the non-linearity. Since the objective function is defined only on line segments, it is necessary to add upper and lower bound constraints on x_1, x_2 and x_3 to assure that these variables stay within their regions of definition. Furthermore, it is necessary to impose the constraint

$$\sum_{i=1}^{3} y_i \leq 1$$

so that only one of the three segments is selected in any solution.

* For a discussion of convex costs see Gray (1968).

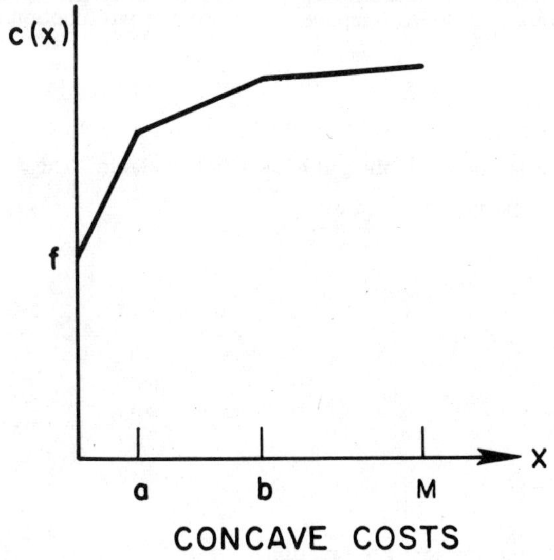

Fig. 1 Piecewise linear approximation to cost function of one variable

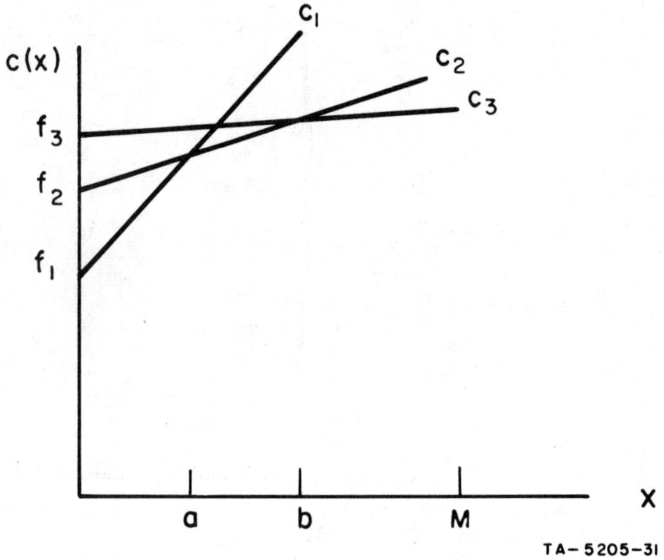

Fig. 2 Replacing a concave cost function by several linear cost functions

The foregoing discussion for a concave cost function in one variable generalizes at once to separable concave objective functions. Suppose that the objective function is

$$\sum_{j=1}^{n} [\phi_j(u_j) + f_j v_j]$$

where $\phi_j(u_j)$ is a concave function in one variable and v_j is a 0-1 variable. The terms of the objective function can be represented as

$$\phi_j(u_j) = \sum_{i=1}^{n_j} c_i{}^j x_i{}^j, \qquad m_{i-1}{}^j \leqslant x_i{}^j \leqslant m_i{}^j$$

and

$$f_j v_j = \sum_{i=1}^{n_j} f_i{}^j y_i{}^j, \qquad y_i{}^j = 0, 1$$

where

n_j = Number of segments in the jth concave function,
$m_i{}^j$ = Upper end of segment i in the jth concave function,
$m_0{}^j = 0$

Substituting into Problem III yields

Minimize

$$\sum_{j=1}^{n} \sum_{i=1}^{n_j} [c_i{}^j x_i{}^j + f_i{}^j y_i{}^j]$$

subject to

$$\left.\begin{array}{l} \mathbf{A}'\mathbf{X}' \geqslant \mathbf{B} \\ m_i{}^j y_i{}^j - x_i{}^j \geqslant 0 \quad j = 1, \ldots n; i = 1, \ldots n_j \\ -m_{i-1}{}^j y_i{}^j + x_i{}^j \geqslant 0 \quad j = 1, \ldots n; i = 2, \ldots n_j \\ \sum_{i=1}^{n_j} y_i{}^j \leqslant 1 \quad j = 1, \ldots n \\ x_i{}^j, y_i{}^j \geqslant 0 \\ y_i{}^j \text{ integer,} \end{array}\right\} \text{Problem V}$$

where $\mathbf{x}'$ is the column vector $(x_1^1, x_2^1, \ldots x_{n_j}^1, x_1^2, \ldots x_{n_j}^n)$, and $\mathbf{A}'$ is the matrix whose columns are $(A_1, \ldots A_1, A_2 \ldots A_2, \ldots A_n, \ldots A_n)$, each column A_j being repeated n_j times, and b is the same as in Problem III.

Examination of Problem V shows its structure to be that of Problem III and hence the algorithm for solving that problem can be applied here. Computationally, Problem V is much larger than Problem III because each variable has been replaced by several variables to remove the non-linearity in the objective function and because constraints have been added. It has been shown that the constraints $- m_{i-1}{}^j y_i{}^j + x_i{}^j \geqslant 0$ never enter into the solution and do not need to be carried along computationally [Gray (1968)]. This reduces the size of Problem V. Although it has been shown that the constraints $\sum_i y_i^j \leq 1$ are extraneous at optimality, it is still necessary to carry these constraints along computationally to ensure that no more than one segment is considered.

3. COMPUTATIONAL METHODS

The decomposition algorithm presented in Section II involves two major procedures:

(1) Finding an eligible vector $\mathbf{y}_1$, and

(2) Solving a linear program in $\mathbf{x}$ given $\mathbf{y}_1$.

In this section, methods for finding the vectors $\mathbf{y}_1$ in an efficient manner and for solving the resulting linear programs in $\mathbf{x}$ are presented.

An eligible vector $\mathbf{y}$ may be thought of as a 0-1 string, that is, a sequence of 0's and 1's corresponding to the values* of the components y_i. If n is the number of integer variables, there are of course, 2^n such 0-1 strings and it rapidly becomes infeasible to examine each of them. At this point, we take recourse to the structure of the problem. As was pointed out in Section II, for fixed-charge problems with positive cost coefficients, simplex solutions of linear programs can be used to find both the continuous optimum and an initial feasible solution in integers. Furthermore, it is possible to find an upper and a lower bound (**FMAX, FMIN**) on the fixed cost **fy**. These bounds serve to reduce the number of 0-1 strings that need to be examined. For example, for a typical nine-integer variable problem imposition of these bounds reduced the number of allowable 0-1 strings from 512 to 85. Thus, for small problems (say, ten to twelve integer variables) it may be possible to enumerate all 0-1 strings between the **FMAX** and **FMIN** limits, particularly if the **FMAX** bound is tightened as better solutions are found.

Mere enumeration, however, ignores a large part of the information available from the structure of the problem. Integer-programming algorithms which successively bound the values of integer variables given the values of the previous variables are particularly attractive for making use of the problem structure. These algorithms, generically known as branch-and-bound algorithms, were developed for all-integer programming problems. One such algorithm, the bound-and scan algorithm, developed by F. S. Hillier (1966), was adapted for the present work.

The basic idea, then, is to use the Hillier algorithm to generate 0-1 strings which, it turns out, satisfy the constraints that are binding (constraints satisfied with equality) if the problem is solved as a linear program and which, furthermore, yield values of the fixed cost that fall within the **FMAX** and **FMIN** bounds. Since upper bounds are known for the components of the $\mathbf{x}$ vector, a simple algebraic test for feasibility of the sub-problem (Problem II of Section II) can be performed if all elements of the **A** matrix are non-negative. This test, which is necessary (but not sufficient in all cases) involves setting those components of $\mathbf{x}$ that correspond to $y_i = 0$ to zero and setting all the other components of $\mathbf{x}$ to their upper bounds. If any of the constraints are not satisfied, the 0-1 string does not lead to a feasible solution. If the algebraic test is passed, a linear program is solved which involves only the components of $\mathbf{x}$ that correspond to $y_i = 1$. This linear program leads either to an infeasible solution or to an optimal solution of the subproblem in $\mathbf{x}$. This optimal solution, if it exists, is then substituted back into the objective function of the original problem together with $\mathbf{y}_1$ to determine whether an improved solution has been found and whether tighter bounds can be imposed on **FMAX.**

The computational details of the algorithm for linear and separable non-linear costs are discussed at length in Gray (1968).

---

* Physically, 0 represents a closed warehouse and 1 an open warehouse.

4. APPLICATION

The algorithms described here have been coded for the Burroughs 5500 computer. The linear objective function algorithm has been applied to the problem of site selection for missile defense. Here a large number of sites, many more than required, are available. A maximum interceptor capacity and a fixed cost are associated with each site. Each site provides a known coverage of part of the area to be defended and overlaps the coverage of other sites to a certain extent. The stockpile requirements depend on defense level and combination of sites chosen. The problem is to select that combination of sites and stockpiles which will minimize the total cost and provide a uniform level of defense across the entire area to be defended.

A second military application [Morey (1968)] has been to the problem of attacking a country with ballistic missiles so as to minimize the total attack size and yet destroy specified fractions of key resources. In this case, the fixed charges arise from the price that the offense has to pay to negate the defense of particular sites or combination of sites. Because of overlapping coverages, several area defenses (i.e., several fixed charges), may have to be negated before a particular group of cities may be attacked. Furthermore, the payoff once the fixed charges are paid is a concave, non-decreasing function of the attack size (first attacker gets more than the second). The problem, then, is to select the cities to be attacked and the attack level at each city.

ACKNOWLEDGMENTS

This work was supported under Naval Research Contracts Nonr-225(53) and Nonr-225(89), National Science Foundation Grant GP-7074, and U.S. U.S. Army Contract DA-49-092-ARO-10.

REFERENCES

1. BENDERS, J. F. (1962) 'Partitioning Procedures for Solving Mixed-Variables Programming Problems,' Numerische Mathematik, Vol. 4, No. 3, pp. 238-252.
2. EFROYMSON, M. A. and T. L. RAY (1966) 'A Branch-Bound Algorithm for Plant Location,' Operations Research, Vol. 14, No. 3, pp. 361-368.
3. GRAY, P. (1968) 'Mixed Integer Programming Algorithms for Site Selection and Other Fixed Charge Problems Having Capacity Constraints,' Stanford University, Stanford, California, Ph.D. Dissertation.
4. HILLIER, F. S. (1966) 'An Optimal Bound-and-Scan Algorithm for Integer Linear Programming,' Department of Industrial Engineering, Technical Report No. 3, Stanford University, Stanford, California.
5. HIRSCH, W. M. and G. B. DANTZIG (1954) 'The Fixed Charge Problem,' RAND Corporation Memo P-648, The RAND Corporation, Santa Monica, California.
6. MOREY, R. C. (1968) 'Imbalancing Attacks Against Defensive Deployments, A Multi-Valued Analysis,' Technical Note TN-OAP-31, Stanford Research Institute Menlo Park, California.

SECTION 7 Integer Programming Applications

W. J. Sullivan, E. Koenigsberg
C. J. Beattie

Mixed Integer Programming Applied to Ship Allocation

WILLIAM J. SULLIVAN, ERNEST KOENIGSBERG
Matson Research Corporation, U.S.A.

The research project reported here arose as a result of a preliminary internal study by Matson Research Corp. of the application to maritime situations of new developments in linear programming. The success of the preliminary study of a specific trade encouraged us to develop linear programming applications for freighter fleet planning and medium-range scheduling. Fleet planning and scheduling require complicated allocations of cargoes to ships and ships to routes in order to optimize some measure of performance for a given forecast period.

1. BACKGROUND

The shipping company operates a liner service between one 'effective'* western port and three effective eastern ports. The operating fleet at present consists of eleven ships, of which four are conventional general-cargo and seven special-purpose. There are four types of special-purpose ship:

All container	1 ship
Container and bulk	2 ships
Container, bulk, and auto (garage)	2 ships
Container and auto	2 ships

There is a degree of interchangeability in the special-purpose ships; some space can be used for either bulk cargoes or containers and autos can be shipped in compatible 'frames' or in containers as well as in special auto space.

The company carries most of the trade in the area and, as a matter of policy, accepts all cargoes offered even if they must be carried at a loss. It has established a set of minimum frequencies of service as constraints at each port to ensure a high level of service to its customers. The trade is not subsidized and operates with American crews and American ships. A portion of the cargo carrying—basically bulk sugar—is under contract.

In this trade area there are seven possible round-trip routes: three two-port shuttles, three triangles connecting two of the eastern ports with western ports, and one route connecting all four ports. Both the longest three-port triangle and the four-port route have been discarded as highly uneconomic. Two other routes have been added because carrying sugar eastbound results in both additional steaming time and additional port time. These additional routes with sugar are a shuttle between the sugar ports and a triangle combining the sugar ports and one other eastern port. The resulting seven routes are the ones actually operated.

For each ship type, we indicate cargo capacities for the four main space types:

- Container space (empty and full)
- Auto slots (garage)
- Bulk liquid (molasses and fuel oil)
- All other space (bulk dry and conventional)

---

* Each port call includes a number of berths. We treat the port area as a single effective port.

These capacities are further restricted. On all eastbound voyages with sugar, weight restrictions limit the number of full containers that can be carried. There is also a trade-off between bulk dry and conventional cargo when both are carried in 'other cargo' space of conventional vessels. When autos are carried in this space, or in containers, loading factors are used to convert the number of autos into the number of measurement and weight tons of space required.

Because of the seasonality of some cargoes, port-to-port cargo forecasts must be made for each quarter of the year on the basis of agricultural reports, customer surveys, economic indicators, and other information. Forecasts are made for nine cargo classes for each applicable port pair:

Westbound containers

Eastbound containers

Westbound conventional cargo

Autos

Sugar

Molasses

Fertilizer

Fuel oil

Eastbound empty containers

Since not all cargoes move between all ports, there are only 26 port-pair cargo forecasts. As might be expected, there is considerable imbalance in segments of the trade. Container cargoes westbound exceed those eastbound, and empty containers must be returned; bulk sugar and molasses move only eastbound; and so on. Movements of empty containers are not forecast; they are directed to return to ports to maintain a suitable balance. The space available for the return of empty containers is not an extreme constraint; empty containers can be stacked on deck higher than full containers without making the ship unseaworthy.

We can readily determine the lay-up costs, steaming and port times, and operating costs (crew, fuel, etc.) for each ship type on each route, as well as net-to-vessel revenues for every cargo class assigned to a given type of space. The objective function used is the difference between net-to-vessel revenues and the sum of operating costs and lay-up costs. This function, though related to profit, is not the total profit generated. All costs which do not depend on the class of cargo, type of space used, and variable operating costs are omitted from our computations.

Net-to-vessel revenue for a cargo class is the difference between the average tariff and the handling costs (including wharfage and claims) associated with that class. The net-to-vessel revenues for the nine cargo classes are based on the space types in which the cargo usually moves. Autos, for example, can be carried in specialized garage space, in containers, or in auto 'frames'; for each space type there is a different revenue. Similarly, sugar revenues depend on the ship used to carry the sugar. There is a different tariff mix of container cargo, and thus a different net-to-vessel for containers, from each port area.

The linear programming model developed to assign cargoes to ships and ships to routes is intended as a planning model rather than an operating model. The model does not produce a sequence of voyage assignments for each ship; it does produce the number of voyages of each type to be made by each ship in a given time period. This is not a serious limitation; for the client's short voyages it is not necessary to determine exact schedules for several quarters in advance.* The LP results are

---

* The maximum round trip voyage time is 30 days, the average is less than 20 days.

meant to indicate basic strategy and to provide inputs for a computerized vessel scheduling model used by the company. *

The number of voyages in a quarter (for each ship type) has been restricted to integer values because the fleet is small and there are only a few ships of each type. Solutions without this restriction will yield values of say, 0.30 voyage of one type, 0.15 voyage of another, and 4.32 voyages of a third. Rounding off to the nearest integer can produce infeasibilities or longer lay-up periods than necessary. It is known that simple round-off does not necessarily yield optimal solutions of the integer variables. We cannot avoid integer variables by extending the time period because the demands vary by quarter.

We have formulated two related models, the first for a single quarter and the second for annual runs of four quarters. The link between quarters is the total number of ship-days for a ship type. To avoid solutions which make non-optimal selections to use up the ship-days, we can vary the number of ship-days in each quarter over a limited range; but the total number of ship-days in a year is fixed. The variation of ship-days in a quarter can also be applied to the single-quarter model to test its sensitivity to the amount of cargo available.

2. FORMAL DESCRIPTION OF THE MODEL

Subscripts

In order to define the terms of LP equations, we will employ the following subscripts:

i = ship type

j = voyage type

k = space type

l = cargo class

m = port-pair combination and direction (westbound or eastbound)

Variables

The following variables are defined:

x_{ijklm} = amount of cargo of class l carried in space type k on voyage type j of ship type i between the ports of pair m during the quarter

y_{ij} = number of voyages of type j made by ship type i during the quarter (integer)

l_i = number of days of lay-up for ship type i during the quarter

Coefficients

The following factors appear in terms on the left-hand side of the equations:

r_{ijklm} = net-to-vessel for a unit of cargo of class l carried in space type k on voyage type j of ship type i between ports m

c_{ij} = operating costs for ship type i on a voyage type j

c_i = costs for one day of lay-up for ship type i

a_{ijklm} = loading factor indicating number of units of space type k on ship type i on voyage type j required by one unit of cargo class l between ports m

d_{ijkm} = amount of space type k available on ship type i on voyage type j for port pair m

b_{ij} = number of operating days required for ship type i on voyage type j

s_{jm} = number of port calls (0 or 1) made on a voyage of type j to an eastern port area, implied by port pair m

* See Olson, Sorensen and Sullivan (1967)

Right-hand sides

The following numbers appear in the right-hand sides of the equations:

t_{lm} = total amount of cargo of class l for port pair m during the quarter

e_i = total number of days available for ship type i during the quarter

g_m = minimum number of calls per quarter required at the eastern port implied by port pair m

h_m = maximum allowable difference between the total number of containers carried in one direction for port pair m and the total number carried in the opposite direction over the whole quarter

n_{ij} = minimum number of voyages of type j a ship of type i is required to make during the quarter

p_{ij} = maximum number of voyages of type j a ship of type i is allowed to make during the quarter

3. EQUATIONS FOR ONE-QUARTER FORMULATION

3.1 Profit equation

The 'criterion function' or profit equation is the difference between the revenue earned and the total direct costs. The algebraic terms representing the revenue can be written as:

$$r_{ijklm} \cdot x_{ijklm}$$

This is the net-to-vessel revenue for a unit of cargo of class l carried in space type k on voyage type j of ship type i between ports m, multiplied by the number of units of that cargo class carried in the same space type k on the same voyage type j of the same ship type i between ports m.

$$\sum_{ijklm} r_{ijklm} \cdot x_{ijklm}$$

is the sum of all the revenues earned for all the cargo carried during the quarter.

The cost terms are:

$c_{ij} \cdot y_{ij}$ = cost of operating a ship type i for the number of voyages y_{ij} of type j

$c_i \cdot l_i$ = cost of laying up a ship of type i for the number of days l_i

Therefore, the total profit is the sum of all revenues minus all costs:

$$\sum_{ijklm} r_{ijklm} \cdot x_{ijklm} - \sum_{ij} c_{ij} \cdot y_{ij} - \sum_{i} c_i \cdot l_i$$

It is this profit that is maximized by the LP program, subject to the following constraint equations.

3.2 Available cargo constraints

$$\sum_{ijk} x_{ijklm} \leq t_{lm}$$

This constraint states that the total amount of cargo class l carried in all spaces on all voyages of all ships between ports m cannot exceed the total amount of cargo of class l available to be carried between ports m in the quarter.

3.3 Ship capacity constraints

$$\sum_{lm} a_{ijklm} \cdot x_{ijklm} - d_{ijk} y_{ij} \leq 0$$

This constraint states that the sum of all the capacity requirements for all cargo carried in space type k for a voyage type j of ship type i cannot exceed the capacity of space type k for ship type i on voyage type j.

We also have provision for a selected subset of cargo types to be constrained within a certain space type k. For example, the capacity of one ship for westbound containers is 690; but we can place a lower limit, say 500 per voyage, when concerned with containers from a particular port.

For sugar space, the inequality is not allowed; there must be enough sugar to fill the sugar space, or the ship will not make a sugar voyage.

3.4 Ship time available

$$\sum_{j} b_{ij} y_{ij} + l_i = e_i$$

This constraint states that the total number of days of operation for all voyages of a ship of type i, along with the number of days of lay-up for ship type i, must equal the total number of days available as input for that ship type during the quarter.

3.5 Port call requirements

$$\sum_{ij} s_{jm} y_{ij} \geq g_m$$

This constraint states that for all ships on all voyages the total number of visits to a port area implied by port pair m must be at least as great as the input number g_m for that port area.

There are separate frequency-of-service constraints of this form for conventional ships and specialized ships.

3.6 Container inventory balance

$$\sum_{ijk} (x_{ijkCm} - x_{ijkXm} - x_{ijkCm}) \leq h_m$$

This constraint states that the total number of full containers (subscript C) shipped from an eastern terminus of port pair m minus the total number of empty containers (subscript X) returned to this port area, minus the total number of full containers shipped to the area, cannot exceed a maximum value h_m. This constraint achieves container inventory balance between ports of a pair by forcing empty containers to make up the deficit of westbound fulls over eastbound fulls.

A similar constraint applies to the western terminus to ensure a balance between the total flow of containers in and out.

3.7 Minimum number of voyages

$$y_{ij} \geq n_{ij}$$

This constraint states that a ship of type i must make at least n_{ij} voyages of type j during the quarter.

3.8 Maximum number of voyages

$$y_{ij} \leq p_{ij}$$

This constraint states that a ship of type i cannot make more than p_{ij} voyages of type j during the quarter.

4. FOUR-QUARTER FORMULATION

For the four-quarter model we have added one further subscript to the variables and coefficients described for the one-quarter model. This subscript denotes a quarter, one of the four possible for an annual run. The four-quarter model thus contains four times as many equations and right-hand sides as the one-quarter version.

In addition, we have added one new set of constraints to add up the number of ship operating days over the four quarters:

$$\sum_{q=1}^{4}\left[\sum_{j} b_{ij}y_{ij}^{q}\right] + L_i = E_i$$

where

y_{ij}^{q} = number of voyages made in quarter q

b_{ij} = number of days required for voyage type j of ship type i

L_i = total year's lay-up for ship type i, not counting scheduled lay-up

E_i = total time available for ship type i during the year, not counting scheduled lay-up

The profit equation for the year is obtained by summing the profit equations for each quarter.

5. RESULTS

We carried out our computations using the mixed integer (MILP) code in CEIR's LP 90/94 program. The code used a branch and bound algorithm in a search for the optimum. A one-quarter problem includes about 500 variables, of which 25 are integer and is subject to about 200 constraints. Computation times ranged from about 15 minutes to several hours for one quarter. We understand that improvements in the code will reduce the time somewhat.

Eight problems were examined; we will discuss all but one of them here. We made two computer runs, using historical data for a peak quarter, to establish base points for comparisons. The first was a relaxation of the integer constraints to determine the unrestricted optimal ship assignments. (Since the MILP code first computes a non-integer optimum, no additional computation time was required.) The second was a run in which the number of voyages made was fixed at the values actually planned by the client for the period. In both cases and also in all later runs, the conventional ships were restricted to the historical voyage pattern, because of the requirement that all cargo must be carried even if a loss is incurred. Without this condition, conventional ships would make fewer voyages and leave conventional cargo on the dock.

The unconstrained solution gives a profit contribution of \$9, 658, 000 on 42. 2 voyages and 104 inactive days for the specialized ships, 86 days of which represent the inactivity of the smallest specialized ship (which is assigned to 0. 2 roundtrip voyage). The solution restricted to planned operations gives a profit contribution of \$9, 522, 000 on 47 voyages and 45 inactive days for specialized ships. In the unconstrained solution a large ship could pick up more cargo than the port could generate between arrivals.

Because the least constrained problem formulation with integers did not converge in two hours of 7094 time, we were forced to examine other conditions by introducing new voyage variables a few at a time, to the set of fixed voyages. In this way we could work from a previous integer solution and reduce computer time. In the first of these runs we added four ship-voyage types and widened the bounds on voyage variables. One of the new types was a triangle voyage of the largest special ship type. Both the non-integer and integer solutions required 43 voyages to meet demands. The difference in the value of the objective function between the two solutions was $1, 000 ($9, 620, 000). There were 95 inactive days for specialized ships, of which 15 were charged to the smallest ship. The largest ship, although severely limited by input constraints on the container and auto cargo that could be loaded at one of the three eastern termini, made one triangle voyage to that restricted port anyway, because the ship could profitably pick up molasses and autos in garage space there.

In the next run three additional ship-voyage types were added. The number of voyages made was again 43; the difference in the value of the objective function between non-integer and continuous solutions was $2, 000 in $9, 628, 000. Again there were 95 inactive days for specialized ships, of which 65 were charged to the smallest ship. The largest ship-type again called at the port with restricted cargo.

In the last one-quarter run the frequency-of-service requirement was relaxed. This produced a profit of $9, 743, 000, required only 39. 2 voyages for the continuous solution, and gave 171 inactive days for specialized ships. All cargo requirements were met, and the smallest ship did not sail. The integer solution did not converge after more than ninety minutes of computer time.

Because of the excessive time required to obtain solutions for some of the single-quarter problems, the four-quarter problems were run in the continuous mode only. The most important continuous run was made to determine the time periods in which the ships should go into scheduled lay up. This run indicated that scheduled lay ups can be spread over the year without reducing profit contribution or the ability to meet demands. For a second four-quarter run, using the output of the continuous run, and our earlier experience to guess the integer values for the whole year, we forced integer results on the voyage variables and obtained a profit contribution equal to about 99. 2 per cent of the value for the previous continuous run.

REFERENCES

OLSON, C. A., SORENSEN, E. E., and SULLIVAN, W. J., 'Freighter scheduling model', Bulletin of the Operations Research Society of America, 15, B-49, 1967.

LAND, A. H. and DOIG, A. G., 'An automatic method of solving discrete programming problems' (unpublished report, London School of Economics and Political Science, (1957).

DANTZIG, G. B., Linear Programming and Extensions, Princeton University Press, (1963) pp. 514 ff.

DISCUSSION

M. G. Shaw

The C-E-I-R mixed integer program (and general branch-and-bound procedures) can be used to exhibit several good but non-optimal integer solutions. Did you consider using this facility, bearing in mind the fact that the shadow prices do not have their usual significance as indicating the effects of changes in the right hand sides in integer programming?

E. Koenigsberg

We recognize that near optimal solutions can be satisfying. We have examined a few near optimal solutions as well as the optimal solutions. In the cases where we did not attain optimal solutions in two hours of machine time, the best solution attained was too far from the non-integer solution (based on other runs) to be of interest. In the runs in which optimality was attained, the largest difference between integer and non-integer solutions was of the order of 0. 8 per cent; differences as large as, say, 1. 5 per cent in non-optimal solutions were not examined. We would like to see a program option allowing a print out of all feasible solutions within E of the optimal non-integer solution and a cutoff within an E'.

We have not used the printout shadow prices for detailed analysis; but we have examined them in some problems as rough indicators of tradeoffs.

Allocating Resources to Research in Practice

C. J. BEATTIE
B.I.S.R.A., U.K.

1. THE PROBLEM

The research evaluation and selection system described in this paper is the system used in the Operational Research Department at BISRA to evaluate and select its research programme. The problem that is solved by the system is that of allocating a fixed quantity of men and capital to competing research projects.

2. THE NEED FOR A FORMAL ALLOCATION SYSTEM

The problem facing most research managers can be stated simply as this:- given a number of projects, find those to be selected, and find the quantity of resources which should be allocated to each selected project. The resources referred to are usually men, space, equipment and money. It is most important that the allocation is such that a high return is obtained from the investment in the resources.

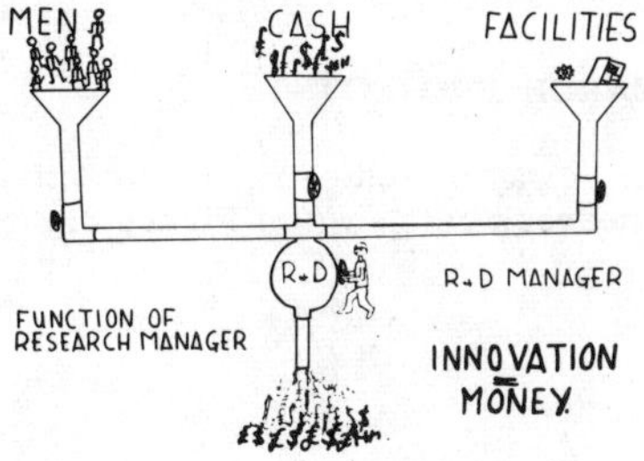

Fig. 1

It was realised in the Operational Research Department at BISRA that this choice of projects and the subsequent allocation of resources to them was being made on very little information. An intuitive assessment of projects is all very well when few projects are competing for the resources on hand. However with more and more competing projects it was thought that some work should be started in order to find the 'best' strategy for management to obtain a high pay-off from research; the immediate problem tackled was that of developing methods for selecting a beneficial research programme which was optimal in some sense.

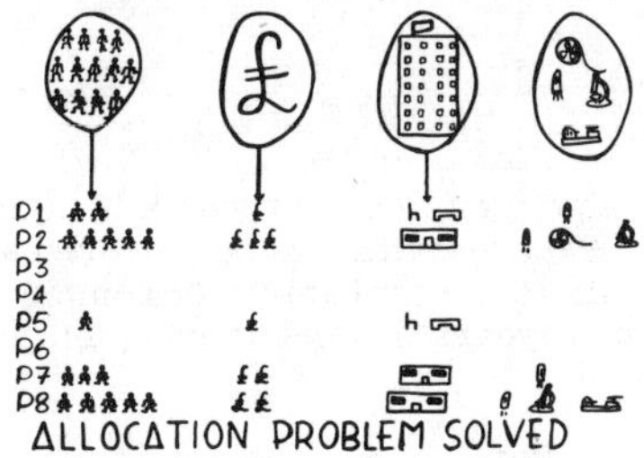

Fig. 2

3. HOW PROJECTS IDEAS ARE INITIATED

There are in general two ways in which an industrial project idea is initiated, namely the industry realises it has or will have a problem, and asks Research to find an answer, or else projects are thought up to solve hypothetical problems, which may be real or not.

Ideally, most research should closely relate to the objectives and plans of the industry it serves. In this way goal-orientated research is started which has practicable applicable results. These results are usually highly beneficial to the industry, and little effort is needed to persuade industrial management of their value, as management helped to formulate the research in the first place.

However most research associations are not in the happy position of being closely integrated into the industries that they serve, and projects are usually merely suggested by industry, or are the ideas of research staff.

In the O.R. Department at BISRA it was found necessary to have a formal system for recording project ideas. For this purpose a form was devised on which would be recorded information such as the aims of the proposed project, the problems to be solved, the amount of work envisaged and so on.

In order to ensure a steady flow of ideas incentives are offered for creativity and small committees consider how research could help industry. The project ideas so produced are vetted by Department Management, who act as an initial coarse sieve. The better project ideas then go forward to the formal evaluation system.

4. THE EVALUATION OF THE COSTS OF RESEARCH PROJECTS

The project to be evaluated will have an objective. A way of achieving the objective can be proposed and planned out in the form of a network as shown in Figure 3.

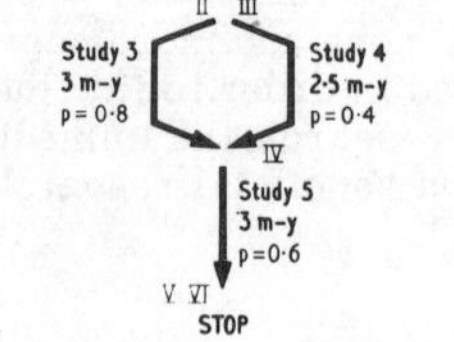

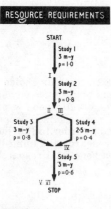

Fig. 3

Only the most important activities, or studies, are shown on this plan, which is called the project strategy diagram. The arrows represent the activities to be performed. Other important information contained in the diagram is the probability of commencing each activity given that it is reached and the man-years of work to complete it. The work content is estimated assuming a team-size of three.

The strategy diagrams are exploded to show each possible outcome of the project and the probability of reaching that outcome. The time taken to reach each outcome is obtained by scheduling on a bar chart, again assuming a team size of three. For other team sizes the time required for completion is calculated using a model which allows for the relative inefficiency of larger teams. As teams grow larger the time required for inter-communication increases, so that while the time to completion decreases, it does not decrease proportionally.

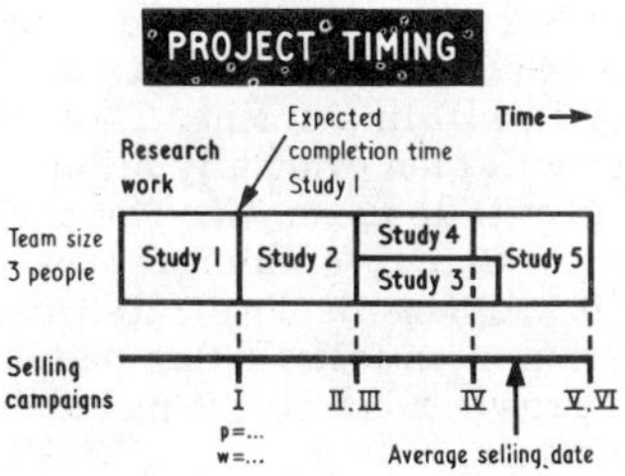

Fig. 4

5. THE EVALUATION OF THE BENEFITS OF RESEARCH PROJECTS

In order to evaluate projects, some criterion for assessment is necessary. As our criterion for success we chose to maximise the benefits accruing to the industry from our research.

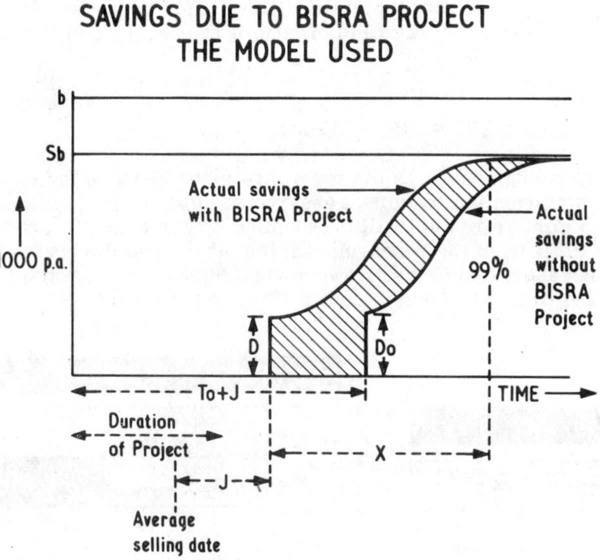

Fig. 5

The model used to describe the mechanism by which benefits are obtained by industry research is shown in Fig. 5. Since the project has some aim, there will be costs to be saved, or a profit that can be made. From a consideration of the sum at risk and the total number of units involved, an expected value of the benefits (b) for the most successful outcome can be calculated under the assumption that the innovation would be immediately and universally adopted. These benefits are represented by the line having value b in Figure 1. For the less successful outcomes of the project a fraction of the savings will be obtained, and this is shown by the line marked Sb. The project will have an expected duration depending on the team size, but the results will not necessarily all come at end of the project. According to the relative importance of the results an average selling date is estimated as the time when on average the results will become available. For example, if results of equal importance are forthcoming throughout the life of a project, the average selling date will be half-way through the project.

When the results are available to industry there will still be an implementation lag (J) before industry can fully use them. This lag is due to factors such as the time taken for re-organisation, or installing new equipment. Only now can benefits be obtained from the project, when a fraction of the industry D, will implement. This step function is shown in Figure 2. The rest of the industry will then gradually follow suit over a period of time, and an S shaped pattern of growth is assumed. The curve adopted is called the logistic growth curve, and is the growth curve which represents the spread of a disease through a closed community. It has been demonstrated, particularly by Mansfield (1961) that the spread of innovation often follows this pattern. In order to fully specify the logistic curve, the rate of growth must be estimated as the time until 99% of the industry has adopted the innovation.

If BISRA does not do the research, then sooner or later someone else will. For our evaluation purposes we try to compare what would happen if BISRA did the project with what would have happened if not. Hence the time (T) when implementation would be possible as a result of someone else's research and the fraction of the industry to implement immediately (Do) are estimated. A logistic growth curve is assumed as before.

The benefits from the research are represented by the shaded area between the two curves in Figure 5, and are discounted to the present time. This procedure is used to evaluate each possible project for every field of benefit at every possible team size.

6. THE SELECTION OF A RESEARCH PROGRAMME

Projects tend to show increasing benefit as the team size rises because the model used for evaluation places a premium on quick research results. If the results are not available quickly, then similar results will be supplied by someone else, and BISRA can claim little benefit from the research. On the other hand the costs of the project also rise with increasing team size. However as long as the benefits of the project rise faster than the costs, the benefit/cost ratio also increases.

PROJECT SUMMARY TABLE

TEAM SIZE	EXPECTED DURATION (YEARS)	TOTAL COSTS (£000)	DISCOUNTED BENEFITS (£000)	BENEFIT/COST RATIO
1	5.6	12.0	190	16
2	3.4	14.1	570	41
3	2.5	15.7	850	54
4	2.0	16.4	1000	64
5	1.7	17.0	1200	69

Fig. 6

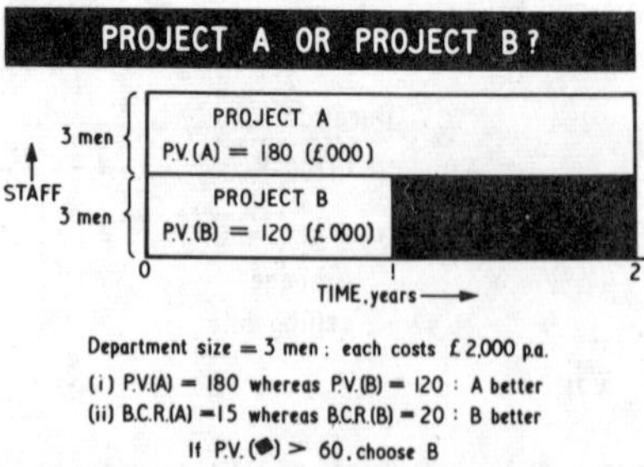

Fig. 7

This does not necessarily mean that projects should be picked with that team size which gives a maximum benefit/cost ratio, as this procedure would not take account of other opportunities. Furthermore, if the projects were selected on the basis of their present discounted values alone, this would bias the selection in favour of the longer projects. For example, consider two projects A and B, each with identical annual costs, but Project A lasting two years with benefits of 180 units, and project B lasting 1 year with benefits of 120 units. If the projects were selected by maximising the benefit/cost ratio, then project B would be selected, but if selected by maximising present value, then project A would be selected. The answer to the dilemma is to examine what the resources would be used for after 1 year, if project B had been selected and had finished. If the present value of the research that would be done is more than 60 units, then project B should be selected.

For this reason a 'planning horizon' is picked, which is at least as long as the longest project. Then an average value of research is assumed for men released from the present selection. This value rises from that of the marginally rejected project in the present selection to the average value of benefits of the present selection at the planning horizon. This hypothetical research is called 'fill-in' research. The benefits of a project are now taken to be the present discounted benefits of the project plus the present discounted value of the fill-in research from the end of the project up to the planning horizon.

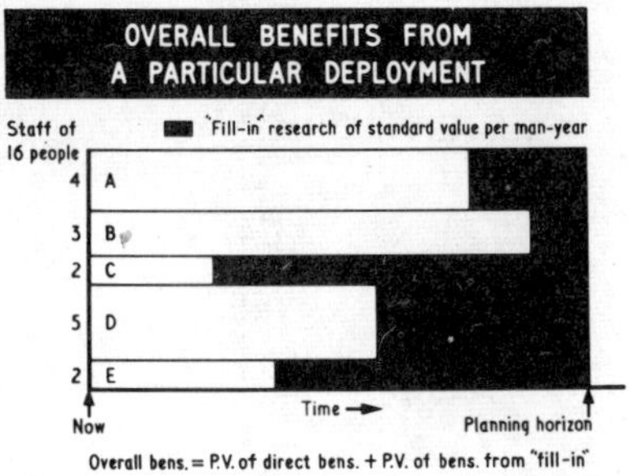

Fig. 8

Most research managers have a good idea of the level of resources that will be made available to them, and this applies to the manager of BISRA's O.R. Department. Accordingly, the selection procedure is to assume several likely levels of resources, and to allocate these resources to the competing projects in such a way as to maximise the expected benefits from the selection. Hence the decision-makers can see what the effect is of varying the budget for the Department.

This selection procedure has been formulated as a mathematical program. Since the benefits are not linear with team size, and furthermore the team sizes are integers, a linear programming approach is inappropriate, and an integer program is used for the selection. The integer program is formulated as follows:

Given

(1) large O.R. Department with certain number of staff;

(2) a number of different projects;

(3) the expected benefits evaluated for each project at each feasible manning level of the project;

(4) the expected capital cost of each project at each team size;

(5) constraints such as:-

(i) some projects must be selected;
(ii) if one project is performed, so must be another;
(iii) limitations on the capital available;
(iv) limitations on personnel of various specialities;
(v) minimum and maximum team sizes allowable on projects;

then the integer programming problem is that of determining which projects must be selected at which manning levels, subject to all the constraints, so as to maximise the benefits from the research programme.

PROJECT		A				B		C					D	
TEAM SIZE		1	2	3	4	2	3	1	2	3	4	5	1	--
INTEGER VARIABLE		X_1	X_2	X_3	X_4	X_5	X_6	X_7	X_8	X_9	X_{10}	X_{11}	X_{12}	--
Constraint (1)	1 ⩾	1	1	1	1	0	0	0	0	0	0	0	0	--
Constraint (2)	1 ⩾	0	0	0	0	1	1	0	0	0	0	0	0	--
Constraint (3)	1 ⩾	0	0	0	0	0	0	1	1	1	1	1	0	--
Constraint (4)	1 ⩾	0	0	0	0	0	0	0	0	0	0	0	1	--
	⋮	⋮	⋮	⋮	⋮	⋮	⋮	⋮	⋮	⋮	⋮	⋮	⋮	
Constraint (5)	22 ⩾	1	2	3	4	2	3	1	2	3	4	5	1	--
Constraint (6)	5 ⩾	1	1	1	2	0	0	1	1	2	2	2	0	--
Constraint (7)	1 ⩽	1	1	1	1	1	1	0	0	0	0	0	0	--
Constraint (8)	C ⩾	C_{A1}	C_{A2}	C_{A3}	C_{A4}	C_{B2}	C_{B3}	C_{C1}	C_{C2}	C_{C3}	C_{C4}	C_{C5}	C_{D1}	--
MAXIMIZE:		B_{A1}	B_{A2}	B_{A3}	B_{A4}	B_{B2}	B_{B3}	B_{C1}	B_{C2}	B_{C3}	B_{C4}	B_{C5}	B_{D1}	--

Fig. 9 Project selection by integer program

The actual matrix formulation is as shown in Figure 9. The projects are shown at the top of the figure marked A, B, C and D with their possible team sizes. For example project B cannot be considered with a team size of one, nor with any team size greater than three. The integer variables denoted by X_n can take the values of 0 or 1 only. The objective function at the bottom is the benefits, $B_{\alpha N}$, of each project at every manning level. The constraints shown are as follows:-

Constraints (1)(2) & (4)	—	Each project can have only one team size.
Constraint (3)	—	Project C must be selected with only one team size.
Constraint (5)	—	There are only 22 members of staff available for the projects under consideration.
Constraint (6)	—	There are only five members of staff with some skill (e.g. ergonomics) available for the selection.
Constraint (7)	—	At least one or both of projects A and B must be selected.
Constraint (8)	—	A possible capital cost constraint.

The solution of the integer program is by a branch and bound method. However if there are no effective capital cost constraints, as is usually the case in operational research, then the solution can be obtained using a dynamic programming method known as the 'Knapsack Algorithm'. This has been programmed in Fortran on an I.C.T. 1905 computer, and is very quick and easy to use.

7. THE UNCERTAINTY ANALYSIS

The input data to the evaluation model is estimated on the basis of present knowledge. The estimates are often little more than informed guesswork, and a common criticism of project evaluation systems is that the answers obtained are only as good as the guesses on which the answers are based. Hence, we are told, the results are valueless.

To test this hypothesis an analysis was made to find two things. Firstly, to find whether the evaluations based on point estimates gave the expected value of the benefits in view of the uncertainty in the actual values of the parameters being estimated. Secondly to find whether or not the fluctuation over time of the actual estimates made was sufficiently large to cause serious differences in the resulting selections, and hence invalidate any selections made.

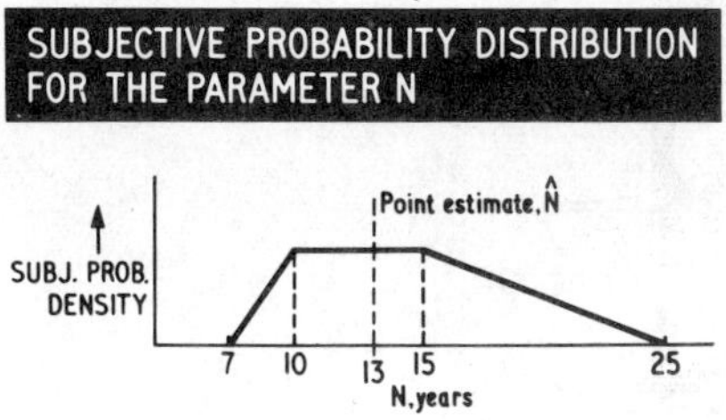

Fig. 10

The first analysis was made by making a subjective probability distribution estimate for every input parameter and finding the actual distribution of benefits that could occur for each project. The expected value of these distributions gave the same selection as the selection based on point estimates, so it is meaningful to use point estimates. However, it was found that most estimators preferred to show the extent of their uncertainty by estimating using subjective probability distributions, so the system was reprogrammed to give this facility.

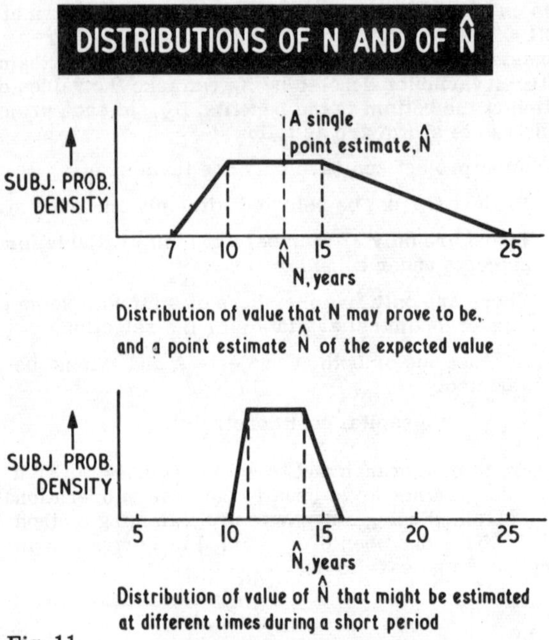
DISTRIBUTIONS OF N AND OF N̂
A single point estimate, N̂
SUBJ. PROB. DENSITY
7 10 N̂ 15 25
N, years
Distribution of value that N may prove to be, and a point estimate N̂ of the expected value
SUBJ. PROB. DENSITY
5 10 15 20 25
N̂, years
Distribution of value of N̂ that might be estimated at different times during a short period

Fig. 11

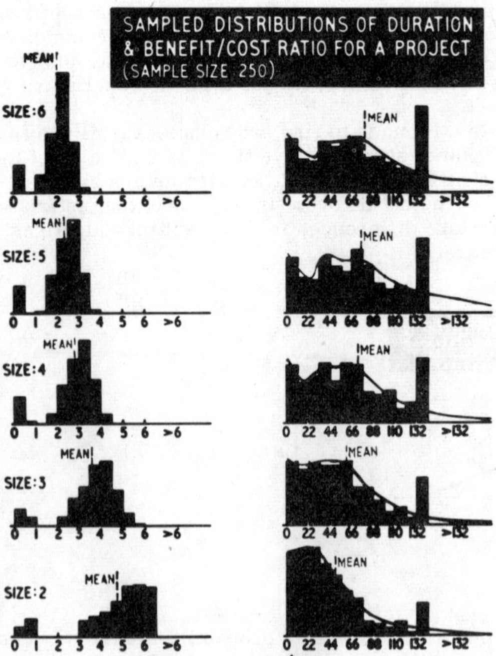
SAMPLED DISTRIBUTIONS OF DURATION & BENEFIT/COST RATIO FOR A PROJECT (SAMPLE SIZE 250)
TEAM SIZE: 6
TEAM SIZE: 5
TEAM SIZE: 4
TEAM SIZE: 3
TEAM SIZE: 2
TEAM SIZE: 1
MEAN
0 1 2 3 4 5 6 >6
0 22 44 66 88 110 132 >132
DURATION, years
BENEFIT/COST RATIO

Fig. 12

The second analysis was made by estimating subjective probability distributions representing the estimator's uncertainty in what the actual estimates should be. Selections were made by taking samples from the parameter distributions. When evaluated, using the expected values of the benefits for each project, it was found that 99% of the sampled selections had values within 10% of the optimum. Hence it was concluded that the differences between projects were greater than the uncertainty in the evaluated benefits, and that it was reasonable to make selections based on subjective estimates.

FREQUENCIES OF SELECTION OF DIFFERENT TEAM SIZES, AS PERCENTAGES

(Sample size: 250 selections: No. of staff available: 22)

Project \ Team Size	0	1	2	3	4	5	6	'Optimal' Selection Team Size
A	1	0	0	99	*	*	*	3
B	8	0	0	2	24	66	*	5
C	15	0	0	4	24	57	*	5
D	43	0	57	*	*	*	*	2
E	48	0	0	4	23	22	3	0
F	85	2	13	*	*	*	*	0
G	100	0	0	0	0	*	*	0
H	100	0	0	0	0	*	*	0
I	100	0	0	0	*	*	*	0
J	100	0	0	*	*	*	*	0
K	*	*	100	0	0	*	*	2
L	*	*	100	0	*	*	*	2
M	*	*	*	100	*	*	*	3

* inadmissible team size

Fig. 13

8. USING THE SYSTEM

The annual evaluation and selection of the research portfolio proceeds in several stages. Firstly, an estimator is appointed for each candidate project. He completes a form describing the aims, objectives, benefits and strategy of his project. These are then discussed and amendments agreed with his formal superior. In the second stage a form is completed on which all the detailed estimation of parameters is made. At the same time critics are appointed who are asked to make independently one or two key estimates. All the estimates can be made in the form of subjective probability distributions, and the form has facilities for specifying the range and type of each distribution.

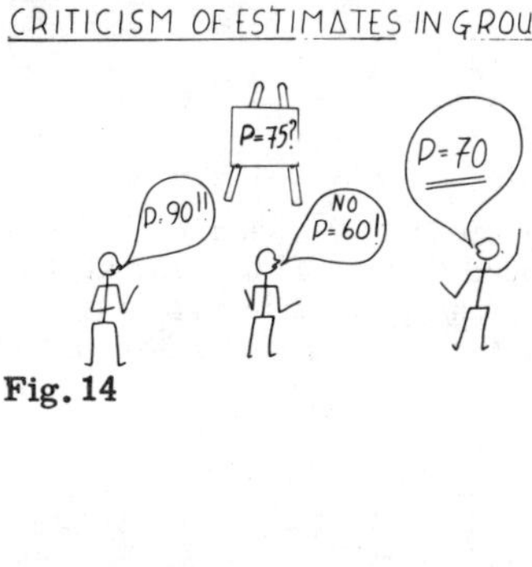

Fig. 14

Then in the third stage, projects are presented to the critics in groups of three or four by their evaluators. The estimates are discussed, compared, and probably altered, and the whole basis of each project is thoroughly reviewed. This process ensures that projects have viable aims, and that the estimates are comparable project to project.

Estimates have sometimes been very difficult to make, and the system has been reprogrammed to make the estimation easier and less tedious. Redundant estimation has been cut, the estimates being asked for have been made as unambiguous and meaningful as possible, and the use of distributions has allowed the estimators to express their uncertainties.

The estimates produced are used as input data to the suite of evaluation and selection programs. The first program takes the requisite number of samples from the input distributions, and these samples are transferred by magnetic tape to the second program which produces the evaluations. The third program in the sequence produces histograms of benefits and costs for each project at each manning level, and finds the means and standard deviations. The mean benefits and costs are used as input data to the integer program, which produces the allocations for a variety of resource levels. The selections are examined by management and with some modification form the basis of the research programme.

9. PRESENTATION OF RESULTS

The results of the selection are analysed by a series of programs according to the requirements of the selections. In general a variety of selections can be obtained which are either optimal or close to the optimum. The benefit figures are calculated for each selection. In addition the marginal value of men on the projects is presented. This gives Department Management information on all the alternative courses of action and hence great flexibility in choosing its research programme.

10. BENEFITS AND COSTS OF USING THIS SYSTEM

The benefits of using the system are intuitively high. Initially projects which had been thought to be well worthwhile were found to be uncompetitive, and hence dropped. However the principal benefit seems to be a better orientation of work. The aim of projects used to be stated in terms of achieving academic objectives, but now they are invariably stated in terms of reaching practical objectives. Another result of using the system has been that projects are now manned by much larger teams, and reach completion more quickly. This means that research results have now rarely been bypassed by events when they are produced.

The costs of the operation have not been inconsiderable, and amount to some 2-3% of the costs affected. This however is a very small figure when compared with the enormous benefits to be obtained by making research more useful.

11. FUTURE DEVELOPMENTS

A formal system for selecting the research programme has now been used at BISRA for the fourth consecutive year. Management has found it to be so useful that it is expected to be operated in some form indefinitely. Any change that occurs is likely to be one of either simplification or the provision of different information because of changing circumstances.

PROJECT SELECTION FOR 1967

	Team size constraints			Optimal Team Sizes							
Project Number	Absolute Max.	Policy Min.	Policy Max.	Unconstrained and 22 men	Constrained, and with department size:- 21	22	23	24	25	26	27
1	7	0	5	6	4	4	4	4	4	4	4
7	5	0	5	0	0	0	0	0	4	4	4
8	6	0	5	6	5	5	5	5	5	5	5
10	4	0	4	2	2	3	2	3	2	3	2
12	6	0	3	6	3	3	3	3	3	3	3
13	3	2	3	0	2	2	2	2	2	2	2
14	3	3	3	0	3	3	3	3	3	3	3
15	2	0	2	0	0	0	0	0	0	0	0
16	3	0	3	0	0	0	0	0	0	0	0
18	2	0	2	0	0	0	0	0	0	0	0
19	4	0	4	0	0	0	0	0	0	0	0
21	4	2	4	0	2	2	2	2	2	2	2
22	2	0	2	2	0	0	2	2	0	0	2
Total benefits (£00, 000)				121	97	101	105	110	114	118	123

Fig. 15

ACKNOWLEDGEMENTS

The author would like to acknowledge the many others concerned with the project, but notably C. K. C. Metz and R. D. Reader.

REFERENCES

1. BEATTIE, C. J., and READER, R. D., Research Management: Some Techniques for the Initiation, Evaluation, Selection and Control of Research Projects. Methuen, London (1969).
2. COLLCUTT, R. H. and READER, R. D., Applying O.R. to the Management of the O.R. Department at BISRA—BISRA Open Report OR/14/66.
3. MANSFIELD, E., Technical Change and the Rate of Innovation, Econometrica, Vol. 29, No. 4 (October 1961).
4. READER, R. D., JAMES, M. F., and GOODSMAN, R. W., The Evaluation and Selection of Research Projects—A Progress Report, BISRA Open Report OR/43/67.

DISCUSSION

P. J. Bannister

I have two questions:-

1. Do you have any system for auditing the benefits that accrue from your research once the results are put into practice?
2. What policy do you have to cater for the problem of allocating men, money, and facilities to unexpected research arising after a year's research programme has been planned and is being implemented?

C. J. Beattie

In answer to the first question, we have no formal system, and it is difficult to see what the information could be used for if it were collected. Because of the long time scale of obtaining benefits and the numerous interactions with other factors it would be very difficult ever to assess the absolute benefits of a completed project. However many of the parameters of our benefit estimation model are subject to post-project checks and have proved to be sufficiently accurate. Furthermore the analysis of uncertainty has shown that our errors of estimation do not invalidate the procedure.

Coming on to the second question, we have about 10% of slack in our allocation. The extra men work on project feasibility studies and ad hoc research. Should the department become completely saturated with pressing new work, then we would consider terminating one of the less beneficial ongoing projects, subject, of course, to our outside commitments.

SECTION 8 Geometrical Programming

M. Avriel

Fundamentals of Geometric Programming

MORDECAI AVRIEL
Mobil Research and Development Corporation, U.S.A.

SUMMARY

Geometric programming is a recently developed mathematical optimization theory for dealing with functions which typically arise in engineering design. The theory provides: (i) algorithms for finding optimal solutions (i.e., most profitable designs); (ii) existence theorems (Is the design possible, subject to the restrictions laid down?); and (iii) characterization theorems (Which are the important parameters and data, and what are the relations among the optimal design parameters at optimum?).

The main theorem of geometric programming is stated, along with indications of its practical applications. Finally, a simple numerical example of how one may analyze an engineering design problem by geometric programming is presented.

This paper is expository. Its purpose is to serve as a point of departure for possible applications and further theoretical work.

1. INTRODUCTION

Mathematical programming is a discipline which treats the problem of maximizing or minimizing a function of one or more variables that are related through specified constraints. The best known and most widely used branches of mathematical programming are linear programming and its nonlinear extensions to convex programming.

In applying mathematical programming to the operation of, say, a refinery, a mathematical model is first set up in which (i) the operating variables are identified, (ii) relationships between the variables and the output of the refinery are established, and (iii) the costs of manufacturing are expressed in terms of the operating variables. The mathematical programming problem is to determine those values of the operating variables for which the required output is manufactured at minimum cost. In other words, mathematical programming chooses the best of all operating possibilities according to the representation by the model. In addition, mathematical programming deals with such questions as under what conditions an optimal solution exists, how it can be characterized, and how sensitive it is to changes in the parameters (data) of the model. Clearly, such a method is of general and widespread interest. But how far does the power of mathematical programming extend in solving optimization problems? In the present state of the art there are quite a few limiting factors, both theoretical, such as mathematical properties of the functions involved, and practical, such as being able to construct a realistic model and being able to carry out sometimes formidable computations. At present, mathematical programming is most successful in cases where the functions appearing in the problem are all linear (linear programming) or have a certain smooth curvature (convex programming).

Linear programming deals with maximizing or minimizing a linear function subject to linear inequality or equality constraints. The main features of linear programming are the extremely efficient computational scheme of the simplex method and a duality theory which provides important and useful economic information in the form of dual variables, or shadow prices. An important feature of convex programming is that it can always be approximated as closely as we please by a linear program.

Although optimal scheduling of refinery operations, as well as many problems of industrial production, transportation and inventory, have a structure that can be represented by a mathematical model involving linear functions only, there are other important areas where the relations between the variables cannot be represented or approximated by linear functions. Such an area is engineering design.

Many engineering design problems can be expressed as optimization problems in which the engineer seeks to design a device or system to perform a certain function with an optimal characteristic, e.g., minimum cost or maximum yield. The design problem usually consists of fixed parameters, which may include material prices, design specifications, etc. and decision variables, such as dimensions, temperature, or pressure, which can be set by the engineer. The objective of an optimal design is to choose those values of the adjustable decision variables which yield an optimal performance criterion. Although this notion may appear to be straightforward, the fact is that modern optimization methods are not widely used in engineering design. Until recently there has been no mathematical method for treating the functions which typically arise in engineering design problems. Many of these are not only nonlinear, but nonconvex as well, i.e., linear approximations are not generally valid. For example, a widely used function in engineering design is one which relates the capital cost of a unit in a certain plant to its size or capacity. Such a function usually has an exponential form represented by the general equation $C = kQ^n$, where C is the capital cost of the unit, Q its capacity, k is a positive constant, and n is a positive exponent, usually about 0.6 to 0.7. Optimization problems involving functions of this type require special methods.

A recently developed theory, called geometric programming fulfills this need to some extent. In its present form geometric programming requires that all engineering characteristics be expressed as 'generalized positive polynomials,' or, briefly, posynomials. A posynomial is a function consisting of a sum of terms, each term being a positive coefficient multiplied by a product of variables, each variable raised to an arbitrary power. The capital cost-capacity relation described above, is a posynomial in one variable, consisting of one term. Other examples of posynomials, frequently encountered in engineering design, are the dimensionless numbers obtained by dimensional analysis, e.g., the Prandtl and Reynolds numbers.

In contrast to the time-honored approach to optimizing engineering design by differential calculus and/or some kind of an exhaustive search over the design variables, geometric programming offers a more efficient method of solution, especially when constraints are involved. This computational efficiency is a result of a duality theory, which associates with each primal geometric program a 'dual' programming problem, which is usually easier to solve than the primal problem. In addition to the computational convenience, the dual variables of geometric programming are important for the design engineer, because they provide information on the relative importance of the various design elements. The precise relationship between the economic interpretation of dual variables in geometric programming and the corresponding one in linear programming is yet to be examined.

In this paper we present some of the highlights of geometric programming, both from the theoretical and practical point of view.

As a preliminary, the general idea (as opposed to the technical details), of geometric programming will be illustrated by a very simple example, similar to one appearing in Wilde (1965). This example, while couched in engineering terms, is of course, much too simplified to come even close to representing any real problem. Its sole purpose is to illustrate, with a problem that can be solved by the differential calculus, the kind of reasoning upon which we will expand in the remainder of this report.

Consider a hypothetical chemical plant in which raw materials are brought up to pressure p (atm) in a compressor at a cost (including both amortized capital cost and operating expenses) of $1000\,p^{0.6}$ \$/year, mixed with a recycle stream r (10^3 lb mole/hour), and fed to a reactor whose operating cost is given by $\frac{4000}{pr}$ \$/year. Product is removed in a separator and the unreacted material is recycled at a combined cost of

10000 r, \$/year. Our objective, then, is to find values for p and r which minimize the total annual cost, given by

$$g = 1000\, p^{0.6} + \frac{4000}{pr} + 10000\, r \tag{1.1}$$

Since there are no constraints in this simple example, we may obtain the optimal pressure p* and recycle r* by setting the first partial derivatives equal to zero:

$$\left(\frac{\partial g}{\partial p}\right)^* = \frac{(0.6)(1000)}{p^{*0.4}} - \frac{4000}{p^{*2}r^*} = \frac{1}{p^*}\left((0.6)(1000)\, p^{*0.6} - \frac{4000}{p^*r^*}\right) = 0 \tag{1.2}$$

$$\left(\frac{\partial g}{\partial r}\right)^* = -\frac{4000}{p^*r^{*2}} + 10000 = \frac{1}{r^*}\left(-\frac{4000}{p^*r^*} + 10000\, r^*\right) = 0 \tag{1.3}$$

The values of p* and r* can be found by solving these nonlinear equations. The method of geometric programming does not require the solution of nonlinear equations, however. Instead we begin by finding the distribution of the minimum cost g* among the three cost terms. Let δ_1, δ_2, and δ_3 be respectively, the fractions of the total minimum cost represented by the compressor, the reactor, and the separator-recirculation. Then

$$\delta_1 = \frac{1000\, p^{*0.6}}{g^*} \tag{1.4}$$

$$\delta_2 = \frac{4000}{p^*r^*g^*} \tag{1.5}$$

and $$\delta_3 = \frac{10000\, r^*}{g^*} \tag{1.6}$$

Since the weights must sum to unity, we have

$$\delta_1 + \delta_2 + \delta_3 = 1. \tag{1.7}$$

Substituting (1.4), (1.5), and (1.6) into (1.2) and (1.3), we get

$$0.6\,\delta_1 - \delta_2 = 0 \tag{1.8}$$

$$-\delta_2 + \delta_3 = 0 \tag{1.9}$$

The last three equations constitute a system of three linear equations in three unknowns which has the unique solution

$$\delta_1 = 5/11,\ \delta_2 = 3/11,\ \delta_3 = 3/11. \tag{1.10}$$

Hence in the optimal design, the compressor contributes $5/11$ of the total cost, the fraction of the reactor cost is $3/11$ of the total expenditure, etc. Note that to find this optimal cost distribution we did not need to know the optimal values of the pressure and recycle stream.

Now we can find the minimum cost, still without knowing the optimal policy. Since the sum of the weights is equal to unity, we can write

$$g^* = (g^*)^{\delta_1}(g^*)^{\delta_2}(g^*)^{\delta_3} = \left(\frac{1000\, p^{*0.6}}{\delta_1}\right)^{\delta_1}\left(\frac{4000}{p^*r^*\delta_2}\right)^{\delta_2}\left(\frac{10000\, r^*}{\delta_3}\right)^{\delta_3} \tag{1.11}$$

Rearranging the last expression we get

$$g^* = \left(\frac{1000}{\delta_1}\right)^{\delta_1}\left(\frac{4000}{\delta_2}\right)^{\delta_2}\left(\frac{10000}{\delta_3}\right)^{\delta_3} p^{*(0.6\,\delta_1-\delta_2)}r^{*(-\delta_2+\delta_3)} \tag{1.12}$$

But by (1.8) and (1.9) the exponents of p* and r* in (1.12) vanish and

$$g^* = \left(\frac{1000}{\delta_1}\right)^{\delta_1}\left(\frac{4000}{\delta_2}\right)^{\delta_2}\left(\frac{10000}{\delta_3}\right)^{\delta_3} \tag{1.13}$$

Substituting (1.10) into (1.13), we get

$$g^* = \left(\frac{(11)(1000)}{5}\right)^{5/11}\left(\frac{(11)(4000)}{3}\right)^{3/11}\left(\frac{(11)(10000)}{3}\right)^{3/11} = 7950 \tag{1.14}$$

i.e., the minimum annual cost is $7950.

Now we can find the optimal design variables. Taking the logarithm of both sides of (1.4) and rearranging, we get

$$\ln p^* = \frac{1}{0.6}\ln\frac{(5)(7950)}{(11)(1000)} = 2.141 \tag{1.15}$$

$$\text{or} \quad p^* \cong 8.5 \text{ atm} \tag{1.16}$$

Similarly, from (1.5):

$$r^* = \frac{(3)(7950)}{(11)(10000)} \cong 0.217 \cong 217 \text{ lb mole/hour} \tag{1.17}$$

The reader can verify that these values do indeed satisfy the derivative conditions (1.2) and (1.3). The clear advantage of the geometric programming approach is that at no time during the analysis of this example has it been necessary to solve simultaneous nonlinear equations. For simplicity, this example was chosen so that: (i) The set of linear equations (1.7)-(1.9) has a unique solution. (ii) No constraints were imposed.

Realistic engineering problems are not usually so limited, however. Geometric programming is the extension of the method illustrated to cases in which there may not be unique solutions to the linear cost-weight equations and in which there may be constraints on the decision variables. This will be discussed in the forthcoming sections.

2. BASIC CONCEPTS

Geometric programming was first formulated by Zener (1961) as a method for optimizing engineering design. He considered the problem of minimizing the sum of component costs of a unit, when each cost depends on products of the design variables, each variable raised to an arbitrary but known power. In addition, it was required that the number of components in the unit exceed the number of design variables by one, as in the example of the preceding section. Zener showed that the solution of such a problem can always be obtained simply by solving a square system of linear algebraic equations.

Zener's method was subsequently extended by Duffin and Peterson to more general optimization problems, including problems with inequality constraints. In their rigorous mathematical treatment they rely on a generalization of the classical inequality of the arithmetic and geometric means as well as other geometric concepts, e.g., orthogonality of vectors. Geometric programming derives its name from its intimate connection with these geometric concepts. The most comprehensive reference on geometric programming is the recent book of Duffin, Peterson, and Zener (1967), where the mathematical derivations and some engineering applications are described in detail. Application of geometric programming to a particular engineering design problem was also discussed by Avriel and Wilde (1967a).

We now turn to a presentation of the basic mathematical concepts of geometric programming. A real valued function g is called a posynomial if it is given by

$$g(t) = \sum_{j=1}^{n} P_j(t) \tag{2.1}$$

where

$$P_j(t) = c_j \prod_{i=1}^{m} t_i^{a_{ij}} \tag{2.2}$$

and each coefficient c_j is positive. The exponents a_{ij} are arbitrary real numbers, and the variables t_i are restricted to take on positive values, i.e., the domain of a posynomial is the positive orthant.

In geometric programming we deal with minimizing posynomials subject to a certain type of posynomial constraint. This minimization problem we call the primal problem of geometric programming, or simply, the primal program.

Primal Program.

Find a vector t that minimizes the function $g_0(t)$ subject to the constraints

$$t_1 > 0, t_2 > 0, \ldots, t_m > 0 \tag{2.3}$$

$$\text{and } g_1(t) \leq 1, g_2(t) \leq 1, \ldots, g_p(t) \leq 1. \tag{2.4}$$

Here

$$g_k(t) = \sum_{j=M_k}^{N_k} P_j(t), k = 0, 1, \ldots, p \tag{2.5}$$

where

$$P_j(t) = c_j \prod_{i=1}^{m} t_i^{a_{ij}} \tag{2.6}$$

and

$$M_0 = 1, M_k = N_{k-1} + 1, k = 1, \ldots, p. \tag{2.7}$$

The exponents a_{ij} are arbitrary real numbers and the coefficients c_j are positive. Thus the functions $g_k(t)$ are posynomials.

An important feature of geometric programming is the central role played by the terms P_j in the posynomials g_k. Instead of focusing on determining the optimal t, the approach is to concentrate on evaluating the minimum of $g_0(t)$ and the relative contributions of the terms $P_j(t)$ to this minimum. Only after they are determined do we find the optimal t. This approach will become clear after we study the relations between the above primal program and its corresponding dual program.

Dual Program.

Find a vector δ that maximizes the product function

$$v(\delta) = \left[\prod_{j=1}^{N_p} (c_j/\delta_j)^{\delta_j}\right] \prod_{k=1}^{p} \lambda_k(\delta)^{\lambda_k(\delta)} \tag{2.8}$$

where

$$\lambda_k(\delta) \equiv \sum_{j=M_k}^{N_k} \delta_j, k = 1, \dots, p \tag{2.9}$$

subject to the constraints

$$\delta_1 \geq 0, \delta_2 \geq 0, \dots, \delta_{N_p} \geq 0 \tag{2.10}$$

$$\sum_{j=1}^{N_0} \delta_j = 1 \tag{2.11}$$

$$\text{and} \sum_{j=1}^{N_p} a_{ij}\delta_j = 0, i = 1, \dots, m. \tag{2.12}$$

Here a_{ij}, c_j, M_k, N_k are the same as in the primal program.

Let us note in detail how this dual program is obtained from the primal program. The constants c_j appearing in the dual function $v(\delta)$ are the coefficients of the posynomials g_k appearing in the primal objective function and constraints. We may say that δ_j is associated with the jth term, P_j, of $g_k, k = 0, 1, \dots, p$, so that each δ_j is associated with one and only one posynomial term P_j. Moreover, each $\lambda_k(\delta)$ is associated with the kth primal constraint $g_k(t) \leq 1$. The λ's, therefore, are related to Lagrange multipliers. The normality condition (2.11) is imposed on the dual variables associated with the primal objective function only, while the orthogonality conditions (2.12) apply to all dual variables. Finally, it should be noted that the ith constraint in (2.12) is associated with the ith primal variable t_i through the exponents a_{ij}.

We say that a program (either primal or dual) is consistent if there exists at least one vector that satisfies its constraints. A primal geometric program is superconsistent if there exists at least one vector $\hat{t} > 0$ such that

$$g_k(\hat{t}) < 1, k = 1, \dots, p. \tag{2.13}$$

A vector t is called primal feasible if it satisfies the primal constraints (2.3) and (2.4), while a vector δ is called dual feasible if it satisfies the dual constraints (2.10), (2.11), and (2.12). Instead of primal feasible or dual feasible, we sometimes say simply feasible.

A primal program is said to be soluble if there is a feasible vector, t^*, such that $g_0(t^*) \leq g_0(t)$ for all feasible t. Similarly, a dual program is said to be soluble if there is a feasible vector δ^*, such that $v(\delta^*) \geq v(\delta)$ for all feasible δ.

Although more general duality theorems can be found in Duffin, Peterson and Zener (1967), we state here the following duality theorem of geometric programming:

Theorem 1. Suppose that a primal geometric program is superconsistent and soluble. Then

(i) The corresponding dual program is soluble.
(ii) The constrained minimum of the primal program is equal to the constrained maximum of the dual program, i.e.,

$$g_0(t^*) = v(\delta^*). \tag{2.14}$$

(iii) The relations between optimal primal and dual variables are given by

$$\delta_j^* = \begin{cases} P_j(t^*)/g_0(t^*), \; j = 1, \ldots, N_0 \\ \lambda_k(\delta^*)P_j(t^*), \; j = M_k, \ldots, N_k, \quad k = 1, \ldots, p. \end{cases} \tag{2.15}$$

and

$$\lambda_k(\delta^*)[1 - g_k(t^*)] = 0, \qquad k = 1, \ldots, p. \tag{2.16}$$

The importance of this duality theorem is quite clear. If we assume that a given primal geometric program is superconsistent and soluble (a reasonable assumption for most engineering design problems), then Theorem 1 tells us that, instead of solving a primal geometric program, we can solve the corresponding dual program and by (2.14) the maximum of the dual is equal to the minimum of the primal. As a matter of fact, one can obtain upper and lower bounds on the solution of the primal and dual programs by simply evaluating $g_0(t)$ and $v(\delta)$ for respectively feasible vectors t and δ. Clearly, for such t and δ we have $g_0(t) \geqslant g_0(t^*) = v(\delta^*) \geqslant v(\delta)$.

This duality theorem not only enables one to find the minimum value of the primal objectives function, without actually solving the primal program, but relations (2.15) also give a method to find the minimizing vector t* from the knowledge of a maximizing vector δ^*. From (2.15) it follows that

$$c_j \prod_{i=1}^{m} t_i^{*a_{ij}} = \delta_j^* v(\delta^*), \; j = 1, \ldots, N_0 \tag{2.17}$$

and for $k = 1, \ldots, p$, such that $\lambda_k(\delta^*) > 0$

$$c_j \prod_{i=1}^{m} t_i^{*a_{ij}} = \delta_j^*/\lambda_k(\delta^*), \; j = M_k, \ldots, N_k. \tag{2.18}$$

Taking the logarithm of both sides of each equation in (2.17) and (2.18), and rearranging them, we obtain

$$\sum_{i=1}^{m} a_{ij} \log t_i^* = \log (\delta_j^* v(\delta^*)/c_j), \; j = 1, \ldots, N_0 \tag{2.19}$$

and

$$\sum_{i=1}^{m} a_{ij} \log t_i^* = \log \left[\delta_j^*/(c_j\lambda_k(\delta^*))\right], \; j = M_k, \ldots, N_k. \tag{2.20}$$

The optimal primal variables t_i^* are thus found by solving the above system of linear equations in the variables $\log t_i^*$.

As we mentioned before, each term P_j in the primal objective function g_0 is associated with one of the dual variables. The first equation of (2.15) now allows an economic interpretation of the optimal dual variables. It tells us that each optimal dual variable δ_j^*, $j = 1, \ldots, N_0$, represents the weight or relative contribution of the term P_j to the minimum of $g_0(t)$. Thus, by solving the dual program one obtains first the minimum of the primal objective function and then the relative contribution of each term P_j to the optimal solution.

Since the feasible t are restricted to be positive, t* is also positive; it follows then, that each term $P_j(t^*)$ in the objective function is positive. By (2.15), therefore, those dual variables which correspond to terms in the objective function are positive, i.e., $\delta_j^* > 0$ for $j = 1, \ldots, N_0$. The remaining δ's, i.e., those δ's corresponding to terms in the constraints are zero or positive, according as the particular constraint is loose ($g_k(t^*) < 1$) or tight ($g_k(t^*) = 1$) at primal optimum. More precisely, relations (2.9), (2.10), and (2.16) imply: (i) whenever $g_k(t^*) < 1$, the optimal dual variables δ_j^*, $j = M_k, \ldots, N_k$ vanish; (ii) if $\delta_j^* > 0$, for some $j = M_k, \ldots, N_k$, then $g_k(t^*) = 1$. Moreover, by the second equation of (2.15) and the positivity of t^*, we can conclude that $\lambda_k(\delta^*) > 0$ for some k, implies $\delta_j^* > 0$ for $j = M_k, \ldots, N_k$.

The duality theorem cited above is probably the most important from a practical viewpoint. For other duality theorems, which can also handle degenerate cases, the reader is referred to Duffin, Peterson and Zener (1967).

3. SOME PROPERTIES OF GEOMETRIC PROGRAMMING

The dual problem of geometric programming consists of maximizing a given function subject to linear equations and nonnegativity constraints on the variables. A particularly easy instance arises when the linear equations have a unique solution; this means the number of equations in the dual constraints is the same as the number of dual variables. If we look at just how the dual is constructed from the corresponding primal, we see that this case occurs when the total number of terms, P_j, in the primal objective function <u>and</u> constraints exceeds the number of primal variables by one, i.e.,

$$N_p = m + 1. \tag{3.1}$$

Zener's original treatment, as mentioned at the beginning of Section 2, was restricted to this case. In fact, it was restricted further to the case in which there were no constraints, i.e., all the P_j were terms of the objective function.

If the number of variables in the dual exceeds the number of equations by one (i.e., the number of terms in the primal program exceeds the number of primal variables by two), the dual optimization is still not very hard. In this case we may solve the dual constraints explicitly in terms of only one variable, thus reducing the problem to a maximization over a single variable. The next case, i.e., the number of primal terms exceeding the number of primal variables by three would lead to a dual program in which a two-variable maximization needs to be carried out and the problem is correspondingly more difficult. In fact, we can define the 'degrees of difficulty' of a geometric program by

$$\text{Degrees of Difficulty} = N_p - m - 1. \tag{3.2}$$

Thus, in the above cases we have zero, one and two degrees of difficulty. Generally, however, well-formulated geometric programs can have arbitrary nonnegative degrees of difficulty.

The linear dual constraints (2.10)-(2.12) have the important property that they are independent of the primal coefficients c_j. Hence the dual optimal solution for a problem with zero degrees of difficulty is <u>invariant</u> in the sense that no matter what the numerical values of the coefficients c_j are, the optimal dual variables are the same, since they are uniquely determined by solving the dual constraints. When the primal objective function represents a cost to be minimized, the optimal dual variables measure the relative contribution of the various cost items to the minimum cost. In the case of zero degrees of difficulty each term in the primal objective function at optimum has an <u>invariant weight</u>, represented by the unique solution of the linear dual constraints, thus providing insight into the economic or engineering structure of the problem. An analysis to find the relative importance of the terms in the primal objective function can be made, in the zero degrees of difficulty case, without prior knowledge of the numerical values of the coefficients. From a computational point of view the invariant weights enable one to evaluate the optimal primal variables from (2.19) and (2.20) for any positive values of c_j without resolving the programming problem. Thus the optimal primal variables are easily adjusted for any change in the coefficients, reflecting market fluctuations or altered design parameters.

The concept of invariant weights in the zero degrees of difficulty case has been extended by Avriel (1966) to the general case of geometric programming with an arbitrary positive number of degrees of difficulty. In fact, for any geometric program one can find an invariant range of weights for the terms of the optimal primal objective function. This can be accomplished conveniently by solving the following $2N_0$ linear programs:

(i) For $j = 1, \ldots, N_0$, maximize δ_j, subject to the constraints (2.10)-(2.12).

Similarly,

(ii) for $j = 1, \ldots, N_0$, minimize δ_j, subject to the constraints (2.10)-(2.12).

The solutions to these programs yield feasible upper and lower bounds on the dual variables, $\delta_1, \ldots, \delta_{N_0}$, associated with the terms of the primal objective function, which in turn define a range for the above dual variables. In many engineering problems the exponents a_{ij} represent technological interdependence of the primal variables. Accordingly, we call the matrix of coefficients appearing in (2.12) the technological matrix. We can say, therefore, that for a given technological matrix one can find the invariant range of weights of the various terms in the objective function without knowing the coefficients c_j, which usually represent unit costs or fixed design parameters. The case of zero degrees of difficulty is thus seen to be a special case of the invariant range of the weights, where the range is only a single point.

Complete solution of a geometric program requires either the minimization of a posynomial, subject to posynomial inequality constraints (primal program) or the maximization of a nonlinear product function, subject to linear inequality and equality constraints (dual program).

Certainly, if the degree of difficulty is low, i.e., zero or one, solving the geometric program by solving the dual rather than the primal is preferable, since, as has been pointed out, the optimization is trivial, or nearly so. For higher degrees of difficulty, the question is not so clear cut, although it does appear that even in this case, the dual problem offers advantages. Let us examine what is involved.

Since posynomials are generally nonconvex functions, straightforward solution of a primal program is usually difficult. It is possible, however, to introduce a change of variables and transform the primal program into a convex nonlinear program.

Consider the posynomial

$$g(t) = \sum_{j=1}^{n} P_j(t) = \sum_{j=1}^{n} c_j \prod_{i=1}^{m} t_i^{a_{ij}}, \; t > 0 \tag{3.3}$$

and make a change of variables by letting

$$t_i = e^{z_i}, \qquad i = 1, \ldots, m. \tag{3.4}$$

The z_i now range over all real numbers.

The transformed function

$$g(z) = \sum_{j=1}^{n} c_j e^{\sum_{i=1}^{m} a_{ij} z_i} \tag{3.5}$$

is easily shown to be convex in z. Thus we can transform the primal program of geometric programming into the following transformed primal program.

Transformed Primal Program.

Find a vector z that minimizes the function $g_0(z)$, subject to the constraints

$$g_1(z) \leqslant 1, g_2(z) \leqslant 1, \ldots, g_p(z) \leqslant 1. \tag{3.6}$$

Here,

$$g_k(z) = \sum_{j=M_k}^{N_k} c_j e^{\sum_{i=1}^{m} a_{ij} z_i} \tag{3.7}$$

$$M_0 = 1, M_k = N_{k-1} + 1, k = 1, \ldots, p, \tag{3.8}$$

and the c_j are positive.

The above transformed program involves minimizing a convex nonlinear function subject to convex nonlinear inequality constraints and thus is amenable to solution by well known convex programming methods.

Let us now consider the dual geometric program. We will show that the dual, also, may be transformed to produce a convex programming problem; but for the dual the constraints will be linear.

In order to obtain a convex program from a dual geometric program, first we note that the natural logarithm is a monotone increasing function of its argument. Thus if a point x* maximizes a function f(x), it will also maximize ln f(x). Thus to find a δ*, which solves a dual program we can maximize ln v(δ), instead of maximizing v(δ), and then we have the following transformed dual program.

Transformed Dual Program.

Find a vector δ that maximizes

$$\ln v(\delta) = \sum_{k=0}^{p} \sum_{j=M_k}^{N_k} \delta_j \ln[c_j \lambda_k(\delta)/\delta_j] \tag{3.9}$$

where

$$\lambda_0(\delta) = 1, \lambda_k(\delta) = \sum_{j=M_k}^{N_k} \delta_j, k = 1, \ldots, p, \tag{3.10}$$

subject to the constraints

$$\delta_1 \geqslant 0, \delta_2 \geqslant 0, \ldots, \delta_{N_p} \geqslant 0 \tag{3.11}$$

$$\sum_{j=1}^{N_0} \delta_j = 1 \tag{3.12}$$

$$\text{and } \sum_{j=1}^{N_p} a_{ij}\delta_j = 0, i = 1, \ldots, m. \tag{3.13}$$

Here a_{ij}, c_j, M_k, N_k are the same as in the primal program. It can be shown that ln ln v(δ) is concave, thus the transformed dual program is a convex programming problem with linear constraints.

Inspection of the above transformed dual program also reveals an interesting correspondence between geometric programming and chemical thermodynamics, which we discuss in the Appendix.

Here we mention only that geometric programs with positive degrees of difficulty can be solved numerically by a computer code, developed by Clasen (1965) to find the equilibrium composition of a reacting chemical system. The numerical results of the next section were obtained by use of this code.

4. ENGINEERING DESIGN EXAMPLE

In this section we present a simple optimal engineering design problem, which can be formulated as a geometric program. The methods and features of this new mathematical approach, described in the preceding sections, are demonstrated in this example. Other techniques, helpful in the solution of problems, are also developed as the need arises. The problem is a variation on an example of optimal chemical reactor design of Levenspiel (1962).

Consider the problem of manufacturing $F_R = 50$ gram moles/hour of product R from a feed consisting of a solution of reactant A in a continuous stirred tank (backmix) reactor. The chemical reaction is

$$2A \rightarrow R \tag{4.1}$$

with rate equation, based on unit volume of reacting fluid

$$-r_A = \frac{-dC_A}{dt} = 2.0\ C_A^2 = 2.0\ C_{A0}^2(1 - x_A)^2 \left(\frac{\text{gm mole}}{\text{liter-hr}}\right) \tag{4.2}$$

where

C_A = concentration of A in the reactor $\left(\frac{\text{gm mole}}{\text{liter}}\right)$

C_{A0} = concentration of A in the feed $\left(\frac{\text{gm mole}}{\text{liter}}\right)$

x_A = conversion-fraction of reactant converted into product.

Suppose the feed solution is available at a continuous range of concentrations of A and its unit cost is given by the empirical relation

$$\alpha_A = k_A C_{A0}^{1.4}\ (\$/\text{liter}) \tag{4.3}$$

where $k_A = 4$, is a constant.

The cost of the backmix reactor including installation, auxilary equipment, labor, depreciation, etc., is assumed to be of the exponential form mentioned in the Introduction and in the proper units it is given by

$$\alpha_B V^n = 0.4\ V^{0.6}\ (\$/\text{hr}) \tag{4.4}$$

where V liters is the capacity (volume) of the reactor. Our problem is to determine the rate of feed solution F_{A0} (liter/hr), its concentration C_{A0}, the volume of the reactor V, and the conversion x_A for optimum operations, i.e., for minimizing total cost/hour, given by

$$g = \alpha_A F_{A0} + \alpha_B V^{0.6}\ (\$/\text{hr}). \tag{4.5}$$

We consider first the case in which the unreacted A is discarded, i.e., there is no recycle.

Material balance around the reactor yields for reactant A:

$$\text{Input of A} = \text{Output of A} + \text{Disappearance of A by reaction} \tag{4.6}$$

$$\text{or} \quad F_{AO}C_{AO} = F_{AO}C_{AO}(1 - x_A) - r_A V. \tag{4.7}$$

Noting that the rate of production of R is

$$F_R = \tfrac{1}{2} F_{AO} C_{AO} x_A \tag{4.8}$$

we get from (4.2), (4.7), and (4.8):

$$V = \frac{F_R}{C_{AO}^2(1 - x_A)^2} = \frac{50}{C_{AO}^2(1 - x_A)^2}. \tag{4.9}$$

Combining (4.3), (4.8) and (4.9), we get

$$\alpha_A F_{AO} = \frac{8\, F_R C_{AO}^{0.4}}{x_A} = \frac{400\, C_{AO}^{0.4}}{x_A} \tag{4.10a}$$

$$\alpha_B V^{0.6} = \frac{(0.4)(50)^{0.6}}{C_{AO}^{1.2}(1 - x_A)^{1.2}} = \frac{4.183}{C_{AO}^{1.2}(1 - x_A)^{1.2}} \tag{4.10b}$$

and we have to minimize

$$g = \frac{400\, C_{AO}^{0.4}}{x_A} + \frac{4.183}{C_{AO}^{1.2}(1 - x_A)^{1.2}} \tag{4.11}$$

Observing the second term on the right hand side of (4.11), we conclude that g is not a posynomial and geometric programming cannot be applied to this problem in a straightforward manner. Fortunately, however, only a simple change of variables is required to bring our problem into a form amenable to geometric programming analysis: Introduce the additional variable w, which satisfies

$$w \geqslant \frac{1}{(1 - x_A)} \tag{4.12}$$

and consider the following related geometric programming problem of minimizing the posynomial

$$g_0 = \frac{400\, C_{AO}^{0.4}}{x_A} + \frac{4.183\, w^{1.2}}{C_{AO}^{1.2}} \tag{4.13}$$

subject to the posynomial constraint, obtained by rearranging (4.12) and given by

$$g_1 = \frac{1}{w} + x_A \leqslant 1. \tag{4.14}$$

Clearly, (w, C_{AO}, x_A) solves this related problem if and only if (C_{AO}, x_A) solves the original problem of minimizing g. Such techniques of constructing related geometric programs from optimization problems involving functions which are not posynomials are readily available for a large class of functions, see Duffin, Peterson and Zener (1967).

Now we turn our attention to the geometric programming problem. First we note that the problem has three variables and it consists of four terms. Hence by (3.2) it has zero degrees of difficulty.

The dual constraint equations are written as

$$\begin{aligned} \delta_1 + \quad \delta_2 \qquad\qquad &= 1 \\ 1.2\delta_2 - \delta_3 \qquad &= 0 \\ 0.4\delta_1 - 1.2\delta_2 \qquad\qquad &= 0 \\ -\delta_1 \qquad\qquad\qquad + \delta_4 &= 0 \end{aligned} \tag{4.15}$$

and the solution is $\delta_1^* = 0.75, \delta_2^* = 0.25, \delta_3^* = 0.3, \delta_4^* = 0.75$. Consequently,

$\lambda_1(\delta^*) = 1.05.$

Recalling the interpretation of the optimal dual variables, we can immediately conclude that in an optimal design the cost of the reactant A and of the backmix reactor are invariably 75% and 25% of the total cost, respectively. This optimal cost distribution will not be affected by any changes in the unit cost of the reactor α_B, reflecting market fluctuations, or the coefficient k_A, appearing in the expression of α_A. Similarly, this distribution is invariant under changes in the required production rate, F_R.

The dual objective function in our case is

$$v(\delta) = \left(\frac{400}{\delta_1}\right)^{\delta_1}\left(\frac{4.183}{\delta_2}\right)^{\delta_2}\left(\frac{1}{\delta_3}\right)^{\delta_3}\left(\frac{1}{\delta_4}\right)^{\delta_4}(\lambda_1)^{\lambda_1}. \tag{4.16}$$

Substituting the optimal dual variables we get for the minimum cost

$$v(\delta^*) = g_0^* = g^* = 420.7 \text{ \$/hour} \tag{4.17}$$

Now we can easily find the optimal design variables: From (2.16) and (4.14):

$$w^* = \frac{\lambda_1(\delta^*)}{\delta_3^*} = \frac{1.05}{0.3} = 3.5$$

$$x_A^* = \frac{\delta_4^*}{\lambda_1(\delta^*)} = \frac{0.75}{1.05} = 0.7143$$

Similarly,

$$\begin{aligned} C_{A0}^* &= 0.238 \text{ gm mole/liter} \\ F_{A0}^* &= 588 \quad \text{liters/hour} \\ V^* &= 10800 \text{ liters} \end{aligned}$$

and the solution is complete.

Of course, our main purpose in presenting the above simple example is to demonstrate the method of geometric programming, rather than discussing the question of whether or not geometric programming is more efficient in solving this particular optimal reactor design problem than conventional methods of calculus. In any case, it offers a systematic approach to design problems, provides insight into the structure of the solution and, in more complex problems, especially when inequality constraints are present, it becomes extremely efficient. To see this we note that classical methods of optimization, based on differentiation as well as most multidimensional search methods, become very cumbersome in solving inequality constrained problems. On the other hand, the difficulty in solving a geometric program (by the dual program) does not depend on the presence of constraints but on the difference between the total number of terms in the primal problem, including the objective function and the constraints, and the number of primal variables.

So far we have discussed a design problem which could be represented as a geometric programming problem with zero degrees of difficulty. Next we demonstrate a problem with a positive degree of difficulty. Inspecting the optimal cost distribution of our reactor design problem reveals that 75% of the manufacturing cost is spent on the raw material A. This suggests that it might be economically beneficial to use a separation process to recover the unreacted A from the product stream leaving the reactor, to bring the reactant up to the feed concentration C_{A0}.

Suppose that the unit cost of this separation process depends on the feed concentration C_{A0}, as well as the fraction of A converted into R, x_A, and can be expressed empirically as

$$\alpha_S = \frac{k_S C_{A0}^{0.5}}{(1 - x_A)^{1.5}} \text{ (\$/liter)} \tag{4.18}$$

where k_S is a constant, assumed to have the value $k_S = 0.1$. Our modified problem is to find the optimum design with this reclaimed A as a recycle stream. Let F_{AI} liters/hour denote the feed rate to the reactor, consisting of F_{A0} liters/hour of fresh feed (make-up) and $F_{AI}(1 - x_A)$ liters/hour of recycle stream, containing the unreacted A, recovered by the separation process. Both the fresh feed and the recycle contain A at a concentration C_{A0}. The total cost now is given by:

$$g^1 = \alpha_A F_{A0} + \alpha_B V^{0.6} + \alpha_S F_{AI}(1 - x_A) \text{ (\$/hr)}. \tag{4.19}$$

Material balances yield

$$F_R = \tfrac{1}{2} F_{AI} C_{A0} x_A = \tfrac{1}{2} F_{A0} C_{A0} \tag{4.20}$$

and

$$F_{AI} C_{A0} = F_{AI} C_{A0}(1 - x_A) - r_A V. \tag{4.21}$$

Combining (4. 2), (4. 3), (4. 4), (4. 18), (4. 19), (4. 20), and (4. 21), we obtain the cost function to be minimized:

$$g^1 = 400\, C_{A0}^{0.4} + \frac{4.183}{C_{A0}^{1.2}(1 - x_A)^{1.2}} + \frac{10}{C_{A0}^{0.5}(1 - x_A)^{0.5} x_A} \tag{4.22}$$

Using the change of variables introduced in (4. 12), we transform (4. 22) into the following geometric programming problem:

Minimize

$$g_0^1 = 400\, C_{A0}^{0.4} + \frac{4.183\, w^{1.2}}{C_{A0}^{1.2}} + \frac{10\, w^{0.5}}{C_{A0}^{0.5} x_A} \tag{4.23}$$

subject to

$$g_1 = \frac{1}{w} + x_A \leq 1. \tag{4.24}$$

First we note that this problem has three variables and 3 + 2 = 5 posynomial terms. Hence, by (3. 2), it has one degree of difficulty. The corresponding dual program is written as follows:

Maximize

$$v(\delta) = \left(\frac{400}{\delta_1}\right)^{\delta_1} \left(\frac{4.183}{\delta_2}\right)^{\delta_2} \left(\frac{10}{\delta_3}\right)^{\delta_3} \left(\frac{1}{\delta_4}\right)^{\delta_4} \left(\frac{1}{\delta_5}\right)^{\delta_5} \lambda_1(\delta)^{\lambda_1(\delta)} \tag{4.25}$$

subject to

$$\begin{aligned}
\delta_1 + \quad \delta_2 + \quad \delta_3 \qquad\qquad &= 1 \\
0.4\delta_1 - 1.2\delta_2 - 0.5\delta_3 \qquad\qquad &= 0 \\
1.2\delta_2 + 0.5\delta_3 - \delta_4 \qquad &= 0 \\
- \quad \delta_3 \qquad + \delta_5 &= 0 \\
\delta \geq 0 &
\end{aligned} \tag{4.26}$$

As a first step in solving this problem, we determine the invariant range of the weights, δ_1, δ_2, and δ_3, corresponding to the various cost items in the primal objective function. First we solve the linear program: maximize δ_1, subject to (4.26); then we solve: minimize δ_1, etc. The results are

$$\begin{aligned}
0.556 \leq \delta_1^* &\leq 0.75 \\
0 \leq \delta_2^* &\leq 0.25 \\
0 \leq \delta_3^* &\leq 0.444
\end{aligned} \tag{4.27}$$

Thus, no matter what numerical values the unit costs may take on, the cost of reactant A in an optimum solution is between 55.6 and 75% of the total cost. Similarly, the costs of the reactor and the separation process cannot exceed 25 and 44.4% of the total cost, respectively. Comparing this range of weights with the unique weights obtained for the zero degrees of difficulty case we conclude that without a separation process, the optimal (and unique feasible) design is attained when the weights of reactant and reactor costs are at their maximum. Optimal design with separation process and recycle will therefore reduce the weight of one or both of these costs and if the separation is relatively inexpensive, the new total cost will be lower.

Next, we get a complete solution to our problem by solving the dual geometric program given in (4.25) and (4.26), using Clasen's code for the chemical equilibrium problem. The optimal dual variables are as follows: $\delta_1^* = 0.665$, $\delta_2^* = 0.141$, $\delta_3^* = 0.194$, $\delta_4^* = 0.266$, $\delta_5^* = 0.194$, $\lambda_1(\delta^*) = 0.460$ and $v(\delta^*) = g^{1*} = 334.3$ \$/hour. It is clear that the proposed separation and recycling can reduce considerably the manufacturing costs of product R. The new design for this case is different from the process without recycle:

$$\begin{aligned}
x_A^* &= 0.422 \\
C_{A0}^* &= 0.230 \text{ gm mole/liter} \\
V^* &= 5650 \text{ liters} \\
F_{A0}^* &= 430 \text{ liter/hour} \\
F_{AI}^* &= 1030 \text{ liter/hour}
\end{aligned}$$

Analysis of these results indicates that for this example an improved design can be achieved without extensive changes in the feed rate and concentration, by the use of the proposed recycle and the reactor volume. These results also suggest that the unit cost k_S of this separation process is an important factor in determining an optimal design. Suppose that the function form of α_S is fixed by the separation process conditions, but the coefficient k_S in (4.18) can vary. Since the corresponding geometric program has one degree of difficulty, the optimal distribution of the total cost does depend on the numerical values of this coefficient. The sensitivity of the optimal design to the numerical value of k_S can be demonstrated by selecting a few values for k_S and solving the corresponding geometric programs. The following table shows the optimal design configurations for various values of k_S:

k_s	g^{1*} (\$/hour)	δ_1^*	δ_2^*	δ_3^*	x_A	V liters	F_{AO} liter/hr	C_{AO} gm mole hour	F_{AI} liter/hr
0.025	274.2	0.706	0.194	0.100	0.262	6910	610	0.163	2340
0.05	297.9	0.688	0.171	0.141	0.338	6460	530	0.188	1570
0.1	334.3	0.665	0.141	0.194	0.422	5650	430	0.230	1030
0.2	391.1	0.636	0.104	0.260	0.506	4410	330	0.304	650
0.25	415.6	0.626	0.091	0.283	0.531	3910	290	0.341	550
0.3	438.4	0.618	0.080	0.302	0.550	3460	265	0.378	480
0.4	480.1	0.605	0.064	0.331	0.577	2780	220	0.449	390

It is interesting to note that changing the value of k_s from 0.025 to 0.4, i.e., multiplying by a factor of 16 yields only a 75% increase in the total cost. This is accomplished by changing the cost distribution. The range of δ_1^* in which it can lie is quite narrow, as reflected in (4.27), and the relative change in δ_1^* is small. On the other hand, as k_s in the unit cost of separation increases, also the weight of the separation cost term increases gradually at the expense of reducing the weight of the reactor cost. It is also clear that for $k_s > 0.25$ the addition of separation and recycle to the process becomes uneconomical, since the total cost of manufacturing without recycle will then be lower.

A discussion of sensitivity analysis, i.e., the perturbation of the optimal design variables due to variations in the coefficients can be found in Duffin, Peterson and Zener (1967). Avriel and Wilde (1967b) formulated and analyzed geometric programming proproblems in which the coefficients c_j are random variables. A survey of these results is, however, beyond the scope of the present paper.

APPENDIX

During the last decade great research effort has been directed into areas involving large and complex reacting chemical systems. One aspect of this research was the mathematical and numerical analysis of chemical equilibrium in a complex reacting system, and consequently the need for developing a general and efficient algorithm for calculating the equilibrium composition became apparent. Conventional equilibrium calculations used the mass action laws and required the iterative solution of a set of simultaneous linear and nonlinear equations. The theory of chemical equilibrium of gases and dilute solutions was initiated by Gibbs, who showed that a chemical system attains an equilibrium state (at a fixed temperature and pressure) when the total free energy of the system is at a minimum.

White, Johnson and Dantzig (1958) developed a method for calculating equilibrium composition of a chemical system at constant pressure and temperature by direct minimization of the free energy function, subject to mass balance constraints. Their original mathematical programming formulation of the chemical equilibrium problem was extended to multiphase ideal gas systems by Shapiro and Shapley (1965) and can be stated as follows: Minimize

$$\hat{F}(\delta) = \frac{F(\delta)}{RT} = \sum_{k=0}^{p} \sum_{j=M_k}^{N_k} \delta_j\{F_j^0/RT + \ln P + \ln[\delta_j/\lambda_k(\delta)]\} \tag{A.1}$$

where

$$\lambda_k(\delta) = \sum_{j=M_k}^{N_k} \delta_j, \; k = 0, 1, \ldots, p \tag{A.2}$$

and

$$M_0 = 1, M_k = N_{k-1} + 1, k = 1, \ldots, p \tag{A.3}$$

subject to the nonnegativity constraints

$$\delta_j \geqslant 0, j = 1, \ldots, N_p \tag{A.4}$$

and the linear mass balance equations

$$\sum_{j=1}^{N_p} a_{ij}\delta_j = b_i, i = 1, \ldots, m. \tag{A.5}$$

Here

$F(\delta)$ = total free energy of the system

R = gas content

T = absolute temperature

P = total pressure in atmospheres

δ_j = no. of moles of molecular species j

$F_j{}^0$ = standard free energy of species j

b_i = no. of moles of element i

m = no. of elements present in the system

a_{ij} = no. of atoms of element i in a molecule of species j

$0, 1, \ldots, p$ = phases in the system

Although the parameters appearing in relations (A. 1) to (A. 5) have real physical meanings, in an abstract mathematical model, they may take on numerical values which may have no physical meanings in conventional physical chemistry. In particular, the mathematical treatment of the chemical equilibrium problem includes the case where the number of atoms i in a molecule of species j, a_{ij}, and the number of element i, b_i are arbitrary real numbers rather than restricted to be nonnegative in the conventional sense. It is to be understood that in the following we refer to the chemical equilibrium problem in its extended and somewhat unconventional form. It was shown by Shapiro and Shapley that $\hat{F}(\delta)$ is convex. Thus the above chemical equilibrium problem is a convex program.

Now we can establish an equivalence between geometric programming and the above chemical equilibrium problem. Let

$$\ln(1/c_j) = F_j{}^0/RT + \ln P. \tag{A.6}$$

Comparing (3. 9) with (A. 1) and noting that if δ^* maximizes $\ln \hat{v}(\delta)$, then it also minimizes $-\ln v(\delta)$, we conclude that $-\ln v(\delta)$ is a special case of $\hat{F}(\delta)$, with $\lambda_0 = 1$. The nonnegativity constraints (3. 11) are identical to (A. 4) and the linear constraints (3. 12) and (3. 13) are a special case of (A. 5) with a right hand side $b = (1, 0, 0, \ldots, 0)$. Thus, for every geometric programming problem there exists an equivalent abstract 'chemical equilibrium problem,' and conversely, it can be shown that for every chemical equilibrium problem there exist corresponding primal and dual 'geometric programs.'

The value of this equivalence is two-sided: (i) The theory of geometric programming can be useful in the study of chemical systems, or the chemical model may suggest new lines of research in mathematical programming, and (ii) Numerical solution of geometric programming problems can be obtained by a computer code originally developed for the Gibbs free energy minimization problem by Clasen (1965).

ACKNOWLEDGMENT

I wish to thank Dr. A. C. Williams for his valuable advice and comments during the preparation of this paper.

REFERENCES

1. AVRIEL, M. (1966), 'Topics in Optimization: Block Search; Applied and Stochastic Geometric Programming,' Ph. D. Thesis, Stanford University.
2. AVRIEL, M., and D. J. WILDE (1967a), 'Optimal Condenser Design by Geometric Programming,' I&EC Process Design and Development, 6, 256-263
3. AVRIEL, M., and D. J. WILDE (1967b), 'Stochastic Geometric Programming,' presented at the International Symposium on Mathematical Programming, Princeton, N.J., August 1967.
4. CLASEN, R. J. (1965), 'The Numerical Solution of the Chemical Equilibrium Problem,' RAND Corporation Research Memorandum RM-4345-PR.
5. DUFFIN, R. J., E. L. PETERSON, and C. ZENER (1967), Geometric Programming, Wiley, New York.
6. LEVENSPIEL, O., (1962), Chemical Reaction Engineering, Wiley, New York.
7. SHAPIRO, N. Z., and L. S. SHAPLEY (1965), 'Mass Action Laws and the Gibbs Free Energy Function,' J. Soc. Indust. Appl. Math., 13, 353-375.
8. WILDE, D. J. (1965), 'A Review of Optimization Theory,' Ind. Eng. Chem., 57, 19-31.
9. WHITE, W. B., S. M. JOHNSON, and G. B. DANTZIG (1958), 'Chemical Equilibrium in Complex Mixtures,' J. Chem. Phys., 28, 751-755.
10. ZENER, C. (1961), 'A Mathematical Aid in Optimizing Engineering Designs,' Proc. Nat'l. Acad. Sic., 47, 537-539.

DISCUSSION

Dr. R. Razani

In many engineering design problems, the objective function is relatively a simple function. There are two types of constraints. The most important of them are the behaviour constraints. Specifying the limits to the behaviour of the system under operating conditions or loads. For example, for a structural design problem one can specify that the stresses or deflections at some point of the structure should not exceed the given allowable values. The difficulty is that the behaviour (stresses or deflection) cannot be expressed explicitly as function of design variables except for very simple case The behaviour is generally obtained from the solution of analysis equations which lead to the inversion of flexibility or stiffness matrices. The relationship between behaviour values and design variables is not expressable in closed form. There are othe types of constraints which are called specification constraints. These are generally simple constraints representing specification limitations or arbitrary architectural or rule-of-thumb constraints on the size of design variables.

Does the geometric programming method have the capability to take care of these two types of constraints and how efficiently can it do it?

Dr. M. Avriel

As I mentioned in my talk, geometric programming can handle only a certain type of inequality constraints, consisting of posynomials, as given by formulas (2.4)-(2.7). Recently, the theory of geometric programming has been extended to include constraints with 'generalized polynomials,' in which the coefficients c_j are unrestricted in sign.

It seems to me that in many engineering design problems one should be able to approximate quite general constraints by either a posynomial or a generalized polynomial inequality constraint and then apply geometric programming.

SECTION 9 Strategic Deployment Problems

F. L. Hassler
M. Arbabi, J. G. Fisher, H. M. Horowitz, D. F. Kocher
W. F. Yondorf, J. C. Grimberg, H. I. Ottoson
P. M. Jenkins, Mary J. O'Brien, Justin C. Whiton
C N. Beard, C. T. McIndoe
M. G. Shaw
J. M. Lodal

Problems of Strategic Mobility

F. L. HASSLER
Special Assistant to the Joint Chiefs of Staff, U.S.A.

1. OBJECTIVES OF THE SPECIAL ASSISTANT TO THE JOINT CHIEFS OF STAFF FOR STRATEGIC MOBILITY (SASM)

The Special Assistant to the Joint Chiefs of Staff for Strategic Mobility (SASM) is the focal point within the Department of Defense on all matters relating to strategic mobility (See Figure 1). It is the responsibility of the SASM to know what the strategic mobility of any military system is, how it can be measured, how it can be changed, what it will cost to change it, and what the value of it is in any given situation.

The critical role of mobility in the theory of warfare has always been apparent. However, a precise, analytical formulation of theory that will permit the SASM to carry out his assigned responsibilities has yet to be developed.

In establishing the SASM organization, the Secretary of Defense directed that this new organization strongly emphasize the role of systems analysis in the execution of his assigned responsibilities. The SASM has, therefore, adopted a formal plan of research and development which has guided the application of systems analysis techniques from the time that his organization was first created. This research and development program plus the definition of the technical support program necessary to apply systems analysis techniques to his internal operations is described in the Plan of Technical Support to the SASM which is published and updated annually.

The Technical Support Plan relies heavily upon the techniques of automation and is divided into five major categories, computer model development, computer data systems development, research, studies, and operational support. The first four categories constitute the formal research and development program of the SASM.

The objectives of the SASM's research and development program are to provide the SASM with the capability to answer those questions for which he is primarily responsible and to be able to demonstrate an analytical basis for such answers.

2. THE CONTEXT

Consideration of strategic mobility problems can be broken down into two major categories of concern. The first category treats military operations planning and execution where the emphasis is on a particular situation and the problem of selecting an appropriate course of action in that situation. The second major category deals with programming and budgeting the mobility resources necessary for the conduct of military operations under all likely circumstances in the future. Because of his responsibilities to the Joint Chiefs of Staff, because of the uncertainties that exist in this area at the present, and because of the sensitive dependence of the entire second area on the results of the first, SASM has chosen to give highest priority to research and development programs in the first category defined. Only with an appropriate understanding of the first category, will we feel confident in building the bridge between the conduct of specific military operations and the problems of programming and budgeting of mobility resources.

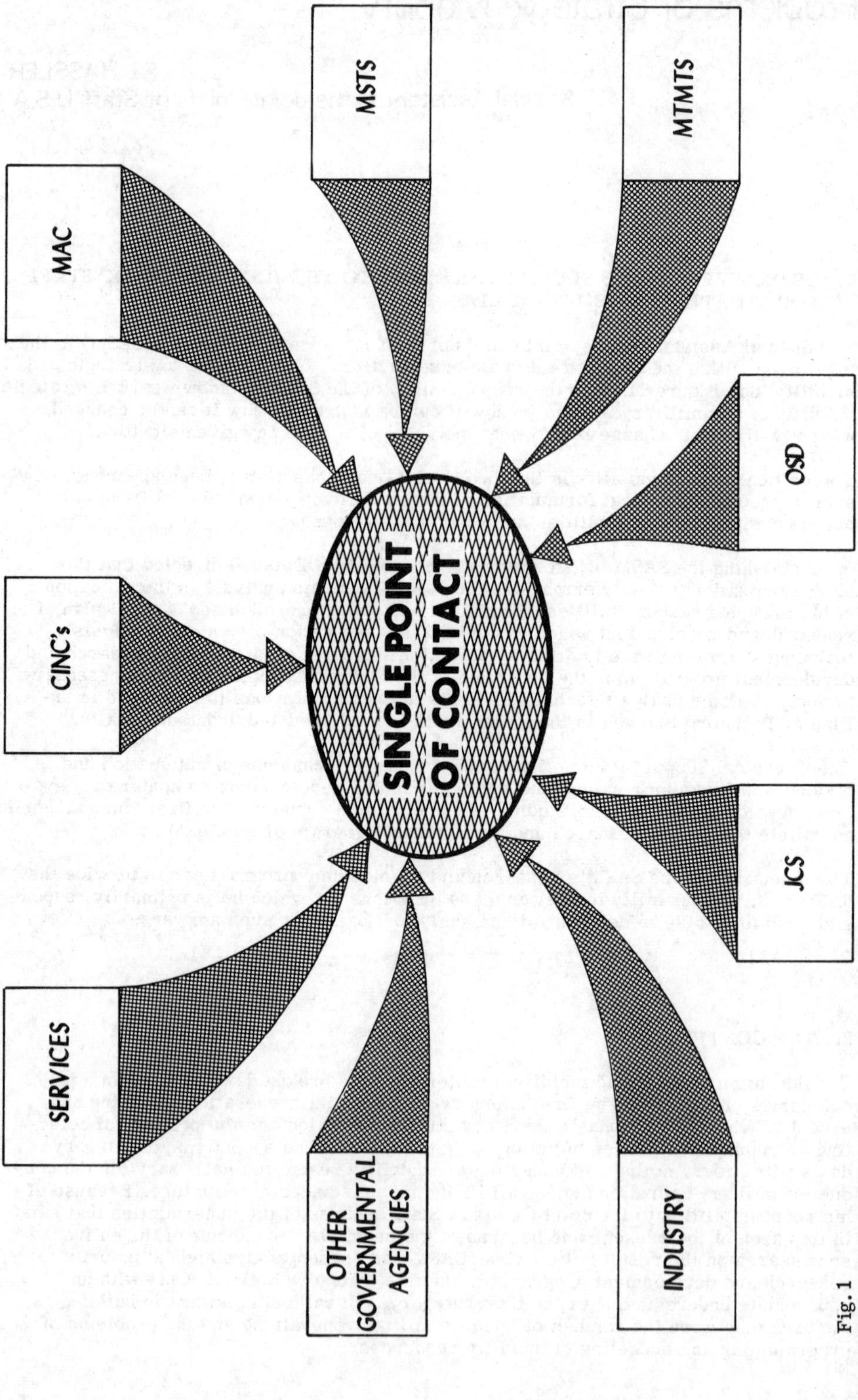
MAC
MSTS
MTMTS
CINC's
SINGLE POINT OF CONTACT
OSD
SERVICES
OTHER GOVERNMENTAL AGENCIES
INDUSTRY
JCS

Fig. 1

2.1 Military Operations Planning and Execution

During the first two and one-half years of the SASM organization's program, most of the research and development has been oriented toward the development and understanding of the theory of mobility analysis. An explicit and analytical theory of mobility analysis is the key ingredient in improving the planning function of the mobility system. The need to make major improvements in the planning function had been highlighted in studies that led to the formation of SASM.

Principal research has been focused around the development of computer models that address different aspects of the mobility problem. The long range objective of the ADP development program is to create a system of ADP models that interact with one another and with the internal data files of a data management system (See Figure 2). The data files are created either by external reporting system inputs connected to other elements of the Department of Defense or from data locally entered by SASM analysts.

The number of different factors which could conceivably enter into strategic mobility analysis is enormously large. However, there is a basic sequence of strategic mobility planning which tends to repeat itself in various guises in most problems. That basic sequence is shown in Figure 3. It begins with a statement of movement requirements (e.g., 20 tons of unit equipment); specifies the movement resources considered available and the routes to be taken; and then allocates movement resources over routes to things to be delivered. The outcome of this process is a movement table or transportation schedule together with information on late arrivals or goods not moved and data on the use of facilities and vehicles.

Mobility planning for emergency or near term commitments must deal with current available resources and dispositions. Planning and analysis for future situations may deal with programmed resources or procurement alternatives.

If economic analysis is desired, the approach is to consider all equally feasible solutions or all solutions optimal from some point of view (e.g., minimum deviation from desired delivery dates) and determine the mix of vehicles that could implement the move with minimum procurement cost or operating expense.

2.2 Analysis for Resources Programming and Budgeting

The principal technique currently employed in the analysis of requirements for mobility resource programming and budgeting is closely related to the procedures employed in movement planning. Demands are established for simultaneous shipment to several world areas. The principal difference lies in the level of aggregation of the things to be moved.

Allocations of resources to demands are made, and for equal effectiveness levels for delivery, alternate resource procurements are evaluated in the standard cost/effectiveness framework.

Because the scope of the analysis in programming and budgeting is much broader, allocation logics must deal with more aggregated representations of the resources and their characteristics. Allocation logics are further complicated by the increased influence of cost-related factors and thus additional aggregation is required for given program size limits.

Accordingly, the subject of the effects of aggregation on the accuracy and reliability of mobility analysis results in receiving great attention in SASM work. Our findings to date indicate that aggregation may seriously effect the answers derived from the models we now employ.

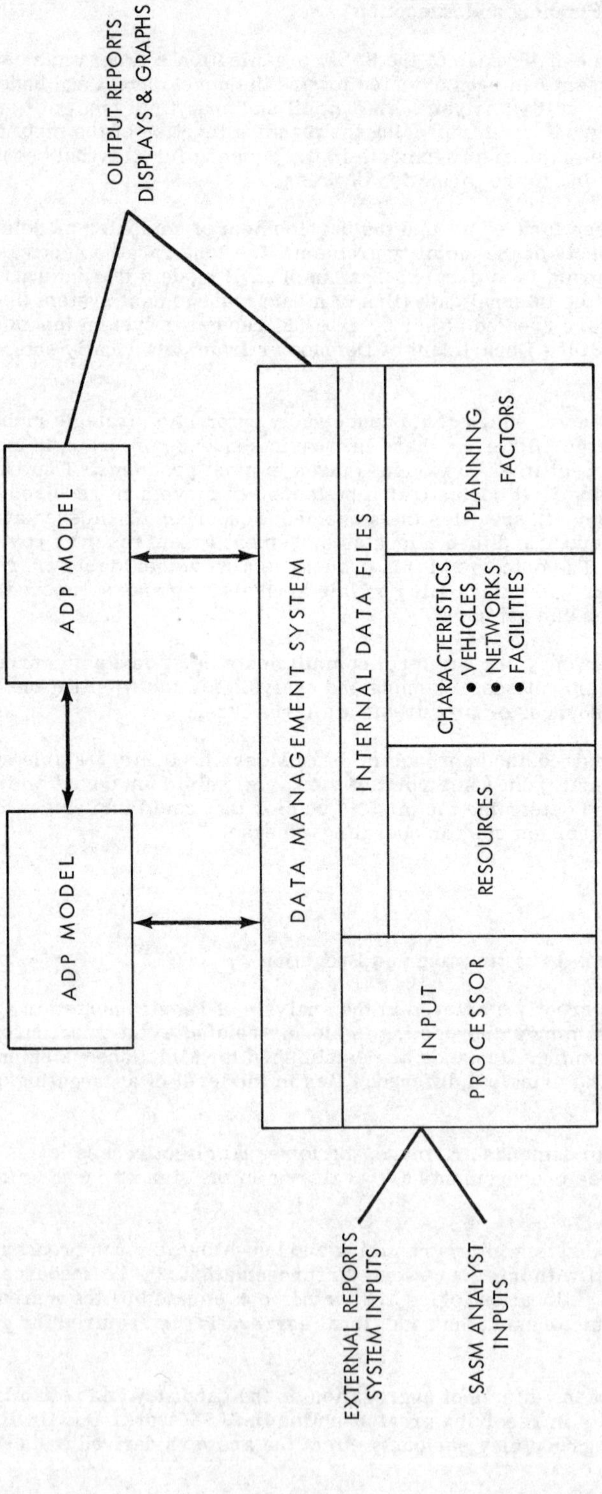

SASM ADP CONCEPT

Fig. 2

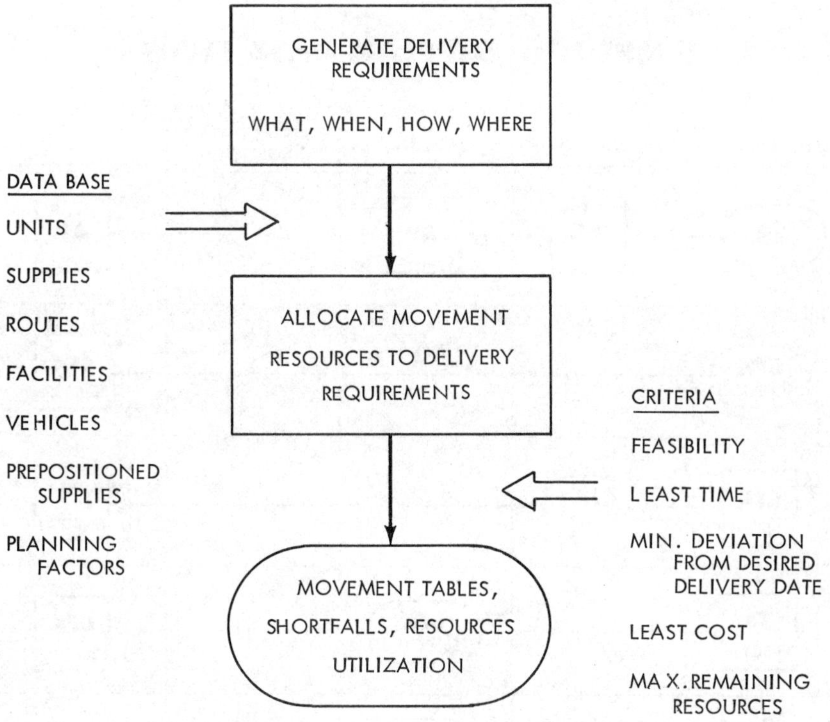

Fig. 3 BASIC MOVEMENT PLANNING SEQUENCE

3. APPROACH TO THE DEVELOPMENT OF A MOBILITY PLANNING SYSTEM

In collaboration with the Commanders of the Unified and Specified Commands, the Services, and the Transportation Agencies of the Defense Department, the SASM has defined the information content that must be generated during the planning of a major force deployment. This information has been defined in a sequence of card formats and a procedure has been specified whereby each significant agent in the preparation of a deployment plan specifies the appropriate information and transmits it to the other agents.

This set of card formats and procedures is referred to as the DEPPLAN System or alternatively the Deployment Reporting System (DEPREP). A simplified card flow and description of the contents of the major cards is shown in Figure 4.

The information content of the DEPPLAN System also defines an internal data file which is used as a principal medium of exchange of information between ADP models of SASM. Figure 5 illustrates this process of communication of information between major computer models and data systems. For example, the Movement Requirements Generator (MORG), a computer model consisting of three processing modules, accepts force requirements statements for a deployment specified in A and B card formats of the DEPREP System as input. MORG then associates information on status of forces, unit characteristics, resupply, and supply build-up factors and generates information appropriate to DEPREP C, F, G, J and K cards. Therefore, SASM has an automated procedure for the generation of DEPREP information which can be employed for special internal analysis, as a consistency check on information prepared externally, or in a cooperative effort with other agencies in the preparation of appropriate DEPPLAN information.

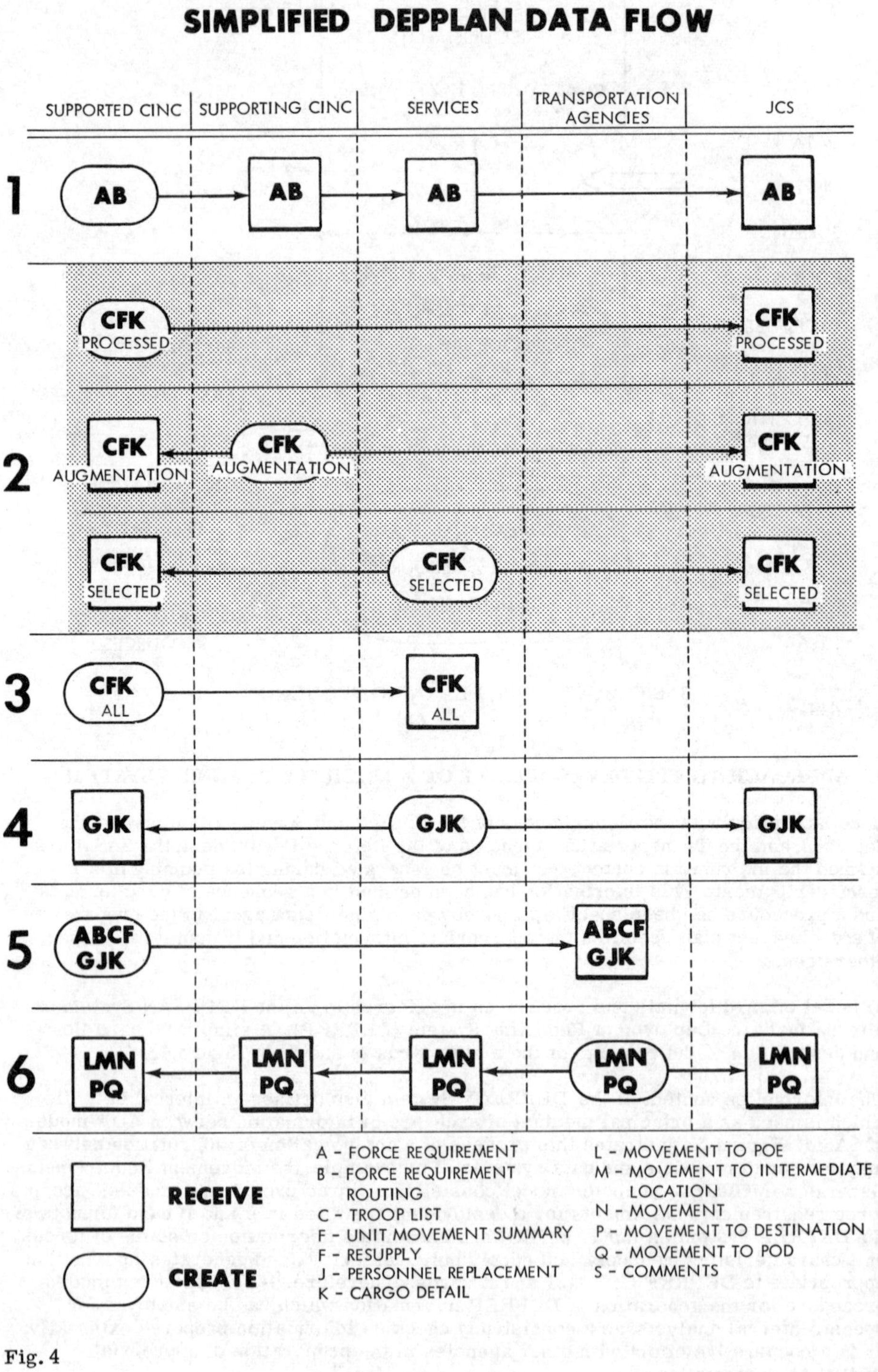
SIMPLIFIED DEPPLAN DATA FLOW
SUPPORTED CINC
SUPPORTING CINC
SERVICES
TRANSPORTATION AGENCIES
JCS
1
AB
AB
AB
AB
2
CFK PROCESSED
CFK PROCESSED
CFK AUGMENTATION
CFK AUGMENTATION
CFK AUGMENTATION
CFK SELECTED
CFK SELECTED
CFK SELECTED
3
CFK ALL
CFK ALL
4
GJK
GJK
GJK
5
ABCF GJK
ABCF GJK
6
LMN PQ
LMN PQ
LMN PQ
LMN PQ
LMN PQ
RECEIVE
CREATE
A - FORCE REQUIREMENT
B - FORCE REQUIREMENT ROUTING
C - TROOP LIST
F - UNIT MOVEMENT SUMMARY
F - RESUPPLY
J - PERSONNEL REPLACEMENT
K - CARGO DETAIL
L - MOVEMENT TO POE
M - MOVEMENT TO INTERMEDIATE LOCATION
N - MOVEMENT
P - MOVEMENT TO DESTINATION
Q - MOVEMENT TO POD
S - COMMENTS

Fig. 4

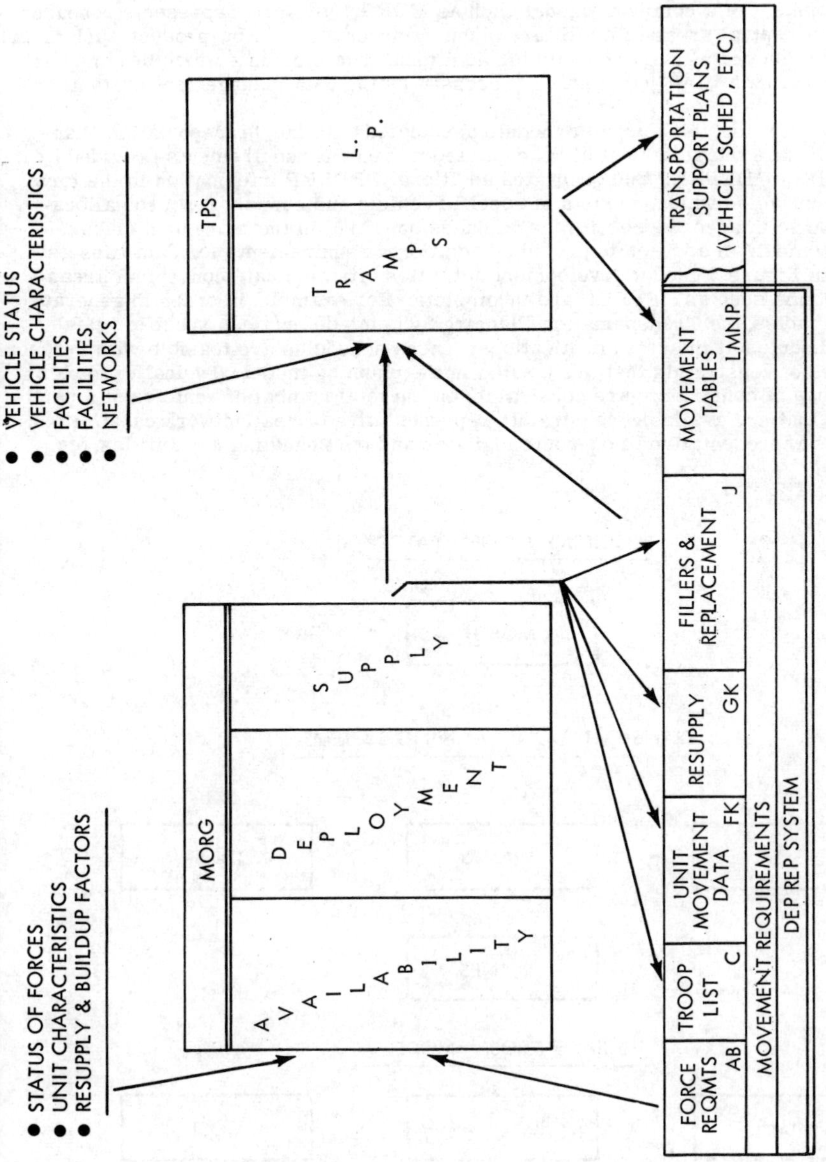

Fig. 5 SASM PLANNING/ANALYSIS PROCEDURES

The development of a computer model such as MORG, therefore, represents research in major conceptual areas of a military planning process. As a by-product, such model development generates requirements for additional data elements which become part of the internal data files that must be processed by the data management system.

In a similar fashion, the computer models associated with the Transportation Planning System take the statement of time-phased movement requirements provided by either MORG or DEPREP and generates additional DEPREP information in the form of movement tables which consists of specific vehicle and support resource allocations prepared in accordance with the demands specified in the movement requirements statement. In addition to providing additional requirements for data files as indicated in Figure 5, model development activities often indicate conceptual areas where systems such as DEPREP are incomplete. For example, in order to generate movement tables, the Transportation Planning Systems determines specific vehicle arrival schedules at indicated destinations. These schedules are feasible with respect to the types of constraints that are treated in the planning model. Typical examples of such types of constraints are constraints on the total number of vehicles of a particular type that are available, constraints representative of real network capabilities, constraints representative of on-route logistics and port handling capabilities, etc.

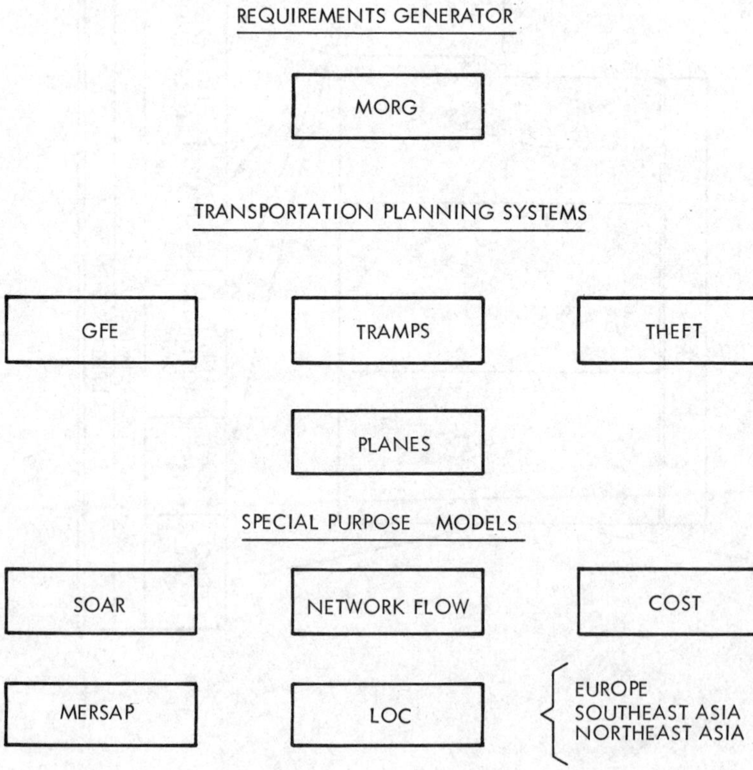

TRANSPORTATION MODELS DEVELOPED BY SASM

Fig. 6

In order to develop movement tables, the scheduled capacity created by such transportation operations is then allocated to the various types of demand in accordance with priorities and time-phased objectives. If the DEPREP system were to provide all the information required to analyze the feasibility of a given deployment, it would contain information on scheduled transportation operations. However, in its current form it does not. Therefore, in order to analyze plan feasibility or the impact of changes in the resources made available, it is essential that alternative movement schedules be provided as part of the deployment plan or else that computer models such as the Transportation Planning System be used to analyze the various alternatives possible. It is through the development of systems like MORG, Transportation Planning System, and DEPREP and the attempt to tie them together in an automated environment, that such conceptual interrelationships are forcefully pounded home. And it is through the effort that we invest in these programs that we are slowly and systematically gaining the analytical insight into the nature of the problems we are dealing with. The nucleus of models and data systems around which the SASM planning and analysis procedures are being developed are those DEPREP, MORG, Transportation Planning System and the related data files. Figure 6 shows some of the other models in the Development Plan.

4. SASM TRANSPORTATION MODELS

The largest and most complex SASM models have been developed to treat various aspects of the transportation planning process depicted in Figure 7a. While all of the models are conceptually similar in these terms, they do not all perform all of the steps, and not necessarily in the order shown, and usually to differing levels of detail. For example, the TRAMPS model (Figure 7b) directly accepts DEPPLAN data, and processes it at the level of detail, along with similar detailed data from stored files, through its six phases, ending in a day-by-day movement schedule for each DEPPLAN line item. TRAMPS is designed to handle 3-4000 line items, over as many as 100 individual inter-theater routes, for periods up to 180 days at one pass. The largest problem handled to date consists of over 6, 000 line items (implying more than 14, 000 discrete commodities), 82 long-haul channels, for 180 days. It took two passes, more than 40 hours, on an IBM 360/50.

By contrast, the Gross Feasibility Estimator (GFE) shown in Figure 7c does not have permanent data files and only provides gross feasibility estimates with some vehicle schedule information. It is designed to run rapidly, usually on aggregated or small problems, and includes some special features such as a convoying routine not included as capabilities in other models.

Linear programming models now in use, such as THEFT and others under development by IBM (See Figure 7d), concentrate heavily upon the subject of allocation as a function of varying constraint formulations and objective specifications. The insights provided by the study of the allocation problem is critical to an understanding of mobility, and, increasingly, we are becoming involved in this area. However, practical limitations on LP techniques presently constrain the size of problem that can be handled. Some of our problems with sizes of 600-2000 rows and 5-6000 columns, . 5-1% dense have run for 1-5 hours on IBM 360/65 and CDC 3600 equipments. We have, therefore, become involved again with the need to understand the quantitative effects of aggregation.

5. OPERATIONAL EMPLOYMENT OF LARGE SCALE MODELS

The selection of models to use in the analysis of strategic mobility aspects of military operations presents some difficulties to SASM analysts. These models have taken 1½ to 2 years to develop, they have not been fully evaluated, they represent first generation developments, and as such are sometimes difficult to use. They often take long times to run, require extensive data preparation prior to running and extensive analysis to interpret results. Automatic model interface, flexible common report generators with high power graphic outputs are second generation capabilities now under development.

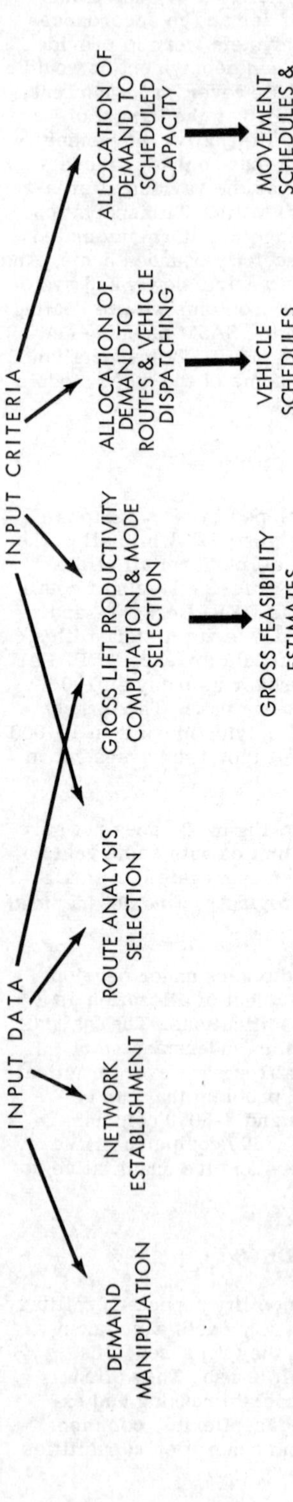

Fig. 7a Transportation planning process

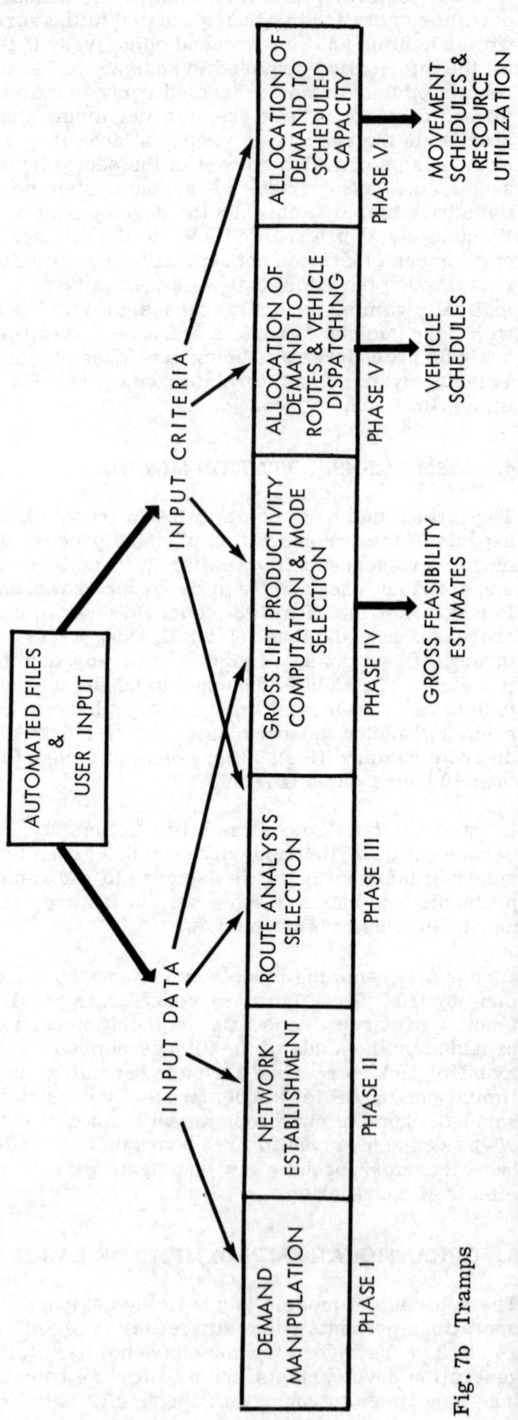

Fig. 7b Tramps

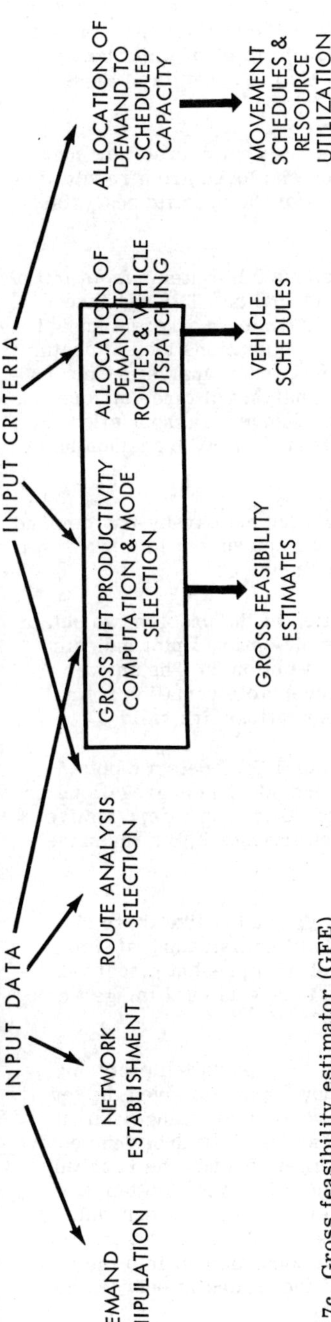

Fig. 7c Gross feasibility estimator (GFE)

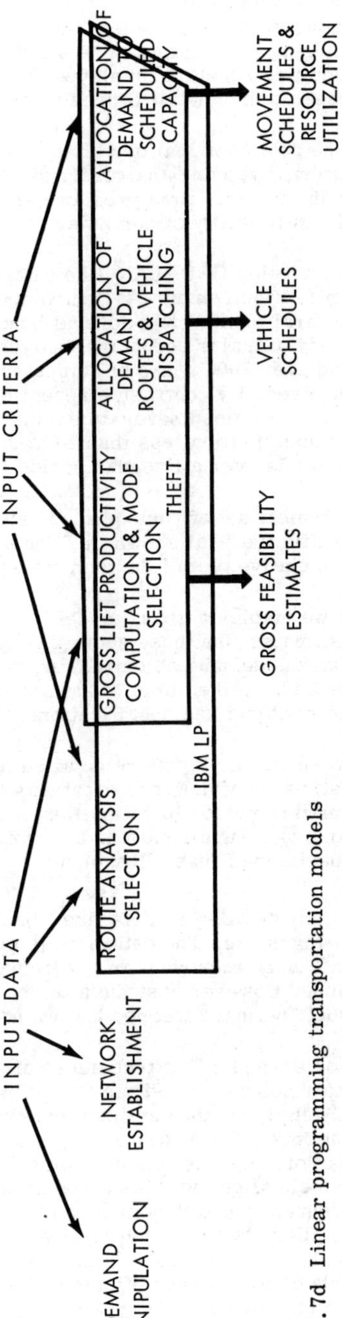

Fig. 7d Linear programming transportation models

However, through experience we are developing a set of procedures for defining and analyzing our problems that begin to exploit the potential offered by our models. At the beginning of analysis, rough problem sizing runs are made with gross feasibility models. As the problem takes shape, the LP models are employed with parametric ranging to investigate the sensitivities of the factors thought to be of importance. Detailed runs are employed for accuracy checks or to establish results in those cases where detail proves to be critical.

The decisions that establish how many factors are varied to what degree, how many alternatives are analyzed, and what level of detail will be employed are a result of many factors: time pressure, the number and capabilities of the specific analysts, the availability of computer time, data, etc.

To develop DEPPLAN-like data on a movement of some 3, 000 line items from initial force requirement (A, B card sets) specifications using MORG and TPS can take several months, and several hundred hours of IBM 360/50 time. In a recent annual study of major scope, the analysis of 4 plans in several combinations took 6 months and over 2000 hours of computer time with several man-years of analyst's time involved. By contrast, a recent emergency development analysis of three or four units, examined several combinations of forces for each of seven transportation options totaling less than fifty hours of computer time, less than two man-months of analysis over a five-day period.

A typical set of time-phased delivery requirements might have a day-by-day demand profile like that shown in Figure 8a. We often find it more convenient to work with the cumulative plot of the demand profile like that shown in Figure 8b.

If we employ a gross model to an aggregate representation of the problem we obtain estimates like those shown in Figure 8c. However, does the option 3 plot indicate that the demand objective can be feasibly met? How about option 2? The answer depends on the effect of aggregation and the degree of acceptable variation from exact object use specifications, assuming the model computations are valid.

When we employ more detailed models such as the various LP models, we obtain estimates of delivery capability like those shown in Figure 8d. The aggregations are smaller, more allocation flexibility allows for surging of utilization rate, effective time varying allocations that optimize delivery for the entire move. But the same questions of feasibility occur.

Highly detailed models like TRAMPS provide movement estimates like those shown in Figure 8c. The detail, while perhaps impressive, is costly, misleading at times, and, more important, of questionable significance, at least with present model versions. However, just the attempt to compute it alone has taught us vital things we would not have learned had we not been forced to try.

For example: The techniques of solving large dimension transportation problem matrices used in Phase 6 of TRAMPS were necessary developments given the level of details we then anticipated. We expect the requirement for increasing detail to continue. The allocation logic of earlier TRAMPS phases often introduces the equivalent of preprocessor bias which distorts the demand profiles through the resulting vehicle allocation. As a result, the fine structure schedules in Phase 6 often cannot adequately match demand and schedule capacity. Thus, even though the schedules are detailed, they may not represent effective solutions to the problem. However, in the analysis of TRAMPS detailed allocations, we have gained some insight into the magnitude of non-integer effects of LP allocation methods and the tradeoffs between high load factors and short fall on low volume shipping routes.

As a result of the lessons learned in the use of TRAMPS and our LP models, we are now engaged in designing LP allocation logics into early phases of TRAMPS, with options to halt if aggregate movement estimates are sufficient.

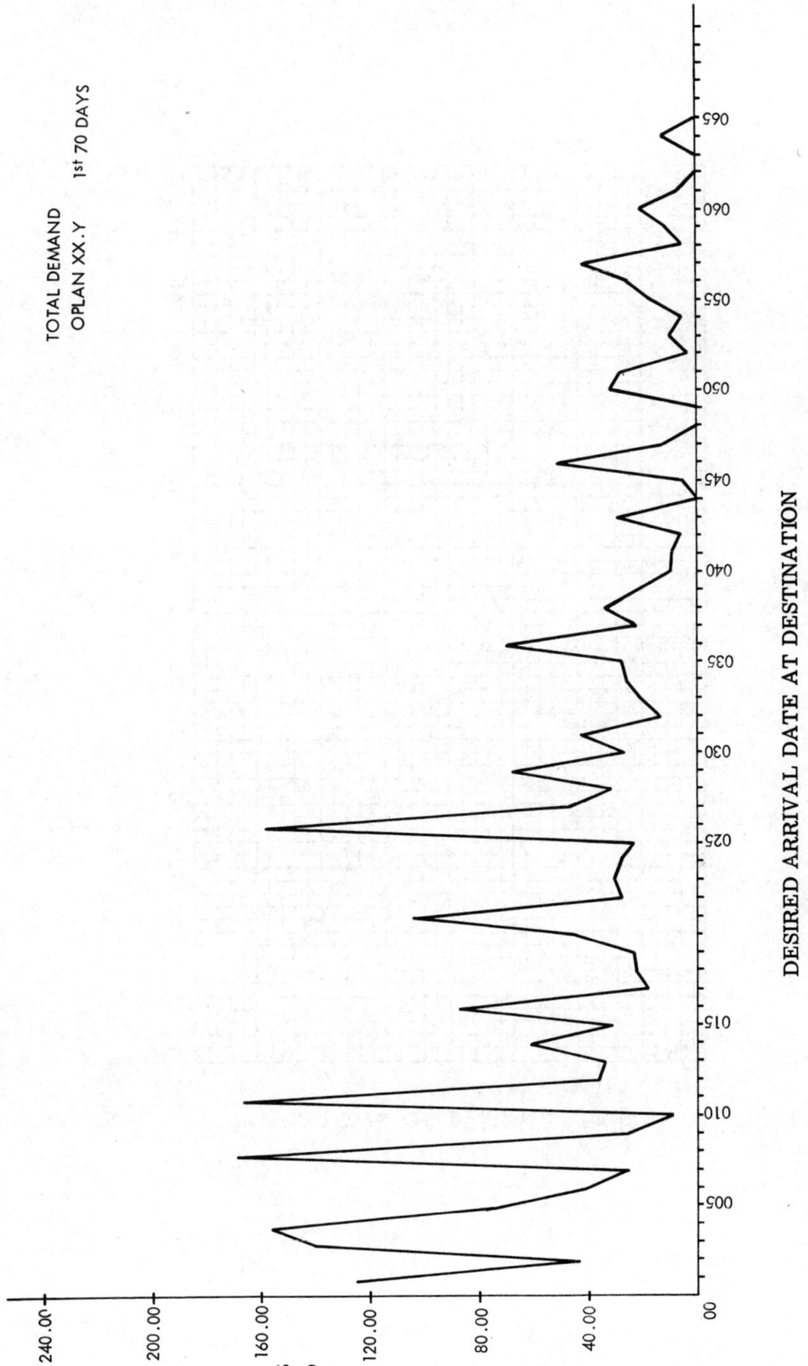
TOTAL DEMAND
OPLAN XX.Y
1st 70 DAYS
SHORT TONS
OF DEMAND
(x10²)
240.00
200.00
160.00
120.00
80.00
40.00
00
005
010
015
025
030
035
040
045
050
055
060
065
DESIRED ARRIVAL DATE AT DESTINATION

Fig 8a

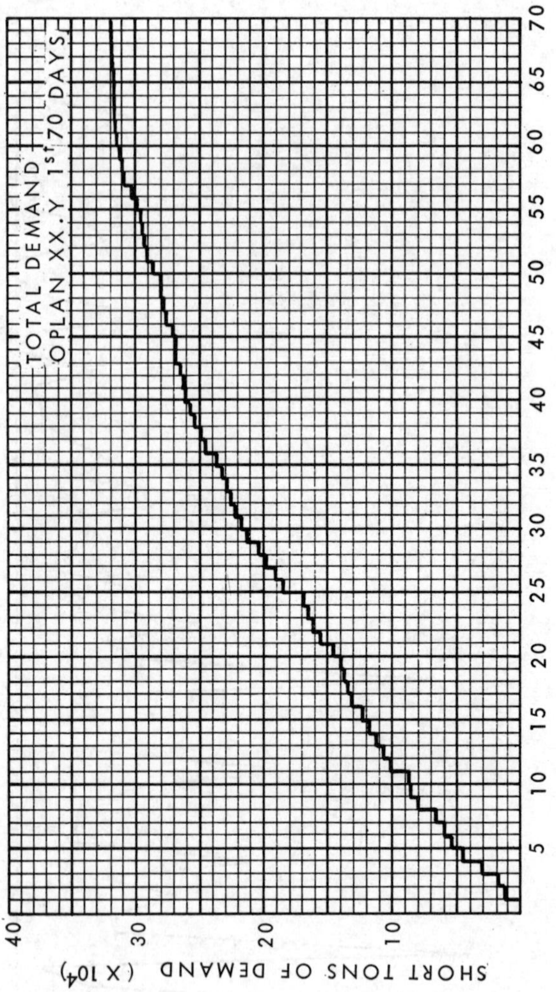
TOTAL DEMAND
OPLAN XX.Y 1st 70 DAYS
40
30
20
10
SHORT TONS OF DEMAND (X 10^4)
5
10
15
20
25
30
35
40
45
50
55
60
65
70
DAYS

Fig. 8b

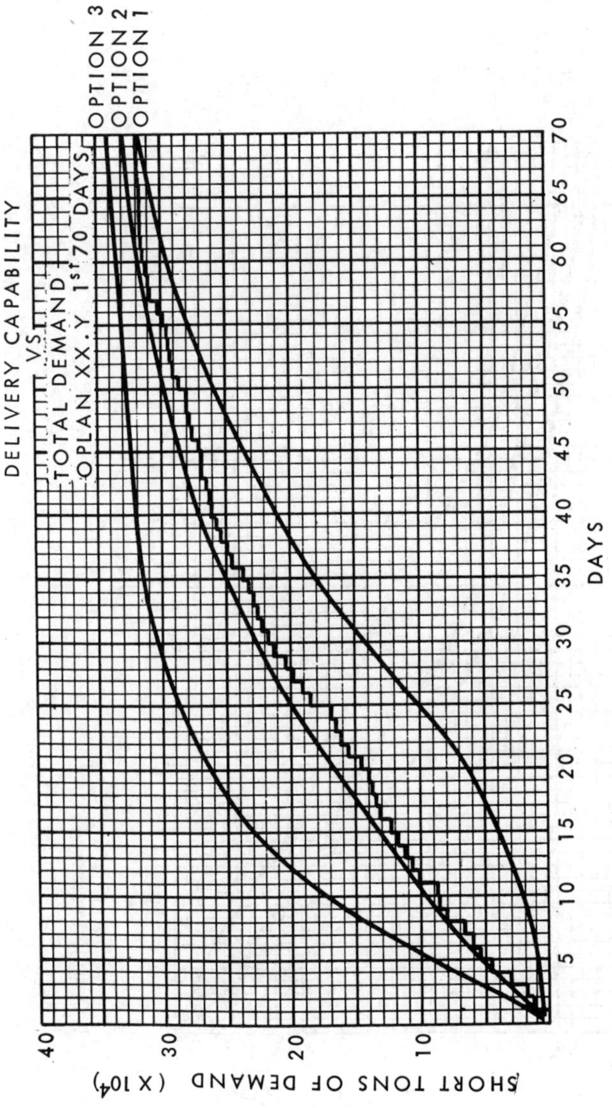

Fig. 8c Gross feasibility estimates

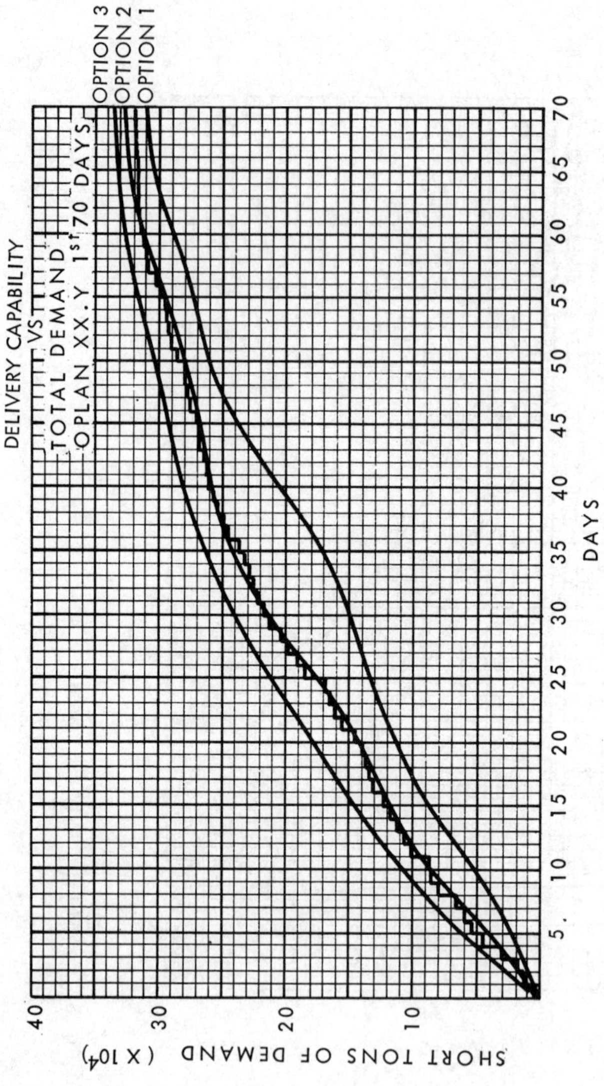

Fig. 8d Optimizing models

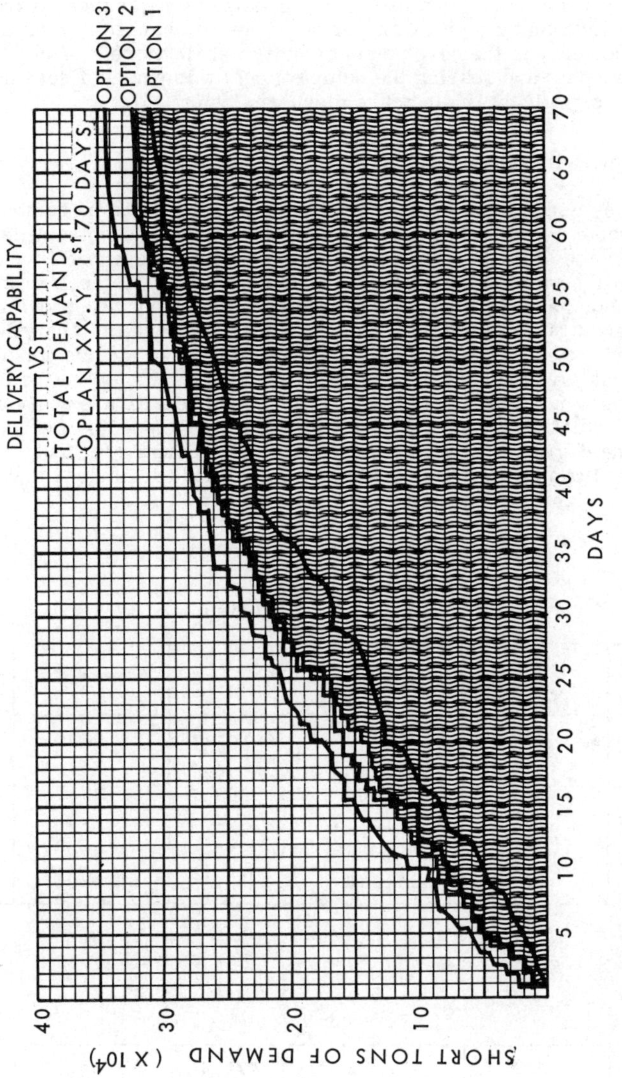

Fig. 8e Detailed scheduler

6. RESULTS TO DATE

SASM now has almost one year's worth of experience using large scale models in the analysis of strategic mobility problems. Early attempts were frustrated by logic bugs and deviations from design expectations. More effort was required than anticipated in the preparation of data and use of the models. Some of these troubles persist. We have learned that a significant period of operational test, evaluation, and validation is a necessity in the development of large scale models. At this time, the first generation models are nearing the completion of this phase of development. Some of the more significant findings are discussed below.

6.1 Vehicle Productivity Computations

Our models employ variants of a basic approach that calculates the number of sorties over a specific route per time interval based upon number available, utilization rate, block speed, and route distance. This, times the allowable cabin load is used as the productivity. While several of our models can accept ACL as a function of route, our grosser calculations have not exercised this feature. As a result, errors in productivity on the order of 10% can be expected in some of our results. Analysis of actual loads carried during recent deployments has shown that load factors considerably less than ACL for a given route can result from factors that apply to the loading of given unit types on a particular type of aircraft. These results are qualitatively in agreement with predictions of specific loading models developed by Lockheed Corp. (See Table 1). The degree to which this effect can be accounted for and corrected within the present framework of our models is unclear, but it is clear that we must, for this effect is a significant source of error.

C - 5
PL = 256,000

DIVISION	No OF SORTIES	AVG PL	AVG VEH PL	AVG PASS PL	AVG. No OF PASS
AIRMOBILE	208	122,489	77,968.7	21,075.2	81.1
AIRBORNE	195	151,611	153,247.6	17,751.1	68.3
ARMORED	569	218,131	211,909.0	7,333.6	28.2
INFANTRY	390	197,424	184,329.9	10,587.5	40.7
MECHANIZED	470	218,337	210,532.2	8,896.1	34.2
FORCE A	835	236,333	223,372.1	11,643.3	44.8

C - 141
PL = 85,000

AIRMOBILE	656	33,539	21,701.4	6,692.8	25.7
AIRBORNE	232	45,893	38,421.5	5,183.6	19.9
ARMORED	1823	63,098	65,643.5	2,287.2	8.8
INFANTRY	1179	47,660	42,995.	3,501.8	13.5
MECHANIZED	1185	56,915	53,004.1	3,540.3	13.6
FORCE A	3009	53,935	47,440.3	3,327.6	12.8

Table 1

Treating aircraft as though prime (rational) aircraft were continuously cycling at their round trip period on a route with even spacing is a misrepresentation of fact. During a recent deployment the trip frequency of aircraft was measured. The results are shown in Table 2. Our models, given the same number of aircraft available would have used an average of about 3 trips per plane. We have no comparable information on large scale deployments. Even though the total number of missions flown over the deployment for a given number of aircraft can be varied by changing the utilization rate, we are still concerned because of the obvious interest in predicting how many missions we can actually extract from the fleet under varying conditions. Hence, our project to develop an Airlift Simulation Model for detailed analysis of airlift operations. Most of our models treat ships in a similar fashion, accounting for repositioning times. As yet, we have no comparable data on the errors introduced, but clearly the effects should be more pronounced. The closely related questions of low load factor on low volume channels is treated below, and will give some insight into the magnitude of the problem.

No OF A/C	No OF MISSIONS PER A/C	No OF MISSIONS
2	5	10
13	4	52
44	3	132
53	2	106
69	1	69
181		369

$$\frac{\text{TOTAL MISSIONS}}{\text{TOTAL A/C}} = 2.0$$

FREQUENCY OF A/C APPEARANCE IN AN AIRLIFT DEPLOYMENT.

Table 2

6.2 Non-Integer Effects and Aggregation

In the analysis on detail of a multi-million ton deployment recently conducted for a particular lift force, approximately 5% of the cargo and a higher percentage of passengers failed to move by our computations. Of this amount, about 75% failed to move because vehicles would have been required to move with load factors of less than .5, and model logic suppressed dispatching. The remainder failed because of some other scheduling maladjustment. The effect of such small scattered tonnage shipments is lost in route aggregation, which makes such shipments undistinguishable from the demands on high volume channels. The small scattered tonnage effect is more pronounced the more the channels, or the smaller the problem, and sets in when meaningful demand aggregates are of the same magnitude as the vehicle capacity. It is, therefore, a more serious concern in shipping, but is of growing concern with the advent of vehicles like the C5A.

To the analyst, non-integer effects are similar but more difficult because they are implicit in the results. Non-integer effects resulting from linear program allocations, like the small scattered tonnage effect, are also more pronounced when the number of routes involved is greater, or the number of vehicle types employed is larger. In a worst case, one vehicle dispatched per time period per vehicle type, could appear as a fractional vehicle allocation on each route of the problem. For very large problems the problem might not be great (of the order of a few percent at most). The operational questions involved are identical—should more vehicles be allocated and allowed to work at low load factors? Or should the load be left or rerouted?

6.3 Optimal Allocations

In the SASM analysis to date, optimality has usually been defined as either minimum cumulative deviation of delivery date from desired delivery date for each specified force, or else as minimum time to close the entire demand. We have on occasion adopted minimum cost or pseudo-cost criteria to meet stated values of deviation or total closure. We are experimenting with other criteria such as maximum residual capacity in ton miles or vehicle days, minimum number of sorties, etc.

Most of these criteria, particularly the minimum deviation areas, optimize over the total period of the deployment, without respect to the uncertainties of the real world that grow the further into the future one tries to project demand. Further, one can question the significance of demand peaks like those shown in Figure 8a. During the rapid deployment phase, unit moves certainly create such peaks, but do they occur during the supply build-up and resupply phase? Probably not nearly as much, but we have had unit deployment spikes throughout the current conflicts in Southeast Asia, some of which could have been predicted—some not.

We are now experimenting with partial, or short-range optimizing procedures as a way of simplifying our overall analysis, more adequately treating time varying uncertainties on demand, and freeing up our computers to analyze detail in the immediate future when it is most significant. But uncertainty on demand and the role it plays or should play in resource allocation is still a largely unexplored area.

6.4 Implications for Resource Programming and Budgeting

From the work done to date, certain observations follow with respect to current techniques in resource programming and budgeting. Current practice uses highly aggregate models, closely related to the SASM models, multiple theater demand scenarios, and heavy emphasis on economic factors, procurement costs and credits. From the above discussions, it is clear that inaccuracies exceeding twenty percent can easily be introduced in the estimates of resources required to do a given job and some times the inaccuracies can be much greater. Usually, the inaccuracies lead to underestimates of required capacity. Experience would indicate that underestimates of the job required are far more serious than that.

One clear implication for resource programmers and budgeteers is that great care should be exercised in using current techniques to differentiate between alternative system procurements that vary from one another by 10-20% in total cost. Although relative costs may be a better measure than absolute costs, even these can be easily miscalculated.

A second implication is that expenditures of resources for improvements in the planning system offers great potential of significantly altering the real productivity of the mobility system. We feel confident that the programs that have been established and are being conducted in SASM represent real progress towards the development of the analytical basis for such planning.

Applications of Linear Programming and Transportation Problem to Strategic Mobility Problems

M. ARBABI, J. G. FISHER, H. M. HOROWITZ, D. F. KOCHER
I.B.M. Corporation, U.S.A.

INTRODUCTION

This paper presents applications of linear programming to strategic mobility problems. The system of which the linear programming is a part is the Transportation Planning System (TPS). The system is being developed to support the transportation management responsibilities of the Special Assistant to the Joint Chiefs of Staff for Strategic Mobility (SASM).

Essentially, TPS is a transportation simulation planning tool which in computer software design, represents a generalized DOD world-wide logistics transportation network. This network consists of transportation resources (vehicles), origins and destinations, ports of embarkation (POE) and debarkation (POD), and the short haul and long haul routes which serve these facilities. TPS will assist the transportation planner with the laborious and time-consuming, data processing and computational aspects of the planning process while, at the same time, affording him the decision-making prerogatives. Wherever practical, the model is being designed to use on-line communication between the Staff Officer and the system.

The initial increment of the system is presently operational. It is known as Transportation Movement Planning System (TRAMPS).*

The emphasis of this paper is on linear programming applications. In order to bring this emphasis into focus, we shall briefly describe the present TRAMPS system; then show where linear programming is used or envisioned to be used.

The paper is divided into three sections: Section 1 describes the TRAMPS System Description, Section 2 describes the Application of Linear Programming to Movement Scheduling and Section 3 describes Application of Linear Programming to Resource Allocation for Transportation Planning.

1. TRAMPS SYSTEM DESCRIPTION

For management flexibility in automatic data processing (ADP) scheduling, as well as transportation planning, the TRAMPS ADP model has been designed in modular form. Six modules are provided which represent logical tasks or functional phases of the basic transportation management function. If alternate courses of action are possible, the user is afforded the ability to direct the proceedings between phases of processing. These six phases are:

(a) Functional Phase 1—Requirements Selection

(b) Functional Phase 2—Find Connecting Channels

(c) Functional Phase 3—Rank and Select Channels

(d) Functional Phase 4—Assign Mode and Channel

(e) Functional Phase 5—Allocate Resources and Dispatch

(f) Functional Phase 6—Movement Scheduling

---

* This contract is based on DCA Contract No. 100-68-C-0011.

The overall tasks of TRAMPS are basically to: select and compile time-phases movement requirements from the JCS Deployment Planning System (DEPPLAN), JCS Force Requirements and Deployment Monitoring System (DEPMON), locally generated plans, or other planning support systems; identify and select logical POEs and PODs, modes of transportation, and associated short haul and long haul transportation routes; rank and select preferred transportation modes and routes; assign requirements to transportation routes; allocate and dispatch resources; and schedule the movement requirements of the stated problem. The relationship of these tasks to the six functional module phases of TRAMPS is depicted in Figure 1-1.

In the performance of these tasks, TRAMPS determines possible routes from origin to destination for the passengers, general cargo, and outsize cargo included in the problem statement. The possible routes are then ranked by such system criteria as: preferred lift mode, resource availability, required arrival dates, etc. In a continued refinement of the transportation problem, the system also evaluates pertinent route associated transportation network features, (e.g. available transportation facility transshipment capacity, available lift resources, and possible arrival times) for the selection of specific routes for each stated movement requirement. Transportation resources are then apportioned to these routes in such a manner as to accommodate the maximum number of requirements that could be moved. The final scheduling of movement requirements is performed through the employment of a modified version of the classic linear programming transportation problem, i.e., the solution for the most economical shipment from a set of production points to a set of consumption points. The TRAMPS version of this solution uses appropriate penalty factors for early or late delivery to ensure that shipments arrive within the stated bounds of the required arrival date, and to ensure that their desired arrival sequence is preserved whenever possible. Requirements may be scheduled for delivery on a preferred date, scheduled for delivery within a fixed number of days from the preferred arrival date (controlled slack time), or scheduled any time to the end of the problem period (uncontrolled slack time).

The first five phases of TRAMPS do not utilize any standard technique, they simply simulate the movement of loads over the network constraint by vehicle availability and route structure. The last phase, Movement Scheduling utilize the transportation problem technique. This phase is described fully in the next section.

2. FUNCTIONAL PHASE 6—MOVEMENT SCHEDULING

This section describes the use of the Transportation Problem to perform detailed scheduling on multi-commodities through a network. The network consists of nodes and the routes or channels which connect these various nodes (ports). Within each channel there exist many subchannels which represent the different transportation resource activity between any two nodes. These transportation resources can be thought of as types of vehicles each with individual speeds, commodity carrying capabilities and capacities. The multi-commodities can be thought of as requirements each with individual quantities, priorities, availability for shipment time and preferred delivery time. The problem to be solved is to schedule requirements on vehicles in such a manner as to minimize the deviation from preferred delivery time without violating any problem constraint such as maximum vehicle capabilities or requirement shipment prior to availability, etc. The initial model described in this paper optimizes the scheduling on one channel of the total network at a time. Reference should be made to section Future Developments for a discussion on a Total Network simultaneous solution.

Conceptually, the input parameters for this application of the Transportation Problem are requirement quantities and vehicle capacities by time period. It is desired that a requirement be delivered in a preassigned time and this preferred eligibility for scheduling is presented in the cost matrix by the assignment of weight penalty numbers. The highest preference delivery periods are represented by zero weight penalties and the penalty weight increases as preferred delivery decreases. This

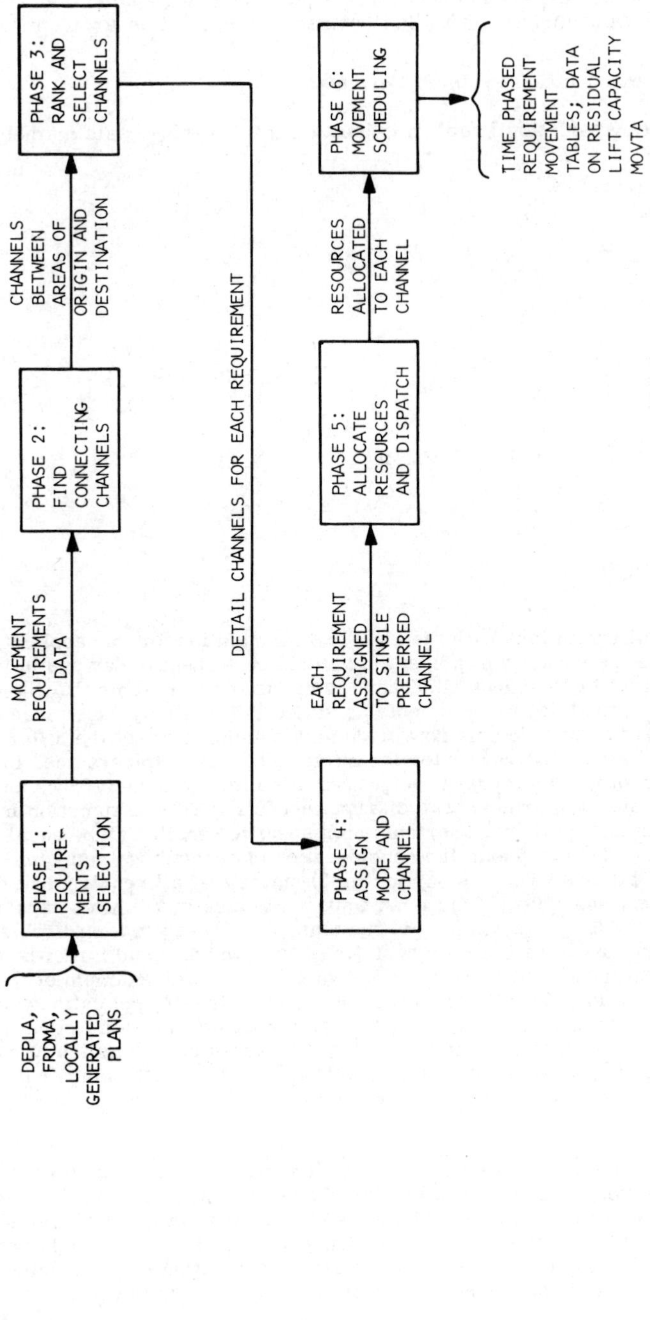

Fig. 1.1 Phase interface diagram

rate of increase is a function of requirement priority, i.e., the highest priority has the highest rate of increase of weight penalty. Of course, as in the classical Transportation Problem the objective function is to minimize the use of these weight penalties, i.e., to deliver requirements as close to preference as possible.

In order to develop this concept into a model which was practical to implement, two major problems had to be resolved. First, the classical Transportation Problem formulation had to be revised and second, detailed scheduling application had to be designed to work in conjunction with this revised Transportation Problem.

2.1 Revised Transportation Problem Formulation

The classical Transportation Problem is built on the following basic mathematical description:

$$\sum_{i=1}^{m} a_i = \sum_{j=1}^{n} b_j \tag{1}$$

$$\sum_{j=1}^{n} x_{ij} = a_i; i = 1, \ldots, m \tag{2}$$

$$\sum_{i=1}^{m} x_{ij} = b_j; j = 1, \ldots, n \tag{3}$$

$$x_{ij} \geqslant \phi; i = 1, \ldots, m \text{ and } j = 1, \ldots, n \tag{4}$$

minimize

$$z = \sum_{i=1}^{m} \sum_{j=1}^{n} c_{ij}x_{ij} \tag{5}$$

There are several limitations with the classical formulation for our application. Assume that a_i is the capacity of a specific vehicle on a specific day and b_j is the requirement amount to be scheduled. There is no reason to assume that the total capacity will equal the total requirements as stated in (1). Early experimentation with the incorporation of a dummy row or dummy column to offset the difference between the row and column vector totals proved to be very cumbersome. In addition, the selection of the appropriate weight penalties for the dummy rows and columns was a difficult process in order not to disrupt scheduling. The assumption in (2) requires that scheduling of all requirements on a specific vehicle day be equal to that vehicle's capacity. This prevents the consideration of scheduling requirements on partially loaded vehicles. The assumption in (3) states that all requirement quantities must be completely scheduled. This is an impossible task in a limited capacity condition. Another interesting shortcoming of the classical Transportation Problem is that it does not offer a guaranteed infinity cost, i.e., a cost which would force the corresponding element in the solution matrix to be equal to zero. An example of the use of infinity weights (the largest number in the cost matrix) in our application would be not scheduling a requirement until available. An obvious conclusion is that a Transportation Problem formulation which considered inequalities in addition to equalities is required for this specific application. The basic addition to the classical formulation which leads to this revision is the incorporation of a permanent fallout row (a_{m+1}) with a capacity of infinity.

This fallout row would accept requirements which could not completely schedule in $1 \leqslant i \leqslant m$. These requirements would be considered as unscheduled. The range of weight penalties assigned to the fallout row is greater than the normal weight penalties but less than infinity weight penalties. The guaranteed infinity weight penalty assures that a lower objective function is obtained by creating positive values in the fallout row rather than in an element corresponding to an infinity cost.

The revised Transportation Problem formulation which allows for inequalities is as follows:

$$\Sigma x_{ij} \leqslant a_i; i = 1, \ldots, m \tag{6}$$

$$\Sigma x_{ij} \leqslant b_j; j = 1, \ldots, n \tag{7}$$

$$x_{ij} \geqslant \phi; i = 1, \ldots, m \text{ and } j = 1, \ldots, n \tag{8}$$

minimize

$$z = \sum_{i=1}^{m} \sum_{j=1}^{n} c_{ij}x_{ij} \tag{9}$$

The revised Transportation Problem offers a less rigid formulation and hence opens the door to many new applications. Some interesting observations about this formulation can be made. Recall that the permanent fallout row had a capacity of infinity. From this fact and from equations (6) and (7) the following relationship exists between the two input vectors:

$$\sum_{i=1}^{m+1} a_i \geqslant \sum_{j=1}^{n} b_j \tag{10}$$

Without considering the permanent fallout row, it could be concluded that the totals of the input vectors (a and b) in no way affect the solution process

The permanent fallout row introduces another interesting by-product of the formulation for the scheduling application. Within the constraints of the problem and without violating the basic objective function (9) the fallout concept maximizes the scheduling of requirements (or minimizes the total unscheduled requirements).

minimize

$$z' = \sum_{j=1}^{n} x_{m+1,j} \tag{11}$$

After the solution is obtained, the capacity vector (a_i') contains the residue or unused capacity, if any. Therefore from (6) the following is true after scheduling:

$$a_i' = a_i - \sum_{j=1}^{n} x_{ij}; \quad i = 1, \ldots, m \tag{12}$$

The technique selected for solving this revised Transportation Problem was based on the shortest path approach presented by Dr. A. J. Hoffman and Dr. H. M. Markowitz. Mr. J. J. Lagemann modified and generalized this approach and produced a computer program which solved the classical Transportation Problem using shortest path techniques. Mr. Lagemann's analysis and program was used as a basis for the revised Transportation Problem that was developed.

2.2 Scheduling Application

The system described in this paper is presently operational and solving complex large scale transportation movements. This section presents some of the salient techniques used in the development of the model.

<u>Multi-Commodity Conversion</u>

Since any transportation problem requires that its two input vectors be of the same commodity, a commodity conversion algorithm is required to transform all requirements and capacity to one base commodity. This technique is an approximation of a Multi-Commodity Movement through a channel.

Conversion factors are computed to convert from commodity A to commodity B, using the following formula:

$$Q = \sum_k \frac{u_k w_k}{w_t} \cdot Q'_k \tag{13}$$

Q is the channel conversion factor for conversion of commodity A to commodity B;

u_k is the total number of appearances of vehicles of type k during the entire scheduling period;

w_k is the capacity of vehicle k in terms of commodity A;

w_t is the total capacity of all vehicles of all types in terms of commodity A; and

Q'_k is the specific vehicle conversion factor for vehicle k for conversion of commodity A to commodity B.

Prior to scheduling all commodities other than the base commodity are converted to the base commodity as follows:

$$E_{j(B)} = Q \cdot b_{j(A)} \tag{14}$$

where

$E_{j(B)}$ is the equivalent weight of requirement j when considered as commodity type B.

$b_{j(A)}$ is the requirement j quantity of commodity A.

After scheduling re-conversion takes place using the following relationship:

$$S_{j(A)} = \frac{x_{ij(B)}}{Q} \tag{15}$$

where

$S_{j(A)}$ is a scheduled portion of requirement j of the original commodity A.

$x_{ij(B)}$ is a scheduled portion of requirement j on vehicle day i of base commodity B.

Multi-Speed Tableau Design

Prior to discussing the Multi-Speed Tableau Design, it is advantageous to compare a one-speed to a multi-speed problem from an operational point of view.

In a one-speed model, requirements not only have to be assigned to a channel, but to a speed as well. If a channel were to transport three commodities and were to consider five speeds, it would be necessary to perform 15 separate transportation problems. A multi-commodity and multi-speed tableau solves all 15 of these transportation problems simultaneously and does not require a previous algorithm to specifically designate requirements to speeds.

The primary objective of the multi-speed approach is to deliver as much of a requirement in its preferred delivery period or span of periods as possible. This may be accomplished by splitting a requirement among several vehicle types, and having the varying speeds compensate for the staggered dispatching from a port of embarkation (POE) and experience a simultaneous arrival at the port of debarkation (POD).

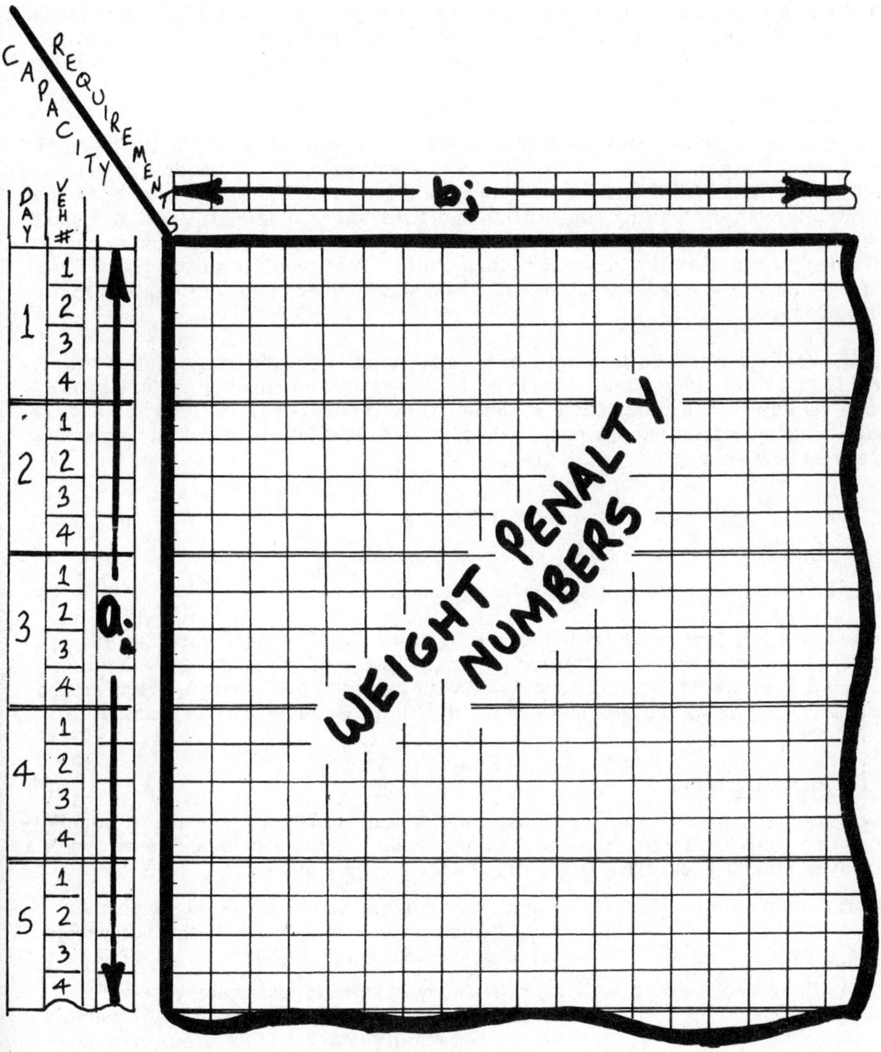

Fig. 2.1 Multi-speed tableau structure

The multi-speed tableau is completely flexible as to the number of vehicle types and the number of time periods one wishes to consider at one time. Consider the following relationship:

$$m' = \frac{m}{v} \tag{16}$$

where

m is the total number of desired vehicle time periods (m + 1 is the fallout row)

m′ is the total number of time periods considered for all vehicle types.

v is the total number of different vehicle types.

For any vehicle time period i, the associated vehicle type k is:

$$k = \{(i - 1)\,\text{MODULO}\ v\} + 1;\ i = 1, \ldots, m \tag{17}$$

In other words, if the limit on m were 150, one could utilize all 150 capacity periods for one vehicle type, or 30 capacity periods for 5 vehicle types, and so on. The fallout period (m + 1) does not represent any particular vehicle but rather the total fallout.

Slack Concept

The Slack Concept determines the limits of allowable scheduling and is the basis of the entire application. A slack is defined as those vehicle-time periods i in which scheduling can occur. Slack is computed considering the different vehicle types, their different commodity carrying capabilities and differences in transit time. If a slack shrinks negative, it is assumed that all the available vehicles are too slow to carry a requirement within its allowable scheduling range. The weight penalties are a function of the slack of a requirement and all scheduling preferences are implicit in the weights produced.

The slack concept requires the fixing of scheduling at one of the two nodes of the channel, e.g., at the originating node (POE). Since one way transit times are known for each vehicle type, all activity at the other node (POD) is easily computed and the slack (allowable scheduling periods) is determined. The following relationships delineate the slack concept:

let

- t_0 be the most preferred scheduling period.
- t_1 be the earliest period that scheduling can take place.
- t_2 be the latest period that scheduling can take place.

The slack of a requirement falls between $t_1 \leqslant t_0 \leqslant t_2$. The weight penalties are a function of this range and the following generalization can be stated about them: $c(t_0) \leqslant c(t_1) \leqslant c(t_2)$. The weight associated with the most preferred delivery period is represented by $c(t_0)$, the weight associated with the earliest delivery period is $c(t_1)$, and etc.

Weight Penalty Function

The weights generated for the scheduling application can be as primitive or sophisticated as one desires. During the development cycle, we actually experimented with a hierarchy of weight generation that started with a binary system.

- o Binary Weights — a primitive system where all the slack periods are represented by zero weights and all other periods are represented by one weights.
- o Arithmetic Progression — a preliminary system where preference within slack is considered as a zero weight and weights are incrementally added as this preference is deviated from. Outside the slack a large weight is generated.
- o Priority — a continuation of the Arithmetic Progression but with different incremental slopes for different requirement priorities.
- o Multi-Speed — a continuation of Priority but with the specialized Multi-Speed Tableau Structure and Multi-Commodity vehicle considerations.
- o Guaranteed Infinity — a continuation of Multi-Speed but with a permanent fallout row containing weight penalties less than the so called infinity weights in the matrix.

This evolution of weight generation led to the sophisticated system which is presently in use for our scheduling application. Other applications would require a weight generation which was considerably different. The following represents a summary of the intent of the weights generated for this application.

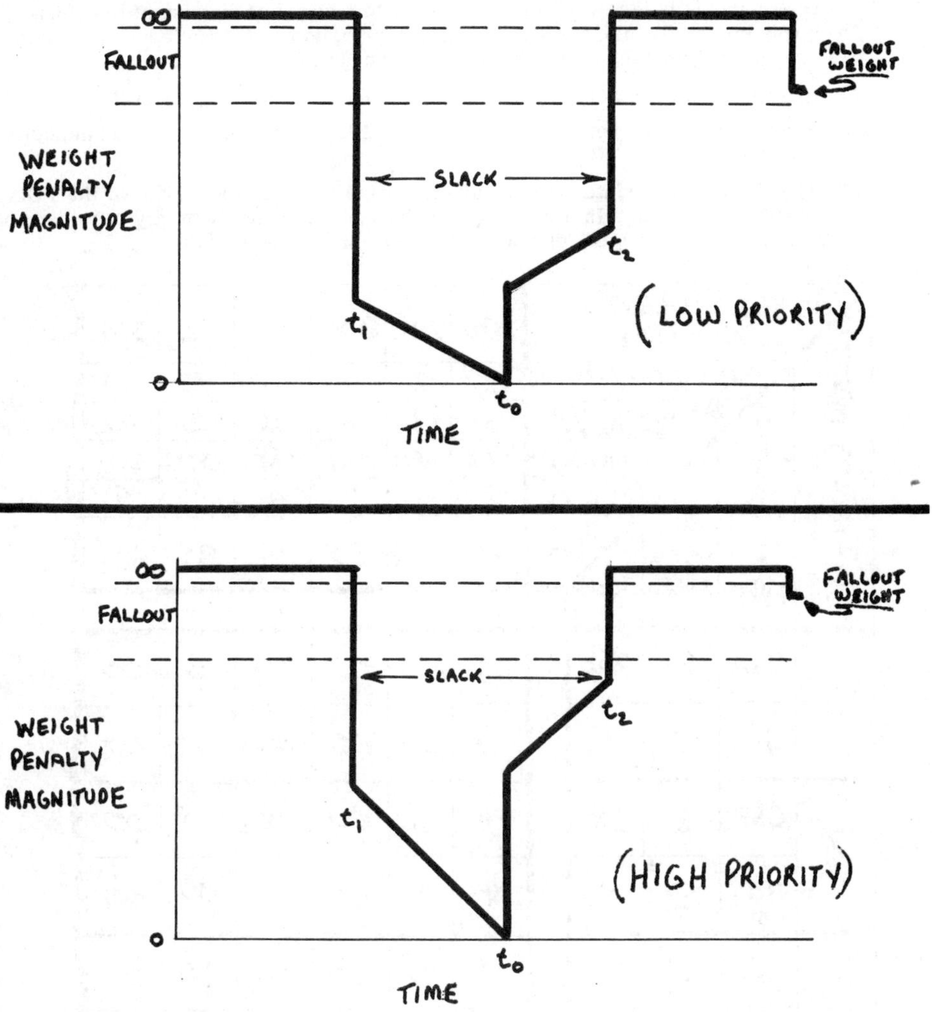

Fig. 2. 2 Priority weight penalties

The weight penalty function is the liaison between the basic movement requirements and the scheduling algorithm. All optimization criteria, such as preferred shipping periods, priorities, and slack periods, are implicitly supplied to the scheduling algorithm via the weight penalty function. The basic scheduling concepts which dictate the weight penalty function output are:

(a) The primary objective is to schedule as much of the total requirement as possible regardless of priority.

(b) The weight function attempts to ship all requirements such that they arrive at the destination in the preferred delivery period or span of periods.

(c) If concept (b) cannot be achieved, low priority items deviate from their preference before high priority items.

(d) If deviations from the preferred delivery date or preferred span of periods are necessary due to capacity constraints, they occur in three stages. The algorithm first attempts to schedule a requirement during the allowable time periods prior to the preference. If this does not dispose of the entire requirement, the allowable late delivery periods are attempted; if this is not enough, the remainder is considered unscheduled.

(e) Requirements are not moved before they are available.

(f) Requirements are not moved by vehicles that cannot handle their commodity designation.

(g) The shipping schedule utilizes adjacent time periods for arrival at the POD whenever possible. In a limited capacity condition, high priority requirements tend to utilize adjacent periods before low priority items do.

			REQUIREMENT QUANTITY	100	240	300	100	72	524
			EARLIEST POE DEPARTURE	1	2	4	8	1	10
			PREFERRED POD DELIVERY	3	3	6	20	2	22
			COMMODITY	PAX	GEN	PAX	PAX	OUT	GEN
			PRIORITY	2	1	2	4	1	5
			TONS	15	240	30	15	81	524
TIME PERIOD	VEHICLE TYPE	NBR. APPEARANCES	TONS						
1	CSA	2	200	8	∞	∞	∞	5	∞
	707	0	0	8	∞	∞	∞	∞	∞
2	CSA	1	100	4	5	∞	∞	0	∞
	707	2	50	4	∞	∞	∞	∞	∞
3	CSA	3	300	0	0	∞	∞	10	∞
	707	3	75	0	∞	∞	∞	∞	∞
4	CSA	4	400	12	10	8	∞	15	∞
	707	5	125	12	∞	8	∞	∞	∞

Fig. 2.3 Sample problem setup

3. RESOURCE ALLOCATION MODEL

As an integral part of Transportation Planning System (TPS), a linear programming formulation of a resource allocation model has been designed. This capability was first developed by MITRE Corporation under the name THEFT, which is an acronym for; To Help Evaluate Feasible Transportation. This initial model has been successfully used in gross transportation planning and gross mobility planning. The model allocates movement resources to movement requirements to meet either of the following criteria:

Minimum deviation from a desired delivery date

Minimum closure time

Secondary models, subject to one of the above criteria, can be used to:

Minimize the number of vehicles employed

Minimize operating cost.

THEFT model is presently being extended by IBM Federal System Division to include several additional features, such as; Allocation of requirements to channels, Capability to reroute ships, and in general, extending the size of the LP problem which can be solved.

It is also expected that this model, upon its perfection will replace early phases of TRAMPS, specifically Phases 2, 3, and 4. (See Figure 1-1).

3.1 Problem Description

The problem consists of a number of demands each of which is to be transported from its source to a destination. A demand is defined by its amount, commodity type, availability date at source, span of preferred delivery at destination, allowable deviation from preferred span, mode restrictions (i.e., air or sea), and priority.

The movement of demands is accomplished by one or more vehicle types whose availability is limited and which operate on a user defined network which connects the sources and destinations. The vehicle types are defined by their loading time, speed and allowable cabin load.

The network is composed of channels which connect Ports of Embarkation (POE) with Ports of Debarkation (POD) via possible enroute ports. The capacity of any port may be limited. The movement of demands is from POE to POD.

The sources and destinations for all demands must be strictly partitioned into regions. Typically, the Eastern and Western U.S. would each be a source region and Korea and Vietnam would each be a destination region. Each of the demands can now be associated with a source region and a single destination region which will be referred to as the region pair for that demand. The demand can embark only through the POEs belonging to its source region and can debark only through the PODs in its destination region.

The time domain in which the problem is framed is composed of time periods which can be of variable length.

This problem is solved by determining the optimum allocation of resources and demands to channels so as to meet some desired schedule of deliveries while neither violating port limitations nor exceeding vehicle availability.

The desired schedule may have one of two possible primary aims or objectives. One of the primary objectives is to achieve closure in the minimum possible amount of time. The other primary objective is to deliver each demand in its preferred span if possible. If not possible, then the demand will be delivered early but still as close as possible to its preferred span. If a demand cannot be delivered on time or early then it is delivered late but still as close to the preferred span as possible. If there is not enough capacity to deliver a demand then as a last resort, it will become shortfall. High priority demands tend to be delivered in their desired period before lower priority demands are considered.

A secondary objective is included which attempts to minimize the number of vehicle days which are consumed by the movement.

A linear optimization procedure is used to find the optimum allocation for the following reasons:

(1) The constraints of the problem are readily expressed as linear inequalities.

(2) The desired delivery schedule is readily reflected through use of weighing factors as coefficients in the objective function.

(3) The IBM Mathematical Programming System* (MPS-360) is readily available to solve the linear program.

(4) This system is flexible and allows for the subsequent addition of parametric procedures and economic considerations.

(5) The MARVEL* language affords high level data preparation, matrix generation and report writing aid.

It is emphasized that this is a test model and as such it differs in important respects from an operational model. The primary purpose of the test model is to investigate the feasibility of novel formulations. It must remain flexible so that new approaches can be incorporated. As a result, some reduction in efficiency in matrix generation is expected. The model will be fully automatic if there are no hardware or software failures; however, breakpoints and/or recovery procedures will require manual intervention. Several features which have already been validated have not been included in this model.

4. PROBLEM FORMULATION AND EXAMPLE

The problem formulation is presented two ways. A conceptual formulation is given which represents only the basic ideas and does not include subscripts on the variables, fallout vectors or residual capacity vectors. Then a detailed expanded formulation is developed which represents the method which will actually be programmed. The definition of terms is given at the end of this section. These pages should be marked as a ready reference to aid the reader in understanding both formulations.

4.1 Basic Concepts

The basic concepts of the test model are given by the following five expressions:

Minimize

$$\Sigma wx \qquad \text{(objective function)}$$

subject to:

$$\Sigma x = d \qquad \text{(demand constraint)}$$

$$\Sigma n \frac{x}{a} \leqslant q \qquad \text{(resource constraint)}$$

$$\Sigma x \leqslant b \qquad \text{(port constraint)}$$

$$\Sigma \frac{x}{a} - \Sigma \bar{x} = 0 \qquad \text{(balance constraint)}$$

The symbol (Σ) means: sum over all possible values of the given variable.

The value of (x) represents the amount of a demand which is delivered in a given time period. In the objective function, the value of the weighting factor (w) increases as the time period of delivery of (x) deviates from the desired period of its demand. Since

---

*See Refernces

the objective function is being minimized the tendency is for the demands to be delivered as close as possible to the desired period because that is where the coefficients (w) are smallest. As a general rule, high priority demands have (x)'s with higher weights than lower priority demands; therefore, high priority demands have more of a tendency to be delivered on time than lower priority demands.

The demand constraint sums the amount of a demand that is delivered in each period and insures that this sum is equal to the entire demand (d). The resource constraint insures that the vehicle days consumed by all deliveries do not exceed the number of vehicle days available. The expression (x/a) divides the allowable cabin load of a vehicle into the amount delivered to get the number of trips required. The number of vehicle days for one trip (n) is then multiplied by the number of trips (x/a) to get the total number of vehicle days consumed by all of the trips needed for delivery of (x) amount of some demand. The total number of vehicle days available in a period is then given by the length of the period times the number of prime vehicles available.

The port constraint insures that the total amount of demand transshipped through each port does not exceed the capacity of that port in tonnage. If the capacity were given in sorties, then the amount delivered (x) could be divided by (a) to get trips or sorties.

The balance constraints insure that the total number of departures (x/a) are balanced by the total number of returns ($\bar{x}$) at each POE and POD. The purpose of these constraints is to cause ships to reposition on the channel which leads back to their next POE. Without this constraint ships would have to reposition on their outbound channel or on an 'average' channel. Either of these approximations could cause distortion in the schedule.

4.2 The Linear Programming Formulation

This section expands each of the expressions given in the previous section and indicates the ranges over which the summations are to be made.

Objective Function

The objective function is to minimize

$$\sum_{i\in I}\sum_{j\in J}\sum_{k\in K}\sum_{m\in M} w_{i,j,k,m}x_{i,j,k,m} + \sum_{i\in I} w_i' f_i - p\sum_{j\in J}\sum_{k\in K} c_{j,k} = z$$

where

$$x_{i,j,k,m} \geqslant 0,\ f_i \geqslant 0 \text{ and } c_{j,k} \geqslant 0.$$

The values of (w′) are large compared to (w) so that the primary objective is to deliver as much as possible, i.e. minimize the fallout (f). The value of (w) for a given demand is lowest when delivery is in the desired period(s). The value of (w) increases as deviation from the desired period increases, i.e. deliver on time if possible.

Priorities for each demand are honored by making the rate of increase of (w) and the value of (w′) larger for high priority items than it is for lower priority items. The smallest value of (w′) is greater than the largest value of (w). Thus the overall tendency is 1) to deliver as much high priority demand on time as possible, 2) to deliver lower priority demand in other than the desired period if necessary and 3) to make the fallout, if any exists, consist of the lowest priority demand. The units for (x) are the same as the units for the corresponding demand. Correction to short tons is included in the value of (w).

The factor (p) is a secondary consideration used to maximize the number of residual vehicle days but not at the expense of late deliveries. The value of (p) is small (say .01 of the smallest (w)) compared to (w) and (w′) and contains an inherent conversion to the units of (x) and (f). A parametric procedure is used to remove any adverse effect that the value of (p) may have had on the optimal allocation.

Demand Constraints

The demand constraints are

$$\sum_{m \in M_r} \sum_{k \in K_{i,m}} \sum_{j \in J_{i,k,m}} g_{i,k} x_{i,j,k,m} + f_i = d_i;\ i \in I_r,\ r \in R$$

where

$$g_{i,k} = \begin{cases} g_i' \text{ if } i \in I' \text{ and } k \in K' \\ 1 \text{ otherwise.} \end{cases}$$

These constraints insure that each demand is divided among eligible early, on time and late allocation vectors and a fallout vector. It is the objective function which causes the solution to contain the optimum allocation for the overall demand. The units of the demand (d) are measurement tons if it is a sea only demand; the units are short tons otherwise. The factor (g) converts from measurement tons to short tons and is required when the allocation vector is in measurement tons (i.e. for a ship) and the demand row is in short tons (i.e. the demand can be delivered by air or sea).

Vehicle Constraints

The vehicle constraints for aircraft are:

$$\sum_{m \in M_{j,k}} \sum_{i \in I_{j,k,m}} \frac{n'_{j,k,m} x_{i,j,k,m}}{a_{i,j,k,m}} + c_{j,k} = q_{j,k};\ j \in J,\ k \in K''.$$

These constraints insure that the amount of demand that is delivered does not exceed the capacity of the available air vehicles. The expression (x/a) gives the number of trips required to deliver (x). The total elapsed time for one round trip times the number of trips (i.e. n′ x/a) gives the number of vehicle-days consumed to deliver (x) amount of a demand. The number of vehicle days available is the product of the number of prime aircraft available and the length of the planning period in days. Since (c) cannot be negative, the vehicle days consumed can never exceed the number available.

The vehicle constraints for ships are:

$$\sum_{j \in J} \sum_{m \in M_{j,k}} \sum_{i \in I_{j,k,m}} \frac{n_{j',j,k,m} x_{i,j,k,m}}{a_{i,j,k,m}} + \sum_{j \in J} \sum_{m \in M_{j,k}} \bar{n}_{j',j,k,m} \bar{x}_{j,k,m}$$

$$+ c_{j',k} = q_{j',k};\ j' \in J,\ k \in K'$$

where $\bar{x}_{jkm} \geqslant 0$.

These constraints insure that the amount of demand that is delivered does not exceed the capacity of the available sea vehicles. Here the expression (n x/a) gives the number of vehicle days consumed by the outbound trips required to deliver (x) amount of demand.

Each outbound trip causes a return trip. The expression ($\bar{n}\ \bar{x}$) gives the number of vehicle-days consumed by return trips.

Sea vehicles may require more than one period to make a round trip; therefore, a ship delivery in a given period precludes the use of that ship for deliveries in adjacent periods. This effect is accounted for in the ship constraints by summing over all deliveries (i.e., including those in other periods) which affect a given period.

Although the summation is over all time periods, the vectors (n) and ($\bar{n}$) will have a large proportion of zeros in long problems; thus the density of the matrix will usually be much lower than is evident by merely looking at the constraint.

The separation of outbound and return trips enables ships to reposition to the port which can make optimum use of the vehicle capacity.

Since (n) and ($\bar{n}$) have a precision of one day, only the portion of a period actually used is charged against a given delivery. Thus if a ship returns in a given period, it may be able to depart on another trip in that same period.

The timing considerations for ships are POD oriented. Since streaming is assumed, there is no period-to-period interaction for aircraft. The unloading of a ship is assumed to be centered in the delivery period, i.e. a ship is half unloaded at the middle of the delivery period. The outbound time consists of the onload time, cruise time for the trip from POE to POD, stop time at enroute bases and offload time. The return time consists of cruise time from the POD to the POE and stop time at enroute bases.

Port Constraints

The port constraints for aircraft are

$$\sum_{m \in M_u} \sum_{k \in K_{j,m}} \sum_{i \in I_{j,k,m}} \frac{x_{i,j,k,m}}{a_{i,j,k,m}} \leq b_{j,u};\ j \in J,\ u \in U_1.$$

The expression (x/a) gives the number of trips through a given base. Since each trip involves a takeoff and a landing (b) must be expressed as sorties.

The POD constraints for ships are

$$\sum_{m \in M_u} \sum_{k \in K_{j,m}} \sum_{i \in I_{j,k,m}} x_{i,j,k,m} \leq b'_{j,u};\ j \in J,\ u \in U_2.$$

This constraint insures that the total tonnage (x) does not exceed the capacity of the port (w).

The POE constraints for ships are

$$\sum_{j \in J} \sum_{m \in M_u} \sum_{k \in K_{j,m}} \sum_{i \in I_{j,k,m}} h_{j',j,k,m} x_{i,j,k,m} \leq b'_{j',u};\ j' \in J,\ u \in U_3.$$

Although ships must offload within a single period, they may onload over several periods. The factor (h) proportions the amount loaded among the appropriate periods so that the total tonnage loaded in a given period does not exceed the port capacity.

The enroute base constraints for ships are

$$\sum_{m \in M_u} \sum_{k \in K_{j,m}} \sum_{i \in I_{j,k,m}} \frac{x_{i,j,k,m}}{a_{i,j,k,m}} + \sum_{m \in M_u} \sum_{k \in K_{j'',m}} \bar{x}_{j'',k,m} \leq b_{j',u};\ j' \in J,\ u \in U_4.$$

This constraint insures that the total number of arrivals of both outbound (x/a) and return ($\bar{x}$) ships does not exceed the handling capacity of an enroute port.

Balance Constraints

The balance constraints for ships at a POD are

$$\sum_{m \in M_u} \sum_{i \in I_{j,k,m}} \frac{x_{i,j,k,m}}{a_{i,j,k,m}} - \sum_{m \in M_u} \bar{x}_{j,k,m} = 0;\ j \in J,\ k \in K',\ u \in U_2.$$

This constraint insures that the number of departures for return trips ($\bar{x}$) is equal to the number of arrivals (x/a) for each vehicle type in each time period at each port. Since offload time is less than one period, arrival and departure always can occur during the same period.

The balance constraints for ships at a POE are

$$L_{j',k,s}y_{j',k,u} + v_{j'-1,k,u} - v_{j',k,u} + \sum_{m\in M_u} \bar{x}_{j'',k,m} - \sum_{m\in M_u} \sum_{i\in I_{j,k,m}} \frac{x_{i,j,k,m}}{a_{i,j,k,m}} = 0;$$

$$j' \in J, k \in K', u \in U_3$$

where

$y_{j',k,u} \geqslant 0, v_{j',k,u} \geqslant 0, v_{j',k,u} \equiv 0$ when $j' = 0$ and

$$L_{j',k,s} = \begin{cases} 1 \text{ if ship availability increases} \\ 0 \text{ if ship availability remains constant} \\ -1 \text{ if ship availability decreases.} \end{cases}$$

A pictorial representation of this expression is shown in Figure 4.1.

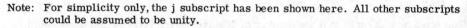

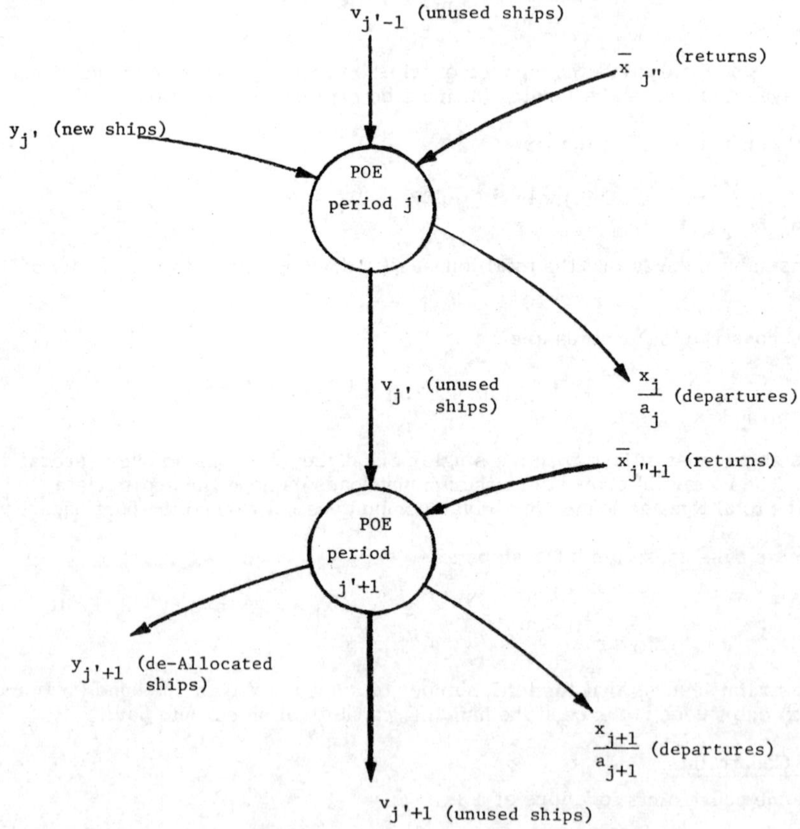

Note: For simplicity only, the j subscript has been shown here. All other subscripts could be assumed to be unity.

Fig. 4.1 Diagram of POE balance constraints

These constraints insure that the number of ships (x/a) departing from a POE does not exceed the number of ships actually at that POE in a given period. The number of ships at a POE in a given period depends on the following items:

(1) The number of ships (+y) newly available in this period.

(2) the number of ships (—y) which are de-allocated in this period.

(3) the number of ships ($v_{j'-1}$) which remain unallocated from the previous period.

(4) the number of ships ($\bar{x}$) which return to the POE in this period.

The number of ships which are unallocated during this period are made available to the next period by means of v_j.

POE to POE Repositioning

Provisions had been made for the inclusion of a pseudo-demand with zero priority and a pseudo-POD with artificial channel lengths, so that the user can realistically simulate the repositioning of ships between POEs in up to three distinct source regions. This feature is data dependent, thus its use and the values of its parameters are entirely under control of the user.

The allowable cabin loads of each vehicle could be assumed to be one tone on each of the artificial channels. Then the demand at the pseudo-POD could be set equal to the number of repositioning trips allowed i.e. one trip for each ton of demand. The values for the artificial distances for two or three region problems are described in the following two paragraphs.

The allowable cabin loads of each vehicle could be assumed to be one ton on each of the artificial distances for two or three region problems are described in the following two paragraphs.

If there are only two regions then the distance from each port in either region to the pseudo-POD is one-half of the average distance between the coastlines of the two regions.

If there are three regions, as shown in Figure 4.2 the distances are given by

$$D_{1p} = \frac{D_{12} + D_{13} - D_{23}}{2}$$

$$D_{2p} = \frac{D_{13} + D_{23} - D_{13}}{2}$$

$$D_{3p} = \frac{D_{13} + D_{23} - D_{12}}{2}$$

where D_{1p} is the value to be used for the distance from ports in region 1 to the pseudo-POD. D_{2p} and D_{3p} are similarly defined for the other two regions.

D_{12} is the average distance between the coastlines of region 1 and region 2. D_{23} and D_{13} are similarly defined for the other combination of regions.

Availability Constraints

The ship availability constraints are

$$\sum_{u \in U_s} y_{j,k,u} = | e_{j,k,s} - e_{j-1,k,s} |; \; j \in J, k \in K', s \in S$$

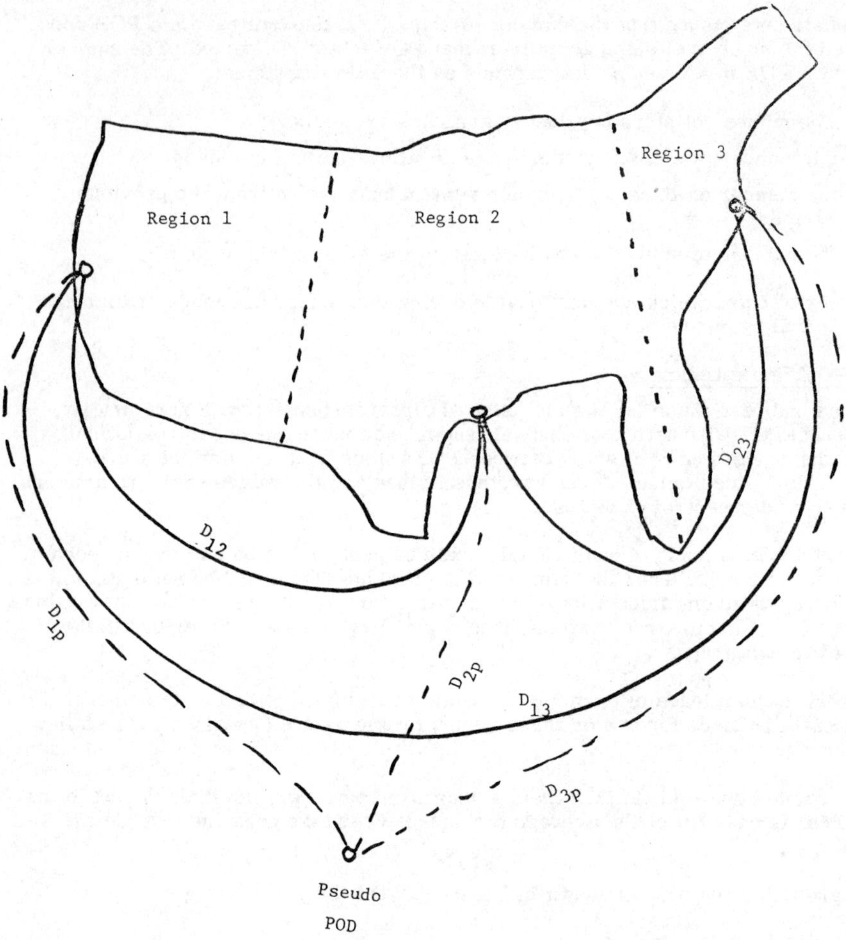

Fig. 4. 2 Construction of a pseudo-POD for repositioning ships between POEs

This constraint insures that number of ships (y) which are made available or are de-allocated at each POE in a region is equal to the total available or de-allocated for that region in each period. This allows the user to control ship availability for each region while allowing the LP to select the best channels serving that region. These rows are generated only when the RHS is non-zero.

4. 3 Definition of Terms

$a_{i,j,k,m}$	is the amount of cargo i which vehicle type k can carry while travelling on channel m in period j.
$b_{j,u}$	is the capacity of port u in period j expressed in sorties.
$b'_{j,u}$	is the capacity of port u in period j expressed in measurement tons.
$c_{j,k}$	is the number of type k vehicle-days which are unused during period j.
d_i	is the amount of the ith demand.

$e_{j,k,s}$	is the total number of type k ships which are available in region s during period j.
f_i	is the fallout for demand i, i.e. the portion which must remain unmoved.
g'_i	is the short ton/measurement ton ratio for demand i.
$h_{j',j,k,m}$	is the fractional part of x which is onloaded in period j but delivered in period j′ on channel m by ship type k.
i	is a demand.
I	is the set of all demands.
I′	is the set of demands which can be delivered by both air and sea.
$I_{j,k,m}$	is the set of demands which may be delivered in period j if carried by vehicle type k travelling on channel m.
I_r	is the set of demands which belongs to region-pair r.
j	is a time period.
j′	is a period in which a ship is in use as a result of a delivery in the reference period j.
j″	is the period of the start of a return trip that causes activity in period j′.
J	is the set of all time periods.
$J_{i,k,m}$	is the set of time periods during which demand i may be delivered by vehicle type k while travelling on channel m.
k	is a vehicle type.
K	is the set of all vehicle types.
K′	is the set of all sea vehicles.
K″	is the set of all air vehicles.
$K_{i,m}$	is the set of vehicles which can carry cargo type i and have routes defined for channel m.
$K_{j,m}$	is the set of vehicles which have routes defined for channel m in period j.
m	is a channel.
M	is the set of all channels.
$M_{j,k}$	is the set of channels defined for vehicle type k in period j.
M_r	is the set of outbound channels which serve region-pair r.
M_s	is the set of all channels serving a source region s.
M_u	is the set of all channels serving port u.
$n_{j',j,k,m}$	is the number of days in period j′ that vehicle type k uses while travelling outbound on channel m to make a delivery in period j.
$\bar{n}_{j',j,k,m}$	is the number of days in period j′ that vehicle type k uses while travelling inbound on channel m as a result of a return trip beginning in period j.
$n'_{j,k,m}$	is the round trip time for vehicle type k while travelling on channel m in the jth time period.
p	is the objective function weight associated with the total number of residual vehicle days.
$q_{j,k}$	is the number of type k vehicle-days available during period j.

r	is a region-pair
R	is the set of region-pairs defined for the problem.
s	is a source region.
S	is the set of all source regions.
u	is a port.
U	is the set of all ports.
U_s	is the set of all POEs in source region s.
U_1	is the set of all air ports.
U_2	is the set of all PODs for ships.
U_3	is the set of all POEs for ships.
U_4	is the set of all enroute ports for ships.
$v_{j',k,u}$	is the number of unallocated ships of type k which remain at POE u at the end of period j′.
$w_{i,j,k,m}$	is the weight penalty attached to delivery of the demand i in period j by vehicle k on channel m.
w'_i	is the weight penalty attached to the amount of fallout (unmoved demand) for demand i.
$x_{i,j,k,m}$	is the allocation vector i.e. it contains the amount of demand i delivered in period j by vehicle k on channel m.
$\bar{x}_{j,k,m}$	is a return trip by ship type k, starting in period j on channel m.
$y_{j',k,u}$	is the increase or decrease in availability of ship type k at POE u in period j′.
z	is the objective function value.

REFERENCES

International Business Machines Corporation, Capability Design Specification (CDS), Rosslyn Facility, TRAMPS Contract.

I.B.M. Corp., Mathematical Programming System/360. (MARVEL) Application Description. Technical Publications Department, White Plains, New York.

I.B.M. Corp., Mathematical Programming System/360. Control Language. User's Manual.

I.B.M. Corp., Mathematical Programming System/360. Linear and Separable Programming—User's Manual.

I.B.M. Corp., MARVEL Primer.

I.B.M. Corp., MARVEL Program Description Manual.

MITRE Corporation, 'An Investigation of the Weighting Function Used in THEFT' Working Paper No. 9321, Washington D.C.

MITRE Corp., 'THEFT—An L.P. Model for Analysis of the Feasibility of Strategic Movement Planning'. Working Paper No. 9164, Washington D.C.

MITRE Corp., 'User's Guide for the 'THEFT' Pre-Processor and Post-Processor' Working Paper No. 9226, Washington D.C.

J. J. LAGEMANN. 'A Method for Solving the Transportation Problem', Naval Res. Logistics Quarterly, Vol. 14 (March 1967), pp. 89-99.

A. J. HOFFMAN and H. M. MARKOWITZ. 'A Note on Shortest Path, Assignment and Transportation Problems', Naval Res. Logistics Quarterly, Vol. 10 (1963), pp. 375-379.

L. R. FORD, Jr. and D. R. FULKERSON. 'Flows in Networks', Princeton University Press, Princeton, N.J., 1962.

G. HADLEY. 'Linear Programming, Addison-Wesley Publishing Co., Inc., Reading, Mass., 1962, pp. 273-330.

G. B. DANTZIG. 'Linear Programming and Extensions', Princeton University Press, Princeton, N.J., 1965, pp. 299-315.

Appraisal of Current Linear Programming Applications to Strategic Mobility Problems

W.F. YONDORF, J.C. GRIMBERG, H.I. OTTOSON,
Mitre Corporation, U.S.A.

1. INTRODUCTION

Two previous papers have characterized transportation planning in the organization of the Special Assistant of the U.S. Joint Chiefs of Staff for Strategic Mobility (SASM), Hassler (1968), and described the mathematical formulation of several linear programming models used in optimizing closure time and scheduling of transportation resources, Arbabi (1968). The models included the SASM THEFT (To Help Evaluate Feasible Transportation) Model, MITRE (1966, 1967, 1968), the last phase of the SASM TRAMPS (Transportation Movement Planning System) Model, and a follow-on test algorithm to be incorporated into TRAMPS.

This paper appraises current applications of linear programs to strategic mobility problems and, in particular, raises the following questions:

What is the essence of the present approach? If one accepts the present formulations, what simplifications and assumptions must be made? What bearing do these simplifications and assumptions have on the solution of strategic mobility problems? Are there alternative approaches which would permit a more effective use of linear programming?

2. THREE TYPES OF MOBILITY PLANNING

It will be recalled that there is a basic sequence of strategic mobility planning which is followed in the solutions of most military transportation planning problems (see Figure 1). The main elements in the process are the specification of units and goods required to be moved, the determination of vehicle resources available, the specification of constraints and criteria applicable to the movement, and, finally, the allocation of vehicle resources to delivery requirements through ports and channels in order to determine feasibility in accordance with the specified criteria. Typical optimizing criteria include least closure time, minimum deviation (in ton-days, for example) from desired delivery dates, maximization of residual resources (unused vehicle or port capacities), or least cost. Typical outputs consist of gross movement tables or detailed shipping schedules, a statement of goods or units not moved (termed 'shortfalls' in mobility planning), and a summary of residual resource capacity.

With some variation, this sequence is applicable to three basic problems of strategic mobility analysis, namely, emergency planning, contingency planning, and resource planning (see Figure 2). Most of the operational research carried out by SASM is directly related to these problems.

<u>Emergency Planning</u>, or operations planning, requires the application of currently available resources to the movement of men and material in the immediate future. Hence, readiness and the disposition of troops and movement resources are given, and the alternatives open are usually severely restricted.

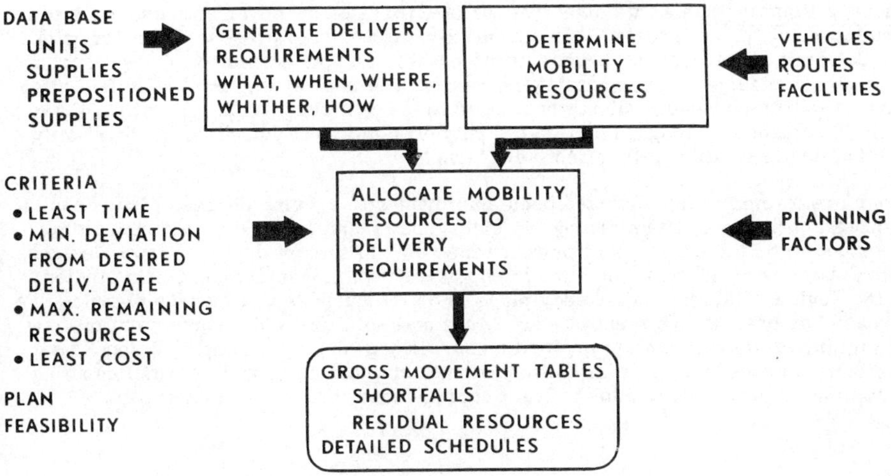

Fig. 1 Basic movement planning sequence

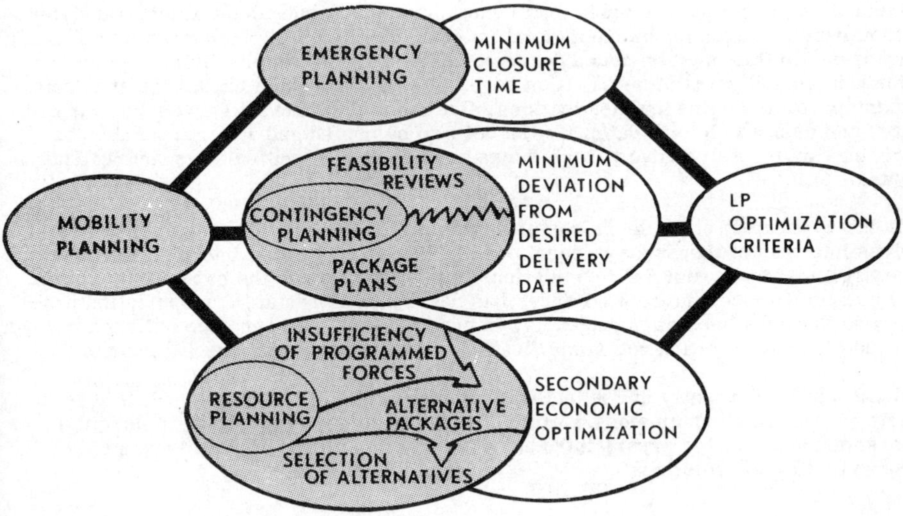

Fig. 2 Mobility planning types

Contingency Planning at the SASM level generally means one of two things: review of mobility plans made by others; or a study to determine the feasibility from a mobility point of view of executing two or more plans simultaneously. The object of the latter is to gauge overall movement capability with existing and programmed movement resources under a wide range of conditions. For this reason, movement resources and contingency plans are held constant; but contextual conditions, such as strategic warning, the availability of commercial shipping resources, and force posture, are varied.

Resource Planning has as its objective the specification of movement resources required to meet future threats. It is essentially a three-step process. The first step is to determine whether and to what extent present and programmed forces fall short of satisfying assumed movement requirements. The second step is to postulate alternative resource packages which, when added to present resources, will satisfy the assumed demands. The third step is to apply economic analysis in order to identify the least costly of all equally effective alternatives.

Linear programming models have been employed with varying degrees of success in all three types of mobility planning. In emergency planning, minimum closure time has tended to be the criterion of primary interest. Minimum deviation from desired closing dates is a criterion primarily associated with movement capability studies and the review of single contingency plans. For the purpose of resource planning, the approach has been to find a set of equally optimal solutions from some point of view (e.g., minimum deviation from desired closure dates) and then to apply a secondary optimization procedure in order to find the mix of vehicles capable of implementing the postulated movements with the least capital investment of operating cost.

3. PRESENT APPROACH TO LP APPLICATIONS

3.1 Problems of Scale

The basic difficulty encountered in applying LP techniques to strategic mobility problems is due to the magnitude of these problems. As stated in the first paper presented, a major contingency plan may contain over 6000 line items which, translated into movement requirements, may total up to 14, 000 items for transportation planning purposes. To this must be added dozens of vehicle types, a multiplicity of ports of embarkation and ports of debarkation, together with throughput capacities and routes linking ports to origins and destinations. One must also take into account availability dates and desired closure dates for the units to be moved and a larger number of planning factors which govern such things as vehicle use, critical legs, and utilization routes.

For the application of LP techniques, all of these factors combined translate themselves into tens of thousands of equations and variables formed into a gigantic matrix. The demands which such LP formulations impose far exceed the capacity of present day computers (see Figure 3). Hence, it is necessary to restate transportation problems so that they become manageable. In practice this has been accomplished through the judicious aggregation and simplification of inputs to the various LP models.

All LP models currently in use by SASM employ such simplification. While it is too early to draw final conclusions, sufficient experience has been gained to permit a characterization of the errors introduced through simplification and to gauge the probable effect on solutions.

3.2 Grouping of Commodities

For purposes of mobility analysis, 'commodity' is viewed as a kind of element to be transported as a physical unit. It is defined by its dimensions, weight, and handling characteristics. A commodity, in the generic sense, includes all those elements that can be treated alike from a mobility point of view. Thus, if all elements had similar handling characteristics and were of such high density that they could load any vehicle to the limit of its weight-carrying capacity without exceeding its maximum cargo space they would be considered a single commodity.

Amounts of cargo in a class rather than individual commodity items are used in mobility analysis because of the need to obtain an estimate of the number of lift vehicle sorties required, without consideration of the actual 'loading' of individual vehicles.

Fig. 3 The sorcerer's apprentice

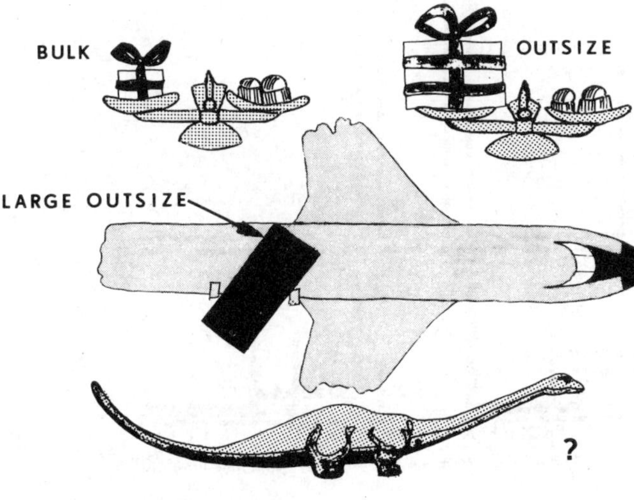

Fig. 4 Grouping of commodities

In moving military units it is most often found that cargo aircraft are space limited rather than weight limited. This is why it is necessary to distinguish at least two classes of cargo: low-density cargo which 'loads' a vehicle without approaching its weight-carrying limit; and high-density cargo which can 'load' a vehicle to its weight-carrying limit (see Figure 4).

The magnitude of the error introduced by the practice of aggregating commodities into a few large cargo classes is difficult to estimate in the absence of a statistical assessment. Statistical analysis, incidentally, is the only means of determining the error introduced in this case, since the number of 'real' loads obtained from a set of requirements would vary from operation to operation even if the same set of vehicles were utilized.

3.3 Aggregation of Time

Aggregation of time into discrete periods greater than a day is required to cover the total deployment period and still keep problem-size within manageable bounds. This means, of course, that all references to time must be made in terms of the defined discrete periods. It follows that the degree of precision with which desired and actual availability and delivery dates may be stated can be no better than the length of the time period selected.

Aggregation of time will generally have the effect of smoothing out peaks in the demand profile created by the fact that planners often state closure times for units not in terms of acceptable periods or time intervals but by a certain date. The sawtooth-like result created by such specifications is indicated in the upper portion of Figure 5. The normally expected smoothing effect of time aggregation is indicated by the dotted line in the lower portion of Figure 5 which groups the same set of delivery requirements into four-day intervals. That this smoothing effect will not always occur is made clear by the case represented by the solid line. It regroups the same delivery requirements into five-day periods, a procedure which should ordinarily result in a further smoothing out of the delivery profile. In this instance this does not occur because time is aggregated into periods which actually reinforce delivery peaks. The lesson is clear: the aggregation of constraining factors in LP formulations will not always automatically skew results in the expected direction. Hence, it is important to inspect problems carefully before formulating them for LP solution and, where necessary, restate them in such a fashion as to minimize the unintended but avoidable effects of error introduced as a result of aggregation.

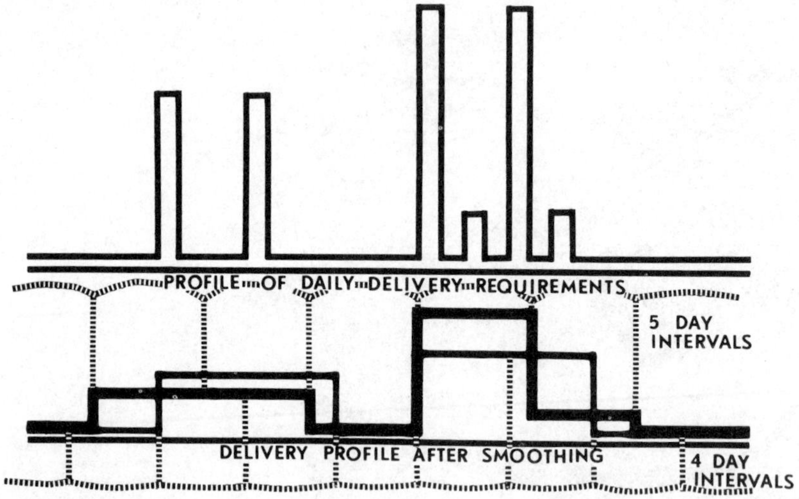

Fig. 5 Time aggregation

3.4 Aggregation of Channels and Ports

Since unique requirements may originate at hundreds of different locations and be destined for delivery to hundreds of different destinations, and since the transportation networks linking origins with destinations may be extremely complex, it is imperative that a simplified transportation net of links and nodes be assumed. In practice this means that ports of embarkation and debarkation are aggregated into composite ports. To some extent, the aggregation of links or channels follows automatically from the compositing of ports; further aggregation of links is possible by reducing the number of alternate routes between pairs of composite ports of embarkation and debarkation (see Figure 6.a and 6.b).

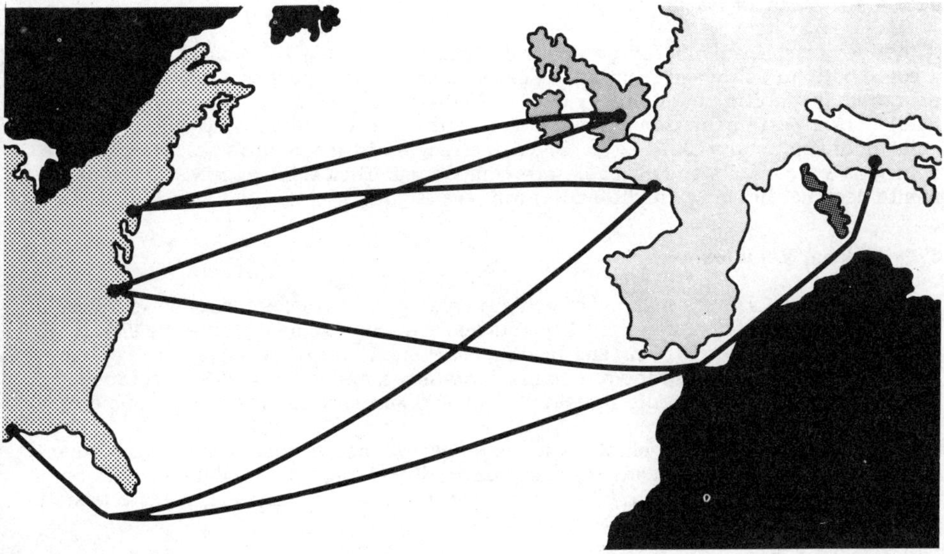

Fig. 6a Aggregation of ports and channels

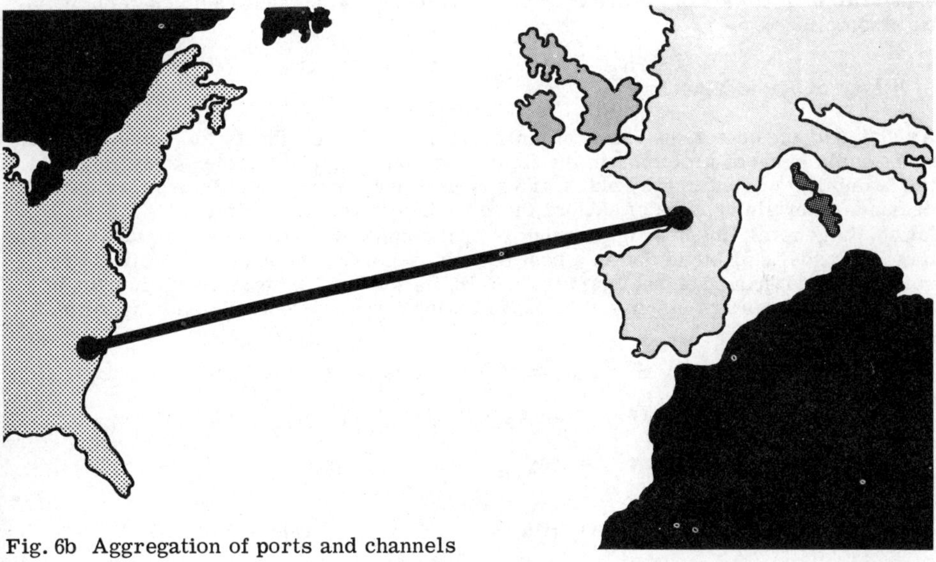

Fig. 6b Aggregation of ports and channels

Port constraints are represented in the problem formulation by assuring that the throughput capacity of composite ports will be equal to the sum of the capacities of the individual ports in each composite. Composite channel distances are normally chosen by determining actual channel distances traversed by each type of vehicle between individual ports of embarkation and debarkation and by applying these distances to travel between composite ports. One effect of such simplification of the transportation network is to neglect the time which ships would spend in the real world in traveling between individual ports of embarkation in order to pick up loads, and between ports of debarkation in order to discharge cargo. In cases where ships normally make several on-loading and off-loading stops, the consequence of this formulation is to understate travel time and to neglect important scheduling problems.

3.5 Aggregation of Requirements

A 'delivery requirement', for purposes of input into present LP models, is defined as a commodity to be moved from a port of embarkation to a port of debarkation within specified availability and delivery dates. The highest possible aggregation in the number of requirements for a given mobility problem is a direct outcome of the number of commodity classes employed, the length of time periods selected, and the degree to which ports and channels are composited. Hence, the aggregation of requirements does not introduce additional error (see Figure 7).

3.6 Notional Vehicles

The reduction of a large number of vehicle types to a reasonable number is inherently related to the selection of commodity classes. Neither can be done independently of the other. The cargo carrying and loading characteristics of vehicles are a primary consideration in the definition of commodity classes, and a knowledge of commodity classes plays an obvious role in establishing notional vehicle types (see Figure 8).

Thus, grouping of several vehicle types into one notional vehicle type presumes consideration of the types of loads (of each commodity class) which unique vehicle types can carry. It is also necessary to assure that the speed of individual vehicle types is reflected in the speed of the notional vehicle type.

With large numbers of vehicle types, it has proven difficult to choose notional vehicle types with characteristics representative of speed as well as commodity carrying capabilities. Analysis of errors introduced by selecting notional vehicles remains to be accomplished.

3.7 Application to Planning

As stated at the outset, the SASM organization has a responsibility for the analysis of three basic kinds of problems in the field of strategic mobility: emergency planning; the feasibility of contingency plans; and resource planning to meet future threats and demands. Certain aspects of each of these problems are well suited for LP formulation. In general, linear programming is best employed where fairly broad trade-offs must be made in order to obtain a best solution according to one of the well-defined optimizing criteria. The peculiarities and limitations of LP applications in each of the planning areas are discussed in the following three subsections (see Figure 9).

$$A_R \le A_{R\,max} = F(A_c, A_t, A_p + {}_c)$$

$$\begin{cases} e_T\,(A_R) = e_T\,(A_{R\,max}) = f(e_{A_c}, e_{A_t}, e_{A_{p+c}}) \\ e_{A_R} \quad = e_{A_{R\,max}} \quad = 0 \end{cases}$$

A_R	AGGREGATION OF REQUIREMENTS	e_{A_R}	ERROR DUE TO A_R
A_c	AGGREGATION OF COMMODITIES	e_{A_c}	ERROR DUE TO A_c
A_t	AGGREGATION OF TIME	e_{A_t}	ERROR DUE TO A_t
A_{p+c}	AGGREGATION OF PORTS & CHANNELS	$e_{A_{p+c}}$	ERROR DUE TO A_{p+c}
$A_{R\,max}$	MAXIMUM A_R COMPATIBLE WITH A_c, A_t, AND A_{p+c}	e_T	TOTAL ERROR IN THE CALCULATIONS WHEN REQUIREMENTS ARE AGGREGATED AS INDICATED INSIDE THE PARENTHESIS

Fig. 7 Aggregation of requirements

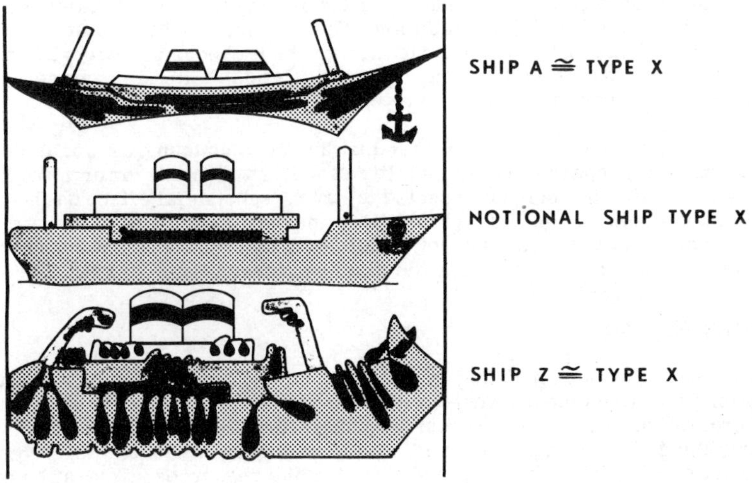

Fig. 8 Notional vehicles

			GOOD	POOR
PLANNING	EMERGENCY OR OPERATIONAL	COMMITMENT OF RESOURCES	✓	
		SCHEDULING		✓
	CONTINGENCY	FEASIBILITY	✓	
		DETAILED ANALYSIS		✓
		STUDY OF STRATEGIC POSTURES	✓	
	RESOURCES	COMPARISON OF ALTERNATIVES	✓	

Fig. 9 LP Applications to strategic mobility problems

3.8 Emergency Planning

In the case of planning an emergency deployment, the simplifications introduced into the LP model during the model formulation phase are apt to be directly transformed into errors of operations planning. The practical consequence of these errors depends largely on the level at which planning is performed.

If the model is only used for broad planning purposes, as would be the case if the problem were to estimate the mobility resources necessary to commit to an operation and to call attention to possible alternatives and limitations in their employment, then results should prove satisfactory—provided they are corrected by an appropriate safety factor. This safety factor should be designed to compensate for the error introduced through simplification and aggregation. Detailed scheduling models might be used to arrive at an initial value for the safety factor, with the idea of adjusting it as experience in actual operations permits a gradual refinement of planning factors.

A second adjustment must be made in order to allow for natural uncertainties whose effect is not represented in the LP formulation. The last point is important: regardless of how accurate and refined LP formulations may gradually become, there will continue to be need for this safety factor, since optimal solutions mitigate against giving proper weight to unforeseen events which are apt to affect any real operation.

If the model were employed in detailed rather than in gross planning, as would be the case if it were used to prepare detailed mobility schedules, then any error introduced for the sake of simplification may be expected to have a substantial effect on the schedules and make them suspect. The limitations imposed by the LP formulation on the amount of detail achievable casts doubt upon the utility of present LP models for detailed scheduling applications in all but the simplest cases.

3.9 Contingency Planning

At the SASM level, contingency planning is primarily concerned with studies to determine the feasibility of implementing one, or several plans simultaneously using only existing and programmed movement resources. The situations considered vary with respect to such things as the availability of units to be moved, status and initial geographic distribution of military and commercial shipping resources, declaration of national emergency, and right of passage through the major canals.

Typical questions raised are: Can deliveries be made by the dates specified in the plans under the various conditions postulated? If it is not possible to deliver all requirements on time, during which periods and where do shortfalls occur? How large are they? What cargo types are involved? Could they be eliminated by relaxing delivery dates, permitting early delivery, disregarding specified modes of transportation, increasing mobility support forces, negotiating for additional load or unload facilities with allies, contracting for more commercial shipping, or declaring a national emergency so as to permit the early mobilization of additional resources?

Most of the problems are susceptible to analysis in gross terms. If a particular scenario results in delivery shortages, then it is a comparatively simple matter to relax constraints in the LP formulation in order to carry out a sensitivity analysis. In that event, as in the case of emergency planning, it will be necessary to include a safety factor in order to compensate for the simplifications built into the LP model and for the uncertainty inherent in the definition of any contingency.

Another aspect of contingency plan review is the detailed analysis of a plan in order to establish its feasibility in detail. LP models, although useful for an initial gross feasibility check, are not suitable for this purpose. Simulation rather than optimization is the proper technique for such an analysis.

Up to this point, contingency planning and emergency planning seem to be alike in their susceptibility to LP model treatment. However, there is one aspect of contingency planning which has no counterpart in emergency planning: the investigation of alternative strategic postures in peacetime for the purpose of identifying the one most effective in the face of likely contingencies. LP models are capable of contributing to such an investigation by providing estimates of the feasibility of implementing a set of contingency plans as a function of alternate strategic postures, including alternative prepositioning and basing policies, thus contributing to the assessment of such postures. The level of detail at which LP models should be used for this purpose is the least detailed possible, provided such grossness does not interfere with the discrimination by the model between alternative postures.

3.10 Resource Planning

The problem of resource or mobility force structure planning involves the application of LP formulations similar to those used in force posture studies. In resource planning, the approach is to define a series of alternative force structures and to play them against a set of contingencies in order to determine their relative effectiveness. Assessments of this type are currently made with LP models capable of examining a very large number of alternative resource packages in the light of a critical and somewhat grossly defined set of contingencies. Cost considerations constitute an additional leading feature in resource planning. Hence, LP models are used first to determine force structures of optimal effectiveness and then, holding effectiveness constant, to determine the least costly of all equally effective force structures.

4. LIMITATIONS OF THE PRESENT APPROACH

The ideas so far discussed were prompted primarily by problems encountered in the use of TRAMPS and THEFT, the two LP models most widely employed by SASM. This section goes further and questions the fundamental adequacy of the present approach in two related respects, the static treatment of time and the handling of shortfalls.

4.1 Static Models for Dynamic Problems

One of the major limitations of the present approach is the static way in which problems are considered. Time is treated in the same way as all other variables, that is, the formulation is made at the outset for all periods of time. This implies equal knowledge of any planned event, with the consequence that late delivery requirements influence results as much as early delivery requirements. However, it is obvious that initial actions planned to meet an emergency have a higher probability of being carried out without modification than actions planned for later implementation (see Figure 10).

This case illustrates once more the fact that dynamic problems cannot be properly analyzed with static formulations.

4.2 Static Consideration of Shortfalls

A static approach also affects the manner in which shortfalls are treated. Under the present system, shortfalls are reported as they occur, and the calculation proceeds just as if such shortfalls had no influence on planning the delivery of the ensuing requirements (see Figure 11). Faced with planning and actual operation, a planner

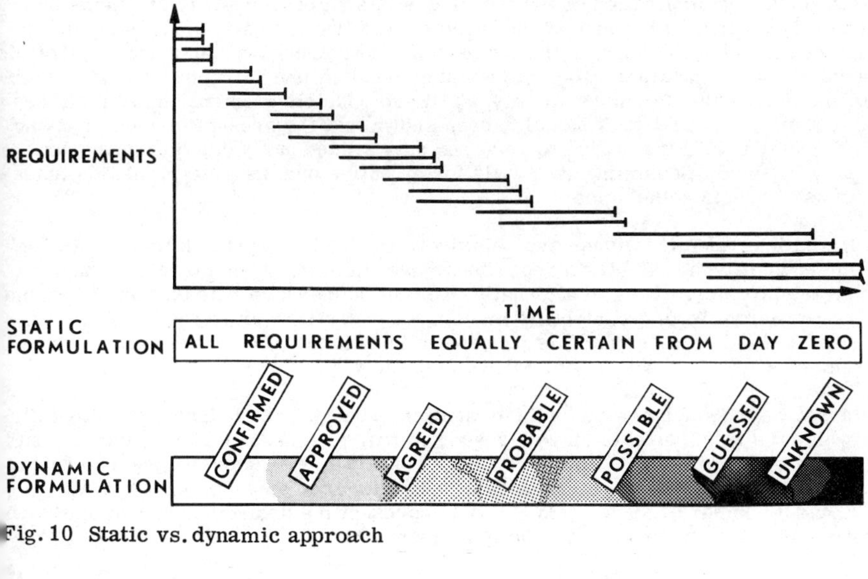

Fig. 10 Static vs. dynamic approach

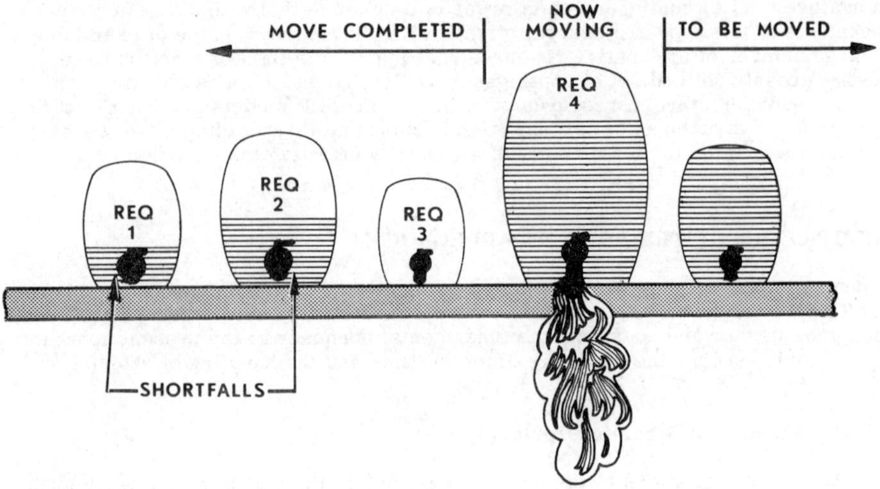

Fig. 11 The treatment of shortfalls

could not possibly proceed in this fashion and would instead find it necessary to decide immediately either to reduce the size of the requirement, or to add extra lift vehicles, or to permit a surge in utilization rates, or to alter closure dates, or to allow earlier deliveries, or to delay or accelerate competing delivery requirements. To make all these choices among alternatives within the restrictions imposed by the present static formulation would require the specification of all possible combinations of all alternatives for all shortfalls at the outset of a problem, a demand which far exceeds the capability of available LP models.

5. ALTERNATIVE APPROACH

As stated above, the application of LP techniques to strategic mobility problems has been severely restricted because of the large size of transportation problems and the difficulty of managing the huge matrix created by their many variables and constraints. It is evident that much larger problems could be handled, and that moderate size problems could be executed far more rapidly, if it were possible to partition LP matrices and solve them sequentially. It has also been shown that the present method of giving equal effect in the LP formulation to requirements needed early and late in the course of a deployment and of summing shortfalls for an entire plan in a static fashion leads to solutions of questionable value.

These limitations and objections have reinforced the many current efforts to develop better methods. At The MITRE Corporation, these limitations and objections have prompted the investigation of an alternative LP approach which takes advantage of the fact, already emphasized, that mobility planning is essentially an iterative, time-oriented process. The nature of this process is in part explained by the need to stop, adjust, and proceed whenever a shortfall comes into view.

The alternative approach now under investigation presupposes (1) that the span of time during which the requirements may be delivered will be explicitly stated, and (2) that the sequencing that units most follow in deployment is given. In each case, two dates, the closure date and a starting arrival date, will determine an allowable arrival interval, and a number of such intervals will represent the desired sequence of arrival for a whole set of requirements, as shown in Figure 12.

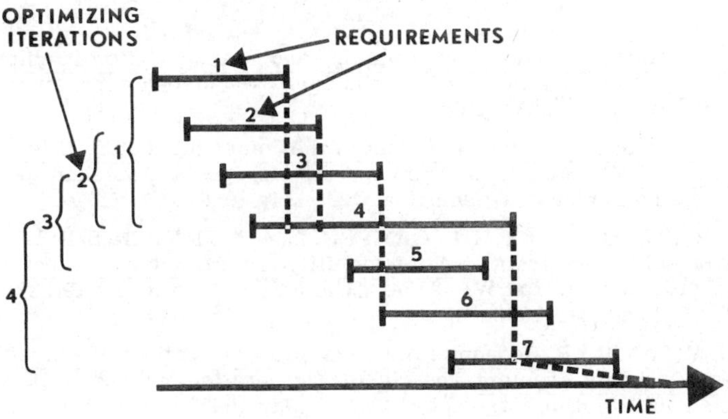

Fig. 12 Alternative approach diagram

In the alternative approach, the calculation would proceed by taking the first requirement, Number 1, and considering with it only those other requirements with overlapping desired arrival intervals, Numbers 2, 3, and 4, in the case of Figure 12. The assignment of resources among these four requirements will be optimized by an LP formulation, and resources will be assigned in a final way to the movement of Requirement 1.

Requirement 2 will be considered next, together with its overlapping requirements, and the result of the calculations will be the assignment of resources in a final way to Requirement 2. The calculation will then proceed in iterations.

If, as in Figure 12, the second requirement does not add new overlapping requirements to those already considered in the case of the first requirement, the assignment of resources in iteration 1 and 2 will be identical, although the second iteration will not include Requirement 1, nor the resources finally assigned to its movement. When new overlapping requirements are added, as in the case of Requirement 3, two iterations involving the same requirement may yield a different assignment of resources to that requirement due to the addition of new demands on the resources. In this case, final assignment of resources is done in accordance with the later iteration.

This procedure will reduce appreciably the size of the matrices generated for the LP model and is expected to reduce the computer time required at each iteration to the point where the system may be operated on-line in a practical way.

On-line operation would allow the user to become aware of shortfalls, if any, at the end of each iteration of the LP calculation, thus permitting him to exercise a series of options to correct the situation. He will then be able to select the option which best fits the overall objectives.

An experimental program to explore this new approach has been initiated on a small scale, and preliminary test programs are now running on the IBM 360/50 computer system at MITRE's Washington facility. Experience to date suggests that on-line operations of the type described are feasible and that small problems can be solved rapidly. Among the many unresolved questions, perhaps the two most important are whether transportation planners will find the new approach sufficiently promising to support its development vigorously, and whether linear programming models are fast enough to make it practical to execute operational size planning problems on-line and in a time-shared mode.

REFERENCES

M. ARBABI, et al, Applications of Linear Programming and Transportation Problem to Strategic Mobility Problems, International Business Machines Corporation, Arlington, Virginia, 30 April 1968. (U)

F. L. HASSLER, Objectives of the Special Assistant to the Joint Chiefs of Staff for Strategic Mobility (SASM), Advisor for System Analysis, Special Assistant for Strategic Mobility, The Pentagon, Washington, D.C., May 1968. (U)

G. E. BENNINGTON, P. P. BUCHYNSKY, H. G. CAMPBELL, A. S. DISTLER, S. H. LUBORE, THEFT—An LP Model for Analysis of the Feasibility of Strategic Movement Planning, The MITRE Corporation, WP-9164. DCA/NMCS Contract AF19(628)5165, Washington, D.C., 1 November 1966. (U)

L. R. GSELLMAN, M. M. SANDLER, Economic Analysis of the Operations Planning Phase of Strategic Mobility Problem, The MITRE Corporation, WP-9251. DCA/NMCS Contract AF19(628)5165, Washington, D.C., 5 May 1967. (U)

L. R. GSELLMAN, M. M. SANDLER, Economic Analysis of the Operations Planning Phase of an Example Strategic Mobility Problem, The MITRE Corporation, WP-9307. DCA/NMCS Contract AF19(628)5165, Washington, D.C., 24 October 1967. (U)

G. E. BENNINGTON, S. H. LUBORE, Linear Optimization Approach to Origin to Destination Movement, The MITRE Corporation, WP-9345. DCA/NMCS Contract AF19(628)5165, Washington, D.C., 29 December 1967. (U)

H. M. WOODALL, Structure of Strategic Mobility Analysis, The MITRE Corporation, MTP-305. DCA/NMCS Contract AF19(628)5165, Washington, D.C., 28 February 1968. (U)

A. S. DISTLER, The Use of the THEFT Model for Strategic Mobility Analysis, The MITRE Corporation, MTR-5040. DCA/NMCS Contract AF19(628)5165, Washington, D.C., 10 April 1968. (U)

DISCUSSION ON THE PAPERS BY HASSLER, ARBABI et al., and YONDORF et al.

P. Bannister

I have the impression that you are attempting to produce operational schedules from all your solutions. In my experience despite all our planning we do not attempt to produce a schedule for more than 30 days ahead, owing to the basic uncertainties that we face. Could you please give the reasons for this difference or is my impression wrong?

Reply

At the JCS level we do not ordinarily develop detailed schedules and stop in fact with the production of gross movement tables. However, 'representative schedules' are on occasion developed in detail in order to explore questions of feasibility. We do not impose our schedules on the transportation operating agencies of the Defense Department. The operating agencies project day-to-day operating schedules 90 days ahead and detail for 30 day periods.

E. L. Leese

In our experience it is sometimes important to take account of the order of delivery of the various loads in an airlift. Is it possible in your model to modify the objective function so as to take some account of the desired order of delivery of the various loads?

Reply

Yes, it is possible. In the usual case we divide the force list up into packages. These packages are then sequenced by a pre-& post-processor where the first package is input, optimal solutions obtained, resource lists updated, the next package brought in, processed, resources updated etc. Within each package, we may or may not consider constraints that are designed to minimize deviation of individual unit deliveries from their specified desired delivery dates. We could, if we so desired, sequence each unit as if it were a package and force complete delivery of that unit before going on to the next unit, but we do not, and probably would not.

C.N. Beard

The last three speakers have implied that aggregation becomes a problem only when one uses linear program techniques. In fact it is a difficulty inherent in the whole process of military planning whether done manually or mechanically.

A further implication has been that aggregation is automatically a bad thing. Hence Dr. Arbabi's statement of the ideal case requiring several thousand rows and billions of vectors. Surely the vagueness of planning assumptions must make such pernickety concentration on minute detail irrelevant. Rigid plans are often bad because acrual events never correspond to the planning assumptions. Is it not better, therefore, to plan with average aggregate values, and rely on the flexibility of officers to adjust the particulars to the circumstances which occur.

Reply

Concerning aggregation for linear programs, admittedly aggregation is done for all our programs in view of the size of our problems. However, many of our other programs can handle detail that our linear programs cannot and hence for us, the use of linear programs has forced, more than anything else, the study of the effects of aggregation.

We do not mean to imply that aggregation is good or bad. We must do it and therefore we must understand as much as we can about the effects one encounters. We have experienced cases where aggregation leads to serious errors resulting in major over estimates and major under estimates of capability. We have also seen cases where it made little apparent difference.

Granting the points concerning rigidity and uncertainty, we feel that detail (or at least the ability to go to detail if desired) is essential to feasibility demonstration. We want our estimates, no matter how grossly rendered, to be reliable in the sense that there is some feasible operation that can be carried out if the occasion should dictate.

There is no intention to make detailed operational plans and impose them on transportation agencies. Operating commands will continue to generate their own detailed plans and schedules for actual deployments.

E. L. Leese

1. Our experience with strategic airlift simulations in Canada is that certain stochastic elements, such as the delays due to repair of aircraft which become unserviceable during the airlift and delays due to weather, have a very important effect on the delivery times for the airlift. Do the models discussed by Dr. Hassler and Dr. Yondorf make at least some aggregated allowance for these stochastic features of the problem?

2. Has any attempt been made to validate the strategic mobility models by comparison with the actual results of airlift services carried out by MATS, or with actual airlift operations conducted by U.S. forces?

Reply

1. Yes. We take variables like aircraft operating characteristics-maintenance requirements etc.-into account through statistical planning factors like utilization rates. Separately we are exploring through detailed simulation techniques, the magnitude of these effects and intend to explore the behaviour of these types of planning factors under a wide range of conditions so as to better the values that we currently employ in our planning models.

2. Yes. We have for certain deployments in which we played a major part in the planning, gathered statistics on the actual operation to compare with our predictions for the sake of validation. The comparison was poor. Planning factors were recomputed on the basis of experience. The problems were rerun and this time the agreement was very close. Whether this will be true for a larger class of problems remains to be seen. We are keenly aware of the problems and are exploring the subject extensively at this time.

Concerning the first question. While it is quite customary by now to carry out sensitivity analyses in order to gauge the effect on results of varying initially assumed conditions, stochastic procedures have only been used to a very limited extent. As far as linear programming models are concerned, the reason is not difficult to find. It is easier and quicker to do right-hand ranging for sensitivity analysis than to attempt to modify an already large LP model for stochastic processing. But we certainly believe that much is to be gained through simulation incorporating stochastic procedures.

D. R. Razani

In connection with your resource planning. Do you use the programs for analysis and optimization of mobility problems to predict the most efficient (or optimum) mode of transportation, vehicle specification for future demand? For example, if I am a ship manufacturer and tell you that I can develop, say, five types of ships with specified amount of tonnage, speed, cost, etc. in the next ten years and ask you which of these ships will optimally fulfil your mobility problem in the future? Can you tell me which type of ship is the optimum type and how many of this type of ship you may need in the next 10 years? The answer to this question can lead to development of the most suitable vehicles in the future.

Reply

The programs analyse the relative cost/effectiveness of any specified vehicle to do a given job or set of jobs. This implies a judgement as to the numbers of vehicles required and their relative effectiveness.

Models used for resource planning analyse various force mixes, taken from a menu of alternative vehicle types and prepositioning options, to determine the most cost-effective combinations for a given set of scenarios. It is, of course, possible to introduce several alternative configurations of a vehicle design in order to determine which would add most to overall effectiveness at given costs—provided one holds constant the performance of all other vehicles on the menu. No doubt, such information would prove to be of interest to shipbuilders. But it must be realized, of course, that the results of any such analysis hold only for the conditions assumed.

P. M. Jenkins

The models as described seem to find optimal or good feasible solutions for a deployment to meet a single threat. As a multiplicity of possible threats is considered within a defence department, do you consider the implications of your single threat solution to the alternative threats, so that a feasible solution is obtained for the various possible threats?

Reply

The models can address single or multiple contingency plans, either with equal or unequal priorities and with or without sequencing relations of the type 'Do plan A and then do B with the left overs.' For large studies we usually do address multiple contingencies—usually for the largest plans we expect to have to implement together, thereby analysing the lift system or requirements for lift systems under conditions of maximum stress.

A Large-Scale Mathematical Programming Model of the U.S. Department of Defense Mobility Resources Acquisition and Allocation Problem

P. M. JENKINS, MARY J. O'BRIEN, JUSTIN C. WHITON
Research Analysis Corporation, U.S.A.

1. INTRODUCTION

DOD has as a major responsibility the continuous review of all major defense programs. These issues are often complex problems of interservice planning, programming, and procurement. In addition to the interservice aspect, the role of the US in international affairs must be considered as well. The development of quantifiable results in the face of such interrelated areas has proven particularly amenable to large-scale linear programming methods.

This paper will focus on the 'Airlift/Sealift Force' program. The series of linear programming models to which the Mobility and Transportation Resources Allocation (MATRA) Model belongs has been developed to aid in the analytical investigation of the problem: What constitutes an optimum airlift/sealift logistic system?

The purpose of this paper is to present some aspects of the development and application of a strategic deployment model that is in the 1000-3000 constraint class. Although the area of application is not the subject of this paper and results in terms of solutions are not of interest, some description of the strategic deployment problem is presented to indicate model formulation questions and types of applications that may be of concern.

The group within DOD for whom these linear programming and strategic deployment applications have been performed is the Strategic Mobility and Transportation Division of the Office of the Assistant Secretary of Defense for Systems Analysis [OASD(SA)]. This group is responsible for recommending to the Secretary of Defense a program of airlift, sealift, and prepositioned materiel which will make up the strategic mobility system. The programmed system must provide an adequate capability to meet logistic requirements for any of a series of limited wars, as well as meet world-wide peacetime demands of DOD and the continuing peacetime demands for the nonwar areas during wartime.

Linear programming has proved to be an ideal technique for generating sensitivity data rapidly and systematically for DOD analysts and planners. It is of special note that such applications do not answer the question but serve to generate extensive sensitivity data in terms of the primal and dual solution information and subsequent postoptimal data.

How is linear programming employed in this field? First and foremost, it is used to generate least-cost solution sets that are subjected to the standard 'What if . . . ?' variations. Typical problems investigated during 1966-1967 have included 2000 constraints and 6000 or more variables describing the transportation of three types of commodities by 10 to 20 vehicle types to five world areas during a total transportation time interval of six months. The use of time intervals is an important feature of this work. By representing requirements (i.e., tons of materiel per time interval, or troops per time interval) in sequential submatrices and by using a set of logical rules to permit the sequential use and reallocation of transportation vehicles, an element of dynamic or time-varying system use is possible. (This is not dynamic programming, of course, but is a set of time-variant assumptions to guide in the deployment profile of the various elements.)

The postoptimal features of mathematical programming are of considerable use in the application of linear programming to the military transportation problem. For example, uncertainty in costs of the various elements that make up the procurement package of thousands of vehicles is to be expected. How sensitive is the optimal solution space to the uncertainty associated with a preliminary cost of heavy jet aircraft (for example, the C-5 cost five years before one aircraft is manufactured)? Parametric cost ranging is used to develop answers to this type of question. Parametric right-hand-side ranging is of similar value in considering the problem of meeting scenario or theater requirements. If Planner A says two divisions are required in 30 days and Planner B estimates three divisions are required in the same time, of what magnitude is this difference in terms of a worldwide, $10 billion dollar deployment system? Parametric right-hand-side ranging of the requirements of concern can answer the question in a few minutes of computer time. Obviously, the military value between the two solutions is outside of this analysis.

Some minor but interesting questions are listed below. These questions have been addressed by the series of models described in this paper. Again, in no case has model use provided 'the answer'; linear programming has merely served as an informative technique augmenting other analyses. The following uses have been included:

Foreign site evaluation

Many of the US extraterritorial sites (allies, possessions, treaty countries) are represented in the models. Their value may be estimated by selective deletion or parametric ranging options.

Maximum capability estimation

Through the use of alternate objective functions (system capability rather than cost) the capability of a lift system or of selected elements may be rapidly estimated.

Rapid deployment cost estimation

By successively increasing requirements in earlier time phases (through parametric programming) the change in cost and composition of rapid deployment systems can be obtained.

Rate of commodity substitution

The tradeoff between systems in the model (i.e., the rate of commodity substitution) is determined by sequentially reducing a system limit and observing the changes in other variables and the total system cost.

It is evident that this is not an exhaustive list of applications. After some experience with such models one conclusion is that large-scale models do not readily lend themselves to a wide range of applications as well as smaller models do. This will be discussed in the final and concluding section of this paper.

2. THE STRATEGIC DEPLOYMENT LINEAR PROGRAMMING MODEL

Strategic deployment poses the well-known military problem of how to arrive first with the most (i.e., how to obtain an early and substantial logistic posture in wartime). In a period of worldwide commitments by the US the difficulty is compounded. 'Flexibility' is the keyword. From Europe to Southeast Asia, from the most highly developed to the most underdeveloped country, from wherever commitments exist that must be met, there are potential requirements for military support.

The problem of simultaneous threat further complicates US policy. A threat in some remote and underdeveloped world area may call for reciprocal and reinforcing action in an alternate corner of the globe; flexibility to meet combinations of simultaneous demands is essential.

Rapidly advancing technology is an additional consideration. Systems existing today will be obsolete by the mid-'70s when current cargo carriers will be overshadowed by the large and efficient vehicles that are currently either on the drawing boards or in the early stages of development. The capability to utilize a rapidly modernizing fleet to meet any of a wide range of shifting contingencies is vital; flexibility must be assured

Linear programming has proven to be an efficient means to measure flexibility by representing most of the pertinent factors and developing the tradeoffs necessary for sensitivity analyses. The linear constraints state the system cost and limitations, system availability, and the meeting or fulfillment of requirements. In addition, linear constraints specify the throughput limitations (e.g., maximum port capacity) by time period. Figure 1 is a simplified representation of the problem, indicating five types of constraints.*

Minimize the objective function:

$$\Sigma C_i X_i \qquad (1)$$

subject to

System limitations, $i = 1, \ldots, I$ (2)

$$X_i \qquad \leq b_i$$

System availability, all feasible (i, k, l, m) combinations (3)

$$X_i - \sum_j P_{ijklm} Y_{ijklm} \qquad \geq O$$

Fulfillment of requirements, all necessary (k, l, m) combinations (4)

$$\sum_{i'}\sum_{j} Y_{ijklm} \qquad \geq R_{klm}$$

Throughput capacity limitations, subset i' of i (5)

$$\sum_{i'}\sum_{j} Y_{i'jklm} \qquad \leq S_{klm}$$

The notation used is

X_i = number of units, system i

C_i = cost per unit, system i

b_i = upper limit of availability, system i

P_{ijklm} = productivity, system unit per tonnage unit of system i from site j to theater k, contingency set l in time period m.

Y_{ijklm} = tonnage unit by system i, site j, theater k, contingency set l time period m

R_{klm} = tonnage required, theater k, contingency set l time period m

Y_{ijklm} = tonnage unit by the subset (i′) of i from site j, etc.

S_{klm} = throughput constraint, tons, in theater k, contingency set l time period m

Fig. 1 Simplified Linear Constraint Set

---

* This arbitrary grouping does not include all of the types of relationships, nor are the logical rules of a real problem indicated by this general formulation. Figure 3 expands this description.

The necessity for describing contingency sets, which are the alternate combinations of theaters represented by the subscript 1 in Fig. 1, is frequently questioned when the problem is first described. However, it should be observed that military planners may present the sets (A, B, and C) or (A, C, and D) or (A, C, and E) as equally likely sets of contingencies†. Why is a system that meets the worst of the three not adequate? Lack of flexibility is the answer. For example if B should prove the worst offender of all the alternates in the group B, D, or E, prepositioning in the vicinity of B could be a part of the optimum solution to (A, B, and C). This could, on the other hand, provide little value in meeting the lower requirements for the geographically distant areas D and E. The linear programming formulation to investigate all sets of equal probability is guaranteed to provide a feasible and least-cost means of meeting requirements. Flexibility is ensured by the formulation.

The linear programming formulation of the total deployment problem is illustrated schematically in Fig. 2. These are the submatrices employed in the most recent formulation, the MATRA Model. The variables are of two basic types: system variables and deployment variables. System variables describe the number of aircraft and tons of shipping required in the total deployment system. Deployment variables describe the fulfillment of requirements in terms of tons of materiel delivered or moved per time interval. The latter variables include in their definition a description of type of carrier, type of load, and origin-destination pair. The constraints are of five basic types, as shown in Fig. 1. First is system cost. This is the simplest because it is only a single constraint. The major difficulty with this constraint is obtaining the cost coefficients, which come from diverse sources and represent complex problems in themselves. Second are the system-limitation constraints. Here

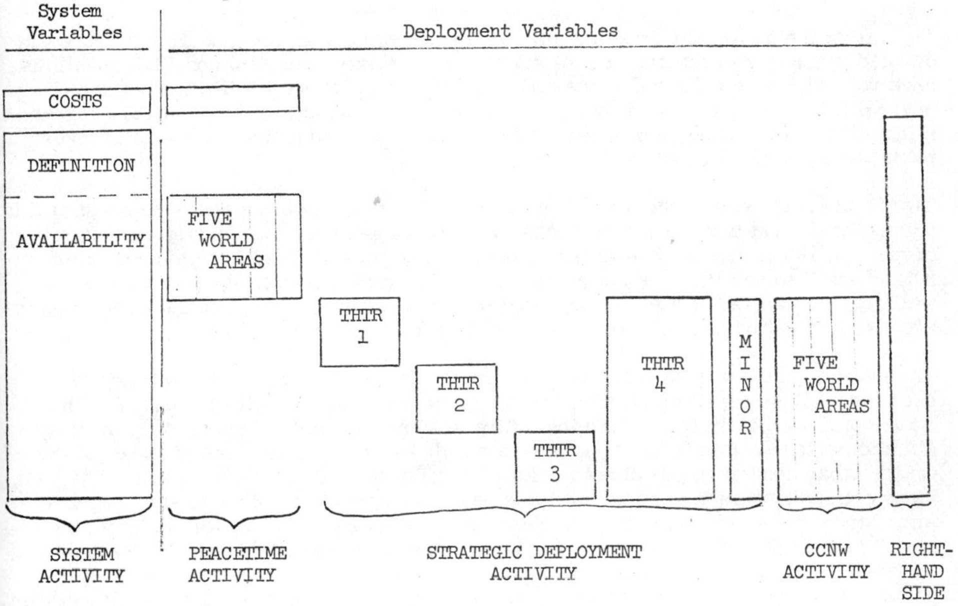

Fig. 2 MATRA model macrostructure

Indicating the four major submatrices: System, Peacetime, Strategic Deployment and Concurrent Nonwar

Numerical designation of theaters is employed to avoid security classification problems.

---

† A, B, C, etc., are used to notionally represent limited wars in countries designated A, B, C, etc.

limits are imposed on the system elements by cost category (e.g., new or inherited asset elements) and by any assumptions as to availability at the time of assumed contingency occurrence. Third, availability or balancing constraints are specified. These constraints establish by contingency area and time period the availability and possible employment of the systems described. Fourth, constraints state how the various military requirements can be met. These are in terms of the theater, contingency set, and time period. The fifth and final type of constraint specifies individual port and airfield throughput limitations as functions of world area and time period.

The models are quite simple and pose considerably less descriptive challenge than do many other linear programming applications. A major difference from many linear programming problems exists in the uncertainty of the 'rim' of the problem (i.e., costs and requirements). This aspect, however, is one characteristic of the problem that makes the postoptimal features of linear programming desirable.

An attractive feature of the linear programming formulation is that it allows the decision maker to consider the important factors of his problem. He is able to develop the tradeoffs between complex transportation systems (e.g., the Air Force C-5, the Navy Fast Deployment Logistic Ship*) and pose the questions regarding sensitivity that he must answer. Obviously, the use of analytical techniques as represented by this linear programming model is considered highly desirable in the current planning, programming, and budgeting system within DOD. A brief description of the models developed for such use in 1966, 1967, and 1968 constitutes the following section.

3. DEVELOPMENTAL HISTORY

The use of mathematical programming for the least-cost airlift/sealift mix has been devoted primarily to the analysis of very large systems. The C-5 and FDL questions have been of major concern. Since either system represents a total expenditure of several billion dollars, the resources required for substantial mathematical programming effort may be justified in part of the basis of the importance of the total investment and potential cost savings.

Two frequent questions regarding the use of linear programming are of interest at this point. First, what has been the experience with acceptance by the decision-making groups? It is not appropriate to indicate details of acceptability in this brief summary, but suffice it to say that complete solution sets with all postoptimal data have been well received and adequately interpreted by government analysts and decision-making groups. In most cases, acceptance and use of results have been good.

The second question is one for which a less direct answer can be given. What have been the visible results in terms of cost benefits or other benefits? Because of the nature of the problem (programming future systems for use in hypothetical military situations), it is not possible to point to a result and say 'X' millions have been saved. On the other hand, it is possible to point to Air Force studies that have altered previous conclusions as a result of analysis with strategic deployment linear programming models. For example, the Air Force programming objective for the C-5 was revised downward. Similarly, Army planners have seen the advisability of requesting additional logistic support from the Navy. A visible result of this type has been to provide an integrated point of view whereby the interdependent service role is better understood as it relates to meeting a team goal. The use and acceptance of linear programming as the technique to study deployment has resulted in common points of departure and improved understanding of the reasons for divergent opinions among the services and the Office of the Secretary of Defense.

The first major model in the series was developed in mid-1965 to provide supporting analysis for the C-5 procurement problem. It contained approximately 300 constraints

* The C-5 is a large USAF cargo aircraft. The Fast Deployment Logistic Ship (FDL) is a proposed cargo ship.

and 500 variables and was structured to include five contingency area requirements in a 30-day time interval. Approximately 1000 hours of IBM 7040 computer time over a full year were expended in generating solutions with this first model.

The use of approximately 1000 computer hours per year has been characteristic. In 1966 the time was mixed 7044 and 7094 and in 1967, 7094 and 360/65 time. The mathematical programming packages used have been the IBM series LPI and LPIII versions for the 7040/7044 computers and the MPS 360 system for the third generation computers.

Two models were developed in 1966, each having approximately 800 constraints and 1500 variables. Size increase in one case was due to attempts to represent more detail, and in the other to an enlargement of the problem to include more areas of interest. All models developed through this period were hand structured and of a type that could be termed experimental in that constraints and variables were not static. Changes in structure were introduced frequently, and numerous runs performed to explore alternate formulations.

The remainder of this discussion concerns the MATRA Model that was developed in 1967. It was a logical outgrowth of the two previous models in that both detail and enlarged areas of interest were included. The MATRA Model ranges in size from 2000 to 3000 constraints and contains approximately 6000 variables. Solution times on IBM 360/65 computers range from one hour for parametrics to twenty hours with no basis.

The most interesting features of the model development from a model-building point of view have been the problems encountered in generating solutions for a matrix of this size and the relative stability of the solutions. These two points are expanded later.

The developmental history is incomplete without some indication and conjecture as to future developments. Experience indicates that larger models are not required in the next year and that detail can be traded off for additional considerations. For example, twice as many system elements (e.g., new aircraft, additional types of ships, an additional commodity) can be added and half of the time periods eliminated, and solutions would not be significantly different. Today, it is not apparent whether larger models will ever be required; it is only clear that continued experience with 2000- to 3000-constraint problems is necessary to fully understand existing formulations.

A continuing trend is to fully automated matrix and report generators. It is desirable to rapidly generate new problems and to extract only selected subsets of solution data. Automated generators are possible today with most of the third-generation computing machines and linear programming software.

4. MODEL DEVELOPMENT

The model structure for the MATRA Model was a logical extension of previously developed models. The constraints, as noted in the discussion of Fig. 1, are of five types. Figure 3 is a more complete description of the structure of the model and is presented to show the size of the problem and the types of variables. Again, secondary details are excluded; for example, subsets of the variables X_{sjklm} (shipping) can contribute to subsets of the X_t (prepositioning), but these options are not shown.

The problem described by Fig. 3 assumes only the system and strategic deployment portions of the model. In actual use, the MATRA Model is composed of four primary submatrices: System, Strategic Deployment, Concurrent Nonwar, and Peacetime.

The System, Peacetime, and Concurrent Nonwar submatrices are relatively simple, consisting primarily of the previously described constraint types: system definition,

Minimize the objective function

$$\Sigma C_r Z_r + \Sigma C_s Z_s + \Sigma C_t Z_t \qquad (1)$$

subject to

$$Z_r \leq b_r \qquad (2)$$

$$Z_s \leq b_s$$

$$Z_t \leq b_t$$

$$Z_r - \sum_j \sum_k P_{rjklm} X_{rjklm} \geq 0 \qquad (3)$$

$$Z_s - \sum_j \sum_k \sum_{m'} P_{sjklm} X_{sjklm} \geq 0$$

$$Z_s - \Sigma Y_{sjklm} \geq 0$$

$$Z_t - \sum_r \sum_k \sum_m X_{rtklm} - \sum_s \sum_k \sum_m X_{stklm} - \Sigma\Sigma\Sigma Y_{stklm} \geq 0$$

$$\sum_{t'} Z_t + \sum_r \sum_j \sum_m X_{rjklm} + \sum_s \sum_j \sum_m X_{sjklm} + \sum_s \sum_j \sum_m Y_{sjklm} \geq R_{klm} \qquad (4)$$

$$\sum_{r'} \sum_j X_{rjklm} \leq S_{klm} \qquad (5)$$

$$\sum_{s'} \sum_j X_{sjklm} + \sum_{s'} \sum_j Y_{sklm} \leq Q_{klm}$$

Notation:

Variables

Z = number of aircraft, ships, or prepositioned tonnage

X = tonnage deployed

Y = shuttle shipping, tonnage

Subscripts

r = aircraft designation, r ~ 10

s = ship designation, s ~ 10

t = prepositioning site designation, t ~ 15

j = site (origin), j ~ 15

k = theater (destination), k ~ 5

l = contingency set, l ~ 3

m = time period, m ~ 10

Coefficients

P = productivity

Right-hand sides

b = system limits

R = requirement;

S, Q = throughput constraint, limits

Comments

(1) Primed subscripts are used to indicate that only a selected subset is of concern in this case.

(2) This formulation shows only one type of cargo, no troop movement, and does not detail the numerous special constraints the MATRA Model contains.

(3) In the MATRA Model, it is always assumed that the same theaters do not have differing requirements or activities by contingency sets, i.e., activities to meet (k_1, l_1) = activities to meet (k_1, l_2).

Fig. 3 Expanded Linear Constraint Set, MATRA Model

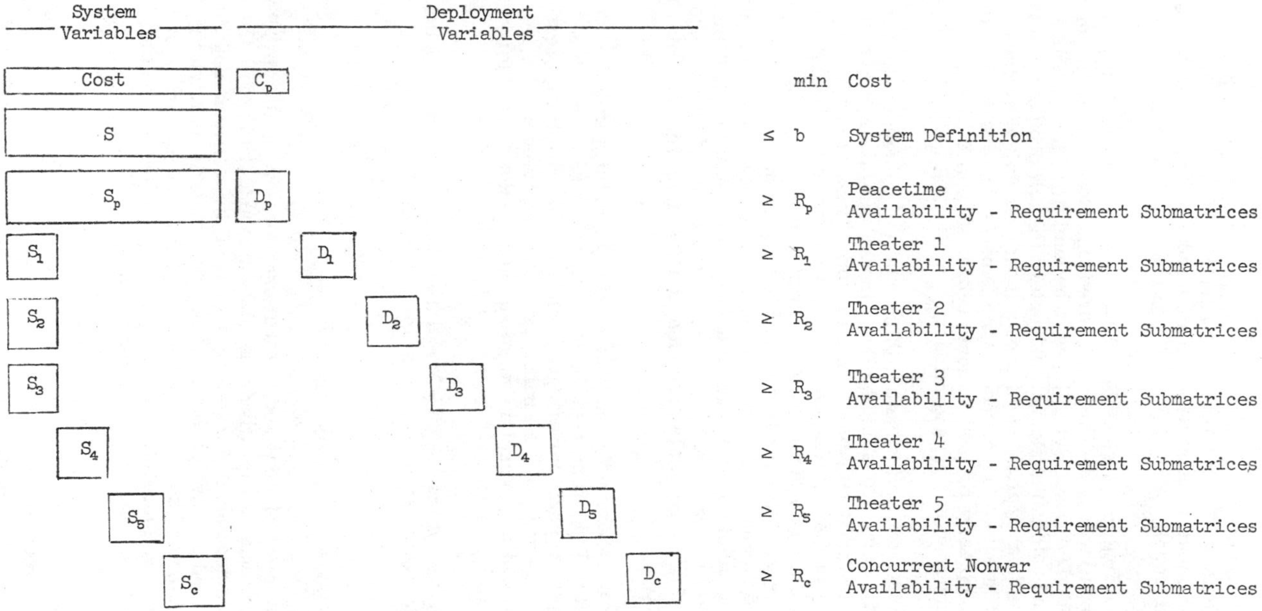

Fig. 4 Expanded MATRA model submatrix schematic

Notation:

S_k = System activities (variables)
D_k = Deployment activities
R_k = Requirement and theater constraints
k = 1, 2, 3, 4, 5 (theater designation)
k = p, Peacetime
k = c, Concurrent Nonwar

availability-utilization balance, and requirement fulfillment. Figure 4 indicates schematically a full-sized structure that includes all submatrices. It is an elaboration of the previous MATRA Model macrostructure, showing in detail how the submatrices are related.

The major formulation complexity occurs in the Strategic Deployment portion of the matrix. This section accounts for 80 or 90 percent of the 30, 000 to 50, 000 coefficients of a full problem. Data manipulation must be automated, and a matrix generator is used for this portion of the model.

The technique used in generating the MATRA Model has been as follows: First, the System and Peacetime submatrices were hand structured: The Strategic Deployment submatrix was then merged via another set of tapes. The total problem, then, is a result of hand-generated and automated means. Ultimately, the full problem was run and checked out as a unit. Actual application of such a model generally involves changes in either cost coefficients or right-hand-side elements. On occasion, either the System Submatrix is completely revised or some subset of coefficients is modified; the full problem has not required regeneration.

One major problem that has arisen as a result of this method of combining the component parts is one of verifying solutions for the large model. How does one 'check' a problem consisting of up to 2, 000 variables? Small solution subsets can be employed to exercise various portions of the formulation. Ultimately, experience through numerous applications leads to the use of a criterion of reasonableness. When this fails (i.e., a solution appears to be unreasonable), the last resort is to trace through the problem. With the MATRA Model, and indeed in all of the strategic deployment models developed thus far, the most frequent problem for the analyst has been that solutions appear unreasonable, or at least defy intuition. In most cases this is resolved by thorough examination of the matrix. For large-scale problems, considerable time is required to conduct confirmatory analysis.

It is concluded that, insofar as performing analyses to verify solutions with a model of the size of the MATRA Model, the time requirements may be excessive. This appears to be true if postoptimal solutions are generated to aid in resolving questionable results or if the analyst is required to further examine the problem. The requirement for large expenditures of time to explain solutions must be anticipated for this size and type of problem.

Additional difficulties in the use of the MATRA Model as related to specific applications and user acceptance are discussed in the next two sections.

5. APPLICATIONS OF THE MATRA MODEL

With the initial applications of a technique such as that used for solving the MATRA Model (i.e., a somewhat novel use of a relatively sophisticated analytical tool), the analyst is faced with two requirements. First, both the use of the technique and the model must be well understood to solve problems efficiently. This is no trivial matter in the case of a large computer and software package and a 2500-constraint linear programming model. Second, the model must be effectively employed for the purposes for which it was constructed: to solve the client's problems.

Determining the most efficient algorithmic/model combination to solve problems is necessary before extensive model use is possible. There are many parameters that can be set either to aid or to guide problem solution when employing a large-scale software package such as the one used here. For example, the software employed with the MATRA Model has been primarily the IBM MPS/360 package that contains the following user-available parameters:

XCYCLESW When XCYCLESW >0, the matrix is read only until p possible candidates are found. Here, p = XCYCLESW.

XPRICE	When the matrix or subset of the matrix is read, the m vectors that give the best rate of change of the objective are selected as candidates to enter the basis. Here, m = XPRICE.
XTOLDJ	When the magnitude of any computed reduced cost is less than or equal to XTOLDJ, it is treated as if it were zero.
XDZPCT	When vectors are chosen to enter the basis on minor iterations, it may be wise to consider only those vectors that will give an objective change at least equal to some fraction of the change in the objective function on the preceding major iteration. XDZPCT is that function.

There are approximately 75 additional parameters that may be set at the user's discretion. There is virtually no guidance available from user experience concerning what values to select for setting these parameters.

Two attempts to investigate better settings for the parameters are detailed in the following tables. First, it appeared that changes in XPRICE and XCYCLESW from the standard values of XPRICE = 5 and XCYCLESW = 50 might produce better solution time. Table 1 summarizes several hours of experimentation in this area. The net result was that the results with the precalculated values could not be improved.

Rate of change of the objective function (in arbitrary units) is listed for eight combinations. This was the best performance achieved* (i.e., the greatest rate of change of objective function).

The test consisted of running the problem from the same point in all eight tests for a fixed number of iterations. Each point required approximately 30 minutes of computer use.

Table 1

XPRICE AND XCYCLESW Variations

	XPRICE value		
XCYCLESW value	2	5	8
50	10.7	11.8[a]	8.0
250	9.4	7.7	9.0
500	7.8	7.6	na[b]

[a]Machine-set value is underlined.

[b]Not available.

Next the XTOLDJ setting was explored. This has been performed twice at different stages of model development with the same results. Table 2 summarizes the experience. Experience not included in this table has indicated that to proceed to a solution without a starting basis and with a machine-set tolerance of 10^{-7} often leads to numerical difficulties. Typically, frequent loss of feasibility in the region of near optimality is encountered. The value of 10^{-4} has been adopted for most of the MATRA Model solutions.

Finally, a note is included relative to XDZPCT. This parameter has been repeatedly explored with no effect on solution history for the MATRA Model.

* The underlined value in Table 1.

A conclusion relative to these experimental applications is that the user and model builder should be offered guidance in the use of these various parameters. It is a most expensive and unsatisfactory experience to be faced with several dozen parameters to manipulate without guidance in addition to attempting to develop a large-scale model. This is one of the authors' principal criticisms of the MPS/360 system. Given unlimited computer use and a well-understood model, the availability of such a range of choices might be desirable. This situation is not one that is frequently faced by the analyst.

Table 2

XTOLDJ Variations

XTOLDJ value	Time to optimality[a]	Objective function[b]
10^{-3}	>10 hours[c]	C_0
10^{-4}	+1 hour	3%
10^{-5}	+5 minutes	<1%
10^{-6}	+5 minutes	≪1%

[a]The values shown as +t are added to the preceding time.

[b]The values shown as X% reflect decrease in the preceding value of the objective function.

[c]Typical solutions for the MATRA Model can generally be described as going through four 'phases.' First, feasibility is obtained in perhaps 5 percent or less of the total time. Next, the objective function is driven to within two times the final value of the optimal solution in perhaps 10 percent of the total time. Going from this higher level to a level within one or two percent of optimality requires greater than 50 percent of the total time. The fourth and final phase is occupied by reaching optimality, frequently losing feasibility, recovering, and making very small changes in the objective function. Figure 5 illustrates the typical solution history.

Use of the series of models of which MATRA is the most recent version has been directed to a large extent toward mapping solution sensitivity to changes in the cost of vehicles and the availability or deletion of system elements. This application history has continued through the use of the MATRA Model. An additional aspect in application is time. One of the purposes in developing the MATRA Model (and also a further version of the strategic deployment class of problem, termed the POSTURE Model) has been the desire to represent time more realistically in linear models. Time phasing is important in this type of problem because early demands require more expensive assets and the later demands can generally be met with less expensive assets. This obvious statement is nevertheless an important basis for developing the set of models used thus far in this area; tradeoffs in the expensive systems are important and must be determined.

To present a graphic demonstration of MATRA Model solutions and the type of information desired, a solution is shown in Table 3.* While this table illustrates the primal solution in terms of levels of the variables in an optimal basis, the analyst must refer to the total solution (i.e., the dual variables) to determine the 'why' of it all. Consider Table 4, in which a partial listing of variables that meet requirements and marginal values associated with requirements is shown. Even the full solution in itself is not completely meaningful since it is in the comparison of the activity levels of the primal and dual variables from solution to solution that the real cause and effect relationships of the activities are developed.

---

* All solutions are hypothetical, extracted from actual runs and subsequently modified for use as unclassified illustrations.

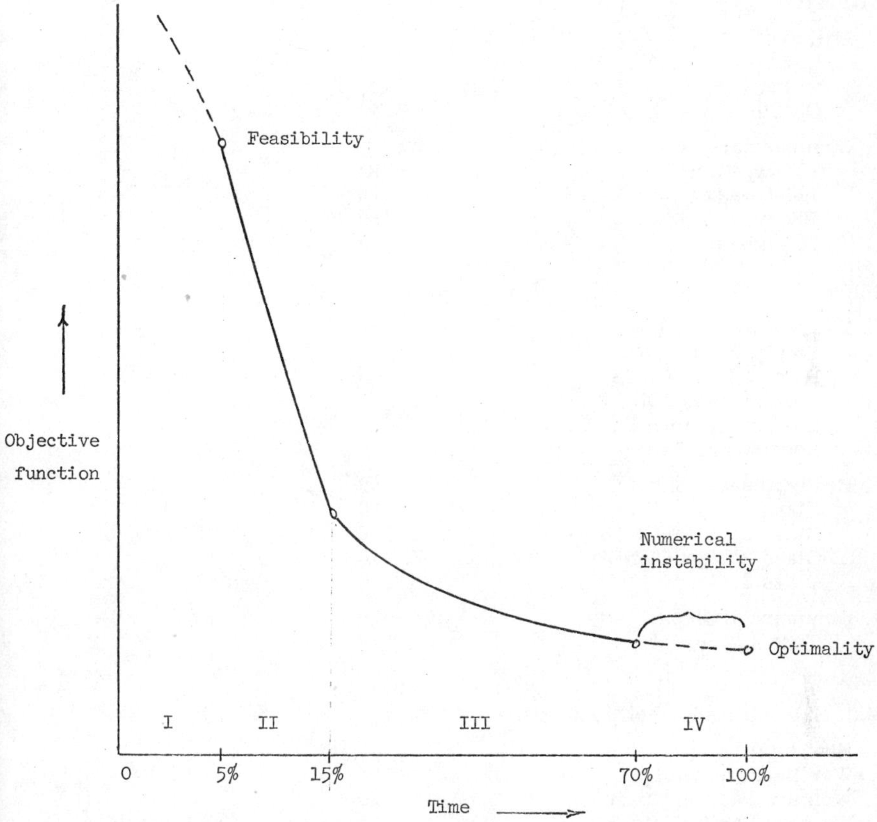

Fig. 5 Typical solution
Objective function decrease with solution time.

Table 3

Typical MATRA Model Solution System Activities

Item	Number
System cost, $ × 10^9 = XX. XXX	
Aircraft	
Military	
C-5	144
C-141	160
C-130	48
Commercial	
707-type (P)[a]	96
747-type (P)	16
707-type (C)[b]	160
747-type (C)	32
Ships	
Fast Deployment Logistics Ships	25
Empty, West CONUS[c]	10
Empty, East CONUS	5
Loaded, East CONUS	4
Loaded, Western Pacific	3
Loaded, Far East	3
Ready Reserve Ships	50
Class 1	0
Class 2, East CONUS	10
Class 2, West CONUS	40
Class 3	0
Commercial Ships	420
Type 1, . . .	.
.	.
Material, unit equipment, 1000 Stons	
New, total	30
War Reserve, total	100
Western Pacific, Site A	30
Western Pacific, Site B	15
.	
Europe, Site X	10
.	.
Material, resupply, 1000 Stons	
New, total	.
.	.
Troops, 1000's	.
.	.

[a](P) = Passengers

[b](C) = Cargo

[c]CONUS = Continental United States

Table 4

MATRA Model Solution

Deployment Activities, Primary and Dual Variables (Partial listing)

(a) Columns Section: Deployment Variables

Theater	Time period	System	Origin	Cargo type	Level of activity
1	1	C-5	West CONUS	Unit	20 kilotons
1	2	C-141	Western Pacific, Site A	Unit/Pax	2 kilotons 700 troops
.	.				
1	7	FDL	West CONUS	Resupply	100 kilotons
.	.				
2	1	C-5	East CONUS	Unit/Pax	4 kilotons 1400 troops
.	.				

(b) Rows Section

Theater	Time period	Description	Net activity in row	Marginal value ($\$ \times 10^6$)
1	1	Unit requirement	30. 0 kilotons	20. 0
2	1	Unit requirement	15. 0	95. 0
3	1	Unit requirement	20. 0	0. 0
.	.			
4	7	Troop requirement	150, 000 pax[a]	0. 0
.	.			
1	2	C-5—balancing	0	4. 0
1	3	C-5—balancing	0	2. 0
2	2	C-5—balancing	0	0. 8
3	2	C-5—balancing	0	1. 5
4	2	C-5—balancing	0	5. 0
.	.			
3	2	Unit materiel	0	0. 2
3	3	balancing, Western	0	3. 5
3	4	Pacific Site B	0	0. 4
.	.			

[a]Passengers.

Table 5 illustrates this comparison. Here, the tabular arrays represent an increased cost in prepositioning at one site from among the 15 possible in the MATRA Model.

A final set of solutions shown in Table 6 illustrates the effect of varying one of the time-phased elements in the model—unit equipment at one site.

An apparent feature of both solution sets in Tables 5 and 6 is the small amount of change in the major system variables over the range of the parameter(s) varied. The total cost may increase considerably, perhaps as much as $\$2 \times 10^9$. However, this amounts to a 15 percent increase in cost, which is not necessarily an appreciable change for driving some parameter from one extreme (maximum level) to the other extreme (minimum level).

Table 5

Parametric Cost Ranging: Partial Solution Set[a] (Cost at Site D)

Cost at Site D	Total system cost, $ × 10^9	Material at Site D, 1000 Stons	C-5 aircraft, number	Marginal value, R_x[b]
C_d	10.010	30	140	94.
2.C_d	10.040	30	140	93.
4.C_d	10.100	30	140	92.
6.C_d	10.170	25	145	97.
8.C_d	10.200	20	150	105.

[a]This table represents one type of commodity substitution in which the cost of material at Site D is parametrically increased and the C-5/Site D material substitution is generated.

[b]This is the marginal value associated with the constraint R_x, which represents the requirement that dominates this solution set.

Table 6

Solution Set: Effect of Time-Phased Equipment Availability

System cost, $ × 10^9	12.5	13.6	13.7	13.8	13.9
L1[a]	X	0	0	0	0
L2	X	X	0	0	0
L3	X	X	X	0	0
L4	X	X	X	X	0
Military aircraft	334	348	348	338	338
Commercial aircraft	300	300	300	300	300
FDLs, total	40	30	32	26	25
CONUS	35	15	17	9	8
Forward	5	15	15	17	17
Commercial ships	200	200	200	200	200
Reserve ships	200	240	240	220	190
Prepositioning, total[b]	280	300	320	400	420
Europe	200	150	150	170	180
Eastern Pacific	40	40	40	75	80
Western Pacific	30	95	110	120	110
Far East	10	15	20	35	50

[a]L1 through L4 are constraints that are applied sequentially to limit material available in CONUS. 'X' means material is unlimited; 'O' means material in the Li period is reduced to the level of feasibility.

[b]Units of prepositioning in 1000 Stons.

These applications serve to illustrate types of solutions, sensitivity, and the meaning of time (as represented in the availability of assets) insofar as results are concerned. Further details as to solution use by the decision-making groups and solution acceptability are contained in the final section.

6. SUMMARY OF MODEL USE

The MATRA Model has been employed by DOD codevelopers of the model, OASD(SA), an Air Force Systems Analysis Group, and a group responsible to the JCS Special Assistant for Strategic Mobility (SASM). Although each of the groups has had a different objective in employing the model, a major effect has been the coordination and integration of the resulting systems.

First, the OASD(SA) group had as a primary objective the use of the model in overall system design (i.e., size and mix). The purpose was to aid in analytical support for the preparation of the annual Draft Presidential Memorandum (DPM) on Airlift and Sealift Force, a document which reviews and makes recommendations for changes in the Airlift and Sealift Force levels. This document sets the major programming objectives of the Secretary of Defense for the coming fiscal year.

The second group, and in fact the most extensive user of the MATRA Model, has been the Air Force Systems Analysis Group. Their primary mission, insofar as the application of the model has been concerned, has been the development of an Air Force position relative to the optimum airlift force size and composition in the strategic deployment context. In this role, the MATRA Model has served as an analytical framework in which Air Force analysts have been able to examine the sensitivity of important questions. In addition, the Air Force undertook to examine the points of view of other services, since the model describes competition between Air Force systems and those of other services for the strategic deployment mission.

The third group, the JCS-SASM users, has been engaged in the development of a joint service position through the appraisal of the total strategic mobility posture and recommendation of airlift and sealift forces required to support postulated strategies. In effect, these recommendations are a consideration in the development of the DPM by OSAD(SA).

In the case of the OASD(SA) group, use of this technique was not new; familiarity with the least-cost linear programming model approach dates back several years. In the OASD case, results of the large-scale model were not substantially different from the results of a smaller model. Aggregated systems of approximately the same composition resulted, and in this sense the results were not different or substantially more desirable than those of earlier and considerably smaller versions.

For the Air Force, there was equal interest in actual employment of the elements of the system and the system composition. As an operating service, one of the major 'owners' of systems represented in the MATRA Model, questions of solution feasibility arise and can be probed. Although solutions have not considerably changed in the Air Force case, the question of feasible operations has been addressed. The detailed 'Why... ?' of solutions is much more addressable by the large-scale model, although answers require interpreting hundreds of variables.

The third user is not only interested in the overall strategic mobility posture but is equally concerned with system composition. However, this group's experience with such analytical tools is not as extensive as that of the two previously noted users. Rephrasing of the JCS-SASM user's questions to take full advantage of the MATRA Model, as well as reformulating the model slightly for this user, is under consideration.

In all cases the matrix of commonality provided by the model allowed the users to address important and related questions. Rather than arrive at diverse solutions to a similar problem and be faced with a major uncertainty in comparison because different techniques are employed, the users are able to argue validity of assumptions on input data. Where previously such groups addressed the same problem with different techniques and wrote different chapters of the same story, a common denominator is now employed to reduce questions to more fundamental issues. This aspect, that of integrating a complex problem within one framework and cutting away the uncertainty of duplicative and costly analytical techniques is a significant achievement in DOD planning.

It must be concluded that the value of using large-scale mathematical programming models for military planning purposes is not quantifiable in the normal sense. However, the benefits seem apparent in terms of user acceptance and continuing interest. It would seem that a continued expansion and development of large linear programming models may occur, primarily as the user realizes the value of employing this device to integrate complex interrelations. It is in the continued expansion to include the more wide-ranging aspects of the user's problems that such large-scale model use will probably continue. It is not difficult to envision problems of 10, 000 to 20, 000 constraints resulting from an expansion of the MATRA-like problems to include other interesting facets of DOD, although within the next two years model use will probably still reside with the 1000- to 3000-constraint versions.

Finally, a conclusion is in order regarding use of such large models from the model-builder's point of view. It is not apparent that large models that include all aspects of potential interest or aggregate models with emphasis only on certain details are the answer to future model development. The OASD(SA) efforts have tended to balance between these two points. Nevertheless, large running times on expensive computers have been necessary throughout this program. The authors' very reserved opinions, after over a year of solving problems of this size, are that for certain problems large models do provide some benefit. In practice, the determination of when to use such a model is not easy. There are instances of months being devoted to formulating and building large models, only to have either failure or such poor solution experience result that no positive results are achieved.* Whether or not to employ a large model is a question that should be answered only after extensive consideration of all techniques and the potential value from a programming model have been weighed, and initial experimentation or small model work performed to test the validity and adequacy of this approach.

ACKNOWLEDGMENTS

This paper represents the authors' experience, interpretations of results, and conclusions. Advice and suggestions have been received from the following: Dr. David C. Dellinger and Dr. Evan Porteus, OSD(SA); Major General Selwyn D. Smith Jr., USA (Ret), RAC; Lieutenant Colonel James T. Brumbeloe and Major John F. Baumann, USAF. Their comments are gratefully acknowledged, particularly regarding the 'Summary of Model Use,' which concludes this paper. The authors assume responsibility for criticisms that may result from their interpretation of the comments and suggestions received.

REFERENCES

The following documents describe earlier model work related to the strategic deployment linear programming problem. All are unclassified.

FITZPATRICK, G. R., and J. C. WHITON, 'A Programming Model for Determining the Least-Cost Mix of Air and Sealift Forces for Rapid Deployment,' RAC-P-19, Research Analysis Corporation, Jul 66.

FITZPATRICK, G. R., J. BRACKEN, M. J. O'BRIEN, L. G. WENTLING, and J. C. WHITON, 'Programming the Procurement of Airlift and Sealift Forces: A Linear Programming Model for Analysis of the Least-Cost Mix of Strategic Deployment Systems,' Naval Research Logistics Quarterly, 14 (2): 241-55 (Jun 67).

JENKINS, P. M., M. J. O'BRIEN, and J. C. WHITON, 'The Mobility and Transportation Resources Allocation Model (MATRA), A Large-Scale Linear Programming Model to Investigate Department of Defense Transportation Problems,' presented at the Twenty-First Military Operations Research Symposium, Colorado Springs, Colo., Jun 68.

* These efforts are not in the strategic deployment linear programming area.

WENTLING, L. G., G. R. FITZPATRICK, M. J. O'BRIEN and J. C. WHITON, 'A Programming Model for the Design of Strategic-Deployment Systems,' RAC-TP-211, Research Analysis Corporation, Sep 66.

Linear programming 'packages' as used in this work are described in the following:

International Business Machines Corporation, '7040-7044 Linear Programming System III (7040-CO-12X) User's Manual,' H20-0195-0, White Plains, N.Y.

—, —, 'Mathematical Programming System/360 (360-CO-14X) Application Description,' H20-0136-1, White Plains, N.Y.

—, —, 'Mathematical Programming System/360 (360A-CO-14X) Linear and Separable Programming—User's Manual,' H20-0476-0, White Plains, N.Y.

DISCUSSION

D. Edmonds

This is a question on your computational experience with MATRA.

Am I correct in thinking that your studies of alternative parameter settings in MPS have not resulted in significant reductions in running time?

J. C. Whiton

That is correct. We have achieved no noticeable reductions in running time with any of the alternate parameter settings we have tried. Such attempts are very expensive with a large model. We have, therefore, not attempted a large exploratory effort regarding alternative settings.

The Determination of the Least Cost Mix of Transport Aircraft, Ships and Stockpiled Material to Provide British Forces with Adequate Strategic Mobility in the Future

C.N. BEARD, C.T. McINDOE
Ministry of Defence, U.K.

1. INTRODUCTION

The problem to which the analysis described in this paper was addressed may be summarized by the following question:

> 'What is the most economical combination of aircraft, ships and stockpiled material which will provide adequate strategic mobility for British Forces in the future?'

where the answer is to be judged against scenarios typifying possible future operations.

The Secretary of State for Defence set up a working party to examine this problem. DOAE were asked to undertake the analysis on behalf of the working party under the guidance of an executive sub-group. The assumptions about scenarios, desirable force build-up rates, the forces likely to be moved, etc. were agreed by this working party upon which the views of the three Services, the Defence Secretariat and DOAE were represented.

2. THE APPRECIATION OF THE PROBLEM

The process of discovering an answer to the question is beset by two types of difficulty. The first results from uncertainty about the future; for example, uncertainty as to the type and scale of operations in which British forces may become involved. The second type of difficulty results from the complexity of the problem; for example, the strong interdependence of all major aspects means that valid conclusions cannot be drawn from a study of one major aspect in isolation. The uncertainty of the future defines certain aspects of an ideal method of solution. Firstly, it should be quick, so that as many different sets of assumptions as possible can be examined within the time available for the study. Secondly, the sensitivity of answers to dubious numerical estimates should be easily determined.

Linear programming obviously provides these desirable features, but also fits the basic logic of the problem. Various operations taken in isolation would define different optimum combinations of aircraft and ships. We needed to derive that combination which would most economically cover a representative selection of possible operations. Thus although no two operations were considered to be taking place at the same time, the analysis needed to take into account the requirements for one operation in determining the requirements for another. It was with these thoughts in mind that the model was constructed.

3. THE CONSTRUCTION OF THE MODEL

3.1 General

In general terms the model was similar to that described in the Research Analysis Corporation's paper, although different in several particulars. The principal difference was that a group of constraints represented the various possible ways of achieving a force build-up rate in an operation by sea and air movement either from UK or from a stockpile. There were therefore four blocks of constraints representing different scenarios. A fifth block set limits on the inventory of aircraft and ships, and the size and composition of the stockpile. The objective function represented the capital and running costs of maintaining aircraft, ships and stockpiles in being for the contingencies considered.

Time phasing of the force build-up in any operation was achieved by using four time periods which could differ in duration from one operation to another. Apart from material balance constraints which ensured that the specified build-up was achieved in any particular time period, the following are a sample of the type of operational limitations that were allowed for:

(a) Saturation of air routes from the UK to the various areas of operation.

(b) Saturation of facilities at destination airfields.

(c) Special load-carrying characteristics of different types of aircraft and ships.

(d) The need for parachute assaults on airfields deemed insecure.

(e) The fact that certain amounts of material must accompany men moving from the UK and so could not be stockpiled.

(f) The movement of men and material forward from a port by air.

(g) The need for aviation fuel to be moved into the area for aircraft involved in the operation.

When four scenarios were embodied in the model, there were 850 constraints comprising about 2000 vectors. Specific aspects of the problem which will be described next defined the need for a mixed integer linear program algorithm. We therefore selected the system devised by CEIR Ltd. which Mr. Shaw will describe in the next paper.

3.2 Selection of an Optimum Stockpile Site

About four possible sites were considered to be available, of which at most one was to be selected. This site had to cover the different operations mentioned earlier. Its selection depended not only on the weight of material to be moved into the various areas, but also on the times of arrival for each operation. These conditions were covered by considering aircraft and ships able to move material from each site to each operation with the following constraints imposed by zero-one integer vectors:

Let Z_1, Z_2, Z_3 and Z_4 be the quantities of material needed on Sites 1, 2, 3 and 4 to cover the various operations.

Let M be a large number, and

$\delta_1, \delta_2, \delta_3$ and δ_4 be zero/one integer vectors,

then the following constraints imposed the necessary conditions

$$\delta_1 M - Z_1 \geqslant 0$$

$$\delta_2 M - Z_2 \geqslant 0$$

$$\delta_3 M - Z_3 \geqslant 0$$

$$\delta_4 M - Z_4 \geqslant 0$$

$$\delta_1 + \delta_2 + \delta_3 + \delta_4 \leqslant 1$$

3.3 Allowance for Fixed Costs

There is a fixed cost incurred in setting up a stockpile irrespective of the amount stockpiled. This fixed cost varied from site to site, depending upon the facilities already existing. It was allowed for by including the following terms in the objective function:

$$\delta_1 F_1 + \delta_2 F_2 + \delta_3 F_3 + \delta_4 F_4 + Z_1 C_1 + Z_2 C_2 + Z_3 C_3 + Z_4 C_4$$

(this must be examined with the constraints given in the previous paragraph)

where F_1 is the set-up cost for Site 1, etc. and C_1 is the cost per ton of material stockpiled on Site 1, etc. Similar allowance was made for port facilities for ships.

3.4 Allowance for Mutually Exclusive Aspects of Operations

In certain operations it was presumed necessary to mount assaults to secure airfields or ports. Where these assaults were required near a coast, they could on occasions be carried out either by parachute troops air-dropped, or by Royal Marine Commandos landing from assault ships. As it was necessary to determine which of these two methods involved least overall expenditure, the condition imposed was that an assault would be carried out by one or other of these methods but not by both. The following constraints expressed this condition:

Let R_p be the number of men needed for the assault;

Let T_c be the number of commandos in an amphibious assault;

Let T_p be the number of parachute troops in a parachute assault;

Let d_5 and d_6 take values of 0 or 1 only;

Let M = any large number.

Then

$$T_c + T_p \geqslant R_p$$

$$d_5 M - T_c \geqslant 0$$

$$d_6 M - T_p \geqslant 0$$

$$d_5 + d_6 = 1$$

3.5 Representation of Ships in the Model

Certain types of ships were available in such small numbers that unknowns representing numbers of ships in the mix could only be allowed to vary by integer amounts (i.e. one needed to avoid the ridiculous situation where half a ship was implied to be available in home waters and the other half in the Far East).

4. COLLECTION AND DERIVATION OF DATA

Officers representing each of the Services were full time members of the team of eight engaged on the study. One of their main responsibilities was to obtain the relevant data and information from their respective Service Departments. These data were then evaluated and processed by the scientific and Service members jointly. After digesting the information in this way and appraising its relevance to the study, the information, assumptions and data to be used were published as working papers and circulated to defence departments and members of the main working party for final approval and comment. In this way many misunderstandings and arguments were eliminated very early in the study.

Before the linear program model was ever run on the computer, it proved valuable in forcusing our attention upon missing data, and so virtually served as a check list of necessary information. The type of preliminary analysis that was done to derive coefficients is illustrated by the project working paper used in the study and attached as an Annex to this paper.

In an analysis covering so wide a field as this, there was some difficulty in ensuring that the data used for different parts of the model were of similar accuracy. One did not want work delayed whilst one part of the model was honed to a sharp edge when one knew that other parts were of necessity relatively blunt instruments. This difficulty has been largely overcome by setting limits on the time available to derive data, and by issuing precise terms of reference to departments deriving information on our behalf stating in exactly what form information was required.

The matrix generator produced for us will be described by Mr. Shaw. From the project team point of view the main requirements were that it should cope efficiently with the mass of data and yet not be too rigidly tied to a particular formulation of the model. This latter requirement was mainly because defence assumptions can change fairly frequently, and so impose changes on the model. If the project team were to maintain a rapid response to questions, it was essential that changes in formulation should not impose inordinate delays on the study.

5. PRESENTATION AND DISCUSSION OF RESULTS

The plan of the study is shown in Figure 1. Two basic policy options were considered. The first was that the UK has commitments, in the period 1970-80, to Internal Security Operations only. The second option was that the UK has commitments, in the 1970-80 period, to both Internal Security and Limited War Operations outside Europe. Within these policy options two alternative rates of force build-up were considered.

		Pre-1975	Post-1975
I.S. Commitments Only	Higher Force Build-Up Rate	Situation A	Situation C
	Lower Force Build-Up Rate	Situation B	Situation D
I.S. and Limited War Commitments	Higher Force Build-Up Rate	Situation E	Situation G
	Lower Force Build-Up Rate	Situation F	Situation H

Fig. 1

The equipment either available or potentially available for operations changes in the mid-1970s. For example:

(a) One type of aircraft became available for purchase in 1975, whereas another was to be phased out of service by 1975.

(b) Certain sites could be considered available for stockpiles before 1975 that were not available afterwards.

It was therefore necessary to study operations before and after 1975 separately, whilst ensuring that equipments available at the end of the first five year period should be considered to be already existing at the beginning of the second five year period. There were therefore eight combinations of the main assumptions to be examined as indicated by Situations A to H in the diagram.

The main results are set out in Table 1 in the general format of the diagram above. For reasons of security they are imaginary, although they illustrate the general points revealed by the actual results. They show the least cost mix of aircraft, ships and stockpiled material required under the various circumstances.

Pre-1975 the change in force build-up rate has a marked effect upon the overall five year cost of transport requirements. The difference in cost is mainly due to the need for a large number of Type C aircraft to achieve the higher performance. Type B aircraft are commonly acknowledged to be the least efficient transport aircraft. When the build-up requirements are relaxed, the analysis shows that they are the aircraft which should be preferentially taken out of service. The stockpile sites shown are the optimum sites selected in the analysis from the alternatives considered.

Table 1

Internal Security Operations

	Pre-1975		Post-1975		
	Number	Cost	Number	Cost	
Ships—Type 1	9	720	—	—	
Type 2	57	2,000	—	—	
Aircraft—Type A	296	4,520	228	2,600	Higher
Type B	27	600	—	—	Build-Up
Type C	81	5,300	40	1,200	Rate
Type D	—	—	45	4,200	
Stockpile Site	Peking	—	Hanoi	—	
Amount Stockpiled	8,000 tons	300	9,000 tons	150	
TOTAL COST		13,440 units		8,150 units	
Ships—Type 1	9	720	—	—	
Type 2	42	1,400	12	300	
Aircraft—Type A	296	4,520	228	2,600	Lower
Type B	—	—	—	—	Build-Up
Type C	20	600	42	1,600	Rate
Type D	—	—	7	1,000	
Stockpile Site	Peking	—	Hanoi	—	
Amount Stockpiled	9,000 tons	350	23,000 tons	700	
TOTAL COST		7,590 units		6,200 units	

Post-1975 a new aircraft, Type D, is available for purchase. The results show the impact of purchasing this aircraft. The same force performance can be achieved more cheaply without any need for ships. In these circumstances a lower build-up rate does not have such a marked effect upon cost as it does pre-1975.

This discussion merely serves to illustrate the way in which such analysis can probe the interrelationships between the various factors which affect a very complex problem. A similar set of results was obtained for situations in which UK also accepted Limited War commitments. Comparison of the two sets of results showed the cost consequence of two possible overseas policies.

5.1 Substantiation of Results

Originally there was some scepticism about the use of a computer in this type of study. This was expressed as a fear that vital supporting evidence, for answers of the type just illustrated, would be lost in the internal workings of the computer. However, apart from these results, we obtained from the computer analysis an outline plan for each operation showing how a particular rate of build-up could be achieved with the ships, aircraft and stockpiles proposed as the solution. Not only was this conclusive in convincing the original sceptics, but also highlighted those areas of each operation which were most critical.

5.2 Sensitivity Analysis

When studying situations so far into the future, there must always be uncertainty about the accuracy of some of the numerical values used. In this study we looked at the effect upon results of uncertainty in:

(a) The costs of Type D aircraft.

(b) The date upon which Type D aircraft could be made available for purchase.

(c) The costs of stockpile facilities.

(d) Restrictions on the rate at which aircraft are permitted to fly along routes from UK.

(e) The specified rate of force build-up.

The insight provided by sensitivity analysis is best illustrated by the results covering the last two points.

5.3 Route Limitations

Figure 2 shows how the overall five year cost of aircraft and ships, needed for intervention operations, changes as the maximum permitted number of aircraft on routes from UK is increased. The scale along the bottom ranges from a low rate to a high rate. Cost is expressed in the arbitrary units shown previously. No benefit is gained for improvements beyond a certain point, for at that point a destination airfield is at

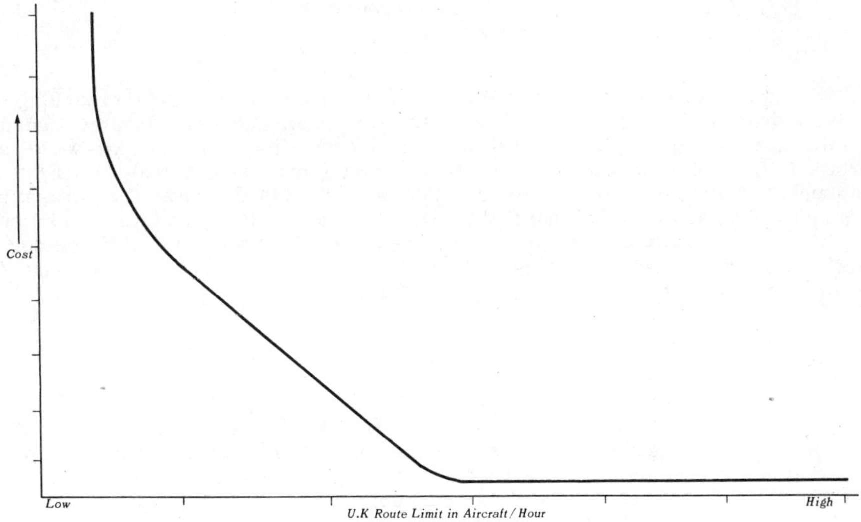

COST v ROUTE LIMITATION

Fig. 2

maximum capacity and so becomes the limiting factor. The reason for the form of the curve is that when route restrictions are severe, expensive aircraft with relatively high payload must be used to stay within the limits; whereas when restrictions are relaxed, cheaper aircraft of lesser payload can be used in greater numbers.

5.4 Force Build-Up Rates

The rate at which intervention forces should be built-up in any operation is an uncertain and controversial assumption. In the main part of the study we considered a higher and a lower build-up rate under several different circumstances, as shown in the plan of the study. It is of interest to know how the cost of intervention forces might vary between these extremes. Figure 3 shows how the five year cost of intervention forces varies as the rate of build-up is progressively reduced from the higher rate to the lower rate. From this it is obvious that minor reductions below the higher level can result in fairly large savings in certain circumstances.

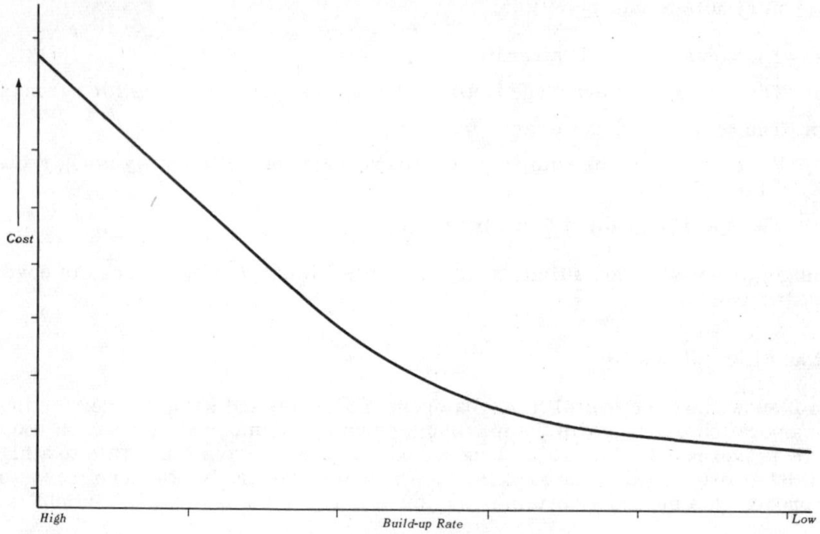

COST v BUILD-UP VARIATION

Fig. 3

Initially there was a feeling in the Ministry of Defence that we were approaching the problem the wrong way around. In a military staff study one would assume a number of ships and aircraft available and then determine what they could do. We were asking for a statement of what needed to be done so that the most economical mix of transport and stockpiles consistent with that statement could be derived. The phrase 'most economical' conjured up the idea that we would produce a cheap and nasty solution which would not really satisfy operational needs. The fact that we could consider a wide range of force build-up rates very quickly and show the cost consequences of the assumptions was decisive in allaying such fears.

6. ACCEPTABILITY OF RESULTS

The process of taking a decision divides into two parts. Firstly, it involves weighing the consequences inherent in various possible courses of action. Secondly, it involves selecting that course of action for which the consequences come nearest to the desired objective. The first part of this process is subject to all the difficulties discussed earlier in this paper. The second part of the process is made difficult in practice because there may be several partly conflicting objectives, the relative merits of which are difficult to assess. We considered that the role of operational analysis was in spelling out the logical consequences of various possible courses of action, and the role of the MOD working party was to bring to bear subjective judgement and considerations of overall national policy to sift these results. This interpretation of roles has been completely accepted and forms the basis of a continuous dialogue between the analysis team and the authority for decision.

The results previously presented as Table 1 provide an example of the sort of question submitted to such a dialogue. Having obtained the results for the high build-up situation pre-1975, there were at least three courses the subsequent analysis could have taken. The first course—the one we took—was to say that the high build-up capability would be required pre-1975, hence the aircraft and ships needed to make it feasible could be considered already available in 1975. The second course was to say that the bill was too high, and that a reduced capability would be accepted pre-1975 which would make requirements for ships and aircraft more consistent with requirements post-1975. The third course was to consider Type D aircraft available for purchase before 1975. By submitting these sorts of issue to the MOD working party, they can thereby direct and control the study, and progressively appreciate the outcome of the analysis.

Apart from the detailed results, we believe that this study has demonstrated the following important general points to the Ministry of Defence decision-making machine:

(a) At a time when the basis for most decisions is becoming broader and more complex, modern analysis techniques coupled with the power of modern electronic computers provide a means of handling both the breadth and the complexity.

(b) If analysts and the responsible authority for decision co-operate in the manner outlined above, then:

  (i) decisions can be founded on an adequate basis of information;

  (ii) full account can be taken of the qualitative aspects of the problem;

  (iii) the responsible authority can retain control over even the most complex issues.

APPENDIX: THE METHOD OF CALCULATING THE NUMBER OF AIRCRAFT OF A PARTICULAR TYPE NEEDED TO LIFT ONE KILOTON ALONG A ROUTE IN A GIVEN TIME

Introduction

The linear programming approach to the problem requires coefficients to be calculated—in the form of a number of aircraft per kiloton—for each type of aircraft, and each route to be considered in the study. The object of this paper is to set out a standard method of calculating these coefficients, and to state the basic assumptions made. The detailed numerical values to be used in specific cases will be set out in a separate working paper.

Derivation of the Method of Calculation

(i) <u>Assumptions</u>

The main assumptions are as follows:-

(a) In an emergency operation Base 1 servicing is done on an opportunity and progressive basis.

(b) At the start of an emergency operation all aircraft have an extension of X hours to the normal number of flying hours between Base 2 servicings.

(c) At the start of an emergency operation the number of flying hours attributed to aircraft since the previous Base 2 service is distributed linearly amongst the available fleet of A_0 aircraft between 0 and H flying hours as illustrated in Figure 4 (where H is the number of flying hours between Base 2 servicings for a particular type of aircraft and A is the number of aircraft with less than h flying hours on the clock initially).

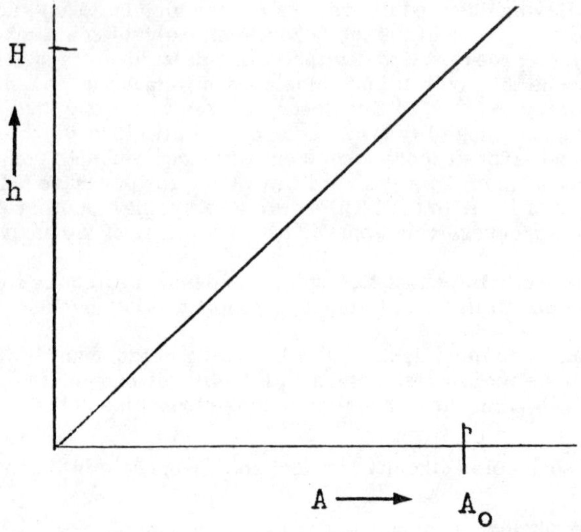

Fig. 4

(d) After the first Base 2 service during the course of the operation, aircraft return for Base 2 servicing each time the normally permitted number of flying hours (H) has elapsed.

(e) The rate at which aircraft return from Base 2 servicing will depend upon routine flying rates, and upon the number of aircraft of a particular type on unit establishment as follows:-

$$\text{Rate of return from Base 2 servicing} - \frac{A_0\, t'}{T'\, H}\, \frac{1}{f}$$

where in routine operations each aircraft of a particular type flies t′ hours in a period of elapsed time T′ hours. (Subsequently the ratio $\frac{t'}{T'}$ is referred to as M.)

A_0 – the number of aircraft of a particular type initially available in the fleet.

H – the number of flying hours between Base 2 services for a particular type of aircraft.

f – the proportion of the total fleet of aircraft of a particular type available at any one time.

(f) Aircraft 'shuttling' on any one route will maintain equal time intervals between landings at the destination, subject to limits corresponding to either

(i) the maximum permissible flow rate on the route; or

(ii) the maximum rate at which aircraft can be handled at the destination.

(ii) Definitions—

In addition to symbols defined in the previous paragraph, the following symbols are used in subsequent calculations:

t — time elapsed from the start of the operation

F — total flying time on a round trip from loading base back to landing base

G — total time spent on the ground during a round trip from loading base back to loading base

T — total round-trip time of an aircraft flying from loading base to loading base

a — mean aircraft turn round time

d — mean delay to an aircraft per turn round

l — mean time taken to load an aircraft

m — mean time taken to unload an aircraft

v — average cruising speed of an aircraft

D — total distance to be flown on a round trip

N — total number of staging posts at which an aircraft lands on a round trip

f — proportion of the total fleet of aircraft of a particular type available at any one time

p — payload carried by an aircraft of a particular type on a particular route

$r = \frac{M}{Hf}$ see paragraph 2e above

A(t) — the number of aircraft of a particular type flying on a route at time t after the start of the operation

B(t) — cumulative build-up at a destination by time t. (Kilotons)

(iii) Aircraft Round-Trip Time

(a) The total flying time during a round trip is given by the expression

$$F = \frac{D}{V} + 0.25\ (N + 2)\ \text{hours}$$

(The second term on the right hand side of this expression allows for time spent in circuit by aircraft at each staging post and base.)

(b) The total time spent on the ground by an aircraft during a round trip is given on average by the expression

$$G = 1 + u + Na + Nd + 0.1\ F\ \text{hours}$$

(the last term on the right hand side of this expression allows for average time spent on replacement servicing.)

(c) The total round trip time for an aircraft is therefore

$$T = F = G\ \text{hours}$$

(iv) Number of Aircraft Available to Fly on a Route at Time t After the Start of the Operation.

(a) The initial number of aircraft A_0 will remain available up to time t_e where

$$t_e = \frac{X}{F}\ T \text{ due to Assumption 2b.}$$

(b) The rate at which the original A_0 aircraft will leave the shuttle for second line servicing will be

$$\frac{A_0\,F}{T\,H} \text{ aircraft per hour.}$$

due to assumptions 2c, 2d and 2f above.

Then if $\frac{F}{T\,H} = S$

all the original A_0 aircraft will have returned to second line servicing by time

$$t_s = t_e + \frac{1}{S} \text{ hours}$$

but none of the aircraft emerging from second line servicing during the period t_e to t_s will have re-entered.

(c) For any time t such that $t_e < t < t_s$ the rate at which aircraft operating decline will be:-

$$-\frac{d\,A(t)}{dt} = A_0 S - A_0 r$$

Therefore

$$A(t) = A_0[1 - (S - r)(t - t_e)]$$

(d) At time $t = t_s$ the number of aircraft available for operations will be

$$A_0\,\frac{r}{S}$$

After time t_s i.e. for $t \geq t_s$ any aircraft emerging from second line service will remain available for a duration 1/S hours before returning for further servicing. During this time the number of aircraft which come from second line servicing will be

$$\frac{A_0 r}{S}$$

these will themselves all return to second line service in a period of time 1/S hours. Therefore after time t_s the rate at which aircraft return to second line service will be $A_0 r$ aircraft per hour which is equal to the rate at which aircraft emergy from second line service. Thus from time t_s onwards the number of aircraft available for operations remains constant at

$$A_0\,\frac{r}{S}$$

(v) <u>The Build-Up of loads at the Destination</u>

(a) For $t < t_e$ the rate of build-up at the destination will be

$$\frac{A_0 p}{T} \text{ due to assumption 2f}$$

$$\therefore B(t) = \frac{A_0 p}{T}\,t$$

(b) For $t_e < t < t_s$ the rate of build-up at the destination at time t will be

$$\frac{A(t)p}{T}$$

$$\therefore B(t) = B(t_e) + \int_0^{t-t_0} \frac{A(t)p}{T}\,dt$$

$$= B(t_e) + \int_0^{t-t_0} \frac{A_0P}{T}[1-(s-r)(t-t_e)]dt$$

$$B(t) = B(t_e) + \frac{A_0p}{T}\left[(t-t_e) - \frac{(S-r)}{2}(t-t_e)^2\right]$$

(c) For $t > t_e$ the rate of build-up at the destination at time t will be

$$\frac{A_0rp}{ST}$$

$$\therefore B(t) = B(t_s) + \frac{A_0rp}{ST}(t-t_s)$$

(vi) Calculation of the Coefficients Required

The coefficients must be expressed as the number of aircraft to be held on unit establishment in order to carry one kiloton over the route in a permitted time. A_0 in the above expressions may be written as:-

$$A_0 = fU$$

where U is the total number of aircraft of the type in question on unit establishment and f is the availability factor for these aircraft.

Suppose the period of time for which a coefficient is to be calculated begins at time x and ends at time y after the start of the operation.

Then there are five particular cases for the calculation of coefficients as follows:-

Case (a)

$$x < y \leqslant t_e$$

$$\text{Then } B(y) - B(x) = \frac{fUp}{T}(y-x)$$

$$\text{But } Q = \frac{U}{B(y)-B(x)}$$

$$Q = \frac{T}{fp(y-x)}$$

Case (b)

$x < t_e < y$

$$B(y) - B(x) = B(t_e) + \frac{fUp}{T}\left[(y - t_e) - \frac{S-r}{2}(y - t_e)^2\right]$$

$$- \frac{fUp}{T} \cdot x$$

$$= \frac{fUpt_e}{T} - \frac{fUpx}{T}$$

$$+ \frac{fUp}{T}\left[(y - t_e) - \frac{S-r}{2}(y - t_e)^2\right]$$

$$B(y) - B(x) = \frac{fUp}{T}\left[(y - x) - \frac{S-r}{2}(y - t_e)^2\right]$$

$$Q = \frac{T}{fp\left[(y - x) - \frac{(S-r)}{2}(y - t_e)^2\right]}$$

Case (c)

$t_e < x < y < t_s$

$$B(y) - B(x) = B(t_e) + \frac{fUP}{T}\left[(y - t_e) - \frac{(S-r)}{2}(y - t_e)^2\right]$$

$$- B(t_e) - \frac{fUp}{T}\left[(x - t_e) - \frac{(S-r)}{2}(x - t_e)^2\right]$$

$$= \frac{fUp}{T}\left[(y - z) - \frac{(S-r)}{2}\{(y - t_e)^2 - (x - t_e)^2\}\right]$$

$$Q = \frac{T}{fp\left[(y - x) - \frac{(S-r)}{2}\{(y - t_e)^2 - (x - t_e)^2\}\right]}$$

Case (d)

$t_e < x < t_s < y$

$$B(y) - B(x) = B(t_e) + \frac{fUp}{T}\left[(t_s - t_e) - \frac{(S-r)}{2}(t_s - t_e)^2\right]$$

$$+ \frac{fUrp}{ST}(y - t_s) - B(t_e)$$

$$- \frac{fUp}{T}\left[(x - t_e) - \frac{(S-r)}{2}(x - t_e)^2\right]$$

$$= \frac{fUp}{T}\left[(t_s - x) - \frac{(S-r)}{2}\{(t_s - t_e)^2 - (x - t_e)^2\}\right.$$

$$\left. + \frac{r}{s}(y - t_s)\right]$$

$$Q = \frac{T}{fp\left[(t_s - x) - \frac{(S-r)}{2}\{(t_s - t_e)^2 - (x - t_e)^2\} + \frac{r}{S}(y - t_s)\right]}$$

Case (e)

$$t_s < x < y$$

$$B(y) - B(x) = B(t_s) + \frac{fUrp}{ST}(y - t_s)$$

$$- B(t_s) - \frac{fUrp}{ST}(x - t_s)$$

$$= \frac{fUrp}{ST}(y - x)$$

$$Q = \frac{TS}{frp(y - x)}$$

Review of Computational Experience in Solving Large Mixed Integer Programming Problems

MAX SHAW
Scientific Control Systems Ltd. (formerly C-E-I-R)

1. INTRODUCTION

General methods for the solution of integer programming problems have been known since the original work of Gomory (1958). More recently, the emphasis in the search for practical computational methods has switched to branch and bound approaches. These were pioneered by Land and Doig (1960), and their potential usefulness became much clearer with the publication of the work of Little, Murty, Sweeney & Karel (1963) on the travelling salesman problem. Other applications of branch and bound have been reviewed by Lawler & Wood (1966) and Agin (1966).

In order to solve practical mixed integer programming problems of any size, it is of course necessary to have a computer program that is integrated with a general linear programming system. Such a program has been developed in C-E-I-R's LP/90/94 System and used on a production basis for the solution of mixed integer programming problems on the IBM 7094 computer since 1966. The methods used are described by Beale & Small (1965), and in more detail by Beale (1968).

The system accepts mixed integer programming problems in SHARE standard input format. Three types of integer variables can be used, and each is identified by the form of the variable name—which like all variable names in the SHARE standard has 6 characters. A general integer variable is identified by the last 2 characters being = =. Since the LP/90/94 system does not provide special facilities for bounded variables, there must be a row associated with each general integer variable defining the upper bound for the variable. Many integer programming problems involve several variables that can only be zero or one, so special facilities were provided for such variables. They are identified by their 6th (i.e. last) character being = but their 5th character not being =, and they do not need an associated row. Another type of zero-one variable has also been recognized; it is identified by the last two characters being / =. Its properties are that when branching on such a variable one first explores the possibility that it is unity, leaving the possibility that it is zero for later consideration. This facility was first developed for 'covering variables', i.e. variables occurring in a constraint of the form $\Sigma x = 1$, where the variables all have to be integers, since one is making much better progress towards an integer solution by driving such a variable to unity, than one is by driving it to zero. So these / = variables are known as covering variables. In practice the facility has not been very useful: it sometimes enables a genuinely feasible solution to be found earlier than it would otherwise have been, but it often makes the value of the objective function at the first feasible solution much worse than it would otherwise have been, and lengthens the overall solution time.

The size of problem acceptable to the system is up to 1023 rows—including rows defining the upper bounds on general integer variables. There is no real limit on the number of columns.

Among the problems solved with the system, those given to C-E-I-R by the D.O.A.E. and described by Beard & McIndoe (1968) are of particular interest. They were of substantial size and involved both zero-one and general integer variables. As often happens with large problems, a number of technical difficulties arose, and the D.O.A.E. has kindly consented to the publication of their joint experiences with C-E-I-R on these problems.

2. NARRATIVE

This section reviews the progress of the work described by Beard & McIndoe from the point of view of programming and computing.

In August 1967, D.O.A.E. approached C-E-I-R with a view to using C-E-I-R's mixed integer system and professional services to solve Beard & McIndoe's model. The model was already completely formulated, and it was established that its dimensions were within the capacity of the system. The first technical task was therefore to prepare the data of the model into a form acceptable to the system.

It would obviously have been possible to punch the coefficients of the problem directly on to cards in SHARE standard format, but this would have been a very tedious process, and would surely have introduced a number of random errors in the formulations which would have been expensive and time-consuming to remove. It was therefore decided to write a matrix generator. The next decision concerned the type of generator. It is usually best with a project of this kind to let the generator form the individual rows and columns of the matrix using specially written programs representing the types of activity and constraint to be considered, and data cards representing the numerical parameters of the individual cases (i.e. the scenarios for this problem). This approach requires that the analysis preceding program writing is concerned with general rules for generating constraints and variables for the type of problem being considered, rather than with the specific formulation of the particular problems to be studied. In fact the D.O.A.E. had developed their specific formulations in some detail, and it was decided to build the matrix generator round these formulations. These decision was justified by the fact that Mr. A. J. Akeroyd of C-E-I-R Ltd. was able to write and debug the generator in 3 weeks. The input to the generator is in the following form which was chosen as being convenient to D.O.A.E.

2.1 Matrix Generator Data

The data for the generator was of 4 types:-

(a) Tables of coefficients

(b) Functions

(c) Rows

(d) Right hand side coefficients.

Each table has a set of coefficients of a particular type. Any of these coefficients could be referenced in data sections b and c by naming the table in which it appeared and a subscript defining the relevant entry. This was a compact way of handling frequently used coefficients and had the advantages of being both easy to prepare and check.

Functions were included to take advantage of the structure of the model. Many of the constraints in the problem were made up from parts of other constraints, e.g.

	$x_1 - x_n$	$y_1 - y_n$	$z_1 - z_n$	$p_1 - p_n$	$q_1 - q_n$	$r_1 - r_n$
Row 1	....			....		
Row 2	....	....			....	
Row 3	....	....	....			....

The set of variables $x_1 - x_n$ has the same entries in all 3 rows, set $y_1 - y_n$ in rows 2 and 3 etc. Sets $p_1 - p_n$ were peculiar to row 1, $q_1 - q_n$ to row 2 etc.

Matrix elements were defined in the row and function data sets by naming each vector and an associated coefficient, the coefficient being either a table entry or an explicit value.

Functions were defined by giving them a name and listing their elements. For example the coefficients of vectors $x_1 - x_n$ in row 1 could be defined as function F1.

```
F1    X1    -£Q(51)        X2       £R(19)
      X3    1.0      .......        XN    -1.0
```

The £ indicates a table entry. For example the coefficient of x_1 is the negative of the 51st entry in the table called Q.

Any function could reference a previously defined function by enclosing its name between solidi and if required appending a minus sign meaning the negative of all the entries was wanted. Thus we had two ways of defining the repetitive part of row 2. Either we could define set $y_1 - y_n$ as function F2

```
F2      Y1     -£Q(52)   Y2......    YN......
```

or alternatively define set $x_1 - y_n$ as function G2

```
G2      /F1/  Y1      -£Q(52)    Y2......      YN......
```

Rows were written in the same way as functions, the first field being the name of the row and its inequality type.

Thus the rows 1 to 3 would be coded either as

```
+R1     /F1/     P1       1.0   ......   Pn   £ABC(2)

+R2     /F1/     /F2/             Q1    1.0 ...... QN.

+R3     /F1/     /F2/             /F3/

        R1    -£Z(1)      R2        3.0 ...... RN.
```

or as

```
+R1     /F1/     P1      1.0 ...... PN    £ABC(2)
+R2     /G2/     Q1      1.0 ...... QN
+R3     /G3/     R1      -£Z(1) ... RN.
```

The right had side data consisted of the name of each row and its value.

As far as practicable the generator was programmed not to fail on data errors. When an error was detected a sutiable message was printed and the coefficient deleted. It could be restored in correct form later by using the LP/90/94 Revise facility.

Examples of the types of error messages are

```
FOLLOWING TABLE ENTRY BEYOND LIMIT OF TABLE
ROW     r
VECTOR v
TABLE  t ENTRY e
ELEMENT IGNORED
```

or

FOLLOWING FUNCTION NOT FOUND

FUNCTION f CALLED FROM ROW r

Besides producing the matrix on a magnetic tape the generator also printed the contents of all tables and optionally

(a) List of functions and their elements

(b) List of matrix elements

(c) List of rows, matrix and rhs

(d) List of row names

(e) List of rhs values.

It should be noted that the generator was in no way specific to the D.O.A.E. model and could be used to generate any matrix for which a special purpose program was unsuitable.

It will be seen that from these input data it is easiest to generate the matrix element by rows. But the LP/90/94 system, or any other production linear programming system using the inverse matrix version of the simplex method, requires that the non-zero matrix elements be sorted by columns. Further, a list of row names must precede the matrix elements, which must in turn precede the non-zero right-hand side elements. On a particular matrix element card, columns 7-12 have a column name identifier, and columns 73-80 are free for sequencing purposes. In the generator output, cards (which were actually card images on magnetic tape rather than physical cards), were numbered in ascending order in columns 76-80, except for matrix elements which were all given the same number. These card images were then sorted on cols. 7-12 within cols. 76-80, giving correct LP/90/94 input. The sort itself was done using the standard sort package on the CDC 3200 computer. The efficiency of the sort packages supplied as part of the basic software of modern computers is such that there is no great objection to generating a matrix by rows where this is most natural, and then using a standard sort to convert it into column order. (On the other hand many problems can be generated just as easily in column order once the analyst gets used to thinking in terms of activities rather than equations.)

As D.O.A.E. wanted to do a first run at the end of September the matrix generator was written and debugged in parallel with D.O.A.E.'s preparation of their data. Statistics of the computer runs done by D.O.A.E. appear at the end of the paper. After doing a few runs with a 200 row version of their model D.O.A.E. did a first run of their 850 row problem. In getting a first integer solution the mixed integer package seemed to take an unexpectedly long time and furthermore suffered from a number of digital difficulties. The digital difficulties were traced to the fact that to force certain vectors produced by the generator not to appear in feasible solutions critical coefficients in these vectors were set to huge numbers.

These huge numbers were intended to prevent these activities from being used, by saying that they required a wholly unreasonable amount of some scarce resource. But the program did try using them in microscopical amounts in the course of iterating towards a feasible solution. Once the difficulty had been identified, the flagging facilities of the LP/90/94 system were used to bar these vectors from the basis, and no further troubles from this source arose.

The other point noticed was that most of the nodes of the search were concerned with finding an integer level for two of the general integer variables able to range between 1 and 20. As these variables were coming into integer solutions at levels between 16-18 it was felt these variables could be treated as continuous (i.e. non-integer) variables.

Tabulation of Runs

	Row	Nodes	O/I Variables	G.I Variables	Time (hours)
E200	200	20	9	6	1.2
F200	200	113	9	6	4.33
G200	200	70	9	6	3.34
E850	850	14	9	6	6.3
F850	850	4	9	4	2.0
G800	800	4	9	4	3.5
H800	800	5	9	4	2.3

NOTE 1. Solution times are times from an optimum continuous solution to a global integer solution, except in case E850 where only a first integer solution was found.

2. G.I. = general integer variable.

Following run E850 shown in the chart it was clear that by treating the two particular general integer variables concerned as continuous variables we would dramatically cut search time. Where run E850 required 6.3 hours to reach a first integer solution; runs F, G and H required about 3 hours each to reach a global optimum. The only significant difference between run E and run F, G, H being the two general integer variables referred to being integer in E and continuous in F, G, H.

In doing a particular set of runs it was the practice to use the matrix generator to produce a basic input tape; and use the data manipulation and revision features of the mixed integer package to change the cost, matrix elements, right-hand sides or vectors to be excluded on a particular run. During the course of the longer runs heavy and continuous use was made of the 729 Mk6 tape units of the 7094 model 2. A considerable number of bad tapes and general tape unit trouble occurred during the runs; and without the presence of skilled production controllers and the fact that the package had rescue facilities to recover in the event of hardware trouble the runs could not have been reliably completed.

In view of the amount of machine time used in mixed integer work; and the necessity to manipulate the data between runs of large problems, I feel to be practical on a production basis a mixed algorithm must be programmed within a generalized LP suite.

3. GENERAL CONCLUSIONS

A wide variety of clients' mixed integer problems have been solved on C-E-I-R's mixed integer scheme and the following simple rules have been found helpful:-

1. Solve the problem as a continuous problem first and examine this solution in an attempt to find any errors in the formulation.
2. If the value of one integer solution, or its approximate value is known beforehand, the mixed package can use this to cut solution time.
3. Only declare those integer variables as integers which have to be integer. If treating a particular variable as continuous instead of integer would make little difference to the solution the particular variable should be in as a continuous variable.
4. For a particular problem formulation after a few versions of the model have been solved a pretty good idea of the value of the best integer solution in a slightly modified re-run of the problem will usually be known. When integer solutions near to this known value have been obtained the search can be terminated.

5. Where the integer solutions represent investment decisions or the closing of industrial plants etc. rather than just the global optimum both the global optimum and several near global optimum integer solutions may be desired. The mixed package normally tends to give these as a by-product of obtaining the global optimum. Study of the tree will usually suggest a variety of near global optimum integer solutions worth investigating.

It would appear from the D.O.A.E. work outlined here and from other work done by C-E-I-R that the branch-and-bound method provides a practical tool for mixed problems. An important aspect of this method is that it can usually be relied on to produce a good integer solution, and quite possibly the optimum solution, fairly quickly, even if a great deal of further exploration is needed to prove that the solution is indeed optimal. With large problems it is therefore often useful to use the program even if one knows in advance that it will be too expensive to complete the search.

In spite of these encouraging experiences, it should be admitted that some integer programming formulations can require considerable amounts of machine time, and that these amounts of time may depend substantially on which of a number of mathematically equivalent formulations is used. When choosing the formulation, it is often helpful to try to anticipate how the branching process will develop.

In summary, integer programming is a practical technique for production mathematical programming work. It needs to be handled carefully, in much the same way that serious linear programming problems needed to be handled carefully 10 years ago. As with linear programming at that time, integer programming codes will undoubtedly become more widespread and easier to use over the next few years.

The author wishes to thank A. J. Akeroyd, E. M. L. Beale and R. Horovitz for their help in preparing this paper.

REFERENCES

N. AGIN (1966) 'Optimum Seeking with Branch and Bound' Management Science **13**, pp B176-185.

E. M. L. BEALE & R. E. SMALL (1965) 'Mixed Integer Programming by a branch and bound technique' Proceedings of the IFIP Congress 1965, pp 450-451. Edited by W. A. Kalenich (Spartan Press, Washington, D.C., and Macmillan, London).

E. M. L. BEALE (1968) Mathematical Programming in Practice Pitmans, London.

C. N. BEARD & C. T. McINDOE (1968) Determination of the least cost mix of transport aircraft, ships and stock-piled material to provide British Forces with adequate strategic mobility in the future. Paper presented at this Conference.

R. E. GOMORY (1958) 'Outline of an algorithm for integer solutions to linear programming problems' Bulletin of the American Mathematical Society 64, pp 275-278.

A. H. LAND & A. G. DOIG (1960) 'An automatic method of solving discrete programming problems' Econometric **28**, pp 497-520.

E. L. LAWLER & D. E. WOOD (1966) 'Branch-and-Bound Methods: A Survey' Operations Research **14**, pp 699-719.

J. D. C. LITTLE, K. C. MURTY, D. W. SWEENEY & C. KAREL (1963) 'An algorithm for the travelling salesman problem' Operations Research **11**, pp 972-989.

DISCUSSION

D. Edmonds

What is C-E-I-R doing to implement this computational system for mixed integer programming on 3rd generation machines?

M. G. Shaw

C-E-I-R is implementing this computational system on the UNIVAC 1108 (to be made available from C-E-I-R on bureau basis) and on ICL system 4 computers for all users of the latter machines.

J. De Buchet

Pouvez vous nous expliquer de quelle façon vous avez pu tirer parti de solutions entières déjà obtenues pour faciliter l'optimisation d'un nouveau problème différant légèrement du premier.

M. G. Shaw

There is no one simple way of using a previous best integer solution to assist in finding the best integer solution to a revised version of the original problem. The previous solution may allow a good 'cut-off' level to be guessed to speed up the new search. A previous integer solution may also be a feasible integer in the revised problem and this would be another way of producing a cut-off for the revised problem.

Lagrange Multipliers, Non-Linear Programming, and United States Strategic Force Effectiveness

JAN M. LODAL
Office of the Assistant Secretary of Defense for Systems Analysis, U.S.A.

1. INTRODUCTION

One of the criteria used in the Department of Defense to design the U.S. strategic force structure is what we call 'Assured Destruction.' Assured Destruction is measured as the maximum amount of damage the U.S. could do to a nation after that nation had attacked first, attempting to destroy as much of the U.S. strategic force as possible. Measuring Assured Destruction is a way of measuring the deterrent capability of our force.

Since the Soviet Union currently presents the only serious threat to the United States' Assured Destruction capability, computing it is reduced to first finding what forces we would have left if the Soviets optimally allocated their entire inventory of strategic forces against our strategic forces, and then computing how much damage these remaining forces would do if they were optimally allocated against Soviet urban-industrial targets. The resulting damage is taken as a measure of the United States' Assured Destruction capability.

Mathematically, an Assured Destruction problem can be framed as a two-sided, zero-sum game. The initial attacker should allocate both his offensive and defensive weapons to minimize the maximum amount of damage the other side could do in return, assuming that the other side would allocate his weapons to maximize the same measure of damage. However, the set of possible strategies is so large that usual methods for solving two-sided games cannot be used. Consequently, we have reduced the problem to two separate resource allocation problems by making use of an intermediate set of values for U.S. weapons, and have used a modified Lagrange multiplier method, described in Section III, to solve the allocation problems. We have thus used Lagrange multipliers to approximate minimax strategies for a two-sided zero-sum game with discrete strategy sets.

We attempt to make the U.S. weapon values a measure of the relative amount of damage various U.S. weapons could do to the Soviets, should they survive a Soviet first strike. We then allocate Soviet weapons to maximize the total U.S. weapon value destroyed, and allocate the remaining U.S. weapons to maximize Soviet urban-industrial value destroyed. Urban-industrial values are assigned to Soviet cities in a way we feel represents the relative values the Soviets would assign to their cities.

If the values assigned to U.S. weapons do not correctly represent their relative destructive power, the Soviet allocation, even though it minimizes the weapon value surviving, will not minimize the resulting damage done by surviving U.S. forces. To ensure that we have used correct values, we compare the relative values originally assigned with the marginal values implied by how much damage the weapons do when they are actually allocated. If the original values and the new values differ significantly, we adjust the assigned values and re-calculate which U.S. weapons survive. Eventually, this process converges to a sufficiently accurate solution.

2. OBJECTIVE FUNCTIONS

In the absence of defenses, the expected value of the damage done by a mixture of weapons allocated to a target located at a single point, such as a missile silo, is given by the following relationship:

$$D_j = V_j\left(1 - P_{1j}^{N_{1j}} \dots P_{mj}^{N_{mj}}\right) \tag{1}$$

Where

V_j = Total value of target j

D_j = Value destroyed

P_{ij} = Probability that target j will survive one weapon of type i allocated to it (the single-shot survival probability)

N_{ij} = The number of weapons of type i allocated to target j

If the target is not located at a single point, but is spread over a large area, such as a city, a different damage relationship holds. Although it is not possible to define an exact relationship whose form is the same for cities with different shapes and different value densities, experience has shown that the following relationship holds for a wide variety of cases:

$$D_j = V_j\left[1 - \left(1 + \sqrt{\sum_{i=1}^{m} K^2{}_{ij} N_{ij}}\right) e^{-\sqrt{\sum_{i=1}^{m} K^2{}_{ij} N_{ij}}}\right] \tag{2}$$

where

V_j = Total value of target j

D_j = Value destroyed

K_{ij} = A fitting factor for weapon i against target j

N_{ij} = Number of weapons of type i allocated to target j

The form of this equation becomes exact for small weapons optimally targeted against large targets with Gaussian (normal) value distributions. The value of K_{ij} can be easily computed if one knows the single-shot survival probability for weapon i against target j because one can then equate this probability to $(1 + K_{ij})e^{-K_{ij}}$.

If defenses and defense penetration aids are present, the above equations become more complicated. However, many defenses can be represented as one of two relatively simple types: random area defenses covering all targets and subtractive terminal defenses covering single targets. There are also only two primary types of penetration aids: decoys that cannot be discriminated from warheads by any defenses (terminal decoys) and decoys that can be discriminated by terminal but not by area defenses (area decoys). Decoys can be handled by assuming that the defense treats them like warheads. Random area defenses simply reduce the probability that a weapon will reach its target. Subtractive terminal defenses destroy weapons allocated to a target until the expected value of the number of weapons plus decoys arriving equals the number of reliable defenders at the target; additional weapons are not destroyed because we assume that the defense has now been exhausted.

Thus, the forms of the above equations are not changed by these two types of defenses. However, the single-shot survival probability must now include the probability of penetrating the area defenses, and N_{ij} must equal the number of weapons of type i allocated to target j minus the 'price', i.e., the number of weapons needed to exhaust the defense.

If the offense has weapons both with and without terminal decoys, it will tend to use the weapons with decoys to exhaust terminal defenses and those without to subsequently destroy the target. Unless the timing of the attack is proper and the knowledge of the defense size is perfect, such an attack policy could entail excessively high risk. Thus, we allow the offense to allocate only a single type of weapon, out of the set of weapons which can be attacked by the defense, to terminally defended targets.

If we let:

N_i = total number of weapons of type i available to be allocated,

N^*_{ij} = number of weapons of type i needed to exhaust the defenses at target j

$$N'_{ij} = \begin{cases} N_{ij} - N^*_{ij} \text{ when } N_{ij} > 0 \\ 0 \text{ otherwise} \end{cases}$$

n = total number of targets,

m = total number of weapons,

P_{ij} = single-shot survival probability including the penetration probability, and

D_j, N_{ij}, K_{ij} be as defined above,

we can describe our resource allocation problem as follows:

$$\text{Maximize } \sum_{i=1}^{n} D_j\text{; where}$$

$$D_j = V_j\left(1 - \prod_{i=1}^{m} P_{ij}{}^{N'_{ij}}\right) \tag{3}$$

for point targets, or

$$D_j = V_j\left[1 - (1 + \sqrt{\sum_{i=1}^{m} K^2{}_{ij}(N'_{ij})}\; e^{-\sqrt{\sum_{i=1}^{m} K^2{}_{ij}(N'_{ij})}}\right] \tag{4}$$

for area targets, subject to the following constraints:

(a) $\sum_{j=1}^{n} N_{ij} \leq N_i, i = 1, \ldots, m$

(b) N_{ij} and N'_{ij} must be non-negative integers.

(c) At a given target, for all i, j such that N^*_{ij} is non-zero, at most one N_{ij} can be non-zero.

3. A SOLUTION USING LAGRANGE MULTIPLIERS

One can solve the above non-linear integer programming problems using a Lagrange multiplier technique, although modifications must be made to the Lagrange technique commonly known. If one were to use the usual technique, he would have to ignore the integer and non-negativity constraints and substitute equality for inequality constraints.

Proofs of the theorems one must use to overcome these difficulties were first published only recently (see Everett, 1963) although they are quite simple. The most important theorem is as follows:

Theorem 1: Let S be the domain of an objective function H(x). If $\lambda^k, k = 1, \ldots, n$ are non-negative real numbers and if $x^* \in S$ maximizes the function:

$$H(x) - \sum_{k=1}^{k=n} \lambda^k C^k(x)$$

over all $x \in S$, then x^* maximizes H(x) over all those $x \in S$ such that $C^k(x) \leq C^k(x^*)$ for all k, regardless of the nature of the set S.

Normally, one solves a Lagrange Multiplier problem by differentiating the Lagrangian function, setting the results equal to zero, and solving for both the λ's and the resource allocation that maximize the objective function, given a set of constraints. The above theorem says that if we maximize the Lagrangian given a set of λ's (i.e., given the shadow values of the resources), the resulting allocation, x^*, implies a set of constraints, $C^k(x^*)$. Thus, we automatically have an optimum allocation, x^*, for a constrained problem, namely the problem with constraints $C^k(x^*)$.

The great advantage of starting with the multipliers and deducing the constraints, rather than deducing the multipliers given the constraints, is that this approach allows one to use the above theorem. This theorem places no restriction on the domain of the objective function. Thus, having an objective function that is neither continuous nor differentiable no longer presents an insurmountable problem.

The disadvantage of this modified method is that unless we choose the correct set of λ's, we will not satisfy all the constraints. However, one can prove that if all but one constraint is held constant, the remaining constraint is a monotone decreasing function of its associated multiplier. Thus, after we have reached a wrong solution, we know which direction to adjust the λ's in order to obtain the desired constraints. Algorithms have been developed which make this adjustment relatively efficiently.

In many cases, one cannot find a set of multipliers which imply the desired constraints. The above theorem ensures only that if one finds a solution, it is a correct solution; it does not guarantee that a solution can be found. The existence of optimum solutions that can be found by adjusting the multipliers depends upon the objective function being approximately concave in the region of the solution. Such concavity is not always present; however, the following theorem (Everett, 1963) allows us to place bounds on the solution to problems where a direct adjustment of the multipliers will not work.

Theorem 2: If x' comes within ϵ of maximizing the Lagrangian, i.e., if for all $x \in S$, where S is the set of all possible strategies,

$$H(x') - \Sigma\lambda^k C^k(x') > H(x) - \Sigma\lambda^k C^k(x) - \epsilon \tag{5}$$

then x' is a solution of the problem with constraints $C^k = C^k(x')$ that is itself within ϵ of the maximum for those constraints.

The proof of this theorem gives some insight into its usefulness. Rearranging the above inequality,

$$H(x) - H(x') + \Sigma\lambda^k[C^k(x') - C^k(x)] < \epsilon \tag{6}$$

Keeping the λ^k fixed, the allocation x' implies constraints $C^k(x')$, and, obviously, x' is a feasible solution to the problem:

$$\text{Maximize } H(x) \text{ subject to } C^k(x) \leq C^k(x'). \tag{7}$$

For any other feasible solution to this problem, $C^k(x) \leq C^k(x')$ by definition, implying $C^k(x') - C^k(x) \geq 0$. Thus, substituting into inequality (6) above, $H(x) - H(x') < \epsilon$ for any x that is a feasible solution to the problem defined by (7) above. No payoff can be more than ϵ greater than the payoff given by the allocation x'.

Thus, this theorem allows us to compute an upper bound to the maximum payoff for any constraints. We pick a set of λ's and compute both the maximum Lagrangian and the implied constraints. We then find a new allocation, x', that implies our desired constraints, and, keeping the λ's fixed, evaluate the Lagrangian at x'. ϵ is then the difference between the value of the Lagrangian at $x = x'$ and the maximum Lagrangian. This is the maximum additional payoff above $H(x')$ we could ever obtain while meeting the desired constraints. Geometrically, Theorem 2 says that the hyperplane with slopes defined by the λ^k, tangent at x_0 to the envelope of the maximum payoff surface in the space of maximum payoff versus resources expended, is an upper bound to H(x) at any point if x_0 maximizes $H(x) - \Sigma\lambda^k C^k(x)$.

We compute actual allocations in the following manner:

(i) An initial, heuristic allocation of weapons to targets is made. The value destroyed by the last weapon of each type allocated is used as an initial estimate of the Lagrange multiplier associated with that weapon.

(ii) Using the initial set of Lagrange multipliers, an upper bound for total value destroyed is computed. If the destruction obtained by the heuristic initial allocation is close enough to the upper bound, that allocation is used.

(iii) If the initial allocation is not good enough, new allocations are computed, the Lagrange multipliers are adjusted, and new upper bounds are obtained until an allocation is found which is close enough (within approximately 1%) to the least upper bound.

4. THE 'CODE 50' COMPUTER PROGRAM

The above described allocations and the resulting target destruction are calculated with a computer program called 'Code 50'. In addition, Code 50 computes the values of the parameters needed to do the weapon allocation. These parameters are:

(i) The number of weapons the attacker has available to allocate to enemy targets.

(ii) The probability that any weapon so allocated will penetrate enemy area defenses.

(iii) The single-shot kill probability (one minus the single-shot survival probability) of each weapon against each target.

(iv) For each target, the number of weapons of each type that must be used to exhaust the terminal defenses at that target.

These four sets of numbers are functionally related to the input variables as follows:

(i) The number of weapons available to be allocated is a function of:

- (a) the number of weapon carriers surviving previous strikes,
- (b) the fraction of carriers either not on alert or withheld,
- (c) the number of weapons per carrier, and
- (d) the probability that a weapon will fail in such a way that the remaining weapons can be re-targeted taking its failure into account (the reprogrammable reliability).

(ii) The probability that a weapon will penetrate area defenses once it has been allocated is a function of:

- (a) the number of area decoys it carries,
- (b) the number of reliable enemy defense units, and
- (c) the probability that the weapon will fail in flight after it is too late to re-target for its failure (the non-reprogrammable reliability).

(iii) The single-shot kill probability is a function of:

- (a) the yield and accuracy (c.e.p.) of the weapon, and
- (b) the hardness of the target (vulnerability number or psi required for destruction).

(iv) The number of weapons needed to exhaust defenses is:

(a) equal to the number of reliable terminal ballistic missile defense (BMD) interceptors, or

(b) equal to the number of reliable surface-to-air missiles (SAMs) that can attack air-to-surface missiles (ASMs) for ASMs on bombers.

These values are computed by the program first; the second part of the program allocates weapons to each target as described above.

As a result of Code 50's computations, one may obtain the following:

1. An annotated presentation of the input data.
2. The number of weapons of each type available to be allocated.
3. The single-shot kill probabilities for each weapon-target combination.
4. The probability of penetrating area defenses for both missiles and bombers on each strike.
5. A summary of target value destroyed on each strike.
6. A summary of the weapon allocation (laydown) used on each strike.
7. The detailed weapon laydown used for each strike. This includes for each target the weapons used, the destruction, the number of weapons destroyed by terminal defenses, and the Lagrange multiplier associated with each weapon used, which is an estimate of the additional destruction one would obtain if he were to use one more weapon of that type.

Other output may also be obtained, but most of it is of use only in helping to locate mistakes in the program.

Several additional options are available in Code 50. For example, on any strike, one may require that a certain percentage of the total value of all defended cities be destroyed. City values are adjusted until either the required destruction is obtained (plus or minus 2%) or until it is shown that the requirement still cannot be met, even if all weapons are sent against defended cities. The allocation and destruction we would obtain if we were to ignore this requirement can also be printed.

Code 50 can also be used to compute the results of force exchanges other than Assured Destruction exchanges. As was mentioned, the relative values assigned to a defender's weapons are used in an Assured Destruction exchange only as an intermediate step. These values are used to convert what is acutally a two-sided zero-sum game into two separate resource allocation problems. In this case, the value of urban-industrial targets to the initial attacker is zero, and the value of military targets to the side retaliating is zero. By changing these zero values, one can evaluate mixed attacks against both military and urban-industrial targets.

In addition, weapon allocations other than the 'optimal' allocation can be evaluated and compared with the optimal allocation. Other allocations may be specified in three ways:

(a) by directly specifying the number of weapons of each type used against each target,

(b) by directly specifying the fraction of available weapons of each type used against each target, or

(c) by computing and saving on magnetic tape an allocation that would be 'optimal' under a different set of assumptions. This allocation can then be evaluated under a new set of assumptions and compared to the allocation that is 'optimal' under the new set of assumptions.

Two programs have been developed to help handle data for Code 50. A program called 'DATAGEN' allows one to create input decks on tape by specifying changes to previously created input decks. Since much of the input data is the same for almost all cases (e.g., target vulnerability; city names and populations; weapon names, yields, c.e.p.'s, loadings, and reliabilities), only one complete deck of input data need be punched on cards. DATAGEN can make all other decks using specified changes.

To illustrate the usefulness of DATAGEN, one can compare the number of cards needed to do the computations in a typical set of Department of Defense Strategic Force and Effectiveness Tables with and without DATAGEN. Results for approximately 250 cases, are included in the tables. A Code 50 input deck contains about 500 cards; 250 decks would contain around 125,000 cards (63 boxes). With DATAGEN, one needs less than 2,000 cards (1 box) for all 250 cases.

A program called STAT summarizes input and output data in a form easier to use and check. Input data are presented by case for each variable. For example, a table can be printed giving weapon inventories for all cases being run at the same time. If, as is common, a set of cases consists of the same basic data for several different years, such a table would be a table of inventories by year. Some output data, such as the number of weapons, megatons, or equivalent megatons used on any one strike, can also be summarized and printed by case.

5. ANTICIPATED IMPROVEMENTS

We are currently working on two primary improvements for Code 50: better defense modeling, and better weapon allocation algorithms. In addition to random area missile defenses, we are including preferential area defenses and overlapping island subtractive area defenses. A preferential defense can be used only if the defender knows what portion of the attack he is seeing with his radars at any one time. If he has sophisticated command and control, he can then allocate an equal portion of his area defenses to minimize the value destroyed by the incoming attack. The resulting defense allocation will 'preferentially' defend certain targets, preventing damage on them, while allowing some targets to go undefended. The mathematical difficulty involved in modeling such a defense is in determining the optimal offense attack in the face of such a defense, especially when certain weapons, such as those carried by aircraft, can completely evade the missile defense and destroy the targets preferentially defended.

A defense interceptor in an island subtractive defense can cover only a certain 'island' of the targets; the interceptors in any one island must be exhausted before targets in the island can be attacked. Mathematically, this type of defense is less difficult to model than the preferential defense because only the attacker has alternative strategies available.

Although our present offense allocation algorithm is usually sufficiently accurate, we are improving it primarily to increase computational efficiency. We are including a new Lagrange multiplier adjusting algorithm. This algorithm approaches the desired constraints by forming a simplex around the point which represents the constraints in the space of maximum pay-off versus resources expended. A unique set of Lagrange multipliers is associated with the hyperplane through the corners of the simplex. The corners of the simplex are moved closer and closer to the constraint point, and, if the constraint point can be reached using Lagrange multipliers, i.e., if the objective function is concave in the region of the solution, the simplex eventually collapses to a point whose associated Lagrange multipliers comprise the desired set. If the solution lies within a 'gap' (convexity), the simplex stabilizes around the gap, and Theorem 2 (the 'epsilon' theorem) can be used to bound the solution. We also intend to include an improved allocation algorithm to be used in this type of situation, i.e., when a solution cannot be found by direct use of the multipliers.

REFERENCES

EVERETT, HUGH III (1963) 'Generalized Lagrange Multiplier Method for Solving Problems of Optimum Allocation of Resources' Operations Research, Volume 11, No. 3, 399-417

DISCUSSION

Prof. E. M. L. Beale

It is of interest to compare Mr. Lodal's treatment of the weapon allocation problem with that described in Beale (1966). My paper describes a linear programming treatment that involves essentially the same approximations as the Lagrange Multiplier treatment. The advantage of the linear programming approach is that the Lagrange Multipliers are computed automatically by the simplex method, as the shadow prices on the resource availability rows, even in the presence of significant local defences that can be distributed between the targets. It also gives some interesting insight into the types of mixed strategy appropriate to both sides. But it is doubtful if this approach can deal effectively with the large number of weapon types considered by Mr. Lodal. In my own numerical work I considered only one weapon type for each side. The method should cope with many problems involving 2 or 3 weapon types; and the use of decomposition methods could extend this range, but possibly not dramatically.

The idea of using Lagrange Multipliers to replace some constraints in a mathematical programming problem is not very new. It is a central feature of the Dantzig-Wolfe Decomposition Principle. But it is very interesting that Mr. Lodal is able to approximate the Lagrange Multipliers effectively for his problems, since we—and others—have had great difficulty with this: as the partial, but patchy, computational success of the decomposition algorithm illustrates.

REFERENCE

E. M. L. BEALE (1966) 'Blotto games and the Decomposition Principle' pp. 64-84 of Theory of Games: Techniques and Applications Edited by A. Mensch. English Universities Press, London.

SECTION 10 Non-Linear Programming

E. L. Leese
H. Stolley

Non-Linear Programming Applied to a Statistical Estimation Problem

E. L. LEESE
Department of National Defense, Canada

1. INTRODUCTION

This paper is concerned with the estimation of production levels for a number of facilities within a region. Each facility has been assigned a certain (unknown) 'daily production quota', which can be assumed to remain constant from day to day. On a daily basis, we have information about the production level of each facility, and the production level of the whole region. However, direct information on the level of production is not given. Instead, the production level of a plant on a given day is given as a percentage of the (unknown) 'daily production quota' for that plant. Similarly the total production of the whole region is given as a percentage of the (unknown) 'daily production quota' for the region.

If such information is available for one day, we can derive a linear relation between the daily production quotas for the individual plants, and the daily production quota for the whole region. If we have such information for n days, we can derive n such linear relations.

Since all the information is in the form of 'percentage of daily quota', it is clear that we cannot use these relations to obtain estimates of the absolute values of daily production quota for the plants. However, we can perhaps make estimates of the daily production quota of each plant relative to that of the whole region.

If there are m plants in the region, and we have information for n days, we can derive n linear relations among the 'relative production quotas' for the various plants. In addition, the sum of the 'relative production quotas' for all plants in the region must be unity. Hence we have available $(n + 1)$ linear relations among the m unknown relative production quotas.

If $n = m - 1$ we can solve these equations. But usually n exceeds $m - 1$, and in such a case it is usually impossible to choose values for the m unknown quotas which will satisfy all the $(n + 1)$ equations.

In this case the only reasonable course is to assume that the information on 'percentage of daily production quota' is subject to error. Moreover, since there is no reason to suspect any one piece of input data more than another, we might as well assume that the error distributions are identical for all input values. If we assume that these error distributions are normal with zero mean and common variance, and are mutually independent, we can apply normal statistical methods to derive estimates of the m unknown relative production quotas.

2. MATHEMATICAL STATEMENT OF PROBLEM

Denote by $100x_{ik}$ the 'observed percentage of daily quota' for day i and for plant k, and by $100y_i$ the 'observed percentage of daily quota' for day i for the whole region. Let the 'true' values of x_{ik} and y_i be denoted by ξ_{ik} and η_i. Let z_k be the 'relative production quota' for plant k, relative to that of the whole region.

Assume that for all (i, k)

(a) x_{ik} is distributed normally about mean ξ_{ik} with variance σ^2,

(b) y_i is distributed normally about mean η_i with variance σ^2,

(c) the distributions of x_{ik} and y_i are mutually independent.

We wish to make estimates of the values of the 'relative production quotas' z_k, on the basis of the observed values of x_{ik} and y_i.

We note that all z_k must be positive or zero, and their sum must be unity. Moreover, for each day i we know that the total region production is the sum of the productions of the individual plants. This leads to the conditions

$$z_k \geqslant 0 \qquad (k = 1, m) \tag{1}$$

$$\sum_{k=1}^{m} z_k = 1 \tag{2}$$

$$\sum_{k=1}^{m} z_k \xi_{ik} = \eta_i (i = 1, n) \tag{3}$$

If we now apply the maximum likelihood method to estimate the unknown parameters z_k, ξ_{ik} and η_i, we find that the required parameter values are those which minimise the value of

$$S = \sum_{k=1}^{m} \sum_{i=1}^{m} (x_{ik} - \xi_{ik})^2 + \sum_{i=1}^{m} (y_i - \eta_i)^2 \tag{4}$$

subject to the constraints (1), (2), (3).

To solve this problem, we first note that for any <u>fixed</u> set of values of the z_k satisfying (1) and (2) we can choose a set of ξ_{ik} and η_i, which will minimise S subject to the constraints (3). The appropriate values of ξ_{ik} and η_i can be obtained by the usual Lagrange method. If we denote the consequent minimum value of S by S*, one can readily show that

$$S^* = \left[\sum_{i=1}^{n} \left\{y_i - \sum_{k=1}^{m} z_k x_{ik}\right\}^2\right] \Bigg/ \left(1 + \sum_{k=1}^{m} z_k{}^2\right) \tag{5}$$

To obtain the maximum likelihood estimates of the parameters z_k, we must now choose the z_k to minimise S* subject to the constraints (1) and (2).

Since S* is a non-linear function of the z_k, this may be regarded as a problem in non-linear programming. We note that the constraints (1) and (2) are linear, and are of very simple form.

3. METHODS OF SOLUTION

For non-linear programming problems with linear constraints, two well-established methods of solution are the 'projected gradient' and 'reduced gradient' methods. These methods are described and discussed in References (1), (2) and (3). Both methods are simple in principle, but involve some complexity in application to a general problem. However, because of the simple nature of the constraints in our problem, both projected gradient and reduced gradient methods can be very simply applied. Several variants of each method were used to solve the problem, with the idea of comparing the effectiveness of the two solution methods, for the particular problem considered here.

The general principles of the two methods are clearly described in articles by Abadie (1967), Wolfe (1963) and Rosen (1960). In both methods we regard a set of values (z_1, $z_2, \dots z_m$) as the coordinates of a point P in m-dimensional space. We start from any arbitrary point P in the 'constraint set' of points defined by the constraints (1) and (2). We choose a 'best direction to move', to reduce the value of the objective function S*. We extend a straight line ('ray') from P in this direction, and choose a new point on this ray, still in the constraint set, for which S* is (approximately) minimum. From this new point we again choose a 'best direction to move', and repeat the process until we can find no direction to move which will cause a decrease in S*. At this point we have solved the problem.

In both methods, if P is inside the constraint set (i.e. if all the constraints (1) are strict inequalities) we choose as 'best direction to move' that of the gradient of S* with respect to the z_k variables. However, if P is on the boundary of the constraint set, the choice of 'best direction to move' is different for the 'projected gradient' and 'reduced gradient' methods.

4. APPLICATION OF PROJECTED GRADIENT METHOD

In the projected gradient method, when P is on the boundary of the constraint set, we obtain the 'best direction to move' by projecting the gradient vector (of S*) on to the boundary of the constraint set. This boundary is defined as the intersection of all surfaces formed by the set of constraints (1) which are strict equalities at P. The direction of the 'projected gradient' vector, which lies on the boundary of the constraint set, is the direction in which the objective function S* changes most rapidly, if we move a short distance from P, keeping (1) and (2) satisfied. This is the direction in which movement is made, in the projected gradient method.

If we get into a position in which P is on the boundary of the constraint set but the 'projected gradient' vector is zero, we try relacing those constraints which correspond to equalities in (1), to see whether we can now make a further decrease in S* by moving into the interior of the constraint set. If this is possible, we make this move. If it is not possible, we have reached the solution.

In our problem, the 'projected gradient' algorithm may be expressed as follows. We start with an arbitrary point P, for which s of the constraints (1) are strict equalities. (We include the case $s = 0$). Denote the set of indices $k = 1, 2, \dots m$ by M, and the set of indices for which the constraints (1) are strict equalities by S. Proceed as follows:-

(I) Calculate S* and its partial derivatives (g_k) with respect to z_k, for all $k \in M$.

(II) Calculate $\bar{g} = \dfrac{1}{(m - s)} \sum_{k \in M-S} g_k$ (6)

$$y_k = \bar{g} - g_k \quad (\text{all } k \in M - S) \tag{7}$$

$$D = \sum_{k \in M-S} {y_k}^2 \tag{8}$$

(The quantities y_k are proportional to the components of the 'projected gradient' vector at P. All components for which $k \in S$ are zero.)

(III) If D is positive go to (IVa). Otherwise, set $g_0 = \bar{g}$ and go to (IVb).

(IVa) Extend a ray from P in the direction defined by the quantities y_k of step II. Find P′, the point at which this ray first hits the boundary of the constraint set. P′ will lie on one of the coordinate planes, $z_{k'} = 0$, say, where k′ does not belong to S.

(Va) Calculate the rate of increase of S* along the 'ray' of step (IVa) at the point P′. This is

$$D' = \sum_{k \in M-S} y_k g_k(P') \tag{9}$$

where the y_k come from step (II). If D′ is positive go to (VIa). Otherwise go to (VIb).

(VIa) Find a point P″ on the ray PP′ at which S* is 'approximately minimum'. Replace P by P″ and return to step (I).

(VIb) Add to the set S the index k′ found in step (IVa). Replace P by P′ and return to step (I).

(IVb) Find t, the member of the set S for which $g_t(P)$ is least. If g_t is greater than or equal to g_0 of step (III), terminate the search at P. Otherwise go to (Vb).

(Vb) Eliminate 't' from the set S and return to step (I).

5. METHODS OF ONE-DIMENSIONAL MINIMISATION

In the projected gradient method, we repeatedly have to find the position of a point P″, on a straight line segment PP′, at which a certain 'objective function' is 'approximately minimum'. We don't need to find a precise minimum, because in most cases P″ is merely a new 'jumping off point' for the next part of the search.

There are various ways of finding P″. One can find the rate of change of S* (the 'gradient') as one moves along the line PP′, and seek to determine a position P″ at which the gradient is approximately zero. Or one can use the values of the objective function itself, and search along PP′ until one finds a point P″ where S* is 'about as small as one can obtain'.

Methods of both these types were tried. The following four methods were used:-

(A) Simple linear interpolation. This just uses the values of the gradient at the two ends of the line PP′, and does a simple linear inverse interpolation to obtain a (crude) estimate of the position P″ at which the gradient is zero.

(B) Repeated linear interpolation. Here we apply (A) a number of times. At each stage we interpolate between gradient values at points P_a and P_b, say, where P_a and P_b are the two closest points at which the gradient values are of opposite sign.

(C) Modified repeated linear interpolation. This is similar to (B), except that in each interpolation stage, we halve the value of the gradient at P_b before interpolating. (Here P_a and P_b are the two closest points at which the gradient values are of opposite sign, and of these two points, P_b is the one which was used in the preceding interpolation step).

(D) 'Golden section' division. This also consists of a number of iteration stages, but involves only the computation of S*. At a given stage, when we know that the minimum of S* lies between points a and (a + h), we compute the values of S* at points (a + rh) and (a + r^2h), where r is the 'golden section' ratio defined by

$$r + r^2 = 1 \tag{10}$$

If S^* is less at $(a + r^2h)$ than at $(a + rh)$, we know that the minimum must lie in $(a, a + rh)$. On the other hand if S^* is less at $(a + rh)$ than at $(a + r^2h)$, we know that the minimum lies in $(a + r^2h, a + h)$. In either case we have reduced the interval in which the minimum must lie, and we can repeat the process. In view of (10), only one new S^* calculation is needed at each stage.

The choice of the 'best' one-dimensional minimisation method involves a compromise. If we use a crude method such as (A), the minimisation process will be quite fast, but we have to do a large number of these steps before solving the whole problem. On the other hand, a 'refined' one-dimensional minimisation method such as (B), (C) or (D) with many iterations, may take longer than (A), but we may need fewer steps before solving the whole problem.

To examine the relative merits of minimisation methods (A), (B), (C) and (D), we applied them to solve a few problems of practical interest, and noted the computation times. The results are given later in this paper.

6. APPLICATION OF REDUCED GRADIENT METHOD

As mentioned earlier, the reduced gradient method differs from the projected gradient method in the selection of the 'best direction to move' when P is on the boundary of the constraint set. In addition, the reduced gradient method looks at the problem in a somewhat different way. For each of the constraint inequalities, a 'slack' variable u_j is defined, as in the simplex method of linear programming, with the object of transforming all the constraint conditions into equalities. For example, a constraint equation

$$\sum_{i=1}^{m} a_{ij} z_i \leqslant b_j \tag{11}$$

is replaced by

$$a_j = b_j - \sum_{i=1}^{m} a_{ij} z_i \geqslant 0 \tag{12}$$

There will be as many slack variables as constraints. At any stage, some of the slack variables will be zero (those corresponding to constraints which are equalities). The other slack variables will be positive.

Since P is a point in m dimensions, we can usually define the position P by specifying the values of only m of the slack variables. These are called 'independent' variables. Values of the z coordinates of P, and of the remaining slack variables (the 'dependent' variables) can be expressed in terms of the independent variables.

Initially we choose our set of independent variables to include all the slack variables which are currently zero. (This is always possible). Then we make a series of 'best direction' moves for P, in the direction of the local gradient vector, as in the projected gradient method.

This is done until P hits a new boundary of the constraint set. At this stage one or more of our 'dependent' slack variables may become zero. In this case we redefine these (currently zero) 'dependent' variables as 'independent' variables and in exchange, redefine some of the 'independent' variables which are currently positive as 'dependent'. This process, which involves the solution of a set of linear simultaneous equations, is similar to the 'pivoting' process in linear programming. By making this exchange we ensure that all the dependent variables remain strictly positive.

When P is on the boundary of the constraint set, the steepest descent direction, given by the gradient, may point outside the constraint set. In such a case, we express the gradient vector in terms of the m 'independent' slack variables, and examine in turn each of the individual gradient components corresponding to a change in just one of the independent variables. Those components which would result in a movement of P outside the constraint set, we replace by zero. The other components we leave unchanged.

The vector formed by these modified vector components is called the 'reduced gradient'. It is in the direction of the reduced gradient that we make our next move.

In our problem the constraints are very simple, and one constraint is always an equality. Only (m – 1) independent quantities are needed to fix the position of P. If we start at a point for which some (arbitrarily chosen) coordinate z_t is positive, we can choose, as our starting set of (m – 1) independent 'slack' variables, the set of all (z_1, $z_2, \ldots z_m$) excluding z_t. Our only dependent slack variable will be z_t.

If we get into a position where z_t becomes zero, we merely make z_t an independent variable and choose one of the other (non-zero) z's as the new dependent variable.

The steps of the reduced gradient method, for our problem, are as follows:-

(I) Calculate S* and its partial derivatives (g_k) with respect to z_k, for all $k \in M$.

(II) Determine the 'reduced gradient' direction, with components

$$\left.\begin{aligned} y_k &= g_t - g_k \quad \text{all } k \in M - \{t\} \text{ for which } g_k < g_t \text{ or } z_k > 0 \\ &= 0 \qquad\quad \text{all other } k \in M - \{t\} \end{aligned}\right\} \tag{13}$$

Calculate $$D = \sum_{k \in M-\{t\}} y_k^2 \tag{14}$$

(III) If D is positive go to (IV). Otherwise, terminate the search at P.

(IV) Extend a ray from P in the direction defined by the quantities y_k of step (II). Find P′, the point at which this ray first hits the boundary of the constraint set. P′ will lie on one of the coordinate planes, $z_{k'} = 0$, say.

(V) Calculate the rate of increase of S* along the 'ray' of step (IV) at the point P′. This is

$$D' = \sum_{k \in M} y_k g_k(P') \tag{15}$$

where the y_k come from step (II). If D′ is positive go to (VIa). Otherwise go to (VIb).

(VIa) Find a point P″ on the ray PP′ at which S* is 'approximately minimum'. Replace P by P″ and return to step (I).

(VIb) If z_t is zero at P′, find the index k = t′ for which z_k is greatest at P′. Replace t by t′ and go to (VII). If z_t is positive at P′ go directly to (VII).

(VII) Replace P by P′ and return to step (I).

In step (VIa) we tried the same one-dimensional minimisation methods as were used in the projected gradient method.

7. COMPUTER RUNS

Each of the above methods of solution was applied to two problems, the input data for which are given in Tables I and II. Computer programs for the projected gradient and reduced gradient methods were written in FORTRAN IV, and were run on an IBM 360 Model 40 computer using Operating System. For each method, we tried using each of the four one-dimensional minimisation methods described above. In the case of minimisation methods (B), (C) and (D), the number of iterations for a single minimisation step was chosen so as to minimise the overall solution time for the whole problem.

Table I

Input Data for Problem A

Value of i	Value of x_{ik} for k = 1	2	3	4	5	Value of y_i
1	0.927	1.034	0.897	0.923	1.007	0.964
2	0.900	1.034	0.897	0.923	1.007	0.964
3	0.940	1.002	0.912	0.951	0.115	0.973
4	0.965	0.976	0.997	0.870	1.093	0.945
5	1.010	0.895	1.094	0.928	1.000	0.991
6	0.971	0.954	0.948	1.001	0.895	0.953
7	1.000	0.913	0.879	0.902	1;082	0.962
8	0.919	0.935	0.880	0.794	0.958	0.910

Table II

Input Data for Problem B

Value of i	Values of x_{ik} for k = 1	2	3	4	5	Value of y_i
1	1.067	1.183	1.502	0.947	0.331	0.936
2	1.104	1.166	1.877	1.038	0.439	0.987
3	1.114	1.123	1.076	1.047	0.486	0.931
4	1.143	1.102	1.329	1.143	0.797	0.099
5	1.220	1.112	1.159	1.161	1.015	1.111
6	1.239	1.328	1.204	1.389	1.727	1.263
7	1.269	1.273	1.280	1.827	1.681	1.175
8	1.089	1.305	1.094	1.873	1.564	1.243

The computation times (for the execution phase of the run) are listed in Table III.

8. DISCUSSION OF RESULTS

For the projected gradient method, Table III suggests that for both problems the fastest solution is obtained by using the 'simple interpolation' minimisation method. In other words, it is better to use a crude, fast method to get a very rough idea of the position of the minimum, and carry out iterations of the projected gradient procedure, than to spend time in getting a more refined minimum position. This simple minimisation method is the one suggested in Ref. (3).

In the case of the reduced gradient method, the simple interpolation method also gives the fastest solution to Problem B. In Problem A the modified repeated interpolation method is fastest, but the simple interpolation method is a close second.

If we compare the solution times for projected gradient and reduced gradient methods, we find that for Problem A the reduced gradient method is faster, but that for Problem B the projected gradient method is faster. The reason for this is not obvious. The solution to Problem A has all z_k values positive, whereas in the solution to Problem B two of the z_k values are zero. The solutions are listed in Table IV.

Table III

Comparison of Computer Times for Various Methods for IBM 360 Model 40

Minimisation method	Execution time (secs) for			
	Problem A using		Problem B using	
	Projected gradient method	Reduced gradient method	Projected gradient method	Reduced gradient method
Simple interp.	19.1	13.9	7.9	11.7
Repeated interp.	31.9	15.6	20.4	19.5
Mod. Repeated interp.	20.8	11.8	20.2	16.4
Golden section division	45.6	17.3	26.9	33.9

Table IV

Solutions to Problems A and B

Plant	Relative Daily Production Quotas	
	Problem A	Problem B
1	0.4391	0.6974
2	0.2909	0.0000
3	0.0799	0.0000
4	0.1199	0.1619
5	0.0701	0.1406

9. CONCLUSION

This very limited study does not show any outstanding advantage for the projected gradient or reduced gradient method, as far as speed of computation is concerned. Which method is faster seems to depend on the particular problem. But in most cases the crudest form of minimisation step seems to produce the fastest solution.

REFERENCES

1. ABADIE, J. (1967) Non-Linear Programming, Wiley, New York.
2. GRAVES, R. L. and WOLFE, P. (1963) Recent Advances in Mathematical Programming, McGraw-Hill, New York.
3. ROSEN, J. B. (1960) 'The Gradient Projection Method for Non-Linear Programming. Part I: Linear Constraints'. Journal of Society of Industrial and Applied Mathematics, Vol. 8, No. 1, pages 181-217.

DISCUSSION

J. N. S. Calis

In this lecture were mentioned 4 methods to minimize a real function of one variable. Simple linear interpolation seemed to be the best. Yet another method is to regard the function locally as a quadratic function and minimize this function instead. For this method 3 evaluations of $f(x)$ are needed as compared with 2 of $f'(x)$ in linear interpolation. But as the linear interpolation is essentially converging linearly and the abovementioned method quadratically, this one seems to be still better suited. Also the abovementioned method does not require the evaluation of $f'(x)$.

E. L. Leese

I agree that the proposed method of one-dimensional minimization is a good one, and I propose to test it in our probelm to see whether it gives better results than the four minimization methods already described.

Application of Quadratic Programming to the Promotion System of Professional Officers in the German Air Force

H. STOLLEY
I.A.B.G., West Germany

1. TASK STATEMENT

For clarification of the task statement, the age and rank structure of the year 1967 is used as a basis (see Fig. 1).

If you consider this distribution by age, you will note that between two maxima (at 53 and 29 years of age) a strong minimum (at 40 years) exists. From this distribution, the difficulties which the Air Force of the Federal Republic of Germany will have to face in the coming decades are easily apparent:

1. In the next 10 years the officers of the upper maximum range must retire. The possibility exists that in some of these years the number of retirements will be higher than the recruitment rate of young officers. As a result, a decrease of the total number of officers must be faced.
2. The subsequent minimum range involves the danger that there is a shortage of candidates for the upper ranks.
3. For the lower age maximum the same holds true as for the upper age maximum.

Our task was to try to aid in the reduction of these difficulties as much as possible by defining a suitable recruitment, promotion and retirement strategy, and pointing out how the present system could be transformed into a system which would no longer have the indicated difficulties.

2. CONSTRUCTION OF A STATIONARY SYSTEM

In the analysis performed, the ranks of Lieutenant-Colonel and Major as well as Second and First Lieutenant were combined and treated as one rank each. The following ranks were analysed:

1.	Generals	($j = 1$)
2.	Colonels	($j = 2$)
3.	Lt. Colonels and Majors	($j = 3$)
4.	Captains	($j = 4$)
5.	Second Lieutenants and First Lieutenants	($j = 5$)

What are the characteristics of a system which does not present the above mentioned difficulties any more? This question shall be answered in two steps; first we shall consider the overall system and then the various ranks.

2.1 Overall system

Figure 2 shows a summary of the factors and conditions determining the overall system.

Figure 3 shows a system satisfying the indicated conditions and having, moreover, the characteristic that in the transition from one year into another the system is maintained; therefore we call it a <u>stationary system</u>.

In this system 5 unknowns exist

(a) NZ the number of new additions

(b) the steps $d(p_1)$, $d(p_2)$, $d(p_3)$ and $d(p_4)$ which indicate the number of retirements in the ranks of General, Colonel, Lt. Colonel + Major, as well as Captain.

These steps in the retirement structure result from the conditions which will be determined for the various ranks later in this presentation. Once these values are determined NZ can be calculated with the help of the total number N.

2.2 Individual Systems

For consideration of the individual systems we introduce the following definitions:

a : age of officers

$h_j(a)$: number of officers of rank j at age a

$\tilde{h}_j(a)$: number of officers of ranks 1, 2, ..., j at age a
$$\tilde{h}_j(a) = h_1(a) + h_2(a) + .. + h_j(a)$$

N_j : number of officers of rank j

$\tilde{N}_j$: number of officers of ranks 1, 2, ..., j

For the analysis it is useful to consider the combined systems, i.e.

(a) Generals

(b) Generals and Colonels

(c) Generals and Colonels and Lt. Colonels + Majors

(i) Generals

The system of the Generals was pre-defined (Fig. 4). According to this system a annual number of BG_1 Colonels are promoted to General at the age of 49 and BG_2 at 55. The total number of Generals is N_1, including

G_1 Lt. Generals
G_2 Major Generals and
G_3 Brigadier Generals.

(ii) Generals and Colonels

Figure 5 summarizes all the conditions specified for this system.

The solution range of the inputted equation and inequation system in the $h_2(a)$ is an n-dimensional polyhedron. For our problem only integer solutions are possible. Later on a quadratic program will be defined over this domain.

Figure 6 shows a system which satisfies the indicated conditions.

(iii) Generals + Colonels + Lt. Colonels and Majors

Figure 7 summarizes all the conditions specified for this system.

Similar to the system of Generals + Colonels the solution range of the indicated equations and inequations for the $\tilde{h}_3(a)$ is an n-dimensional polyhedron which only the integer solutions are of interest. Figure 8 shows a stationary system for Generals down to Majors which satisfies the indicated conditions.

(iv) Captains and First Lieutenants + Second Lieutenants

The still missing conditions are summarized in Figure 10.

These conditions conclude the description of the overall system. Figure 11 shows a system which satisfies all the specified conditions.

3. PROVISIONAL SOLUTIONS

Comparing the stationary system developed here with the basic system, we can see that the present system is very far from the constructed system. Also it will be realized that it will not be possible to materialize this system within the next few years, since the number of officers who at present are 28, 29 and 30 years of age is far above the calculated number NZ.

It will therefore be necessary to find satisfactory provisional solutions for the next 3 decades. Such solutions will be calculated up to the year 2000 in the following pages. From these calculations also the difficulties mentioned at the beginning become apparent.

In this connection the question arises as to which of the solutions in the two indicated Simplexes shall be aimed at. For this purpose the solutions for each year to be analysed will be calculated. These solutions shall satisfy the following criterion function:

$$(\tilde{h}_j(a) - \tilde{h}_{jE}(a))^2 \rightarrow \text{Min } (j = 2, 3)$$

where $\tilde{h}_{jE}(a)$ is the number of officers in the ranks of 1 to j who are of the age a in the year under consideration.

From the mathematical viewpoint the above criterion function constitutes together with the pertaining Simplex a quadratic program which is being solved for each year under consideration. The solutions obtained are not integers, it is true, but they give indications as to the age at which promotions must be effected—as far as this is possible on account of the available officers—in order to approach a stationary solution.

Furthermore it has become obvious that with the progress of the years considered, the calculated solutions converge towards integer solutions.

Figure 11 summarizes all the conditions that apply for the provisional solutions.

Figure 12 shows in a flow diagram the computer calculations that have to be made for each year under consideration.

4. RESULTS

At the conclusion of the presentation a few results are shown which clearly demonstrate the difficulties with which the Air Force is faced in the field of personnel planning:

Fig. 13 (Age and rank structure in the year 1975)

In 1975 a large fraction of rather old officers must be retained in service in order to compensate for the subsequent minimum and to give the younger officers time to gather sufficient experience for the higher ranks.

Fig. 14 (Age and rank structure in the year 1985)

In 1985 it will become obvious that for the subsequent years no officers are available for retirement and that consequently no promotions can be effected either.

Fig. 15 (Total number of professional officers)

One of the biggest problems is the total number of officers. While in the coming years up to 1980 the total number of officers can be kept on the desired level by an appropriate retirement policy, this number will increase as soon as the age group corresponding to the present minimum reaches the age of retirement.

Fig. 16 (Average age for promotion to Major)

This figure shows the professional prospects of the young officers. It is apparent that starting in 1978 the average age for promotion to Major is over 40 years. These prospects can hardly compete with those of similar positions in industry so that it may become difficult to get enough young men interested in the officers career.

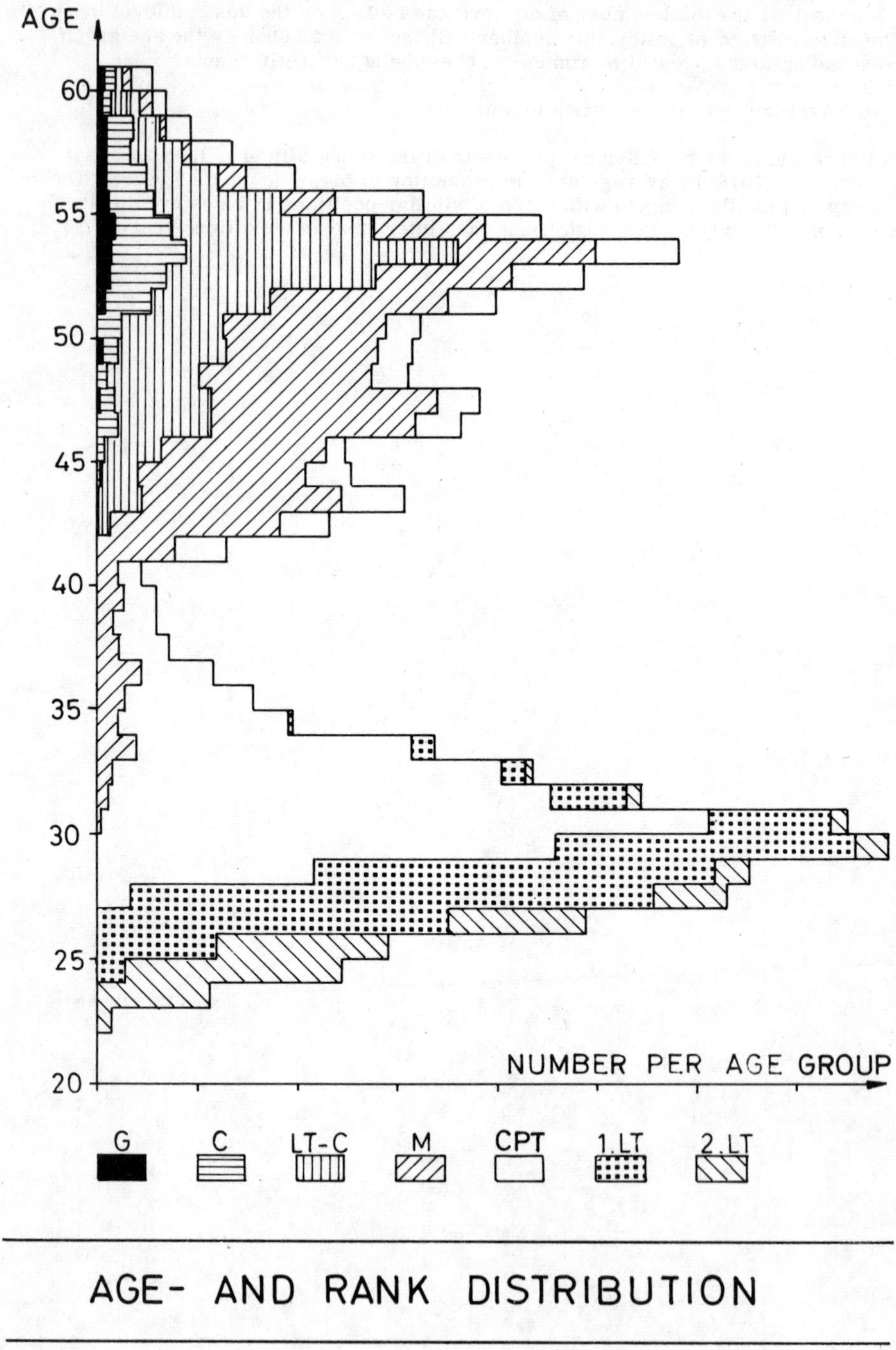
AGE
60
55
50
45
40
35
30
25
20
NUMBER PER AGE GROUP
G
C
LT-C
M
CPT
1. LT
2. LT
AGE- AND RANK DISTRIBUTION

Fig. 1

1 TOTAL NUMBER OF OFFICERS CONSIDERED

$NT_1 \leq N \leq NT_2$

2 RETIREMENT AGES OF THE VARIOUS RANKS

GENERALS	p_1 = 60 YEARS
COLONELS	p_2 = 58 YEARS
LT. COLONELS AND MAJORS	p_3 = 56 YEARS
CAPTAINS	p_4 = 52 YEARS

3 NEW ADDITIONS TO THE SYSTEM

PROMOTIONS TO FIRST LIEUTENANT :

50% AT THE AGE OF 23

50% AT THE AGE OF 24

4 FAILURE RATES

a) EVERY YEAR 1% OF THE OFFICERS QUIT PREMATURELY DUE TO ILLNESS & ACCIDENTS

b) IT IS ASSUMED THAT EACH OFFICER HAS THE SAME PROBABILITY OF FAILURE (THIS ASSUMPTION HAD TO BE MADE DUE TO LACK OF SUFFICIENT STATISTICAL MATERIAL CONCERNING THE AGE STRUCTURE OF FAILURES)

CONDITIONS FOR THE OVERALL SYSTEM

Fig. 2

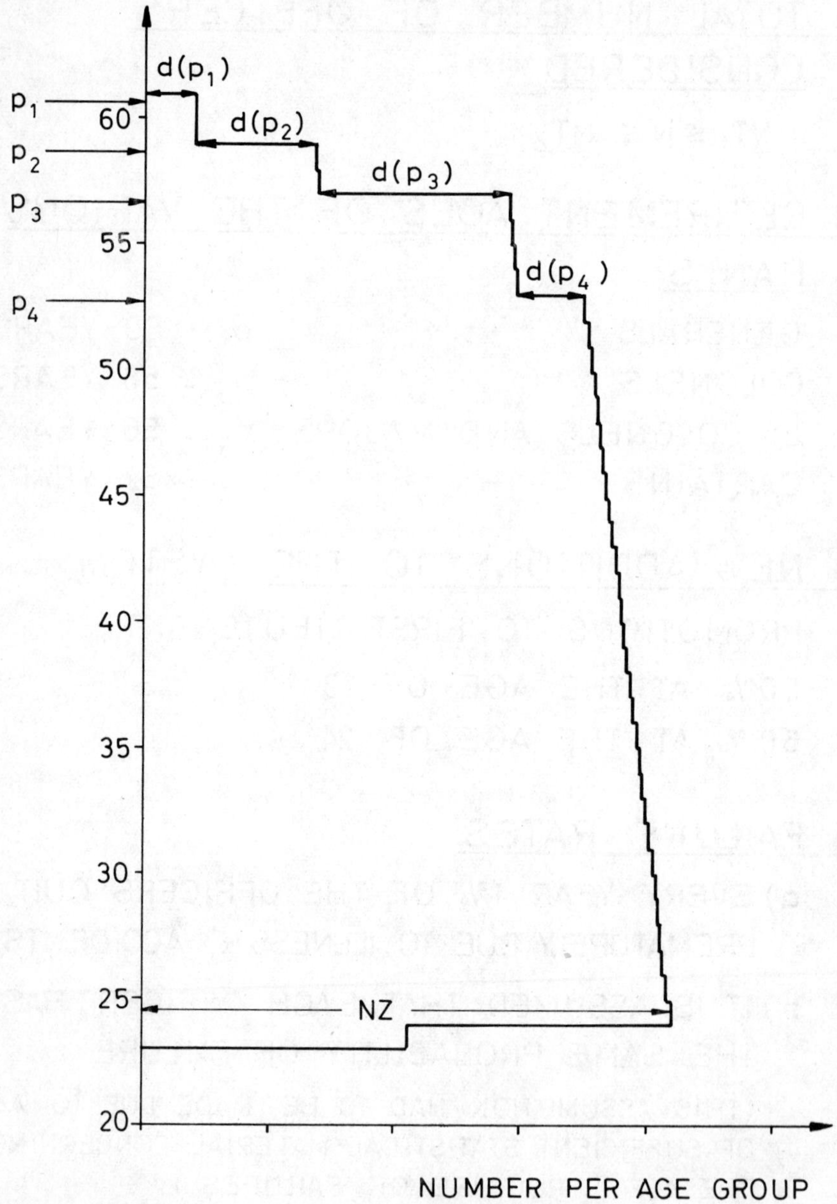

STATIONARY SYSTEM WITH FAILURES

Fig. 3

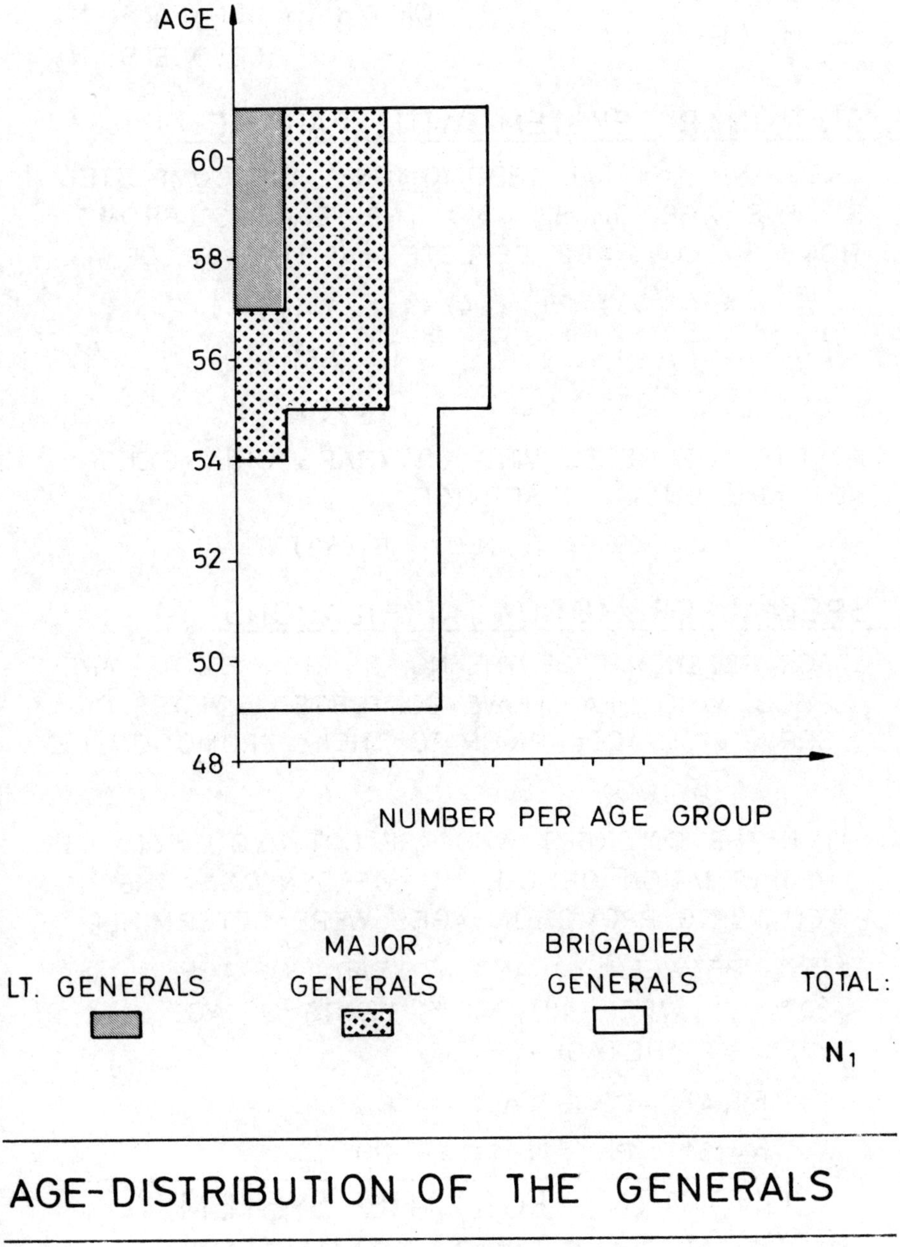
AGE
60
58
56
54
52
50
48
NUMBER PER AGE GROUP
LT. GENERALS
MAJOR GENERALS
BRIGADIER GENERALS
TOTAL:
N1
AGE-DISTRIBUTION OF THE GENERALS

Fig. 4

1. TOTAL NUMBER

$$\sum_{a=43}^{60} \tilde{h}_2(a) = \tilde{N}_2$$

OF WHICH GENERALS: N_1
COLONELS: N_2

2. STATIONARY SYSTEM WITH FAILURES

ASSUMING THAT ALL PROMOTIONS ARE COMPLETED BY THE AGE OF 52 AND THE EARLIEST PROMOTIONS TO COL. ARE EFFECTED AT THE AGE OF 43.

$$0 \le \tilde{h}_2(43) \le \tilde{h}_2(44) \le \ldots \le \tilde{h}_2(52)$$
$$= \tilde{h}_2(53) = \ldots = \tilde{h}_2(55)$$
$$= \tilde{h}_2(56) + 1 = \ldots = \tilde{h}_2(58) + 1$$

AT THE AGE OF 59 AND 60 YEARS ONLY GENERALS ARE STILL IN SERVICE

$$\tilde{h}_2(59) = \tilde{h}_2(60) = h_1(60)$$

3. SPECIAL PROMOTION REGULATIONS

a) FOR REGENERAT. OF THE NO. OF GEN. B COL. ARE REQU. WHO SHALL HAVE COMPLETED 6 YEARS OF SERVICE AS COL. PRIOR TO THEIR PROMOTION, I. E.

$$\tilde{h}_2(43) = B$$

b) FOR THE COLONELS WHO ARE NOT REQU. FOR THE REGENERATION OF THE NO. OF GENERALS THE FOLLOWING PROMOTION AGES WERE DETERMINED:

25 % BETWEEN 43 AND 47 YEARS OF AGE
50 % BETWEEN 48 AND 51 YEARS OF AGE
25 % AT THE AGE OF 52

$$\tilde{h}_2(47) - B = \frac{1}{4}(\tilde{h}_2(52) - B)$$
$$\tilde{h}_2(51) - B = \frac{3}{4}(\tilde{h}_2(52) - B)$$

CONDITIONS FOR THE SYSTEM GENERALS AND COLONELS

Fig. 5

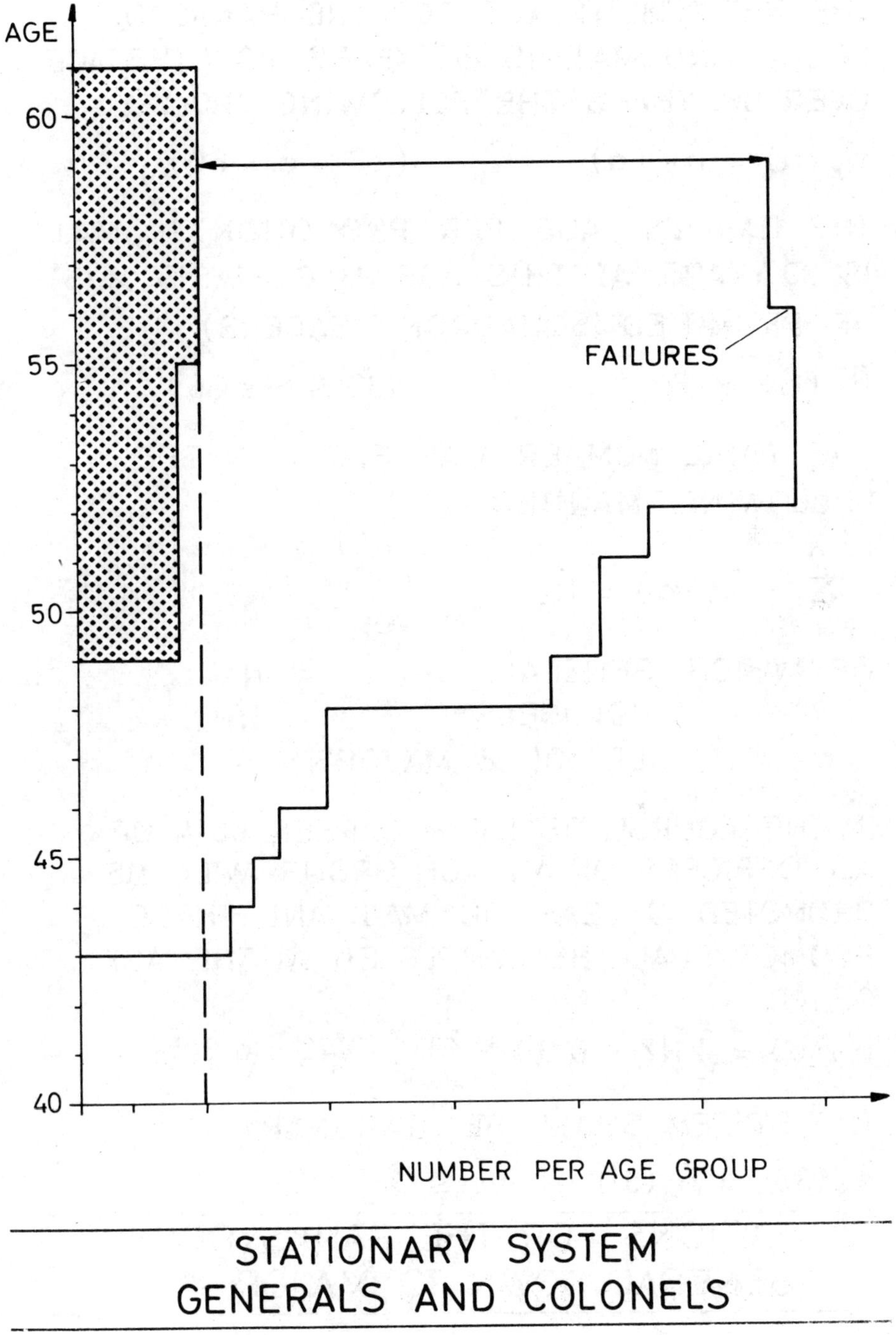
AGE
60
55
50
45
40
FAILURES
NUMBER PER AGE GROUP
STATIONARY SYSTEM
GENERALS AND COLONELS

Fig. 6

1. THE RETIREMENT AGE FOR THE RANKS OF LT. COL. AND MAJ. IS 56 YEARS. FOR THE AGE OVER 56 YEARS THE FOLLOWING APPLIES:

   $\tilde{h}_3(a) = \tilde{h}_2(a) \qquad (57 \leq a \leq 60)$

2. THE EARLIEST AGE FOR PROMOTION TO MAJ. IS 30 YEARS. AT THIS AGE B CAPTAINS MUST BE PROMOTED (SQUADRON LEADERS)

   $\tilde{h}_3(a) \geq B \qquad (30 \leq a \leq 56)$

3. THE TOTAL NUMBER WAS FIXED IN THE FOLLOWING MANNER:

   $$\sum_{a=30}^{60} \tilde{h}_3(a) = \tilde{N}_3$$

   OF WHICH GENERALS : N_1
   COLONELS : N_2
   LT. COL. & MAJORS : N_3

4. IN THE COURSE OF THEIR CAREER 90% OF ALL OFFICERS OF AN AGE GROUP WILL BE PROMOTED AT LEAST TO MAJ. AND THESE PROMOT. SHALL BE COMPLETED AT THE AGE OF 40

   $\tilde{h}_3(a) = \frac{9}{10} NZ \cdot 0.99^{a-24} \qquad (40 \leq a \leq 56)$

5. THE SYSTEM SHALL BE STATIONARY

   $\tilde{h}_3(30) = \tilde{h}_3(31) \leq \ldots \leq \tilde{h}_3(40)$

CONDITIONS FOR THE RANKS OF GENERAL DOWN TO MAJOR

Fig. 7

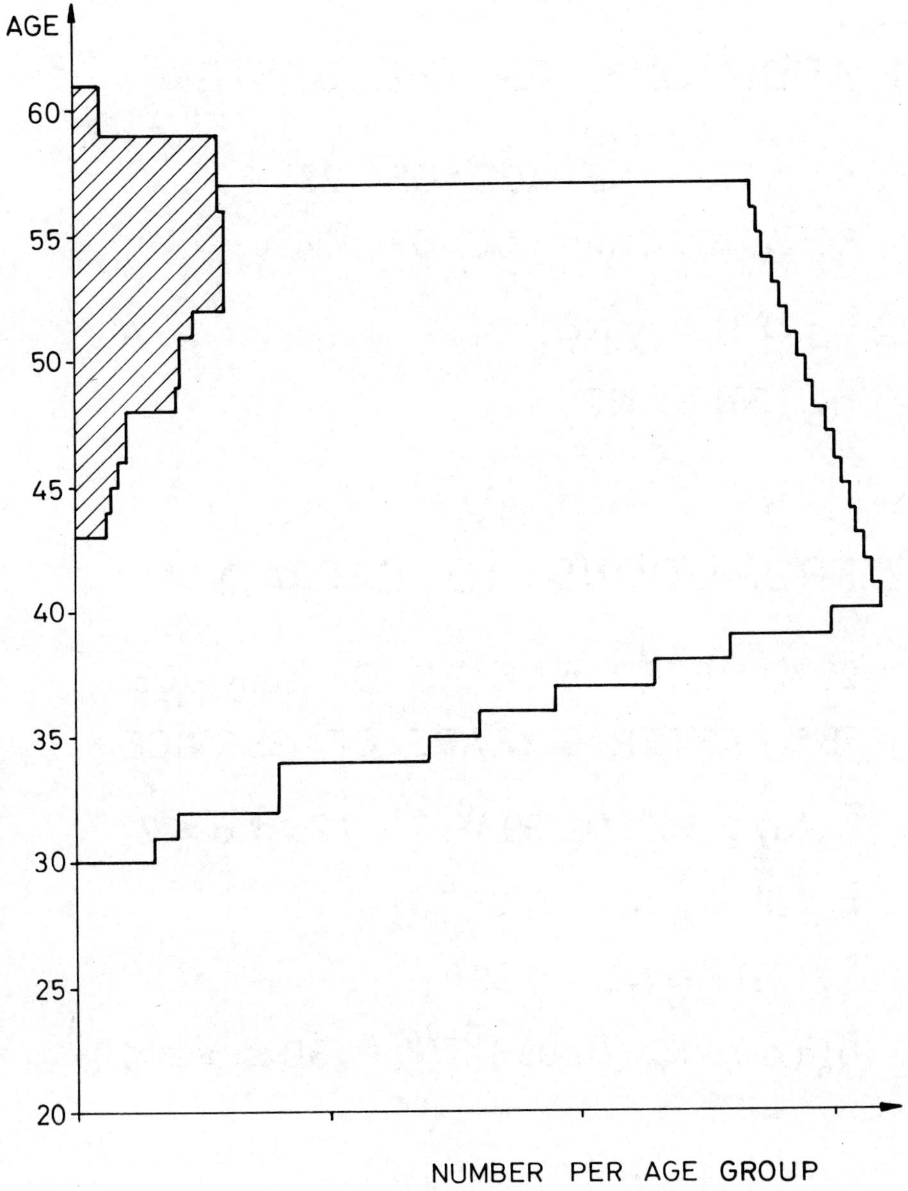

STATIONARY SYSTEM FOR
GENERALS - MAJORS

Fig. 8

1. ADDITIONS TO THE SYSTEM

50 % AT THE AGE OF 23

50 % AT THE AGE OF 24

$$\tilde{h}_5(23) = \frac{1}{2} NZ$$

$$\tilde{h}_5(24) = NZ$$

2. PROMOTIONS TO CAPTAIN

25 % AFTER 5 YEARS OF SERVICE

75 % AFTER 6 YEARS OF SERVICE

$$\tilde{h}_5(a) = NZ \cdot (0.99)^{a-24} \quad (25 \leq a \leq 27)$$

$$\tilde{h}_4(28) = \frac{1}{8} NZ \cdot 0.99^4$$

$$\tilde{h}_4(29) = \frac{5}{8} NZ \cdot 0.99^5$$

$$\tilde{h}_4(a) = NZ \cdot (0.99)^{a-24} \quad (30 \leq a \leq 52)$$

CONDITIONS FOR CAPTAINS, SECOND LIEUTENANTS AND FIRST LIEUTENANTS

Fig. 9

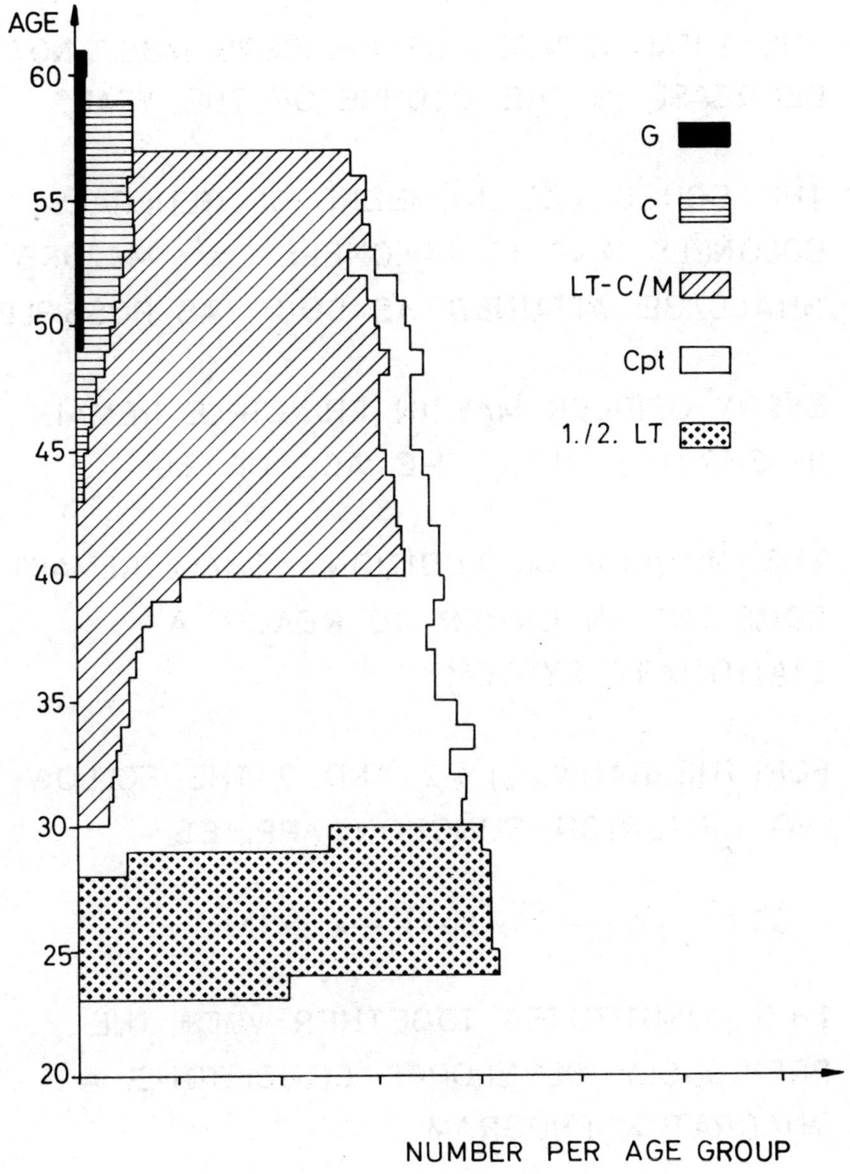

TYPICAL STATIONARY SYSTEM

Fig. 10

1. THE TOTAL NUMBER OF OFFICERS MUST NOT DECREASE IN THE COURSE OF THE YEARS

2. THE SCHEDULED NUMBERS OF GENERALS, COLONELS AND LT. COLONELS + MAJORS SHALL BE ATTAINED AS SOON AS POSSIBLE

3. EVERY OFFICER MAY IN PRINCIPLE REMAIN IN SERVICE UNTIL THE AGE OF 60

4. THE NUMBER OF ADDITIONS SHALL REMAIN CONSTANT IN ORDER TO REACH A STATIONARY SYSTEM

5. FOR THE RANKS j = 2 AND 3 THE FOLLOWING CRITERION FUNCTION APPLIES:

$$\sum_a \left(\tilde{h}_j(a) - \tilde{h}_{jE}(a)\right)^2 \longrightarrow \text{MIN.}$$

THIS CONSTITUTES TOGETHER WITH THE PREVIOUSLY DEVELOPED CONDITIONS A QUADRATIC PROGRAM

CONDITIONS FOR PROVISIONAL SOLUTIONS

Fig. 11

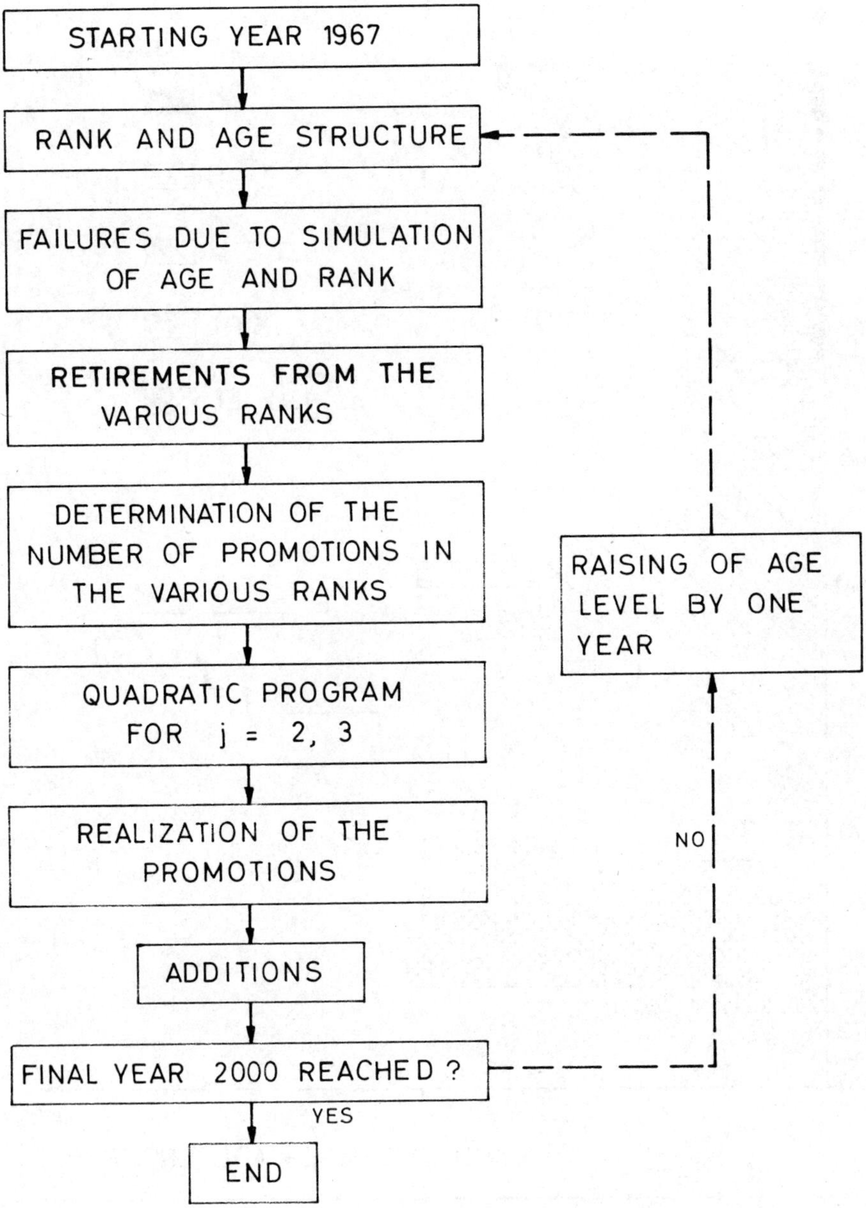

FLOW DIAGRAM OF THE CALCULATIONS TO BE MADE

Fig. 12

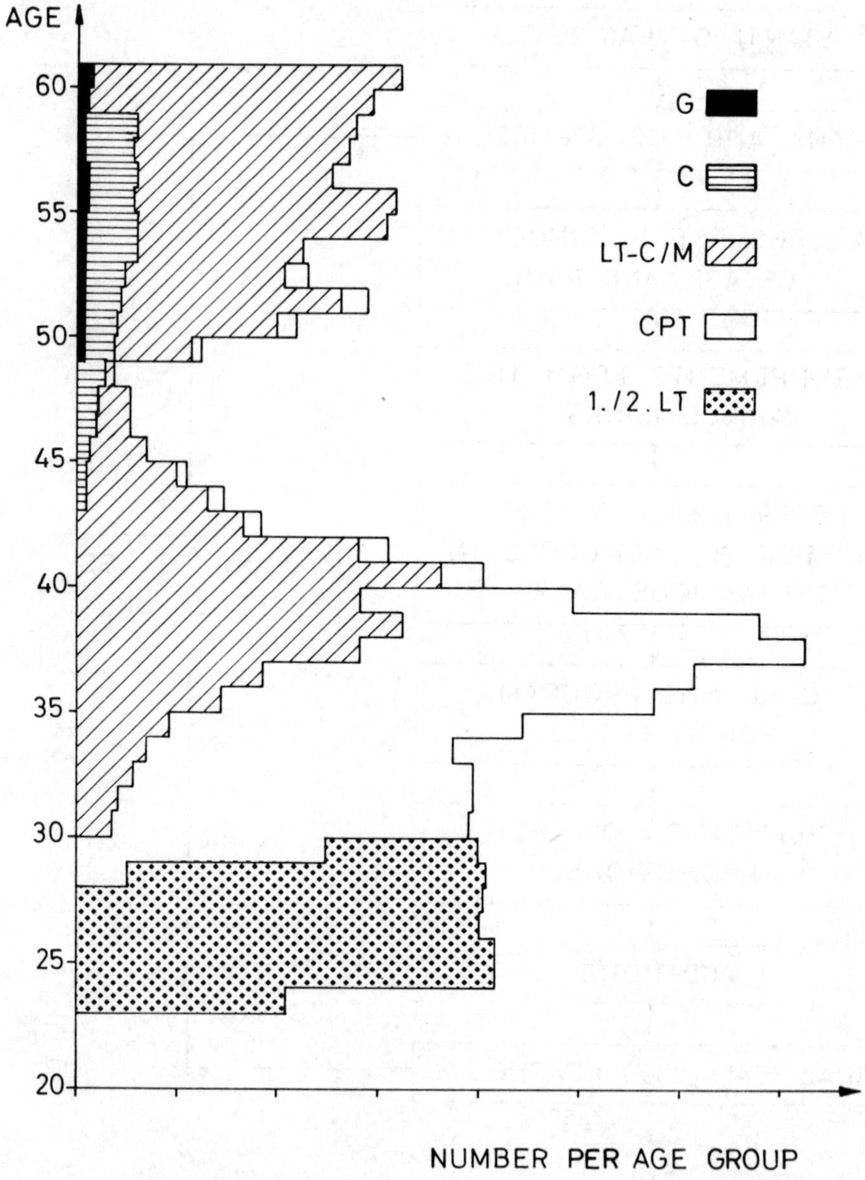

AGE- AND RANKDISTRIBUTION 1975

Fig. 13

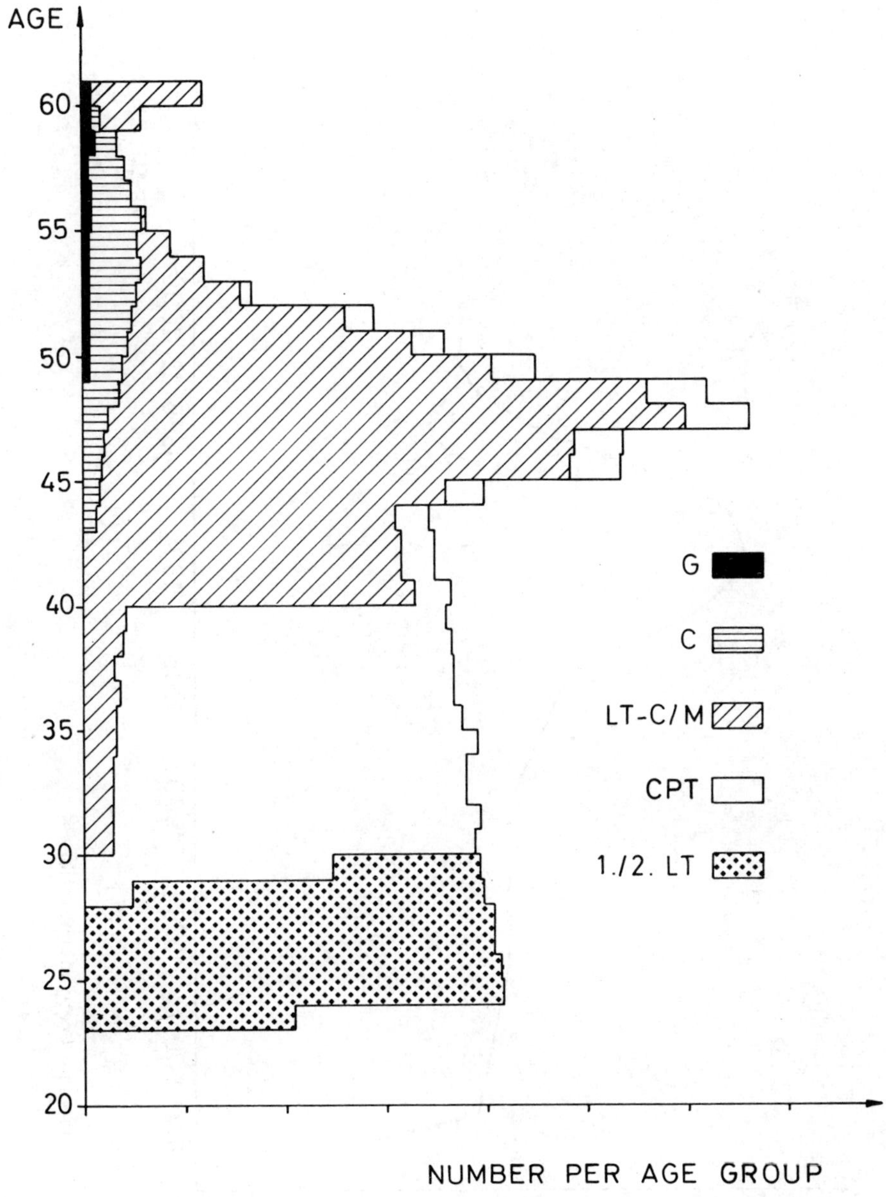

AGE- AND RANKDISTRIBUTION 1985

Fig. 14

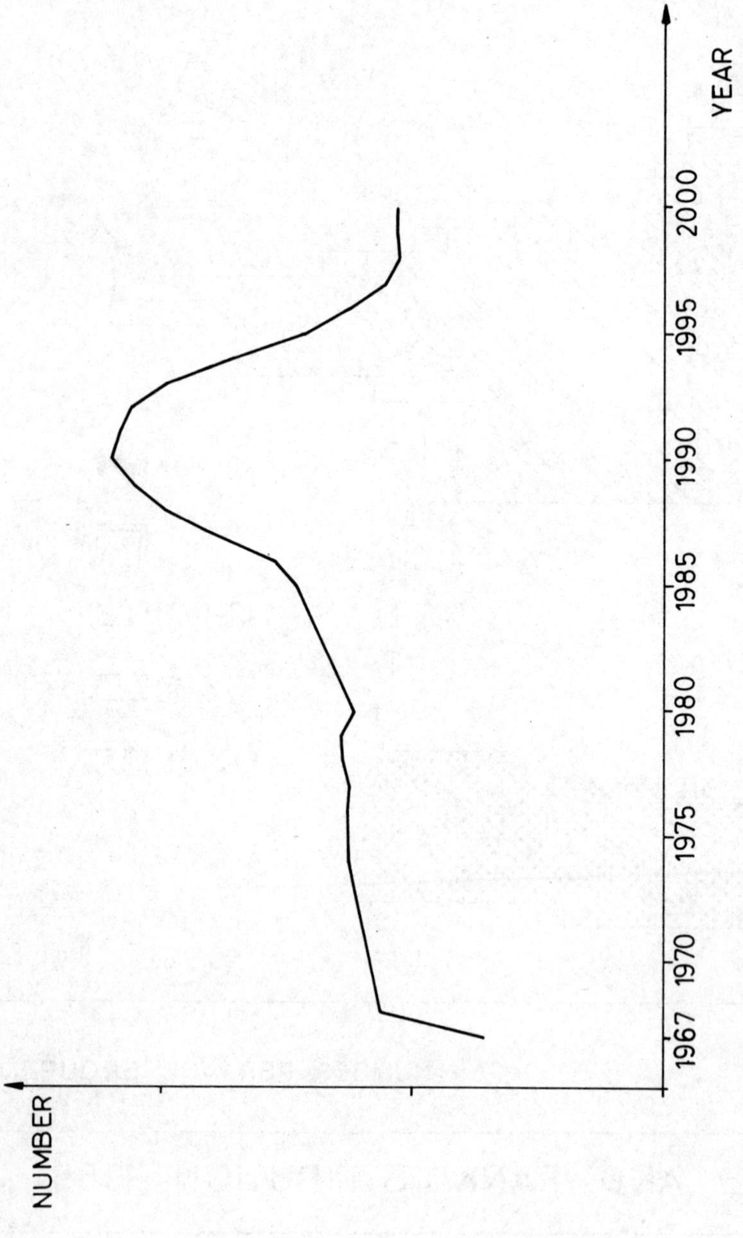

Fig. 15 Total number of professional officers

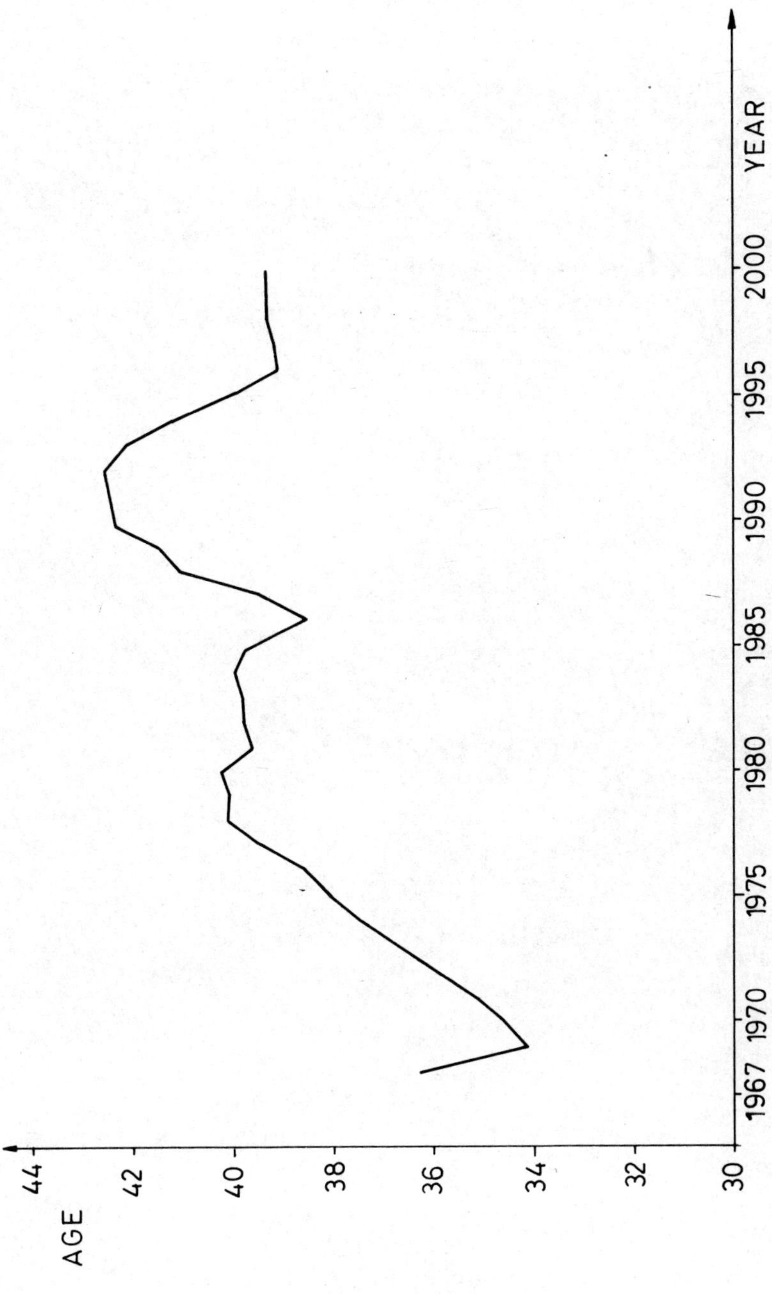

Fig. 16 Average age for promotion to major